The Higgs
Hunter's Guide

The Higgs Hunter's Guide

John F. Gunion,
University of California, Davis

Howard E. Haber,
University of California, Santa Cruz

Gordon Kane,
University of Michigan

Sally Dawson,
Brookhaven National Laboratory

The Advanced Book Program

CRC Press
Taylor & Francis Group
Boca Raton London New York

CRC Press is an imprint of the
Taylor & Francis Group, an **informa** business

First published 1990 by Westview Press

Published 2018 by CRC Press
Taylor & Francis Group
6000 Broken Sound Parkway NW, Suite 300
Boca Raton, FL 33487-2742

Visit the Taylor & Francis Web site at
http://www.taylorandfrancis.com

and the CRC Press Web site at
http://www.crcpress.com

Library of Congress Cataloging-in-Publication Data

The Higgs hunter's guide / John F. Gunion ... [et al.].
   p. cm.
   Includes bibliographical references.
   1. Higgs bosons. I. Gunion, J. F. (John Francis) 1943–
QC793.5.B62H54 1990   539.7'21—dc20   90-38530
ISBN: 0-7382-0305-X

ISBN 13: 978-0-7382-0305-8 (pbk)

This book was typeset by Nora Rogers,   using TEX typesetting language on an
IBM mainframe computer.

Frontiers in Physics

David Pines, Editor

Volumes of the Series published from 1961 to 1973 are not officially numbered. The parenthetical numbers shown are designed to aid librarians and bibliographers to check the completeness of their holdings.

Titles published in this series prior to 1987 appear under either the W. A. Benjamin or the Benjamin/Cummings imprint; titles published since 1986 appear under the Addison-Wesley imprint.

Volumes published from 1974 onward are being numbered as an integral part of the bibliography.

Editor's Foreword

The problem of communicating in a coherent fashion recent developments in the most exciting and active fields of physics continues to be with us. The enormous growth in the number of physicists has tended to make the familiar channels of communication considerably less effective. It has become increasingly difficult for experts in a given field to keep up with the current literature; the novice can only be confused. What is needed is both a consistent account of a field and the presentation of a definite "point of view" concerning it. Formal monographs cannot meet such a need in a rapidly developing field, while the review article seems to have fallen into disfavor. Indeed, it would seem that the people most actively engaged in developing a given field are the people least likely to write at length about it.

FRONTIERS IN PHYSICS was conceived in 1961 in an effort to improve the situtation in several ways. Leading physicists frequently give a series of lectures, a graduate seminar, or a graduate course in their special fields of interest. Such lectures serve to summarize the present status of a rapidly developing field and may well constitute the only coherent account available at the time. Often, notes on lectures exist (prepared by the lecturer himself, by graduate students, or by postdoctoral fellows) and are distributed on a limited basis. One of the principal purposes of the FRONTIERS IN PHYSICS Series is to make such notes available to a wider audience of physicists. A second principal purpose which has emerged is the concept of an *informal monograph*, in which authors would feel free to describe the present status of a rapidly developing field of research, in full knowledge, shared with the reader, that further developments might change aspects of that field in unexpected ways.

The Higgs Hunter's Guide represents just such an informal monograph on a frontier topic in physics. Experiments which establish the properties of Higgs bosons represents the major goal of the next generation of e^+e^- and hadron colliders, hence the authors' title. In the present monograph, the Standard Model for the strong and electromagnetic interactions of elementary particles is used to bring together what is known today about experimental information about Higgs bosons. As leading members of the particle physics community, Drs. Gunion, Haber, Kane, and Dawson, are thoroughly qualified to write an account of the "Higgs sector" which is intended for all practicing members of that community, from advanced graduate students to active researchers, and it is a pleasure to welcome them to the ranks of *Frontiers* authors.

David Pines
Urbana, Illinois
March, 1990

Preface

Despite the enormous success of the Standard Model for the strong and electroweak interactions of elementary particles, the exact mechanism by which the masses of the weakly-interacting vector bosons and the fermions are generated remains uncertain. In the Standard Model, the Higgs boson is the most direct experimental manifestation of this crucial aspect of the theory. More generally, the Higgs sector (or equivalent) can be expected to provide the most detailed tests of the correctness of any electroweak gauge theory. It is generally accepted that the mass scale of the phenomena responsible for electroweak symmetry breaking must not exceed the masses of the W and Z bosons by a large factor. Thus, the ability to either discover or establish the non-existence of Higgs bosons with mass below the TeV scale, and the ability to study WW scattering in the TeV region, are crucial goals for the next generation of e^+e^- and hadron colliders. In addition to exploring in detail the physics of the Standard Model Higgs boson, we also emphasize theoretical approaches which go beyond the Standard Model and consider the extent to which their Higgs sectors can be probed by accelerators now in existence or being planned. Not surprisingly, the appropriate techniques and experiments for fully exploring the Higgs boson spectrum are somewhat model dependent.

This book is intended for practicing particle physicists, both theoretical and experimental, from advanced graduate students to active researchers in high energy physics. While our emphasis is on techniques necessary to explore the Higgs sector experimentally, some theoretical material is included in order to make the discussion self-contained. Chapter 1 provides an introduction and pedagogical material to remind the reader of the details of the Higgs mechanism (although we do not attempt to provide an accurate history of its theoretical development). Chapter 2 summarizes the theoretical properties of the Standard Model Higgs boson. Chapter 3 summarizes existing Higgs mass limits (as of October, 1989), and describes the most effective experimental methods of searching for a Standard Model Higgs boson. Methods for studying the interactions of longitudinal W bosons in the TeV region are also reviewed.

Chapter 4 extends the analysis to the case of the two-doublet Higgs model; particular attention is given to the minimal supersymmetric extension of the Standard Model which requires such a Higgs sector. Chapter 5 examines the plethora of Higgs bosons which occur in non-minimal supersymmetric models inspired by superstring theory, and Chapter 6 discusses various special topics including a number of models with more exotic Higgs and/or gauge sectors. We provide a summary in these chapters of the types of phenomena that might be observable and of the associated experimental requirements. Finally, Chapter 7 provides a brief survey of alternatives to the theory of weakly-coupled Higgs bosons, in which electroweak symmetry breaking is due to the dynamics of a yet-undiscovered strongly interacting sector. Appendix A gives a complete set of Feynman rules for Higgs boson interactions in the minimal supersymmetric extension of the Standard Model; some of the Feynman rules have not appeared before in the literature. In Appendices B and C, we collect various useful formulae for Higgs boson partial decay rates.

Our goal has been to bring together most of what is known today concerning how to get direct experimental information about the Higgs sector of the Standard Model and various extensions thereof. One reason we undertook this project was to find gaps in the theory or calculations and to fill them or stimulate others to do so. Another was to collect all the relevant information for decisions about future facilities and/or future experiments, perhaps making it more likely that optimal choices are made in the pursuit of the Higgs boson. We have included considerable material on approaches to the Higgs sector beyond the Standard Model, since some of the standard lore about Higgs physics is considerably modified in many such models.

We have specifically not tried to report on recent attempts to improve the fundamental understanding of the Higgs mechanism, although we mention some issues when relevant, nor have we tried to argue in favor of one view rather than another. We think it will probably be necessary to have experiments that establish the existence or non-existence of scalar bosons at all masses up to and including the TeV region, and that study WW interactions in the TeV region, before an understanding emerges of why and how the electroweak symmetry of the Standard Model is broken. This book is intended to help inform and to stimulate the planning of those experiments.

This book could not have been completed without the diligent work of our collaborators and the many comments of and discussions with our friends and colleagues. We are happy to acknowledge our many collaborators: R.M. Barnett, R. Cahn, M. Capdequi Peyranere, M. Duncan, J. Ellis, P. Galison, G. Gamberini, K. Griest, J.A. Grifols, P. Irulegui, P. Kalyniak, I. Kani, B. Kayser, Z. Kunszt, A. Mendez, Y. Nir, S.F. Novaes, F. Olness, F.E. Paige, M. Quiros, W. Repko, W. Rolnick, L. Roszkowski, H. Sadrozinski, A.S. Schwarz, A. Seiden, M. Sher, A. Snyder, M. Soldate, D.E. Soper, T. Sterling, A. Tofighi-Niaki, W.-K. Tung, A. Turski, R. Vega, A.J. Weinstein, S. Willenbrock, J. Wudka, D. Wyler, C.P. Yuan, and F. Zwirner. We greatly appreciate their contributions to the material presented in this review. In addition, we gratefully acknowledge fruitful discussions with T. Banks, W. Buchmüller L. Bergstrom, C. Bernard, P. Burchat, M. Chanowitz, S. Chivukula, A. Cohen, J.L. Diaz-Cruz, J. Donoghue, M. Einhorn, H. Georgi, M. Golterman, B. Grinstein, B. Holdom, G. Hou, D. Kaplan, A. Khare, S. Komamiya, J. Kuti, A. Manohar, W. Marciano, P. Nason, A. Nelson, M. Peskin, E. Rabinovici, S. Raby, L. Randall, S. Rindani, M. Shifman, G. Valencia, M. Voloshin, G. West, and R. Willey. We are particularly grateful to Paula Franzini for her numerous comments on the manuscript. Other helpful comments have been provided by S. Bertolini, A. Buras, M. Duncan, M. Golden, P. Langacker, A. Linde, R. Petronzio, P. Roy, S. Sharpe, and P. Taxil. We would like to thank the following experimentalists who assisted us in ascertaining limits on light Higgs bosons: J. Lee-Franzini, Paulo Franzini, L. Littenberg, P. Meyers, H. Nelson, T. Shinkawa, B. Winstein, and M. Zeller.

This project has been partially supported by the U.S. Department of Energy. We would like to acknowledge the hospitality of the Institute for Theoretical Physics (Santa Barbara) and of Stanford Linear Accelerator Center, where large portions of this book were written. We also thank SLAC for computing and other technical support. We are particularly grateful to Allan Wylde and Addison-Wesley for their support during the course of this project; we are especially pleased to acknowledge Jan Benes for his guidance and assistance during the final formatting of the book. Finally, we wish to express our appreciation to Nora Rogers, who did an outstanding job of turning a ragged TeX manuscript into a textbook-quality document.

S. Dawson   December, 1989
J.F. Gunion
H.E. Haber
G.L. Kane

Post-Publication Addendum

In this second printing of *The Higgs Hunter's Guide*, we have taken the opportunity to correct a number of errors and omissions, and have also updated the references at the end of each chapter. Since the initial publication of this book, recent data from LEP have dramatically improved the mass limits for the Higgs boson of the minimal Standard Model and the Higgs bosons of the minimal supersymmetric model. We have summarized the latest LEP limits (as of April, 1991) in the new §1.6 at the end of Chapter 1. In addition, in §1.5, we have included the most recent limits for the top quark mass, which plays a key role in Higgs phenomenology.

S.D., J.F.G., H.E.H., and G.L.K.   May, 1991

Errata for *The Higgs Hunter's Guide*

1. On p. 30, eq. (2.25) should read:

$$\lim_{m_f \to \infty} I_1(\tau_f, \lambda_f) - I_2(\tau_f, \lambda_f) = -\tfrac{1}{3}.$$

2. On p. 30, in eq. (2.29), replace $|\ln x|$ with $\ln x$ (*i.e.*, remove the absolute value signs). Also, replace $|1 - x^2|$ with $(1 - x^2)$, although this will not make a difference since it is implicitly assumed that $1/2 \leq x \leq 1$.

3. On p. 62, in eq. (2.131), the denominator on the right hand side should read $\sigma_0(\nu_\mu N \to \mu^- X)$.

4. On p. 96, eq. (3.9) is not correct. To obtain the correct answer, we must compute the contribution of fig. (3.7) to Δa_μ. Assuming that the interaction is specified in eq. (3.7), where h is now the negatively charged Higgs boson $(H^\pm)$, and F is now a neutral-charged fermion, we find:

$$\Delta a_\mu|_{\text{fig. 3.7}} = \frac{-m_\mu^2}{8\pi^2} \int_0^1 x(1-x)\,dx \frac{\{C_S^2(x + m_F/m_\mu) + C_P^2(x - m_F/m_\mu)\}}{m_\mu^2 x^2 + (m_{H^\pm}^2 - m_\mu^2)x + m_F^2(1-x)}.$$

Our calculation agrees with that of Leveille [6]. (In the results of ref. [6], one must set the charged Higgs boson charge equal to -1 in units of e, since the calculations presented there are for $g - 2$ of the μ^-.) In the two-Higgs doublet model (see §4.1), C_S and C_P depend on the choice of Higgs-fermion couplings. The Model-II couplings [see eq. (4.22) and the discussion following] yield $C_S^2 = C_P^2 = G_F m_\mu^2 \tan^2 \beta/\sqrt{2}$. Inserting these couplings into the equation above (and noting that F is the neutrino), we obtain in the limit of $m_{H^\pm} \gg m_\mu$

$$\Delta a_\mu|_{\text{fig. 3.7}} = \frac{-G_F m_\mu^4 \tan^2 \beta}{24\sqrt{2}\,\pi^2 m_{H^\pm}^2}. \tag{3.9}$$

This result should replace eq. (3.9) at the bottom of p. 96. Note that the above result disagrees with the results of Grifols and Pascual (ref. [8] of Chapter 3) and Donoghue and Li (ref. [20] of Chapter 4) both in magnitude (by a factor of 2) and in sign. We have carefully checked that the sign of Δa_μ in eq. (3.9) is indeed opposite to that of the standard QED one-loop correction. In addition, Grifols and Pascual [8] denote $\kappa \equiv \tan \beta$, but incorrectly state that $\kappa \leq 1$. Thus, the statement below eq. (3.9) on p. 96 should be eliminated.

5. In view of the previous erratum, eq. (3.11) on p. 98 is not correct. It should read:

$$\frac{\Delta a_\mu|_{\text{fig. 3.8}}}{\Delta a_\mu|_{\text{fig. 3.7}}} \simeq \frac{77}{\tan^2 \beta} \,. \tag{3.11}$$

In obtaining this result, we have also made use of the most recent Particle Data Book values for the fundamental constants. Therefore, the two-loop charged Higgs contribution is *larger* than the corresponding one-loop contribution unless $\tan \beta \gtrsim 9$.

6. In chapter 3, there is a typographical error in some of the headlines appearing on odd-numbered pages. Headlines on pp. 109, 111, 113, 115, and 129 should read "Very Light Higgs Bosons (mass $\lesssim 5$ GeV)". Headlines on pp. 137 and 139 should read "Light Higgs Bosons ($5 \lesssim m_{\phi^0} \lesssim 85$ GeV)".

7. On p. 146, the Feynman diagrams in Figs. 3.24 and 3.25 have been incorrectly switched. The two figure captions should remain where they are, while the corresponding diagrams should be interchanged.

8. On p. 155 in the line right above eq. (3.94), the text should read: "Before applying any cuts, reaction (3.92) overtakes reaction (3.93) at roughly".

9. On p. 195, we state that eq. (4.8) is the most general two-Higgs doublet scalar potential subject to a discrete symmetry $\phi_1 \rightarrow -\phi_1$ which is only softly violated by dimension-two terms. This is not strictly correct. There is one additional term that can be added:

$$\lambda_7 \left[\text{Re} \ \phi_1^\dagger \phi_2 - v_1 v_2 \cos \xi \right] \left[\text{Im} \ \phi_1^\dagger \phi_2 - v_1 v_2 \sin \xi \right] \,.$$

However, this term can be eliminated by redefining the phases of the scalar fields. To see this, note that if $\lambda_7 \neq 0$ then the coefficient multiplying the term $(\phi_1^\dagger \phi_2)^2$ in the scalar potential is complex, while if $\lambda_7 = 0$ then the corresponding coefficient is real. Subsequent results presented in §4.1 are not affected by this choice. Moreover, in the minimal supersymmetric model, $\lambda_7 = 0$ (at tree-level). On the other hand, in CP-violating two-Higgs-doublet models, it is important to keep $\lambda_7 \neq 0$ if one wishes to retain the overall freedom to redefine the Higgs field phases.

10. On p. 209, in the text five lines above eq. (4.46), replace the reference to eq. (2.29) with eq. (2.30).

11. On p. 217, in the last sentence preceding the section "The Top Quark and the Charged Higgs", replace the phrase: "Typically, $\tan \beta \lesssim 2$. . ." with "Typically, $\tan \beta \gtrsim 1/2$. . .."

12. On p. 264, the reference to Witten in ref. 106 is incorrect. It should read: E. Witten, *Nucl. Phys.* **B188** (1981) 513.

13. On p. 264, there are some errors in ref. 107. First, A. Kakuto's name was mistakenly omitted. Second, a reference to one paper was accidently left out. The corrected reference should read: K. Inoue, A. Kakuto, A. Komatsu and S. Takeshita, *Prog. Theor. Phys.* **67** (1982) 1889; **68** (1982) 927 [E: **70** (1983) 330]; **71** (1984) 413.

14. A point of clarification regarding the four-point $VVHH$ vertices on pp. 364–8 (where V =vector boson and H =Higgs boson or Goldstone boson). When a charged line is indicated and no arrow is specified, then the given Feynman rule applies (with no change in sign) for both charge states. For example, whereas the Feynman rules for the $\gamma W^+ A^0 H^+$ and $\gamma W^- A^0 H^-$ vertices are opposite in sign, the Feynman rules for the $\gamma W^+ H^0 H^+$ and $\gamma W^- H^0 H^-$ vertices have the same sign.

15. Here is a clarification regarding the Feynman rules for cubic Higgs vertices involving at least one Goldstone field, which appear on pp. 373–375. The rules for cubic and quartic scalar interactions given in Appendix A are specific to the minimal supersymmetric model (MSSM), since the MSSM imposes specific constraints on the Higgs potential. Nevertheless, it turns out that the rules for cubic scalar interactions involving at least one Goldstone boson can be written in a completely model independent way. Three examples of these model-independent rules (which apply to the most general CP-invariant two–Higgs-doublet model) were given in Fig. A.17 on p. 375. In fact, we can obtain similar results for the other six non-zero three-Higgs vertices given on pp. 373–374. For completeness, the Feynman rules in the general CP-invariant two-Higgs-doublet model for all non-zero three-Higgs vertices involving at least one Goldstone field are listed below:

$$g_{h^0 G^0 G^0} = \frac{-ig}{2m_W} m_{h^0}^2 \sin(\beta - \alpha) \,,$$

$$g_{H^0 G^0 G^0} = \frac{-ig}{2m_W} m_{H^0}^2 \cos(\beta - \alpha) \,,$$

$$g_{h^0 G^+ G^-} = g_{h^0 G^0 G^0} \,,$$

$$g_{H^0 G^+ G^-} = g_{H^0 G^0 G^0} \,,$$

$$g_{h^0 A^0 G^0} = \frac{-ig}{2m_W} (m_{h^0}^2 - m_{A^0}^2) \cos(\beta - \alpha) \,,$$

$$g_{H^0 A^0 G^0} = \frac{ig}{2m_W} (m_{H^0}^2 - m_{A^0}^2) \sin(\beta - \alpha) \,,$$

$$g_{h^0 H^\pm G^\mp} = \frac{ig}{2m_W} (m_{H^\pm}^2 - m_{h^0}^2) \cos(\beta - \alpha) \,,$$

$$g_{H^0 H^\pm G^\mp} = \frac{-ig}{2m_W} (m_{H^\pm}^2 - m_{H^0}^2) \sin(\beta - \alpha) \,,$$

$$g_{A^0 H^\pm G^\mp} = \frac{\pm g}{2m_W} (m_{H^\pm}^2 - m_{A^0}^2) \,.$$

In the rule for the $A^0 H^\pm G^\mp$ vertex, the sign corresponds to $H^\pm$ entering the vertex and $G^\pm$ leaving the vertex. [If CP were not conserved, these rules would still apply, except that h^0, H^0 and A^0 would no longer be mass eigenstates.] One can easily check that if tree-level MSSM relations are imposed on the Higgs masses, and angles α and β [see eqs. (A.9)–(A.12) on p. 355], one recovers the MSSM Feynman rules listed on pp. 373–374 in Figs. A.15 and A.16.

16. On p. 375, in the caption to Fig. A.17, on the third line, replace the reference to eq. (A.10) with eq. (A.9).

17. On pp. 404–405, in Fig. A.35(c) and (d), the Feynman rules for the $H^0 h^0 \tilde{q}_{kL} \tilde{q}_{kL}$ and $H^0 h^0 \tilde{q}_{kR} \tilde{q}_{kR}$ vertices are incorrect. The correct Feynman rules are:

$$H^0 h^0 \tilde{q}_{kL} \tilde{q}_{kL} \qquad \frac{ig^2 \sin 2\alpha}{2} \left[\frac{T_{3k} - e_k \sin^2 \theta_W}{\cos^2 \theta_W} - \frac{m_q^2}{2m_W^2} D_k \right],$$

$$H^0 h^0 \tilde{q}_{kR} \tilde{q}_{kR} \qquad \frac{ig^2 \sin 2\alpha}{2} \left[e_k \tan^2 \theta_W - \frac{m_q^2}{2m_W^2} D_k \right].$$

18. On p. 407, in Fig. A.36(f), the Feynman rule for the $G^\pm H^\mp \tilde{q}_{kR} \tilde{q}_{kR}$ vertex is incorrect. The corrected rule is:

$$G^\pm H^\mp \tilde{q}_{kR} \tilde{q}_{kR} \qquad \frac{ig^2}{2} \sin 2\beta \, e_k \tan^2 \theta_W - \frac{ig^2 m_q^2}{2m_W^2} H_k.$$

19. On p. 417, in eq. (C.1), replace $\Gamma(\phi^0 \to \gamma\gamma)$ with $\Gamma(h \to \gamma\gamma)$, in order to be consistent with the notational conventions of Table 1.1 on p. 4.

20. On p. 418, at the beginning of the second line below eq. (C.3), it should read: "$F_{1/2}^{A^0} \to -2$."

21. On p. 421, in the index listing for "Decays of A^0 . . . to Zh", replace 413 with 412.

22. On p. 422, in the index listing for "Decays of neutral Higgs bosons . . . to W^+W^-; ZZ", replace 413 with 412.

Added Note to the Paperback Edition

This paperback edition is a reprinting of the second printing of *The Higgs Hunter's Guide*, which corrected a number of errors and omissions in the original first printing. Errors found subsequent to the second printing are collected and listed on the previous four pages.

Much progress on Higgs physics has been made during the last ten years. An excellent starting point for updating some of the material of this book can be found in the contributions of the Light Higgs Working Group to *New Directions for High-Energy Physics*, Proceedings of the 1996 DPF/DPB Summer Study on High Energy Physics, Snowmass '96, edited by D.G. Cassel, L.T. Gennari and R.H. Siemann (Stanford Linear Accelerator Center, Stanford, CA, 1997). [In particular, see the overall summaries by H.E. Haber *et al.*, pp. 482–498, and by J.F. Gunion *et al.*, pp. 541–587.] Another useful reference is *Perspectives on Higgs Physics II*, edited by G.L. Kane (World Scientific, Singapore, 1997), which reviews many of the most recent theoretical and phenomenological advances in Higgs physics.

As noted in §1.5, the top quark plays an important role in the phenomenology of Higgs bosons. Due to the impressive efforts of experimenters at the Tevatron $p\bar{p}$ collider, the top quark was discovered in 1995. The Particle Data Group currently quotes a top quark mass of $m_t = 173.8 \pm 5.2$ GeV.

Meanwhile, the LEP2 Collider continues to improve the sensitivity of the Higgs boson searches. As it enters its final year of operation, the Higgs mass limits given by the LEP Working Group for Higgs Boson Searches [P. Bock *et al.*, CERN-EP-2000-055 (April, 2000)] are as follows. In the Standard Model, $m_{\phi^0} > 107.9$ GeV. In the minimal supersymmetric extension of the Standard Model, $m_{h^0} > 88.3$ GeV and $m_{A^0} > 88.4$ GeV. In a general two–Higgs-doublet model (in which H^+ decays primarily to either $c\bar{s}$ or $\tau^+\nu$), $m_{H^\pm} > 78.6$ GeV. All limits quoted are at the 95% CL.

During the past decade, precision measurements of electroweak observables have provided some of the most stringent tests of the Standard Model. No significant deviation has yet been observed. A global fit to the electroweak observables in the context of the Standard Model yields an *upper* limit for the Higgs boson mass due to the logarithmic sensitivity to m_{ϕ^0} that arises from virtual Higgs boson corrections to $W^\pm$ and Z gauge boson self-energies. On the basis of such an analysis, the LEP Electroweak Working Group has quoted a 95% CL upper limit of $m_{\phi^0} < 188$ GeV [see, *e.g.*, A. Straessner, presentation at the XXXV Rencontres de Moriond, 11–18 March 2000, Les Arcs, France]. Although this upper limit is not applicable in all approaches that introduce new physics beyond the Standard Model, it is perhaps suggestive that the Higgs boson search may soon reach a successful conclusion.

The Higgs hunters continue their quest as we enter the third millennium.

S.D., J.F.G., H.E.H., and G.L.K.   May, 2000

Table of Contents

Chapter 1

Introduction and Preliminaries

Although the Higgs mechanism [1] was used to introduce mass into the Standard Model [2,3] two decades ago, experimental sensitivity to a Standard Model Higgs boson remains extremely limited. Masses below about $2m_\mu$ can be excluded by a combination of low energy experimental data on nuclear transitions and rare decays of K mesons. Recent results in K and B decays probably rule out masses from $2m_\mu$ to $2m_\tau$. Upsilon decays are potentially sensitive to masses above $2m_\mu$, up to about 5 GeV, but uncertainties regarding the exact magnitude of the expected decay rate to Higgs prevent firm conclusions at this time; although no Higgs bosons have been observed in such decays. Certainly, it will be 1990 (at the earliest) before experiments begin to probe the mass region above 5 GeV, where one might most naively expect to find the Standard Model Higgs boson.

As is often emphasized [4–6], the Higgs sector of the Standard Model is not understood from a fundamental point of view, although it performs technically in an entirely satisfactory way as an effective low energy theory, without conflict or contradiction. The physics that underlies electroweak symmetry breaking is simply not certain, although theorists may have guessed how it functions.

1.1   The Higgs Mechanism

Let us review the Higgs mechanism, to recall how the Higgs boson arises as the direct physical manifestation of the origin of mass in the Standard Model.* The Standard Model is a gauge theory. The SU(2) × U(1) gauge invariance of the theory requires masses of the gauge bosons to be zero, since the presence of a mass term for the gauge bosons violates gauge invariance ($M^2 A_\mu A^\mu$ is not invariant if $A_\mu \to A_\mu - \partial_\mu \chi$ where χ is a function of position in space-time, so M^2 must be zero). The Higgs mechanism circumvents this constraint

---

*More detailed textbook treatments can be found, for example, in refs. 7 and 8.

by beginning with a gauge invariant theory having massless gauge bosons, and ending with a spectrum having massive gauge bosons, after algebraic transformations on the Lagrangian. The physics leading to a gauge boson mass and a physical Higgs boson is contained in the simple Abelian case, which we now review.

Assume there exists a complex scalar boson ϕ and a massless gauge boson A^μ. Assume the Lagrangian of the theory has the form

$$\mathcal{L} = (D_\mu \phi)^* (D^\mu \phi) + \mu^2 \phi^* \phi - \lambda (\phi^* \phi)^2 - \tfrac{1}{4} F^{\mu\nu} F_{\mu\nu}. \qquad (1.1)$$

The parameters are constrained by $\lambda > 0$ (so that the potential is bounded from below), and $\mu^2 > 0$. $F^{\mu\nu}$ is the antisymmetric tensor of the gauge boson field, $F^{\mu\nu} = \partial^\mu A^\nu - \partial^\nu A^\mu$. Invariance of the theory under a local gauge transformation,

$$\begin{aligned} \phi &\to \phi' = e^{ig\chi(x)} \phi \\ A^\mu &\to A'^\mu = A^\mu - \partial^\mu \chi(x), \end{aligned} \qquad (1.2)$$

is guaranteed if in the Lagrangian we use the covariant derivative $D^\mu = \partial^\mu + igA^\mu$, in place of the ordinary partial derivative ∂^μ.

The potential for the scalar field has its minimum value at $\phi = v/\sqrt{2} = \sqrt{\mu^2/2\lambda}$. It is appropriate to expand ϕ near its minimum to find the spectrum of the theory, so write

$$\phi = [v + h(x)]/\sqrt{2} \qquad (1.3)$$

where $h(x)$ is a real field. Substituting this into $\mathcal{L}$, we have explicitly

$$\begin{aligned} \mathcal{L} = \tfrac{1}{2} \, & [(\partial_\mu - igA_\mu)(v+h)(\partial^\mu + igA^\mu)(v+h)] \\ & + \tfrac{1}{2}\mu^2 (v+h)^2 - \tfrac{1}{4}\lambda (v+h)^4 - \tfrac{1}{4} F^{\mu\nu} F_{\mu\nu}. \end{aligned} \qquad (1.4)$$

This contains several important terms. There is a term $(g^2 v^2/2)A_\mu A^\mu$ that should be interpreted as a mass term for the gauge boson. There is a term $-\lambda v^2 h^2$ that is a mass term for the scalar boson. There are interaction terms h^3, h^4, hAA, and $h^2 AA$, with related strengths. The theory with a complex scalar boson and a massless gauge boson has been reinterpreted as a theory with a real scalar boson and a massive gauge boson, because the scalar potential had its minimum at a value of ϕ that was non-zero. This way of giving mass to the gauge boson is called the Higgs mechanism [1].

Four things should be emphasized for our purposes. First there is a real boson, h, that should occur as a physical boson—the Higgs boson. Second, its mass depends on λ and on v. The gauge boson mass determines v, but λ is a parameter characteristic of the scalar potential and no one has ever found a way to calculate or determine λ without finding experimental information about the Higgs spectrum itself. Therefore the mass of the Higgs boson is unknown. Third, the interaction terms (plus those that occur when fermions are given mass) determine the production mechanisms and decays of the Higgs boson (*i.e.*, how it couples to particles that are accelerated or that enter detectors). The self-interaction terms depend on λ but the terms describing the interaction of h with A do not depend on λ, so their strength is known. Fourth, the counting of the number of independent states is consistent. This example began with one complex scalar field ϕ, having two real fields since it is complex. The massless gauge boson had two polarization states, just as a photon would. After the reinterpretation, there is one real Higgs boson, plus the three polarization states $(J_Z = 1, 0, -1)$ of a massive spin-one boson. In both cases the total number of degrees of freedom is four.

Continuing the counting, when we consider the Standard Model we add an SU(2) internal quantum number to the Higgs fields, so there is an SU(2) doublet of complex scalars, $\phi = \binom{\phi^+}{\phi^0}$, with four real fields. There are three massless gauge bosons, $W^\pm$ and Z, with two polarization states each, so the total number of independent fields is ten. Symmetry breaking is initiated by giving a vacuum expectation value $\langle \phi^0 \rangle = v/\sqrt{2}$ to the neutral Higgs field. The result is three massive gauge bosons, with nine degrees of freedom, so there will be one physical Higgs boson that should appear as a real particle.

In a supersymmetric theory the added symmetry implies that two SU(2) doublets of complex Higgs fields are required to give mass to fermions (see §4.2 for a detailed discussion), so there will be eight real scalar fields, plus six massless gauge boson degrees of freedom, fourteen in total. After the Higgs mechanism operates the same nine states are required for the gauge bosons, so five real fields remain, and there should be five spin-zero Higgs fields in the spectrum. This is one of the many experimentally checkable predictions of these different approaches. In this case, three of the scalar bosons are neutral and the other two are a charged pair.

Any theory with additional physics beyond the Standard Model will have a spectrum of spin-zero Higgs fields (one or two or more states, SU(2) singlets or doublets or triplets, etc.) that leads to a specific number of spin-zero bosons. In addition, definite relations (sometimes depending on parameters that have yet to be measured) hold between masses and coupling strengths of the various bosons. We will mention a few more examples in the following. Ultimately it will be necessary to determine experimentally the full spectrum of scalar bosons, from mass $\simeq 0$ through the TeV region, in order to be confident that any particular theory is correct. If history is any guide, only by finding some spin-zero bosons, or by knowing conclusively that they are not present with the necessary couplings over various mass regions, will it be possible to achieve any consensus on what theory is correct; to arrive at a fully valid theory of the Higgs sector will require detailed and complete experimental information.

1.2   Notation

Our notation to describe the various possibilities will be as follows. We use h for any generic Higgs boson, whatever its origin. The single neutral scalar boson of the minimal Standard Model is denoted ϕ^0. The neutral scalars of a minimal supersymmetric theory are h^0 and H^0 for the lighter and heavier scalars respectively, and A^0 for the pseudoscalar. Additional notation will be introduced in context, and is summarized in table 1.1.

Table 1.1

Symbol	Denotes
ϕ	General scalar field
ϕ^0	Single Higgs boson of the minimal Standard Model
h	Any generic Higgs boson
h^0	Lightest CP-even scalar boson of a non-minimal model (*e.g.*, supersymmetry)
H^0	Heavier CP-even scalar boson of a non-minimal model (*e.g.*, supersymmetry) [if needed, such states will be labelled H_i^0 ($i = 1, 2, \ldots$) in order of increasing mass]
A^0	CP-odd scalar boson of a non-minimal model (*e.g.*, supersymmetry) [if needed, such states will be labelled A_i^0 ($i = 1, 2, \ldots$) in order of increasing mass]
$H^\pm$	Charged scalar ("Higgs") bosons
V	Either $W^\pm$ or Z
LSP	Lightest supersymmetric particle

Beyond giving masses to the W and Z bosons, the Higgs mechanism is remarkable in that it can be simultaneously used to generate masses for chiral fermions in the theory. Since left-handed fermions are in SU(2) doublets and right-handed fermions in SU(2) singlets, a mass term for fermion f of the form

$$m\bar{f}f = m(\bar{f}_L f_R + \bar{f}_R f_L) \tag{1.5}$$

is not SU(2) invariant; the product of a doublet and a singlet is not invariant under SU(2) rotations. Introduction of the Higgs SU(2) doublet allows Yukawa-like terms of the form

$$g_f \left[(\bar{f}_L \phi) f_R + h.c. \right] \tag{1.6}$$

to be written. Since f_L and ϕ are doublets, $\bar{f}_L \phi$ is an SU(2) singlet in the Lagrangian. When the Higgs field gets a vacuum expectation value as in eq. (1.3), this term becomes

$$\sqrt{\tfrac{1}{2}} g_f v \left(\bar{f}_L f_R + \bar{f}_R f_L \right), \tag{1.7}$$

which is automatically a mass term for the fermion, with the identification $m_f = g_f v / \sqrt{2}$. Note that g_f was an arbitrary coupling, so m_f is not calculated, but it can be introduced into the theory; since m_f is generally measured, g_f is replaced by $\sqrt{2} m_f / v$. As with the gauge bosons, the Higgs boson couples to fermions with a strength proportional to their mass. The above procedure is repeated separately for every fermion.

1.3   Theoretical Overview

In the Standard Model a single Higgs doublet can give mass to both the gauge bosons and the fermions. In supersymmetric theories [9], although there are two doublets, this aspect is basically unchanged. The two vacuum expectation values contribute in the combination $v_1^2 + v_2^2$ to the gauge boson masses, while down-type fermions have mass proportional to v_1, up-type fermions to v_2. However, in non-supersymmetric models it is not necessary that the same Higgs field(s) give mass to both fermions and gauge bosons, nor is it required that the Higgs fields that give mass to the up and down quarks be different. There has been some exploration [10] of the possibility that two doublets occur, with one giving mass to fermions but hardly any mass to gauge bosons; but if m_t is of order m_W, much of the motivation for such an approach is removed. In a technicolor approach [6], where Higgs bosons are composites of hypothetical fermions, the gauge bosons get mass by a rather elegant and plausible mechanism, while it is necessary to introduce an entirely new sector of the theory ("Extended Technicolor") to give mass to fermions [11].

It is worthwhile looking at the role of Higgs bosons from a different point of view, following the arguments of ref. 12. If the process $f\bar{f} \to W\overline{W}$ is considered, when the produced W's are longitudinally polarized (those W's that arose above by the Higgs mechanism) there are contributions from s-channel gauge bosons and t- or u-channel fermions. If the couplings are in precisely the ratios required for a gauge theory, a term in the cross section that is quadratic in s vanishes, because of cancellations among the contributions. However, there is still a term in the cross section that grows as $m_f^2 s$. This is the piece cancelled by the contribution of an s-channel Higgs boson that couples proportional to m_f (by considering other channels such as $WW \to$

WW the couplings can be uniquely determined). This argument makes clear the fact that a physical Higgs boson, or scalar interaction, must be included along with the gauge bosons to have a sensible theory. Without such a contribution, some amplitudes exceed their unitarity limits at large s. In addition, when tree-level processes which violate unitarity appear as subdiagrams within higher loop diagrams of the theory, infinities result which cannot be removed by renormalization. The theory would, therefore, not be renormalizable. However, it could happen that the scalar interaction is not due to a single fundamental boson, but is generated by non-perturbative behavior of the theory. Whatever happens, some new spin-zero interaction must occur, and it can be discovered experimentally. Later we will go into quantitative details about such questions.

In more complicated theories, such as supersymmetric ones, E_6-based ones, or left-right symmetric ones, several neutral scalars, charged scalars, and even doubly-charged scalars are required in order to give all amplitudes acceptable high energy behavior. Because this physics is in common with that of the Standard Model Higgs boson, we have chosen to call any scalar that is necessary to regulate the high energy behavior a "Higgs boson", whether or not it is directly involved in giving mass to vector bosons and fermions.

It should be emphasized that there are no definitive theoretical upper or lower limits on Higgs boson masses relevant to experiments. A possible lower limit deriving from the requirement that the symmetry breaking vacuum be the absolute minimum is dependent upon knowledge of fermion masses. For example, in the minimal Standard Model, when radiative corrections from gauge boson and fermion loops are included in the Higgs potential, there are two possible minima: the symmetric minimum at $\phi = 0$ and the symmetry-breaking minimum at $\phi = v/\sqrt{2}$. Insisting that the minimum at $\phi = v/\sqrt{2}$ be the lower one gives a condition [13]

$$m_{\phi^0}^2 > \frac{3\left(2m_W^4 + m_Z^4\right) - 4\sum_\ell m_\ell^4 - 12\sum_q m_q^4}{16\pi^2 v^2}, \qquad (1.8)$$

where $v = (\sqrt{2}G_F)^{-1/2} \simeq 246$ GeV. Each particle enters in proportion to its number of charge, spin and color states. We sum over three generations of quarks (q) and leptons (ℓ). If all fermions have mass small compared to m_W, m_Z, eq. (1.8) yields the bound $m_{\phi^0} \gtrsim 7$ GeV [13]. However, if a heavy fermion (e.g., the t quark) with $m_f \sim m_W$ exists, there is no lower limit on m_{ϕ^0}, even in this minimal model. (For $m_f > m_W$, one can make theoretical arguments for a lower Higgs mass bound based on the requirement of stability of the renormalization-group-improved Higgs potential. These issues will be discussed in §2.5.) Beyond minimal models, limits are more model dependent and generally apply only to a combination of scalar masses. For example, in the two-Higgs-doublet model, in the absence of information about scalar masses and mixing angles, there is only a lower limit for the heavier of the two neutral (CP-even) scalars. Any lower Higgs mass limit depends on assumptions about the fermion and boson spectra. Some lower limit will occur in any particular theory, but such lower limits should not be used to bias the manner in which experimental searches for scalar bosons are carried out. Reference 14 reviews what occurs under various assumptions.

The existence of a model independent upper limit on m_{ϕ^0} is less certain. In fact, it is entirely possible that no scalar particle exists which is sufficiently light to prevent some amplitudes for $WW \to WW$ from approaching their unitarity limit, at which point the true theory must begin to differ from the perturbative predictions of the Standard Model. This was first emphasized in ref. 15 and later in refs. 4 and 5. As we will discuss, such behavior is likely to be observable. For large m_{ϕ^0}, the value of $\sqrt{s_{WW}}$ where significant non-perturbative behavior sets in differs from model to model, but is normally around 1.2 TeV (see §2.6). Sometimes this is described as an upper limit on m_{ϕ^0}, but we prefer to view this as simply being the scale where perturbative analysis ceases to be useful. Upper limits on the Higgs boson mass of the Standard Model have also been derived on the basis of the probable triviality of ϕ^4 theory, beginning with the early work of ref. 16. In the strongest version of such limits, where it is assumed that there is no new physics up to some very large energy scale and that the theory remains perturbative at all lower scales, upper bounds on m_{ϕ^0} of order 200 GeV are obtained. However, if there is new physics (e.g., supersymmetry) at a relatively low energy scale, or the theory becomes non-perturbative, clear cut limits have not been derived. As discussed in §2.5, theoretical analyses of the "triviality bound" suggest that $m_{\phi^0} \lesssim 700$ GeV, which is more stringent than the unitarity "bound" quoted above. Nevertheless, at the present state of the art, we believe that it is premature to limit the Higgs mass range over which experimental searches are conducted. Thus, our emphasis will be on how to extract from the experimental data a useful signal for Higgs production for any given Higgs mass in the entire range from ~ 0 to $\gtrsim 1$ TeV. However, we explicitly exclude from this book the phenomenology of a strongly-interacting Higgs sector.

1.4 Higgs Search Overview

A large part of this book will be devoted to the search for the Higgs boson: how the discovery could occur, and what kinds of data are required. We give here a brief survey of when various kinds of data will become accessible, and how large a Higgs boson mass can be probed in a particular kind of experiment. This is summarized in fig. 1.1. Clearly, the years given are lower limits; we have assumed that the construction and commissioning of new accelerators (especially the luminosities that are achieved) will proceed as scheduled. A summary of the machines that are under construction, planned, and under discussion, along with probable luminosities, is given in table 1.2.

It could happen that as LEP accumulates Z decays, no signal appears for a light scalar. We would still know very little about Higgs physics. As the energy of LEP increases it slowly covers an increasing range of Higgs mass, until around 1995 or later when LEP-II will have searched up to about 90 GeV.

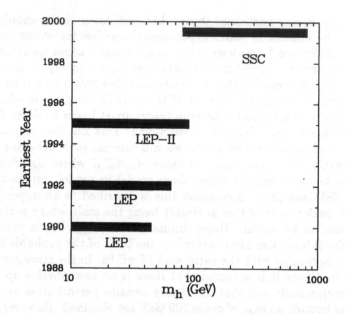

Figure 1.1 Survey of Higgs mass reach based on expected experimental data to be obtained at existing accelerators or colliders presently under construction as an estimated function of year.

Table 1.2

A summary of new and possible future accelerators, energy and luminosity.

	e^+e^-			pp, $p\bar{p}$	
	$\sqrt{s}$	$L(\mathrm{cm}^{-2}\mathrm{s}^{-1})$		$\sqrt{s}$	$L(\mathrm{cm}^{-2}\mathrm{s}^{-1})$
LEP-I	$\sim m_Z$	$\sim 10^{31}$	Tevatron	1.8 TeV	$\sim 10^{32}$
LEP-II	$\sim$180 GeV	$\sim 10^{32}$	LHC	16 TeV	$\gtrsim 10^{33}$
NLC	$\sim$400 GeV	$\sim 10^{33}$	SSC	40 TeV	$\gtrsim 10^{33}$
TLC	0.5–1 TeV	$\sim 10^{33}$	Eloisatron	200 TeV?	$\gtrsim 10^{33}$?
CLIC	1–2 TeV	$> 10^{33}$	UNK	6 TeV?	$\sim 10^{32}$?
			ep		
HERA	0.32 TeV	$\geq 10^{31}$	LHC-ep	1.5 TeV	$\geq 10^{32}$

Note: Event rates will be quoted for $L_{\min} \times 10^7$ s, unless otherwise specified.

It will be a challenge for LEP-II to cover the mass region overlapping m_Z, since this will probably require showing that there is an excess of $b\bar{b}$ events as compared to the number expected from $Z \to b\bar{b}$. Given sufficient luminosity and detector capabilities, it should be possible to rule out $m_{\phi^0} \approx m_Z$ at LEP-II, if $\sqrt{s} = 200$ GeV [17]. A Higgs mass limit of $m_{\phi^0} \gtrsim m_Z$ would imply useful constraints on theoretical approaches and model building, but no decisive conclusions could be drawn. For example, "low-energy" supersymmetric models typically predict a light scalar Higgs boson (h^0), but models are easily constructed where m_{h^0} is somewhat above m_Z. At the same time, suppose no signal has appeared in the charged scalar sector. This would not provide a constraint at all on the Standard Model, which predicts none, or on supersymmetric theories, which normally expect $m_{H^\pm} > m_W$. However, it would be a constraint on composite Higgs ideas. For example, technicolor approaches have been thought [18] to require a charged pseudoscalar with a mass that is about 20 GeV (say to a factor of two either way); although, recent models may yield much higher values for the masses of such pseudo-Goldstone bosons (see chapter 7).

Other colliders (e.g., Tevatron and HERA) lack either the energy or luminosity or detection capabilities to extend the Higgs boson search beyond the mass region that can be probed by LEP and LEP-II. At the hadron colliders mentioned above, Higgs bosons would be produced, if they are not too heavy, via gg fusion and by radiation off W's. But the predicted rates are low, and the QCD backgrounds to detecting the main decay modes of the Higgs bosons are overwhelming as far as anyone is aware today.

A few years after the Higgs mass limits from LEP and LEP-II have been obtained, the SSC and LHC will (hopefully) begin to yield useful data. At the SSC, for instance, without requiring full luminosity, one could search in the channel $h \to ZZ \to \ell^+\ell^-\ell^+\ell^-$. Probably the mass region $2m_Z \lesssim m_h \lesssim 450$ GeV would be rather easily covered with the upper limit extending to about 800 GeV as luminosity and knowledge of the detector improved. If a signal appeared it would be strong evidence against supersymmetric approaches, where the heavier neutral Higgs bosons tend to decouple from W^+W^- and ZZ [19]. As time went by, increased luminosity and experience with detectors would allow some portion of the intermediate mass region $m_Z/2 \lesssim m_h \lesssim 2m_Z$ to be covered using rare decay modes of the Higgs boson, and would allow use of more channels at high masses, so that the upper limit of accessible m_h would approach 900–1000 GeV. Below that mass a complete scan would have been done, and any scalar bosons found. If there were none, it would remain to study the scattering of longitudinal W's around and above a TeV, to see if the cross section deviated from perturbative predictions. It is not yet known how high in mass it will be possible to go, because it is essential to use some W's that decay to $q\bar{q}$, a channel for which the QCD backgrounds are large. Some who have studied the problem [20] are optimistic that it will be possible to study pairs of longitudinal W's having total energy somewhat above a TeV. Eventually, perhaps it will be necessary to have an SSC upgraded in both luminosity and energy, or an e^+e^- collider with several TeV center of mass energy, to completely determine the presence or absence of scalar bosons or anomalous WW interactions at the TeV scale.

At any stage in the above scenario, a signal might have appeared. Any charged scalar is immediate evidence of physics beyond the Standard Model. If LEP or LEP-II were to find a charged Higgs with $m_{H^\pm}$ significantly below m_W it would be a very serious constraint on supersymmetry ideas, eliminating most simple low-energy supersymmetric approaches, and would probably be good evidence for either composite Higgs physics or a more complex (perhaps superstring-based) supersymmetric model, as surveyed in later chapters. At the SSC the production rate for $H^\pm$ is large, but signatures and backgrounds are not cooperative [21]. The decay branching ratios of a charged scalar are necessarily model dependent. Not all of the backgrounds have been computed. Consequently, it is not possible to say confidently that $H^\pm$ could be detected at the SSC, though some choices of parameters seem to allow such a discovery. If a signal is found, the first order of business is to check that the $H^\pm$ has spin consistent with zero, from the production angular distribution ($\sin^2\theta$) in e^+e^-, and the isotropic decay angular distributions. Then the mass dependence of its couplings must be measured. Charged scalars are generally expected to couple proportional to mass, but not strictly so; other factors may enter.

A signal might appear in the neutral Higgs sector. The earliest possibility would be a light scalar at LEP. With 10^6 Z's a signal in $Z \to h f \bar{f}$ could appear for $m_h \lesssim 50$ GeV, and for 10^7 Z's the mass range extends to about 65 GeV. If such a signal is found, its spin and the mass dependence of its couplings must be determined. While not all couplings can easily be measured, it will be possible to check very quickly that it is effectively uncoupled from the light fermions, such as e^+e^-, $\mu^+\mu^-$, $u\bar{u}$, $d\bar{d}$. If a specific state is known to exist, it may be possible to check its couplings to $W^\pm$'s by associated production at the Tevatron Collider.

If a neutral scalar were found in the next few years, many theorists (though not all) would expect it to be part of the supersymmetric spectrum of scalar bosons. Searching for the other members of the spectrum (at least two neutrals and the charged pair) should have the highest priority (along with searches for supersymmetric partners) but may not be possible before the SSC becomes available. A few measurements of the h couplings to other particles are possible that may allow a distinction to be made between the minimal Standard Model case $h = \phi^0$, and the supersymmetric case $h = h^0$. A few methods also exist to distinguish between scalar and pseudoscalar spin-zero bosons [22–26]. If one or more light scalars were discovered, it would still be important to check that the scattering of W's at higher energies is correctly described by the perturbative Standard Model predictions, before we could finally feel that all the relevant data were in hand.

To conclude this survey, we want to emphasize that the success of the Standard Model guarantees that a *complete* search for Higgs physics *must* turn up a positive result. No one knows what that result will be. It could be a single light point-like boson, it could be five scalars all of mass within a factor of two of m_W, it could be a resonant scalar state with mass of a TeV and width of half a TeV, it could be a cross section for $WW \rightarrow WW$ that is different from its perturbative value, or something else. Whatever it is, it can be found by appropriate experiments. Although there are gaps to fill in, we know today essentially what accelerators and detector properties are required to carry out the full search for Higgs physics that is the main topic of this review. The first, and still very relevant, detailed compilation and study of Higgs boson decays and interactions was in ref. 27. Various other reviews of Higgs boson physics have appeared and are valuable additional references [28–37].

The success of the Standard Model has been astonishing. The central problem today in particle physics is to understand the physics of the Higgs sector. Unless a theoretical picture emerges that is overwhelmingly compelling, it seems likely that the only path to an understanding of the Higgs sector is to measure the full spectrum of scalar interactions and scalar bosons. As has happened so often in the past, once the data informs us of the path nature has chosen, understanding should soon follow.

1.5   The Relevance of the Top Quark Mass

Before proceeding with a detailed description of the Higgs bosons of the Standard Model and of extended models, it is important to emphasize that the phenomenological features of any Higgs sector are intimately tied to the still unknown mass of the top quark. This fact will become very apparent as we proceed. Thus it is important to review and place in context the direct experimental limits on the top quark mass that are available at this time. Here, we briefly update the top-quark mass limits as of April, 1991.

The strongest experimental limit on m_t has been obtained by the CDF Collaboration at the Tevatron, under the assumption that the top quark decays via the Standard Model process $t \rightarrow W^{+*}b$ (where the asterisk indicates virtual exchange). The 95% confidence level (CL) limits obtained by the CDF Collaboration are: $m_t > 77$ GeV [38], and a preliminary improved limit of $m_t > 89$ GeV [39].

If non-Standard-Model physics exists such that the assumption of $BR(t \rightarrow W^{+*}b) \simeq 100\%$ is significantly altered, then the top-quark mass limits just quoted may not be valid and must be reconsidered. Most notably, if there is an extended Higgs sector, one must consider the possible influence of a charged Higgs boson, H^+. If kinematically allowed, the two-body decay $t \rightarrow H^+b$ would certainly dominate the three-body decay, $t \rightarrow W^{+*}b \rightarrow f\bar{f}'b$. However, one should not ignore the possibility that there could be other types of new physics that can cause anomalous top-quark decays and make it hard to detect at the colliders. Examples include $t \rightarrow \tilde{t}\tilde{\gamma}$ in supersymmetric models, $t \rightarrow D\ell\nu$, where D is a heavy new charge $-1/3$ quark, and flavor-changing neutral-current top-quark decays. Thus, it is important to obtain limits on m_t that are independent of how the top quark decays.

Based on data from LEP, SLC, and lower energy e^+e^- colliders, the most recent Particle Data Group compilation gives $m_t > 45.8$ GeV at 95% CL [40]. This limit depends very weakly on assumptions about how the top quark decays. One can also obtain a limit on m_t at the Tevatron that is truly independent of the top quark decay mechanism. This can be done by making a precise measurement of [41,42]

$$R \equiv \frac{BR(W \to e\nu)\,\sigma(p\bar{p} \to W + X)}{BR(Z \to e^+e^-)\,\sigma(p\bar{p} \to Z + X)}. \qquad (1.9)$$

By using the known value of $BR(Z \to e^+e^-)$ as measured at LEP, an accurate measurement of R [eq. (1.9)] will be sensitive to the presence or absence of the decay channel $W^+ \to t\bar{b}$. The current CDF limit on m_t based on this technique is $m_t > 44$ GeV at 95% CL [42]. In the future, a model-independent limit of $m_t \gtrsim 75$ GeV may be possible using this technique, when sufficient data are obtained.

It is also possible to obtain an upper bound on the top quark mass based on theoretical analysis of neutral current and charged current data. In 1987, Amaldi *et al.* [43] concluded that $m_t \lesssim 200$ GeV. Since then, the very precise electroweak measurements at the Z mass at LEP have reinforced this theoretical bound, although it still should not be taken with the same confidence as the experimental lower bounds on m_t quoted above. Thus, conservatively, we would conclude that 45 GeV $\lesssim m_t \lesssim 250$ GeV. In the context of the Standard Model, the lower bound on m_t is considerably stronger, based on the CDF limits quoted at the beginning of this section. In this book, we shall usually survey the implications and prospects for Higgs detection as a function of m_t, even allowing for m_t values smaller than the stronger CDF limits quoted earlier. Clearly, the discovery of the t-quark will have a profound effect on the phenomenology of the Higgs boson.

1.6   Post-Publication Update on Higgs Masses

In 1989–1990, the four LEP detectors collected over half a million Z's. In the Standard Model, the Higgs boson can be produced via $Z \to \phi^0 f\bar{f}$ [see fig. (2.12)], with branching ratio shown in fig. (3.19). The four LEP detector collaborations have presented Higgs mass limits [44] based on very detailed analyses which examine many different $f\bar{f}$ final states, and carefully treat the low mass Higgs region where many different Higgs decay modes must be taken into account. In the higher mass region, the search for the decay $Z \to \phi^0 \nu\bar{\nu}$ plays a significant role in establishing the ultimate Higgs mass limit. As of April, 1991, the ALEPH collaboration has obtained the most stringent Higgs mass limit [45]: $m_{\phi^0} > 48$ GeV (at 95% confidence level). The entire

Higgs mass range from 48 GeV down to $m_{\phi^0} = 0$ is ruled out! The success of the LEP experiments in obtaining such restrictive limits is remarkable, and greatly exceeded the expectations of both experimentalists and theorists (described in §3.2). As for the near future, a high statistics search in Z decays with 10^7 Z's should be sensitive to Higgs masses up to $m_{\phi^0} \simeq 60$–70 GeV.

The LEP collaborations have also presented results of searches for Higgs bosons of the two-Higgs-doublet model (described in Chapter 4)—a charged Higgs pair, $H^\pm$, a CP-even neutral scalar h^0, and a CP-odd scalar A^0. The LEP limits on the charged Higgs mass are the least model-dependent [see eq. (4.57)], and depend only on $m_{H^\pm}$ and the ratio of branching fractions: $BR(H^+ \to \tau^+\nu)/BR(H^+ \to c\bar{s})$. All four LEP detector collaborations quote similar limits on the charged Higgs mass [46]. The most conservative limit is $m_{H^\pm} > 36.5$ GeV (at 95% confidence level) if the hadronic decay modes dominate. The limit improves if some fraction of the charged Higgs bosons decay to $\tau\nu$, reaching $m_{H^\pm} > 43$ GeV if $BR(H^+ \to \tau^+\nu) = 100\%$. The LEP limits on the masses of h^0 and A^0 are obtained by searching simultaneously for $Z \to h^0 f\bar{f}$ and $Z \to h^0 A^0$. The ZZh^0 and $Zh^0 A^0$ couplings which govern these two decay rates depend on one particular combination of model parameters, $\sin(\beta - \alpha)$ [see eqs. (4.20) and (4.57)]. One can use the LEP data to deduce limits on m_{h^0} and m_{A^0} as a function of $\sin(\beta - \alpha)$.

Stronger limits can be obtained in the minimal supersymmetric extension of the Standard Model (MSSM), where $\sin(\beta - \alpha)$ is determined (at tree-level) by m_{h^0} and m_{A^0} [see, e.g., eq. (4.80)]. The four LEP detector collaborations have made an extensive search for h^0 and A^0 of the MSSM [47]. For example, the L3 collaboration claims to rule out nearly the entire region of h^0 and A^0 masses up to 41.5 GeV (at 95% confidence level). The results of the other LEP experiments are similar.

Future running of LEP (and LEP-II) can significantly constrain the MSSM. In §4.2, we learn that one of the most important predictions of the MSSM is that $m_{h^0} \leq m_Z$ [see eq. (4.78)]. This seems to imply that experiments running at LEP-II operating at design luminosity could either discover the Higgs boson or rule out the MSSM. (Whether this is possible to do in practice depends on whether these experiments can rule out a Higgs boson with $m_{h^0} \approx m_Z$ [17].) However, the Higgs mass bound quoted above is based on *tree-level* mass relations obtained in eq. (4.75). In particular, $m_{h^0} \leq m_Z$ need not be respected when radiative corrections are incorporated. The size of the radiative corrections are surprisingly large! An important clue can be found in a paper by M. Berger [48], who considered the radiative corrections to the Higgs mass sum rule: $m_{h^0}^2 + m_{H^0}^2 = m_{A^0}^2 + m_Z^2$ [*c.f.* eq. (4.75)]. A number of groups have recently addressed the question of one-loop radiative corrections to the light Higgs scalar mass in the MSSM [49-51]. The results indicate a very large positive mass shift if the top-quark mass is large. For example,

Haber and Hempfling [49] obtain the following upper bound for m_{h^0} (assuming $m_{A^0} > m_Z$) in the limit of $m_Z \ll m_t \ll M_{\tilde{t}}$ [where top-squark ($\tilde{t}_L - \tilde{t}_R$) mixing is neglected]

$$m_{h^0}^2 \lesssim m_Z^2 + \frac{3g^2 m_t^4}{16\pi^2 m_W^2} \left\{ \ln\left(\frac{M_{\tilde{t}}^2}{m_t^2}\right) \left[\frac{2m_t^4 - m_t^2 m_Z^2}{m_Z^4}\right] + \frac{m_t^2}{3m_Z^2} \right\}. \qquad (1.10)$$

For $M_{\tilde{t}} = 1$ TeV, we find positive mass shifts in m_{h^0} of about 20 GeV for $m_t = 150$ GeV, and 50 GeV for $m_t = 200$ GeV. Even when $\tan\beta = 1$ (so that $m_{h^0} = 0$ at tree-level), one finds a large mass shift due to radiative corrections of similar size. Clearly, these radiative corrections will have a significant impact on the search for the Higgs bosons of the MSSM at LEP and LEP-II. Further study of the radiative corrections in the MSSM Higgs sector is in progress, although the dominant effect occurs in the mass shifts such as those just described. These corrections can be incorporated in a straightforward way in the analysis of LEP phenomenology [51,52]. The reader should keep in mind the possible importance of these radiative corrections while reading the discussion of tree-level phenomenology in §4.2.

REFERENCES

1. P.W. Higgs, *Phys. Lett.* **12** (1964) 132, *Phys. Rev. Lett.* **13** (1964) 508, *Phys. Rev.* **145** (1966) 1156; F. Englert and R. Brout, *Phys. Rev. Lett.* **13** (1964) 321; G.S. Guralnik, C.R. Hagen and T.W.B. Kibble, *Phys. Rev. Lett.* **13** (1964) 585; T.W.B. Kibble, *Phys. Rev.* **155** (1967) 1554.
2. S. Weinberg, *Phys. Rev. Lett.* **19** (1967) 1264; A. Salam, *Proceedings of the 8th Nobel Symposium* (Stockholm), edited by N. Svartholm (Almqvist and Wiksell, Stockholm, 1968) p. 367.
3. S. Glashow, *Nucl. Phys.* **22** (1961) 579.
4. M. Veltman, *Acta Phys. Pol.* **B8** (1977) 475.
5. B.W. Lee, C. Quigg and G.B. Thacker, *Phys. Rev. Lett.* **38** (1977) 883; *Phys. Rev.* **D16** (1977) 1519.
6. L. Susskind, *Phys. Rev.* **D20** (1979) 2619; S. Weinberg, *Phys. Rev.* **D19** (1979) 1277.
7. I.J.R. Aitchison and A.J.G. Hey, *Gauge Theories in Particle Physics* (Adam Hilger, Bristol, 1982).
8. T.-P. Cheng and L.-F. Li, *Gauge Theory of Elementary Particle Physics* (Oxford University Press, Oxford, 1984).
9. H.E. Haber and G.L. Kane, *Phys. Rep.* **117C** (1985) 75.
10. H.E. Haber, G.L. Kane, and T. Sterling, *Nucl. Phys.* **B161** (1979) 493.
11. S. Dimopoulos and L. Susskind, *Nucl. Phys.* **B155** (1979) 237; E. Eichten, and K.D. Lane, *Phys. Lett.* **90B** (1980) 125.

12. J.S. Bell, *Nucl. Phys.* **B60** (1973) 427; C.H. Llewellyn Smith, *Phys. Lett.* **46B** (1973) 233; J.M. Cornwall, D.N. Levin, and G. Tiktopoulos, *Phys. Rev. Lett.* **30** (1973) 1268, and *Phys. Rev.* **D10** (1974) 1145.
13. A.D. Linde, *JETP Lett.* **23** (1976) 64; *Phys. Lett.* **62B** (1976) 435; S. Weinberg, *Phys. Rev. Lett.* **36** (1976) 294.
14. R.A. Flores, and M. Sher, *Ann. Phys. (NY)* **148** (1983) 95.
15. D.A. Dicus and V.S. Mathur, *Phys. Rev.* **D7** (1973) 3111.
16. N. Cabibbo, L. Maiani, G. Parisi, R. Petronzio, *Nucl. Phys.* **B158** (1979) 295.
17. Z. Kunszt and W.J. Stirling, *Phys. Lett.* **B242** (1990) 507; N. Brown, *Z. Phys.* **C49** (1991) 657; V. Barger, R.J.N. Phillips and K. Whisnant, *Phys. Rev.* **D43** (1991) 1110; J.F. Gunion and L. Roszkowski, to appear in the Proceedings of the 1990 Snowmass Workshop.
18. E. Eichten, I. Hinchliffe, K. Lane and C. Quigg, *Phys. Rev.* **D34** (1986) 1547.
19. J.F. Gunion and H.E. Haber, *Nucl. Phys.* **B278** (1986) 449.
20. J.F. Gunion, G. Kane, C.P. Yuan, H. Sadrozinski, A. Seiden and A.J. Weinstein, *Phys. Rev.* **D40** (1989) 2223.
21. J.F. Gunion, H.E. Haber, S. Komamiya, H. Yamamoto, and A. Barbaro-Galtieri, in *Proceedings of the 1987 Berkeley Workshop on Experiments, Detectors and Experimental Areas for the Supercollider*, edited by R. Donaldson and M. Gilchriese (World Scientific, Singapore, 1988), p. 110.
22. G.J. Gounaris and A. Nicolaidis, *Phys. Lett.* **102B** (1981) 144; *Phys. Lett.* **109B** (1982) 221.
23. H.E. Haber and G.L. Kane, *Nucl. Phys.* **B250** (1985) 716.
24. J. Pantaleone, *Phys. Lett.* **172B** (1986) 261.
25. H.E. Haber, I. Kani, G.L. Kane, and M. Quiros, *Nucl. Phys.* **B283** (1987) 111.
26. C.A. Nelson, *Phys. Rev.* **D30** (1984) 1937 [E: **D32** (1985) 1848]; J.R. Dell'Aquila and C.A. Nelson, *Phys. Rev.* **D33** (1986) 80, 93; *Nucl. Phys.* **B320** (1989) 86.
27. J. Ellis, M.K. Gaillard, and D.V. Nanopoulos, *Nucl. Phys.* **B106** (1976) 292.
28. A.I. Vainshtein, V.I. Zakharov, and M.A. Shifman, *Sov. Phys. Usp.* **23** (1980) 429.
29. L.B. Okun, *Leptons and Quarks*, (North-Holland, Amsterdam, 1982).
30. A.A. Ansel'm, N.G. Ural'tsev, and V.A. Khoze, *Sov. Phys. Usp.* **28** (1985) 113.
31. J. E. Kim *Phys. Rep.* **149** (1987) 1.
32. J.D. Bjorken, in *Proceedings of the 7th Moriond Workshop on New and Exotic Phenomena*, Les Arcs-Savoie-France (1987), edited by O. Fackler and J. Tran Thanh Van (Editions Frontières, Gif-sur-Yvette, France, 1987).
33. M. Chanowitz, *Ann. Rev. Nucl. Part. Phys.* **38** (1988) 323.
34. R. Cahn, *Rep. Prog. Phys.* **52** (1989) 389.

35. *The Standard Model Higgs Boson*, edited by M.B. Einhorn, (North-Holland, Amsterdam, 1991).

36. D. Callaway, *Phys. Rep.* **167** (1988) 241.

37. M. Sher, *Phys. Rep.* **179** (1989) 273.

38. F. Abe *et al.* (CDF Collaboration), *Phys. Rev. Lett.* **64** (1990) 142; *Phys. Rev.* **D43** (1991) 664.

39. K. Sliwa (CDF Collaboration), in Z^0 *Physics, Proceedings of the XXVth Rencontre de Moriond*, Les Arcs, France, March 4–11, 1990, edited by J. Tran Thanh Van (Editions Frontières, Gif-sur-Yvette, 1990), p. 459.

40. J.J. Hernandez *et al.* (Particle Data Group), *Phys. Lett.* **B239** (1990) 1.

41. F. Halzen and K. Mursula, *Phys. Rev. Lett.* **51** (1983) 857; K. Hikasa, *Phys. Rev.* **D29** (1984) 1939; N.G. Deshpande, G. Eilam, V. Barger and F. Halzen, *Phys. Rev. Lett.* **54** (1985) 1757; A.D. Martin, R.G. Roberts and W.J. Stirling, *Phys. Lett.* **189B** (1987) 220.

42. F. Abe *et al.* (CDF Collaboration), FERMILAB-PUB-90-229-E (1990), to be published in *Phys. Rev. D*.

43. U. Amaldi, A. Bohm, L.S. Durkin, P. Langacker, A.K. Mann, W.J. Marciano, A. Sirlin, and H.H. Williams, *Phys. Rev.* **D36** (1987) 1385.

44. D. Decamp *et al.* (ALEPH Collaboration) *Phys. Lett.* **B245**, 289 (1990); **B246**, 306 (1990); P. Abreu *et al.* (DELPHI Collaboration), *Nucl. Phys.* **B342**, 1 (1990); B. Adeva *et al.* (L3 Collaboration) *Phys. Lett.* **B248**, 203 (1990); **B252**, 518 (1990); M. Akrawy *et al.* (OPAL Collaboration) *Phys. Lett.* **B236**, 224 (1990); **B251**, 211 (1990).

45. D. Decamp *et al.*, CERN-PPE/91-19 (1991), presented at the Aspen, La Thuile and Moriond conferences (Winter 1991).

46. D. Decamp *et al.* (ALEPH Collaboration), *Phys. Lett.* **B241**, 623 (1990); P. Abreu *et al.* (DELPHI Collaboration), *Phys. Lett.* **B241**, 449 (1990); B. Adeva *et al.* (L3 Collaboration) *Phys. Lett.* **B251**, 299 (1990); M.Z. Akrawy *et al.* (OPAL Collaboration), *Phys. Lett.* **B242**, 299 (1990).

47. D. Decamp *et al.* (ALEPH Collaboration), *Phys. Lett.* **B237**, 291 (1990); P. Abreu *et al.* (DELPHI Collaboration), *Phys. Lett.* **B245**, 276 (1990); B. Adeva *et al.* (L3 Collaboration) *Phys. Lett.* **B251**, 311 (1990); M.Z. Akrawy *et al.* (OPAL Collaboration), *Z. Phys.* **C49**, 1 (1991).

48. M. Berger, *Phys. Rev.* **D41** (1990) 225.

49. H.E. Haber and R. Hempfling, *Phys. Rev. Lett.* **66** (1991) 1815.

50. Y. Okada, M. Yamaguchi, and T. Yanagida, *Prog. Theor. Phys.* **85**, (1991) 1.

51. J. Ellis, G. Ridolfi and F. Zwirner, *Phys. Lett.* **B257** (1991) 83.

52. R. Barbieri and M. Frigeni, *Phys. Lett.* **B258** (1991) 395.

Chapter 2

Properties of a Standard Model Higgs Boson

In this book we will repeatedly emphasize that many people who have studied Higgs physics do not expect a single, neutral boson of some definite mass to occur as the only direct manifestation of electroweak symmetry breaking. Nevertheless, for historical reasons, for definiteness, and to provide a standard of comparison we summarize here a number of properties of the Higgs boson of the minimal Standard Model (denoted hereafter by SM). As mentioned earlier, such a boson is characterized by a complete knowledge of its couplings to all quarks, leptons, and gauge bosons, and complete ignorance of its mass. Thus we must consider how the ϕ^0 decays, and, conversely, what primary states can decay to the ϕ^0, as m_{ϕ^0} is varied. In addition, we shall consider in this chapter the limits that can be placed upon the ϕ^0 from radiative corrections at low energy, and from a variety of theoretical arguments.

2.1 Higgs Boson Couplings and Decays

If the SU(2) doublet Higgs field ϕ acquires a vacuum expectation value, couplings to gauge bosons and fermions arise directly from the Standard Model Lagrangian. For example, for one family consisting of left-handed quark and lepton doublets $q_L = \begin{pmatrix} u_L \\ d_L \end{pmatrix}$, $\ell_L = \begin{pmatrix} \nu_L \\ e_L \end{pmatrix}$ and the corresponding right-handed singlets u_R, d_R and e_R, we have

$$\mathcal{L} = (D_\mu \phi)^* (D^\mu \phi) - g_d \bar{q}_L \phi d_R - g_u \bar{q}_L \phi^c u_R - g_\ell \bar{\ell}_L \phi e_R \qquad (2.1)$$

with $\phi^c = i\tau_2 \phi^*$, and

$$D^\mu = \partial^\mu + ig \frac{\vec{\tau}}{2} \cdot \vec{W}^\mu + ig' \frac{Y}{2} B^\mu, \qquad (2.2)$$

where $Y = 1$ for the Higgs field and τ_i are the Pauli matrices with $Tr(\tau_i \tau_j) = 2\delta_{ij}$. Note that the isospin matrices are $\vec{T} = \vec{\tau}/2$, and the electric charge is

$Q = T_3 + Y/2$. By writing

$$\phi \rightarrow \frac{1}{\sqrt{2}} \begin{pmatrix} 0 \\ v + h(x) \end{pmatrix}, \qquad (2.3)$$

we obtain

$$\begin{aligned} m_W &= g\,v/2 \\ m_f &= g_f\,v/\sqrt{2} \\ v &= (\sqrt{2}G_F)^{-1/2} = 246 \text{ GeV} \end{aligned} \qquad (2.4)$$

where $f = u, d$ or ℓ, and the couplings of the SM ϕ^0 given by the Feynman rules of fig. 2.1 are generated. In the above, g is the standard SU(2) gauge group coupling; g' will be used to denote the Standard Model U(1) coupling strength and g_s will be used for the SU(3) coupling strength. We shall also employ $\alpha_2 \equiv g^2/(4\pi)$, $\alpha_1 \equiv g'^2/(4\pi)$, and $\alpha_s \equiv g_s^2/(4\pi)$.

Branching Ratios

$\phi^0 \rightarrow f\bar{f}$

Since the couplings of ϕ^0 to fermions are proportional to the fermion mass, the decay branching ratio to any fermion f is proportional to m_f^2. The partial width to any fermion channel is [1]

$$\Gamma\left(\phi^0 \rightarrow \bar{f}f\right) = \frac{N_c g^2 m_f^2}{32\pi m_W^2} \beta^3 m_{\phi^0} \qquad (2.5)$$

where N_c is 1 for leptons and 3 for quarks, and $\beta^2 = 1 - 4m_f^2/m_{\phi^0}^2$. The one-loop electroweak radiative corrections for fermionic final states have been computed in refs. 2 and 3. The latter reference focuses on the limit of large Higgs mass, $m_{\phi^0} \gg m_Z$. The corrections can approach 10% at $m_{\phi^0} \sim 1$ TeV, but are small for Higgs boson masses such that $f\bar{f}$ final states have a substantial branching ratio.

The QCD one-loop corrections to the tree level result in the case of quark final states have been computed in refs. 4, 5 and 6. Interpretation of the results is not altogether straightforward, so we present a brief summary. Because quarks are confined, there is no "natural" definition of the quark mass. In QCD, one can define a quark mass (like the coupling constant) that runs as a function of the momentum scale of the physical process. Thus we must consider $m(Q)$ where Q is the physical process energy scale. The most consistent procedure for defining the "physical" quark mass M is to require that $m(M) = M$; i.e., the running mass at the position of the quark propagator pole must agree with the location of that pole. In leading-log QCD this leads

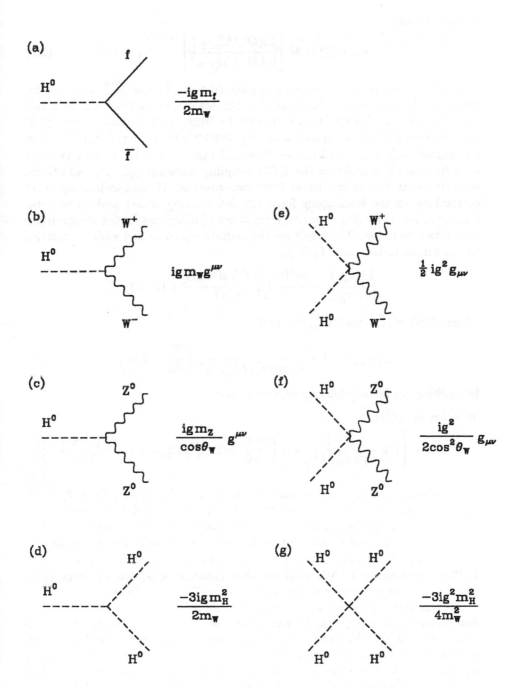

Figure 2.1 Feynman rules for the Standard Model Higgs boson.

to the relation

$$m_{LL}(Q) = M \left[\frac{\ln(Q^2/\Lambda_{QCD}^2)}{\ln(M^2/\Lambda_{QCD}^2)} \right]^{\gamma_0/2\beta_0}, \tag{2.6}$$

where Λ_{QCD} is the QCD scale parameter (defined in the $\overline{MS}$ scheme), for which a value in the neighborhood of ~ 150 MeV is extracted from available data. The leading-log QCD correction to the Higgs width formula of eq. (2.5) consists in replacing the quark mass m_f everywhere by $m_{LL}(m_{\phi^0})$, both in the explicit m_f^2 factor and in the threshold factor. The next stage is to go to next-to-leading order in the QCD coupling constant g_s. The additional corrections at this order derive from two sources: i) next-to-leading order corrections to the leading-log form for the running mass; and ii) next-to-leading order corrections deriving from order g_s^2 diagrams for the Higgs–Higgs correlation function. The result for the radiatively corrected width Γ relative to the zeroth order result Γ_0 is [5]

$$\frac{\Gamma(m_{\phi^0})}{\Gamma_0(m_{\phi^0})} = \frac{m^2(m_{\phi^0})\beta^3[m(m_{\phi^0})]}{M^2\beta^3(M)} C(m_{\phi^0}), \tag{2.7}$$

where $\beta(m) = (1 - 4m^2/m_{\phi^0}^2)^{1/2}$ and

$$C(m_{\phi^0}) = 1 + \frac{1}{\beta_0 \ln(m_{\phi^0}^2/\Lambda_{QCD}^2)} \left(\frac{K_1}{K_0} - 2\gamma_0 \right). \tag{2.8}$$

In addition, in next-to-leading order we have

$$m(Q) = m_{LL}(Q)$$

$$\times \left[1 + \frac{1}{\beta_0 \ln(Q^2/\Lambda_{QCD}^2)} \left(\frac{\gamma_0\beta_1}{2\beta_0^2} \ln\ln(Q^2/\Lambda_{QCD}^2) + \frac{\gamma_0\beta_1 - \gamma_1\beta_0}{2\beta_0^2} \right) \right]. \tag{2.9}$$

In the above equations the various constants are: $K_1 = 5/\pi^2$, $K_0 = 3/(4\pi^2)$, $\beta_0 = 11 - (2/3)N_f$, $\beta_1 = 102 - (38/3)N_f$, $\gamma_0 = -8$, and $\gamma_1 = -(404/3) + (40/9)N_f$. It is perhaps amusing to note that $C = 1 + (17/3\pi)\alpha_s(m_{\phi^0})$ at large m_{ϕ^0}. This should be compared to the much smaller correction to $R = \sigma(e^+e^- \to \text{hadrons})/\sigma(e^+e^- \to \mu^+\mu^-)$ obtained in the same way, where the 17/3 is replaced by 1. We shall see that another unexpectedly large QCD correction emerges when considering radiative $\Upsilon \to \phi^0\gamma$ decays.

Numerically, it is straightforward to evaluate the above corrections. We first focus on the top quark, for which we consider two possible physical mass values, 40 and 80 GeV. Since we are dealing with results above the $t\bar{t}$ threshold, we take $N_f = 6$ when evaluating the various constants appearing in the above equations. Let us consider first the leading-log correction obtained by using $m_f = m_{LL}(\phi^0)$ instead of $m_f = M$ in all locations in eq. (2.5). Let Γ_0 be the width evaluated using $m_f = M$. The ratio Γ/Γ_0 is plotted in this leading-log approximation as the dashed line in fig. 2.2. Obviously, the correction

Corrections to $\Gamma(\phi^0 \rightarrow q\bar{q})$

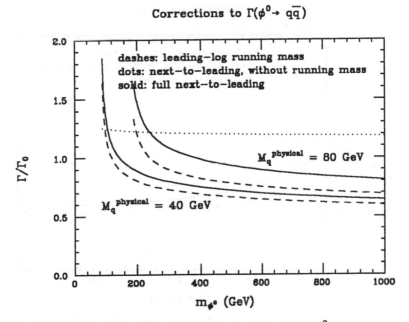

Figure 2.2 Corrections to the $q\bar{q}$ decay width of the ϕ^0 deriving from various sources: a) the leading-log running-mass correction (dashes); b) next-to-leading corrections C [see eq. (2.8)] other than those incorporated in the running-mass correction (dots); and c) the full next-to-leading corrections (solid), incorporating running-mass corrections to both leading and next-to-leading order in addition to those corrections included in b). In all cases we plot $\Gamma(\phi^0)/\Gamma_0(\phi^0)$, where Γ_0 is computed from eq. (2.5) with $m_f = M$, where M is the physical quark mass as defined in the text. We have chosen $M = 40$ and 80 GeV (top-quark-like masses), $N_f = 6$ and $\Lambda_{QCD} = 150$ MeV.

is large and generally negative once away from the decay channel threshold. The reason that the correction is negative is simply that the running mass appropriate to the Higgs mass scale, predicted by eq. (2.6), is smaller than the on-shell mass defined at the lower energy scale M. For higher values of the physical mass M, the correction is smaller at a given m_{ϕ^0} since there has been less evolution in going from scale M to scale m_{ϕ^0}. At next-to-leading order, the corrections from the order g_s^2 graphs for the Higgs–Higgs correlation function, C, are indicated by the dotted line and are generally positive. But full next-to-leading corrections to the ϕ^0 width, relative to Γ_0, are still large and negative, and rather similar to those found in the leading-log approximation. It is these corrections (indicated by the solid line) that should be employed in order to obtain the most reliable available prediction

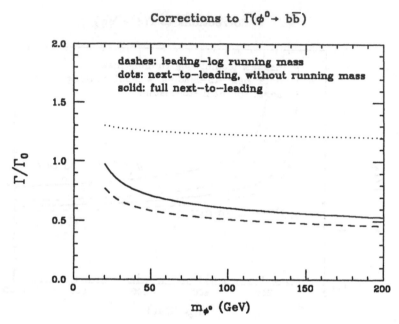

Figure 2.3 Corrections to the $b\bar{b}$ decay width of the ϕ^0 deriving from various sources. The notation for the curves is the same as in fig. 2.2. We have taken $m_b = 4.5$ GeV.

for the decay width.

In most of the discussions to follow we have chosen to use the tree level results for the $q\bar{q}$ decay widths of the Higgs boson. In some instances the QCD corrections could lead to some correction to these discussions. The most important instances in which this will arise are in the considerations involving rare decays. When the $t\bar{t}$ mode is not allowed kinematically, the largest decay channel of the ϕ^0 will be the $b\bar{b}$ mode. By computing the width for the $b\bar{b}$ mode using a b mass that corresponds to the approximate threshold in e^+e^- annihilation, we will be overestimating $\Gamma(\phi^0 \rightarrow b\bar{b})$, and thereby underestimating some of the rare decay rates. In short we will be erring on the conservative side. To illustrate the potential improvement in the rare decay rates due to QCD corrections to the $b\bar{b}$ mode, we plot in fig. 2.3 the QCD corrections to the $b\bar{b}$ mode, assuming the top is too heavy to appear in ϕ^0 decay. We see that the full QCD correction is predicted to cause as much as a factor of two suppression in the $\phi^0 \rightarrow b\bar{b}$ decay rate in the interesting intermediate mass range up to $\phi^0 = 2m_W$, yielding up to a factor of two improvement in the rare decay mode rates to be presented based on the tree level $b\bar{b}$ width. This statement, of course, assumes that radiative corrections to the rare decay widths of interest are not large. We do not anticipate that they will be, but explicit computations are not available. When the $t\bar{t}$ mode is kinematically allowed, the analysis of QCD corrections that we have presented shows that a reasonably good approximate way to include the effects of the corrections in these discussions is to think of the top mass referred to as being the running mass defined at the mass of the ϕ^0, rather than the physical on-shell mass of the top.

$\phi^0 \to WW, ZZ$

For gauge bosons the widths are [7]

$$\Gamma\left(\phi^0 \to ZZ\right) = \frac{g^2}{128\pi} \frac{m_{\phi^0}^3}{m_W^2} \sqrt{1 - x_Z} \left(1 - x_Z + \tfrac{3}{4}x_Z^2\right) \qquad (2.10)$$

where $x_Z = 4m_Z^2/m_{\phi^0}^2$ ($x_Z \leq 1$ for an energetically allowed decay), and

$$\Gamma\left(\phi^0 \to W^+W^-\right) = \frac{g^2}{64\pi} \frac{m_{\phi^0}^3}{m_W^2} \sqrt{1 - x_W} \left(1 - x_W + \tfrac{3}{4}x_W^2\right), \qquad (2.11)$$

where $x_W = 4m_W^2/m_{\phi^0}^2$. The leading term as $x_V \to 0$ ($V = W, Z$) grows proportional to $m_{\phi^0}^3$ because of the presence of the m_W^{-2} factor, which in turn arises from the $1/m_W$ in the longitudinal polarization vector of each of the final V's. The $1/m_W$ factor in a longitudinal polarization vector is a necessary consequence of the Lorentz boost to a moving gauge boson, $\epsilon^\mu = (0; 0, 0, 1) \to (\beta\gamma; \gamma\hat{z})$. An explicit calculation then gives

$$\frac{\Gamma(\phi^0 \to V_T V_T)}{\Gamma(\phi^0 \to V_L V_L)} = \frac{\tfrac{1}{2}x_V^2}{(1 - \tfrac{1}{2}x_V)^2} \qquad (2.12)$$

where $V_T(V_L)$ is a transversely (longitudinally) polarized massive vector boson V. Since the longitudinal gauge boson states arise from the Higgs mechanism (and are in some sense the Goldstone bosons arising from the breaking of a global symmetry when the Higgs field acquired a vacuum expectation value), the dominance of the longitudinal vector bosons in the decay of a heavy Higgs boson is a profound and testable prediction of the Standard Model Higgs mechanism. A useful form to remember is that summed over $W^\pm, Z$,

$$\Gamma\left(\phi^0 \to VV\right) \simeq \tfrac{1}{2}m_{\phi^0}^3 \quad \text{[TeV units]}, \qquad (2.13)$$

where m_{ϕ^0} and Γ are in TeV. This gives $\Gamma \simeq m_{\phi^0}$ when $m_{\phi^0} \simeq 1.4$ TeV, so once the mass is larger than about a TeV it becomes experimentally very problematical (and perhaps not very useful) to separate the Higgs resonance pole from the VV continuum.

The complete set of one-loop radiative corrections to the decay widths for $\phi^0 \to VV$ has been computed in ref. 2.* For $m_{\phi^0} < 1$ TeV, the maximum size of the corrections is of order 15%. The radiative corrections have a minimum at $m_{\phi^0} = 500$ GeV for $\phi^0 \to W^+W^-$ and at 600 GeV for $\phi^0 \to ZZ$. For m_{ϕ^0} less than these critical masses, the radiative corrections decrease with increasing m_{ϕ^0}. Only above $m_{\phi^0} \sim 700$ GeV do the corrections begin to grow rapidly with increasing m_{ϕ^0}.

Some insight into the size of the radiative corrections can be obtained by looking at the large Higgs mass limit. Marciano and Willenbrock [8] have found the leading terms of $\mathcal{O}(G_F m_{\phi^0}^2)$ with the result,

$$\Gamma(\phi^0 \to VV) = \Gamma_0(\phi^0 \to VV) \left[1 + \frac{G_F m_{\phi^0}^2}{\sqrt{2}\pi^2} \left(\frac{19}{16} - \frac{3\sqrt{3}\pi}{8} + \frac{5\pi^2}{48}\right)\right]$$

$$= \Gamma_0(\phi^0 \to VV) \left[1 + \frac{G_F m_{\phi^0}^2}{\sqrt{2}\pi^2} (0.175)\right],$$

$$(2.14)$$

where Γ_0 is the tree level result. Hence for a 1 TeV Higgs boson, the width is increased by 15%. When the radiative corrections are included, a 1.3 TeV Higgs boson has a width equal to its mass.

The effect of large fermion masses on the decay widths can also be easily found [3]. The corrections of $\mathcal{O}(G_F m_F^2)$ (where m_F is a heavy fermion mass) decrease the partial width of Higgs boson decays to vector boson pairs by up to 75% for mass splittings within the fermion doublet of less than 200 GeV. The heavy fermion corrections have the opposite sign from the heavy Higgs corrections and thus the possibility of large cancellations between the radiative corrections exists.

The $f\bar{f}$ and VV decays are the only two-body ϕ^0 decays that occur at tree level. Allowing a loop of gauge bosons and/or fermions, the decays $\phi^0 \to \gamma\gamma$, $\phi^0 \to Z\gamma$, $\phi^0 \to gg$ arise. Since the loop suppression will typically be of order $\alpha_s^2/16\pi$ or $\alpha^2/16\pi$, perhaps enhanced by the contribution of a large number of diagrams, especially those containing heavy particles, these branching ratios will be of order 10^{-4}. That is large enough to be important under some circumstances, so let us examine these rare decays in some detail.

---

*Re(c_{2V}) in eq. (5.4) of ref. 2 has the wrong sign.

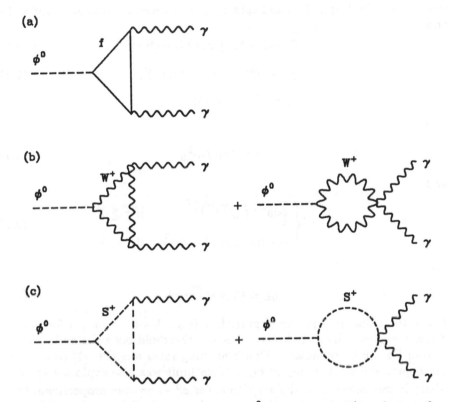

Figure 2.4 Diagrams contributing to $\phi^0 \to \gamma\gamma$ for spin-1/2, spin-1, and spin-0 loops.

$\phi^0 \to \gamma\gamma$

To be completely general, let us include intermediate particles of spin-0, spin-1/2, and spin-1 in the loop. (See fig. 2.4.) The relevant part of the interaction Lagrangian is

$$\mathcal{L}_{\text{int}} = \frac{-gm_f}{2m_W}\,\bar{\psi}\psi\phi^0 + gm_W W^+_\mu W^{\mu -}\phi^0 - \frac{gm_H^2}{m_W}\,H^+H^-\phi^0 + \dots, \quad (2.15)$$

where $H^\pm$ is a charged scalar. The width is given by [9,10]

$$\Gamma\left(\phi^0 \to \gamma\gamma\right) = \frac{\alpha^2 g^2}{1024\pi^3}\,\frac{m_{\phi^0}^3}{m_W^2}\,\left|\sum_i N_{ci}\, e_i^2\, F_i\right|^2, \quad (2.16)$$

where $i =$ spin-0, spin-1/2, and spin-1, e_i is the electric charge in units of e, and

$$F_1 = 2 + 3\tau + 3\tau(2 - \tau)f(\tau)$$

$$F_{1/2} = -2\tau[1 + (1 - \tau)f(\tau)], \tag{2.17}$$

$$F_0 = \tau[1 - \tau f(\tau)],$$

with

$$\tau = 4m_i^2/m_{\phi^0}^2 \tag{2.18}$$

and

$$f(\tau) = \begin{cases} \left[\sin^{-1}\left(\sqrt{1/\tau}\right)\right]^2, & \text{if } \tau \geq 1, \\ -\frac{1}{4}\left[\ln(\eta_+/\eta_-) - i\pi\right]^2, & \text{if } \tau < 1; \end{cases} \tag{2.19}$$

where

$$\eta_\pm \equiv (1 \pm \sqrt{1 - \tau}). \tag{2.20}$$

N_{ci} is the color multiplicity of particle i (e.g., 1 for leptons, 3 for quarks). These formulae fully include the necessary threshold behavior and kinematic factors for arbitrary masses. This branching ratio was first [1] computed in the approximation that $m_i \gg m_{\phi^0}$. This limit was also exploited by ref. 9, where it was shown that the amplitude for $\phi^0 \to \gamma\gamma$ was proportional to the contribution of the heavy particle to the photonic Gell-Mann–Low β function. This allows one to quickly check the signs of the contributions corresponding to loops with particles of different spins. For example, a fermion loop and gauge boson loop contribute with opposite sign. Since the original calculations, the matrix element for this process has been recomputed using the "non-linear R-gauge", which has significantly reduced the labor of the computation [11]. If the couplings of the particles in the loop differ from Standard Model ones, appropriate factors should be introduced. This will be discussed in chapter 4.

It is perhaps useful to note the limits of the three F functions of eq. (2.17) when τ becomes large, $i.e.$, when the particle in the loop is much heavier than the Higgs boson. We find

$$F_0 \to -\tfrac{1}{3}, \quad F_{1/2} \to -\tfrac{4}{3}, \quad F_1 \to 7. \tag{2.21}$$

Note how dominant the W loop contribution is, and how small any scalar loop is likely to be. Note also the opposite sign between fermion and W loops, as mentioned above.

Figures 2.5 and 2.6 show the ϕ^0 branching ratios for $\gamma\gamma$ and its competing channels. $BR(\phi^0 \to \gamma\gamma)$ is most uncertain in the light Higgs region $2m_e < m_{\phi^0} < 2m_\mu$, where the $\gamma\gamma$ decay width depends sensitively on how one treats

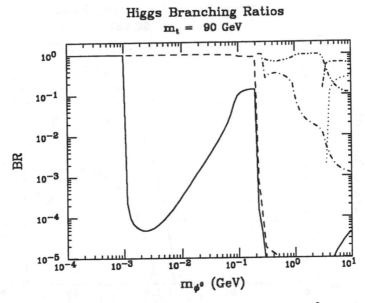

Higgs Branching Ratios
$m_t = 90$ GeV

Figure 2.5 The branching ratios for the decay of a light ϕ^0 to a variety
of channels. We have taken $m_t = 90$ GeV (although this graph is not
sensitive to this choice). The curves for the various channels are: solid =
$\gamma\gamma$; dashes = e^+e^-; dashdot = $\mu^+\mu^-$; dots = $\tau^+\tau^-$; dash-doubledot =
non-charmed hadrons; doubledash-dot = charmed hadrons. The branching
ratio for the non-charmed hadrons is obtained by adding the gg mode and
the light-quark ($u\bar{u}$, $d\bar{d}$, and $s\bar{s}$) modes. Light quark masses have been
chosen to match onto a chiral Lagrangian model below ~ 1 GeV, to be
discussed later in this section.

the light quark masses in the loop [see fig. 2.4(a)].* If we use constituent
quark masses for the u, d and s quarks, then the sum of all fermion loops
actually dominates the W boson loop [fig. 2.4(b)], in this mass region. In
contrast, if we use current quark masses ($m_{u,d} \sim 10$ MeV, $m_s \sim 150$ MeV),
the fermion and W loop contributions nearly cancel! (This would imply that
$BR(\phi^0 \to e^+e^-) \approx 100\%$ for $2m_e < m_{\phi^0} < 2m_\mu$.) In fig. 2.5, we have chosen
light quark masses which are closer in size to constituent quark masses, which
allows $BR(\phi^0 \to \gamma\gamma)$ to be as large as 14%. (See fig. 2.8 and the text which

---

*Leutwyler and Shifman [12] have shown that one can derive a low energy
theorem for the $\phi^0 \to \gamma\gamma$ matrix element which can be applied in the light Higgs
region. Their technique implicitly incorporates the light quark loops; the effects of
the non-zero quark masses are treated in chiral perturbation theory. The end result
is that $\sum N_{ci}e_i^2 F_i$ in eq. (2.16) is replaced by C, where C is a complex number
whose magnitude is of $\mathcal{O}(1)$. (In the chiral limit, $C = -19/9$.) For a graph of C
vs. m_{ϕ^0}, and for further details see ref. 12.

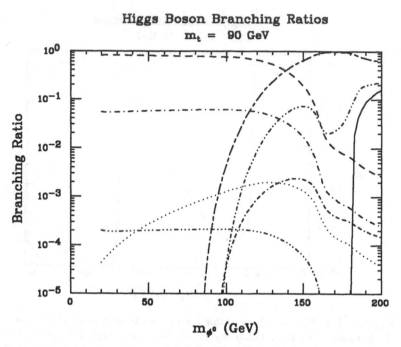

Figure 2.6 The branching ratios for ϕ^0 decay to a variety of channels, for $m_t = 90$ GeV. The curves for the various channels are: solid $= t\bar{t}$; dashes $= b\bar{b}$; dashdot $= \tau^+\tau^-$; longdash-shortdash $= WW$ or WW^* (with no W, W^* branching ratios included); dash-doubledot $= ZZ^*$ (no Z, Z^* branching ratios included); dots $= \gamma\gamma$; doubledash-dot $= Z\gamma$; dash-tripledot $= \mu^+\mu^-$. Since the gg decays are not experimentally useful, they are not plotted. Radiative corrections to $\Gamma(\phi^0 \rightarrow b\bar{b})$ [see fig. 2.2] have been included.

follows for a discussion of our parameter choices.) The QCD corrections to the allowed quark decay modes of the ϕ^0 are not included in fig. 2.5, since the relevant quark masses are themselves rather uncertain. The Higgs branching ratios for $m_{\phi^0} \geq 20$ GeV are depicted in fig. 2.6, for $m_t = 90$ GeV. Here, we have included QCD corrections to the $b\bar{b}$ and $t\bar{t}$ decay channels of the ϕ^0, using the formalism of eqs. (2.6)–(2.9). $BR(\phi^0 \rightarrow \gamma\gamma)$ is quite insensitive to m_t for $m_{\phi^0} < 2m_t$, since in this regime the gauge boson loop contribution is much larger than the top quark loop contribution, as indicated in eq. (2.21). If additional fermions, or new gauge bosons, or supersymmetric partners exist and couple to the Higgs boson, they would yield additional loop contributions. Consequently, combining information on $\phi^0 \rightarrow \gamma\gamma$ with the other loop-mediated processes such as $\phi^0 \rightarrow Z\gamma$ and $gg \rightarrow \phi^0$ can lead to especially valuable constraints. We will discuss the supersymmetric case in §4.2 (see also Appendix C).

$\underline{\phi^0 \to Z\gamma}$

The calculations here are similar to those for $\phi^0 \to \gamma\gamma$. The analytic results were first given in ref. 13. (See also the Appendix of ref. 14.) The decay width can be expressed in the form

$$\Gamma(\phi^0 \to Z\gamma) = \frac{1}{32\pi}|\mathcal{A}|^2 m_{\phi^0}^3 \left(1 - \frac{m_Z^2}{m_{\phi^0}^2}\right)^3,$$

$$\mathcal{A} = \frac{\alpha g}{4\pi m_W}(A_F + A_W),$$

$$A_F = \sum_f N_{cf} \frac{-2e_f(T_f^{3L} - 2e_f \sin^2\theta_W)}{\sin\theta_W \cos\theta_W}[I_1(\tau_f, \lambda_f) - I_2(\tau_f, \lambda_f)], \quad (2.22)$$

$$A_W = -\cot\theta_W \left\{4(3 - \tan^2\theta_W)I_2(\tau_W, \lambda_W)\right.$$

$$\left. + \left[\left(1 + \frac{2}{\tau_W}\right)\tan^2\theta_W - \left(5 + \frac{2}{\tau_W}\right)\right]I_1(\tau_W, \lambda_W)\right\},$$

where θ_W is the electroweak mixing angle. m_{ϕ^0} is the ϕ^0 mass, m_f is the mass of the fermion (with charge e_f, in units of e, and 3rd component of weak isospin, $T_f^{3L} = \pm 1/2$) in a given fermion loop, and A_F, A_W are, respectively the contributions from the fermion and W loops. N_{cf} is the color multiplicity of fermion f. We have defined the ratios

$$\tau_f \equiv \frac{4m_f^2}{m_{\phi^0}^2}, \qquad \lambda_f \equiv \frac{4m_f^2}{m_Z^2}, \qquad \tau_W \equiv \frac{4m_W^2}{m_{\phi^0}^2}, \qquad \lambda_W \equiv \frac{4m_W^2}{m_Z^2}. \quad (2.23)$$

The parametric integrals have been performed explicitly with the result

$$I_1(a, b) = \frac{ab}{2(a-b)} + \frac{a^2 b^2}{2(a-b)^2}[f(a) - f(b)] + \frac{a^2 b}{(a-b)^2}[g(a) - g(b)],$$

$$I_2(a, b) = -\frac{ab}{2(a-b)}[f(a) - f(b)],$$

$$g(\tau) = \begin{cases} \sqrt{\tau - 1}\sin^{-1}(1/\sqrt{\tau}), & \text{if } \tau \geq 1, \\ \frac{1}{2}\sqrt{1 - \tau}[\ln(\eta_+/\eta_-) - i\pi], & \text{if } \tau < 1, \end{cases}$$

$$(2.24)$$

with $f(\tau)$ and $\eta_\pm$ as given in eqs. (2.19) and (2.20), respectively. The value of A_F is small compared to A_W and is usually neglected. Note that, for an infinitely massive fermion, we have

$$\lim_{m_f \to \infty} I_1(\tau_f, \lambda_f) - I_2(\tau_f, \lambda_f) = -\tfrac{1}{2} \qquad (2.25)$$

and A_F never gets too large as m_f varies. It must be kept in mind, however, that if several very heavy fermions exist, their amplitudes add up coherently and may become a significant contribution.

Except for phase space suppression if m_{ϕ^0} is too close to m_Z, this branching ratio is larger than $\gamma\gamma$ by a factor of order $1/\sin^2\theta_W$. However, only a fraction of Z decays provide a usable signature and good resolution in $M_{Z\gamma}$, so this mode is unlikely to be a good one to search for a Higgs boson. This is discussed in more detail in §3.4. If a Higgs boson is detected, so that its mass is known, probably the $Z\gamma$ branching ratio can be measured and provide important information on the couplings of the Higgs boson. When combined with the $\phi^0 \to \gamma\gamma$ and $gg \to \phi^0$ couplings, this decay provides especially valuable information regarding possible loop contributions from new particles. Of course, if $m_{\phi^0} < m_Z$ it is the decay $Z \to \phi^0\gamma$ (more generally, $Z \to h\gamma$) that one studies, but the branching ratios are very small.

$\underline{\phi^0 \to W^*W, Z^*Z}$

These modes can be quite significant, especially as m_{ϕ^0} approaches the real WW and ZZ thresholds. The widths, summed over all channels available to the W^* or Z^* are given by [15]

$$\Gamma(\phi^0 \to W^*W) = \frac{g^4 m_{\phi^0}}{512\pi^3} F(m_W/m_{\phi^0}) \begin{cases} 3 & W^* \to tb \text{ not allowed} \\ 4 & W^* \to tb \text{ allowed,} \end{cases} \qquad (2.26)$$

and

$$\Gamma(\phi^0 \to Z^*Z) = \frac{g^4 m_{\phi^0}}{2048\pi^3} \frac{7 - \frac{40}{3}\sin^2\theta_W + \frac{160}{9}\sin^4\theta_W}{\cos^4\theta_W} F(m_Z/m_{\phi^0}), \qquad (2.27)$$

where

$$F(x) = -|1 - x^2| \left(\frac{47}{2}x^2 - \frac{13}{2} + \frac{1}{x^2} \right)$$
$$+ 3(1 - 6x^2 + 4x^4)|\ln x| + \frac{3(1 - 8x^2 + 20x^4)}{\sqrt{4x^2 - 1}} \cos^{-1}\left(\frac{3x^2 - 1}{2x^3} \right). \qquad (2.28)$$

Note that in the above formulae, $1/2 \le x \le 1$. For the case of ZZ^*, we have assumed that m_t is large enough such that $Z^* \to t\bar{t}$ is not an open channel. As we shall discuss, these modes may be useful for detection of the ϕ^0 at a high luminosity hadron collider, using the final states obtained when both the W and W^* or Z and Z^* decay leptonically [14].

$\underline{\phi^0 \to \psi, \Upsilon, \Theta + \gamma}$

These decays, $\phi^0 \to$ quarkonium $+\gamma$, are examined in refs. 13 and 14. A convenient expression for the width is given (using toponium notation, where Θ is the toponium state) by

$$\Gamma(\phi^0 \to \Theta\gamma) = \frac{3m_\Theta}{2\sin^2\theta_W m_{\phi^0}} \frac{m_t^2}{m_W^2} \left(1 - \frac{m_\Theta^2}{m_{\phi^0}^2}\right) \Gamma(\Theta \to \ell^+\ell^-), \qquad (2.29)$$

where $\Gamma(\Theta \to \ell^+\ell^-) \equiv 4\alpha^2 e_t^2 |R_s(0)|^2/m_\Theta^2$ is the toponium decay width to a single leptonic channel, $R_s(0)$ is the radial wave function at the origin, $e_t = 2/3$, and m_Θ is the toponium mass.

Probably only the decay to the heaviest quarkonium state allowed by energy conservation is large enough to detect, the others being suppressed by form factor effects. Even then, because not all decays of the quarkonium state are detectable, this mode can also be used mainly to measure couplings once a Higgs boson is found.

$\underline{\phi^0 \leftrightarrow gg}$

The $\phi^0 \to gg$ decay [16–19] can be obtained from eqs. (2.16)–(2.20) by keeping only quark loops and replacing $N_{ci}^2 e_i^4 \alpha^2$ by $2\alpha_s^2$. Then,

$$\Gamma(\phi^0 \to gg) = \frac{\alpha_s^2 g^2 m_{\phi^0}^3}{128\pi^3 m_W^2} \left| \sum_i \tau_i [1 + (1 - \tau_i)f(\tau_i)] \right|^2. \qquad (2.30)$$

We note that the expression $\tau[1 + (1 - \tau)f(\tau)]$ approaches $2/3$ when $\tau \to \infty$ (i.e., in the limit of the colored quark becoming much heavier than the ϕ^0). More generally, any colored particle that couples to the Higgs boson enters the loop.

The effective $\phi^0 gg$ interaction is also important because it provides the dominant production mechanism for Higgs bosons at a hadron collider. The production cross section at a hadron collider was derived in ref. 19. A convenient form is

$$\frac{d\sigma}{dy}(AB \to \phi^0 + X) = \frac{\pi^2 \Gamma(\phi^0 \to gg)}{8m_{\phi^0}^3} g_A\left(x_a, m_{\phi^0}^2\right) g_B\left(x_b, m_{\phi^0}^2\right), \qquad (2.31)$$

where

$$x_a = \frac{m_{\phi^0} e^y}{\sqrt{s}}, \qquad x_b = \frac{m_{\phi^0} e^{-y}}{\sqrt{s}}, \qquad (2.32)$$

(y is the rapidity of the ϕ^0) and $\Gamma(\phi^0 \to gg)$ is given in eq. (2.30). The function $g_A(x, Q^2)$ is the gluon distribution in hadron A evaluated at gluon momentum fraction x and scale Q^2.

It will be difficult to measure $\Gamma(\phi^0 \to gg)$ well from Higgs decays, but the $\phi^0 gg$ coupling may be well-determined from the production rate due to

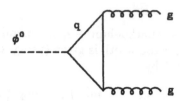

Figure 2.7 Triangle diagram for $\phi^0 gg$ coupling.

$gg \rightarrow \phi^0$. This coupling can provide a valuable sum rule on all heavy colored particles that get mass from the Higgs mechanism. When combined with the decay widths for $Z \rightarrow \phi^0 \gamma$ and $\phi^0 \rightarrow \gamma\gamma$, $gg \rightarrow \phi^0$ provides important constraints on possible heavy particle contributions to the various loops.[†]

Higgs–Nucleon Coupling

The coupling of the Higgs boson to nucleons, $g_{\phi^0 NN}$, is not well determined in the Standard Model. Knowledge of this coupling is crucial, however, in order to derive limits on light ($m_{\phi^0} \lesssim 20$ MeV) Higgs bosons from nuclear physics. In this section, we shall follow the analysis of ref. 20 (also reviewed in refs. 10 and 21). The theoretical underpinnings of this approach will be reviewed in greater depth in §2.2.

In the quark-parton model, the coupling of the Higgs boson to quarks, ψ_i, is given by

$$\mathcal{L} = -\frac{g}{2m_W} \sum_{i=u,d,s} m_i \overline{\psi}_i \psi_i \phi^0. \tag{2.33}$$

We have omitted the heavy quarks (c, b, t) since the heavy quark content of the nucleon is negligible.

The Higgs also couples to gluons via the triangle diagram of fig. 2.7. This is the same diagram responsible for the decay $\phi^0 \rightarrow gg$. For quark masses, m_Q, much larger than m_{ϕ^0}, the effective Lagrangian describing the gluon–gluon–Higgs coupling is

$$\mathcal{L} = \frac{g\alpha_s}{24\pi m_W} N_H G_{\mu\nu}^a G^{\mu\nu a} \phi^0, \tag{2.34}$$

where N_H is the number of heavy ($m_Q > m_{\phi^0}$) quarks in the loop. Note that this coupling is independent of the masses of the heavy quarks. To find the

---

[†]Similar remarks apply to the h^0, H^0, and A^0 of a typical two-doublet model to be discussed in chapter 4.

Higgs–nucleon–nucleon coupling we need

$$\langle N | \mathcal{H} | N \rangle = -\frac{g\alpha_s}{24\pi m_W} N_H \langle N | G^a_{\mu\nu} G^{\mu\nu a} \phi^0 | N \rangle, \qquad (2.35)$$

where in the static limit $\mathcal{H} \simeq -\mathcal{L}$.

The matrix element of $G_{\mu\nu} G^{\mu\nu}$ can be related to the trace of the QCD energy momentum tensor computed in a theory in which all heavy quarks are decoupled. Let us divide the quarks into n_L light quarks and N_H heavy quarks, where N_H is the number of quark flavors heavier than m_{ϕ^0} and Λ_{QCD}. (The u and d are always light, while the s may be either light or heavy.) Then, as a result of the conformal anomaly [22],

$$\Theta^\mu_\mu = -\frac{b\alpha_s}{8\pi} G^a_{\mu\nu} G^{\mu\nu a} + \sum_{i=1}^{n_L} m_i \overline{\psi}_i \psi_i, \qquad (2.36)$$

where $b = 11 - (2/3)n_L$ is the lowest order coefficient of the QCD beta function with only light quarks taken into account. The contribution of the light fermion masses to eq. (2.36) is usually neglected since it is much smaller than the gluon contribution. The matrix element of Θ^μ_μ between nucleon states is known at zero momentum transfer,

$$\langle N | \Theta^\mu_\mu | N \rangle \big|_{q=0} = m_N \langle N | \overline{\psi}_N \psi_N | N \rangle, \qquad (2.37)$$

where m_N is the nucleon mass. Combining eqs. (2.36) and (2.37), we find

$$\langle N | \mathcal{H} | N \rangle = \frac{g N_H m_N}{3 b m_W} \langle N | \overline{\psi}_N \psi_N | N \rangle \phi^0. \qquad (2.38)$$

This yields an effective Higgs–nucleon–nucleon coupling

$$\mathcal{L} = -g_{\phi^0 NN} \overline{\psi}_N \psi_N \phi^0, \qquad (2.39)$$

where

$$g_{\phi^0 NN} = \frac{g N_H m_N}{3 b m_W} \simeq 10^{-3}. \qquad (2.40)$$

As an example, for very light Higgs bosons in the MeV range, we take $n_L = 2$ and $N_H = 4$. Then the numerical value of $g_{\phi^0 NN}$ is approximately that which would result from the Lagrangian of eq. (2.33) with an effective mass of $m \simeq 2 N_H m_W / 3b \simeq 260$ MeV. The manipulations which led to eq. (2.40) are strictly valid only at zero momentum transfer. However, this approximation should be adequate for deriving limits on light Higgs bosons from nuclear physics (see §3.1).

We will often parametrize the Higgs–nucleon–nucleon coupling by,

$$g_{\phi^0 NN} \equiv \frac{g M_N}{2 m_W} \eta \qquad (2.41)$$

where η is a parameter which expresses our uncertainty in the coupling. Equation (2.40) corresponds to $\eta = 2 N_H / 3b \simeq 0.3$ for $n_L = 2$ and $N_H = 4$.

Let us attempt to include the effects of the strange quark mass in eq. (2.36). For example, let us make use of a recent estimate of Kaplan and Manohar [23]

$$\langle p|\, m_s \bar{s}s \,|p\rangle = 334 \pm 132 \text{ MeV} \, \langle p|\, \overline{\psi}_p \psi_p \,|p\rangle . \tag{2.42}$$

Then, we must modify the previous argument by including explicitly the effects of the s-quark in eqs. (2.33) and (2.36). We then find

$$g_{\phi^0 NN} = \frac{g}{2m_W} \left[\left(1 - \frac{2N_H}{3b} \right) \frac{\langle p|\, m_s \bar{s}s \,|p\rangle}{\langle p|\, \overline{\psi}_p \psi_p \,|p\rangle} + \frac{2N_H m_N}{3b} \right] \tag{2.43}$$

which implies $\eta \simeq 0.5 \pm 0.1$. Similar results have been obtained in refs. 24, 25, and 26.

The largest conceivable value for η would be $\eta = 1$ which would follow by naively replacing the quark fields with the nucleon fields in eq. (2.33). Many of the limits on light Higgs bosons will be extremely dependent on the choice of coupling. To be conservative, we will use $g_{\phi^0 NN}$ as defined in eq. (2.41) when analyzing the effects of light Higgs bosons in §3.1 and explicitly show the dependence on η.

Higgs–$\pi\pi$ Coupling

The coupling of the Higgs boson to pions can be found [21,27] in a similar manner as the coupling to nucleons in the previous section. However, here it is certainly not correct to take the s-quark as heavy. Thus, we choose $n_L = 3$ and $N_H = 3$ for the remainder of this subsection. The decay probability into $\pi^+ \pi^-$ pairs is,

$$\langle \pi^+ \pi^- |\, \mathcal{L}_{\text{int}} \,|\phi^0\rangle \tag{2.44}$$

where $\mathcal{L}_{\text{int}}$ is the sum of eqs. (2.33) and (2.34). Using eq. (2.36) we can write $\mathcal{L}_{\text{int}}$ in the following form

$$\mathcal{L}_{\text{int}} = \frac{g}{2m_W} \, \phi^0 \left\{ \left(\frac{2N_H}{27} - 1 \right) \sum_{i=u,d,s} m_i \overline{\psi}_i \psi_i - \frac{2N_H}{27} \, \Theta_\mu^\mu \right\}. \tag{2.45}$$

Using PCAC it is straightforward to find the contribution of the quark mass terms

$$\langle \pi^+ \pi^- |\, \sum_{i=u,d,s} m_i \overline{\psi}_i \psi_i \,|0\rangle \simeq m_\pi^2 . \tag{2.46}$$

On the other hand, the contribution from the matrix element of Θ_μ^μ is not small since it does not vanish in the chiral limit [28,29]. The easiest way to evaluate the matrix element $\langle \pi^+ \pi^- |\Theta_\mu^\mu |0\rangle$ is to make use of the chiral

Lagrangian, which yields

$$\Theta^\mu_\mu = -2\partial_\mu \pi^+ \partial^\mu \pi^- + 4m_\pi^2 \pi^+ \pi^- + \cdots, \tag{2.47}$$

where we have explicitly isolated only those terms quadratic in the charged pion field. It then follows that

$$\langle \pi^+ \pi^- | \Theta^\mu_\mu | 0 \rangle = q^2 + 2m_\pi^2 , \tag{2.48}$$

where $q = p_{\pi^+} + p_{\pi^-}$. For the on-shell $\phi^0 \to \pi^+\pi^-$ decay, $q^2 = m_{\phi^0}^2$; therefore, combining eqs. (2.46) and (2.48) and setting $N_H = 3$ we obtain

$$\mathcal{A}(\phi^0 \to \pi^+\pi^-) = -\frac{g}{9m_W}\left(m_{\phi^0}^2 + \frac{11}{2}m_\pi^2\right). \tag{2.49}$$

Hence the decay of the Higgs boson into pions is enhanced relative to the decay into muons [27],

$$
\begin{aligned}
R_{\pi\mu} &\equiv \frac{\Gamma(\phi^0 \to \pi^+\pi^- + \pi^0\pi^0)}{\Gamma(\phi^0 \to \mu^+\mu^-)} \\
&= \frac{1}{27}\frac{m_{\phi^0}^2}{m_\mu^2}\left(1 + \frac{11}{2}\frac{m_\pi^2}{m_{\phi^0}^2}\right)^2 \frac{\left[1 - (4m_\pi^2/m_{\phi^0}^2)\right]^{1/2}}{\left[1 - (4m_\mu^2/m_{\phi^0}^2)\right]^{3/2}}.
\end{aligned}
\tag{2.50}
$$

For $m_{\phi^0} = 1$ GeV one finds $R_{\pi\mu} = 4.3$. Similar relationships to eq. (2.50) hold for the decay of the Higgs boson into $K\overline{K}$ and $\eta\eta$ pairs with m_π replaced by m_K and m_η in eq. (2.50) and with a statistical factor of 4/3 for $K\overline{K}$ and 1/3 for $\eta\eta$ production.

As m_{ϕ^0} becomes larger than about 1 GeV, the accuracy of the low energy soft-pion relations used to derive eq. (2.50) becomes suspect. However, once m_{ϕ^0} becomes large compared to $2m_\mu$, $2m_\pi$, and $2m_K$, the perturbative spectator approach might become relevant. Indeed, with appropriate threshold factors we can attempt to use it throughout the range of interest. We adopt the approximate form (for $2m_\pi < m_{\phi^0} < 2m_\tau$)

$$
\begin{aligned}
\Gamma_{\mu^+\mu^-} : \Gamma_{u\bar{u},d\bar{d}} : \Gamma_{s\bar{s}} : \Gamma_{gg} = {}& m_\mu^2\beta_\mu^3 : 3m_{u,d}^2\beta_\pi^3 : 3m_s^2\theta(m_{\phi^0} - 2m_K)\beta_K^3 : \\
& (\alpha_s/\pi)^2 m_{\phi^0}^2 \left([6 - 2\beta_\pi^3 - \beta_K^3\theta(m_{\phi^0} - 2m_K)]/3\right)^2,
\end{aligned}
\tag{2.51}
$$

where $\beta_\mu = (1 - 4m_\mu^2/m_{\phi^0}^2)^{1/2}$ and similarly for β_π and β_K. In this form we have explicitly incorporated the p-wave threshold factor for the $\mu^+\mu^-$, $d\bar{d}$, $u\bar{u}$, and $s\bar{s}$ final states. In addition, we have approximated the effect of the quark loops contributing to $\phi^0 \to gg$ (which contribute to a variety of final states) by taking the c, b, and t quarks to be heavy compared to ϕ^0 but allowing a p-wave threshold disappearance of the u, d, and s quark loop contributions once the appropriate thresholds have been passed. The appropriate values of

α_s, m_d, m_u and m_s to match the chiral Lagrangian approach will be specified shortly.

Raby and West [30] have pointed out that final state interactions can considerably enhance the importance of the $\pi\pi$ decay channels relative to the $\mu^+\mu^-$ modes. In a single resonance approximation they would multiply the result for $R_{\pi\mu}$, as computed from the chiral Lagrangian, by an enhancement factor of

$$\Omega^2 \equiv \frac{(M_R^2 + m_\pi \Gamma_R)^2}{(M_R^2 - m_{\phi^0}^2)^2 + k^2 \Gamma_R^2}, \qquad (2.52)$$

where M_R is the resonance mass, Γ_R is its width, and $k = \frac{1}{2}(m_{\phi^0}^2 - 4m_\pi^2)^{1/2}$. As one example they quote the result for a broad S-wave enhancement characterized by $M_R = 850$ MeV and $\Gamma_R = 1.3$ GeV (a crude approximation to the S-wave $\pi\pi$ scattering phase shift results). They find $BR(\phi^0 \to \mu^+\mu^-)$ between 0.1 and 0.04 in the range 300 MeV$< m_{\phi^0} < 1$ GeV, a factor of 5 to 10 below the result without the $\pi\pi$ channel enhancement factor. Presumably matching to the perturbative formulas of eq. (2.51) would be accomplished by modifying these to include QCD 'pre-binding' final state corrections that enhance the $s\bar{s}$ and gg final states when the outgoing physical particles have small relative velocity. Such effects have been considered in many other contexts, most notably the Appelquist-Politzer analysis of the Schwinger formula [31] leading to their prediction of $c\bar{c}$ resonances. The effects of the final state interactions discussed in ref. 30 have been reconsidered in ref. 32 using a coupled-channel approach which includes both $\pi\pi$ and $K\overline{K}$ channels. These latter authors find that experimental data for the broad S-wave resonance are most easily parametrized in such a way that it is strongly coupled to the $K\overline{K}$ channels. As a result, its enhancement of the $\pi\pi$ channel below the $K\overline{K}$ threshold is quite modest. For $m_{\phi^0} \lesssim 950$ MeV they find $BR(\phi^0 \to \mu^+\mu^-) \gtrsim 0.25$; the $\mu^+\mu^-$ branching ratio drops to negligible values in the immediate vicinity of the $K\overline{K}$ threshold, rising again to of order 0.2 by $m_{\phi^0} \sim 1.5$ GeV. The crucial question is clearly the extent to which the $\pi\pi$ and $K\overline{K}$ phase shift data uniquely determine the elasticity of the S-wave resonance. Since the $\phi^0 \to \pi\pi$ branching ratio is important for setting limits on Higgs bosons, it will be important to resolve the question of just how much enhancement of the $\pi\pi$ final state does occur as a function of m_{ϕ^0}.

An alternative approach has been considered by Grinstein et al. [33]. They propose to compute $\phi^0 \to \pi^+\pi^-$ using the following chiral Lagrangian*

$$\mathcal{L} = \frac{f_\pi^2}{4}\left(1 + c_1\frac{\phi^0}{v}\right)\text{Tr}\partial^\mu\Sigma\partial_\mu\Sigma^\dagger + \frac{f_\pi^2}{2}\left\{\text{Tr}\mu M\left(1 + c_2\frac{\phi^0}{v}\right)\Sigma^\dagger + \text{h.c.}\right\},$$
$$(2.53)$$

---

*A term of the form $\phi^0\text{Tr}\Sigma\square\Sigma^\dagger$, originally included in ref. 33, is in fact equal to $-\phi^0\text{Tr}\partial^\mu\Sigma\partial_\mu\Sigma^\dagger$ by employing matrix identities and integrating by parts. In addition, we have modified the mass breaking term by introducing a new constant $c_2 \neq 1$.

where $f_\pi \simeq 93$ MeV, $\Sigma = \exp(2i\pi^a \lambda^a / f_\pi)$, λ^a are the usual Gell-Mann SU(3) matrices, $v = 2m_W/g$, $M = \mathrm{diag}(m_u, m_d, m_s)$, and μ is fixed to give the observed pseudoscalar meson masses. The constants c_1 and c_2 are two real *a priori* unknown strong interaction parameters. This Lagrangian should be valid for $m_{\phi^0} \lesssim 800$ MeV according to ref. 33. From eq. (2.53) it is easy to derive

$$\mathcal{A}(\phi^0 \to \pi^+ \pi^-) = -\frac{gc_1}{4m_W}\left[m_{\phi^0}^2 + 2\left(\frac{c_2}{c_1} - 1\right)m_\pi^2\right] . \qquad (2.54)$$

This result coincides with eq. (2.49) if we identify $c_1 = 4/9$ and $c_2 = 5/3$. The deviation of c_2 from 1 is due to the corrections to the $\phi^0 q\bar{q}$ interaction (where $q = u$, d or s) coming from diagrams in which ϕ^0 couples to two gluons via a heavy quark loop, with the two gluons attached, in turn, to a light quark line [34]. In this chiral Lagrangian approach c_1 and c_2 might be taken as unknown parameters. Grinstein *et al.* note that the $\phi^0 \to \pi^+ \pi^-$ decay width is very sensitive to c_1, and thus cannot be determined reliably without further input. Experimentally, the above considerations are important because of their influence on the branching ratio for $\phi^0 \to \mu^+ \mu^-$. The sensitivity of this branching ratio to c_1 is illustrated in fig. 2.8. In contrast, changes in c_2 have a much smaller effect on the results. The $K\overline{K}$ and $\eta\eta$ modes have been included by using exactly the same formula as for the $\pi\pi$ final states (with appropriate mass replacements and quark counting factors). Also plotted in fig. 2.8 is a dotted curve obtained by using the perturbative spectator model of eq. (2.51). The parameters have been adjusted so as to match with Voloshin's prediction in the vicinity of $m_{\phi^0} \sim 1$ GeV, by taking $m_{u,d} = 40$ MeV, $m_s = 450$ MeV and $\alpha_s/\pi = 0.15$. (Note, in particular, that use of current-quark masses, *e.g.*, $m_{u,d} \sim 5$ MeV and $m_s \sim 150$ MeV, does not successfully reproduce the very sharp drop in the $\phi^0 \to \mu^+ \mu^-$ branching ratio as one crosses the $K\overline{K}$ threshold.) However, no reasonable choice of parameters in eq. (2.51) can produce the much smaller $\mu^+ \mu^-$ branching ratios that are predicted by the chiral Lagrangian approach when a large value for c_1 is chosen.

Voloshin argues that c_1 and c_2 are determined by reliable low-energy theorems (which make use of the heavy quark expansion). Recently, Chivukula *et al.* [35] have re-examined and confirmed Voloshin's derivation of the low-energy theorems. In addition, they have estimated the size of the corrections to eq. (2.49). There are two types of corrections: (i) "high-energy" corrections of order $\alpha_s(m_c)$; and (ii) "low-energy" corrections suppressed by powers of $m_{\phi^0}^2/\Lambda^2$, the most important of such terms having an additional logarithmic enhancement factor of $\ln(m_{\phi^0}^2/\Lambda^2)$. Here, $\Lambda \approx 1$ GeV is the scale of chiral symmetry breaking. The high-energy corrections represent perturbative QCD corrections to eq. (2.45), and are probably small. The leading chiral logarithmic corrections may be large; Chivukula *et al.* find that $\mathcal{A}(\phi^0 \to \pi^+ \pi^-)$ is enhanced by roughly 50% for $m_{\phi^0} \simeq 700$ MeV. Note that the existence of an enhancement of this amplitude due to final state interactions is consistent

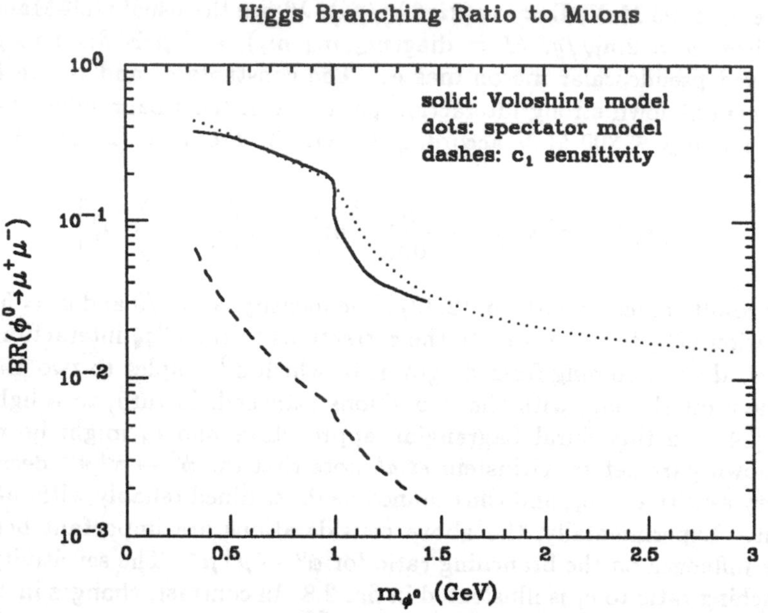

Higgs Branching Ratio to Muons

Figure 2.8 Branching ratio for $\phi^0 \to \mu^+\mu^-$ for a light Higgs using the effective Lagrangian given in eq. (2.53), which depends on strong interaction parameters c_1 and c_2. The solid curve corresponds to $(c_1, c_2) = (4/9, 5/3)$, which is the result of Voloshin (ref. 27) based on low-energy Higgs boson theorems. (All two-body pseudoscalar meson final states have been included.) The dotted curve corresponds to a perturbative spectator model [eq. (2.51)] whose parameters have been adjusted to match with Voloshin's prediction in the vicinity of $m_{\phi^0} \sim 1$ GeV (see text). The dashed curve, corresponding to $(c_1, c_2) = (3, 5/3)$, indicates typical sensitivity to a change in the parameter c_1. (Voloshin's prediction is much less sensitive to changes in c_2.)

with the findings of refs. 30 and 32. In addition, this suggests that the results obtained from the low-energy theorems become untrustworthy for Higgs masses somewhat smaller than previously thought. Finally, we note that similar considerations apply to $K\overline{K}$ and $\eta\eta$ final states, although the quantitative reliability of the chiral Lagrangian approach becomes less certain. Thus, it is not possible to give a firm prediction for $BR(\phi^0 \to \mu^+\mu^-)$ at this time. We have seen that the perturbative model curve of fig. 2.8 represents an accurate estimate before including enhancements for meson-pair modes, but such enhancements are probably significant and can be expected to substantially depress the muon pair branching ratio.

Finally, now that we have reviewed all the components that enter into the computation of the width of a light Higgs boson, it is useful to present results

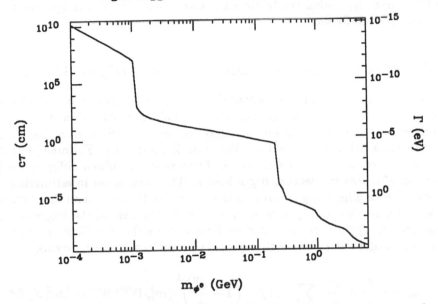

Figure 2.9 The lifetime and width of a light Higgs boson as a function of mass. We have employed the phenomenological spectator approximation above the two pion threshold, as discussed in the text.

for the lifetime and width as a function of mass. In so doing, we shall adopt the spectator approximation of eq. (2.51) above the two-pion threshold, suitably generalized to include charm quark and tau lepton channels. We will not incorporate any $\pi\pi$ channel enhancement. Below the two-pion threshold there are no uncertainties; we compute the photon–photon channel branching ratio including the W boson, quark, and lepton loops according to the formulas given in §2.1 [see eqs. (2.5), (2.16)–(2.20)]. For computations, we have employed $m_u = m_d = 40$ MeV, $m_s = 450$ MeV, and $\alpha_s/\pi = 0.15$, which are the phenomenological values that allow the spectator approximation to successfully match the chiral Lagrangian prediction. The results appear in fig. 2.9. The reader should be reminded that in the regime $2m_\pi \lesssim m_{\phi 0} \lesssim 1.5$ GeV the lifetime could be reduced by as much as a factor of 10 due to final state resonance enhancements. In addition, as noted in ref. 30, there could be a much larger reduction factor in the vicinity of the narrow $f_0(975)$ scalar resonance. For 1.5 GeV $\lesssim m_{\phi 0} \lesssim 4$ GeV the uncertainties could again be as large as a factor of 10 due to the limited reliability of the spectator model in this regime. Nonetheless, it is clear that these uncertainties do not affect the basic conclusion for Higgs bosons with mass in the range $2m_\pi \lesssim m_{\phi^0} \lesssim$ 4 GeV: namely, neither the lifetime nor the width of such a Higgs boson can be experimentally measured.

2.2 Low-Energy Theorems Involving Higgs Bosons

Let us now consider in greater detail the theoretical foundations of the Higgs–nucleon–nucleon and Higgs–meson–meson couplings discussed in the previous section. Low-energy theorems for Higgs boson interactions have been studied in great detail by Vainshtein, Voloshin, Zakharov and Shifman [9,21,27]. These theorems relate the amplitudes of two processes which differ by the insertion of a zero momentum Higgs boson. They are useful in estimating the properties of a light Higgs boson in the same way that soft-pion theorems are used to study low energy pion interactions. In the case of the Higgs bosons, the low energy theorems were derived by observing that the Higgs interactions in the standard electroweak theory could be written in the following form

$$\mathcal{L}_{\text{int}} = -\left(1 + \frac{\phi^0}{v}\right) \sum_f m_f \bar{f} f - \left(1 + \frac{\phi^0}{v}\right)^2 (m_W^2 W^{\mu+} W_\mu^- + \tfrac{1}{2} m_Z^2 Z_\mu Z^\mu),$$

(2.55)

where $v \equiv (\sqrt{2} G_F)^{-1/2} = 2m_W/g \simeq 246$ GeV, and we sum over all fermions f in the theory. Consider a Higgs boson with zero four-momentum $[P_\mu, \phi^0] = i\partial_\mu \phi^0 = 0$. This implies that ϕ^0 is a constant field. From eq. (2.55), it follows that the effect of a constant field ϕ^0 is equivalent to redefining all mass parameters of the theory

$$m_i \to m_i \left(1 + \frac{\phi^0}{v}\right).$$

(2.56)

This immediately implies the following low energy theorem [1,21]

$$\lim_{p_\phi \to 0} \mathcal{M}(A \to B + \phi^0) = \frac{1}{v} \left(\sum_f m_f \frac{\partial}{\partial m_f} + \sum_V m_V \frac{\partial}{\partial m_V}\right) \mathcal{M}(A \to B),$$

(2.57)

where the sum over V includes both the W and Z bosons. This theorem is rather trivial when applied to the elementary particles of the model. But its range of applicability is much wider. As a demonstration, let us derive a low energy theorem for the $\phi^0 gg$ interaction (g = gluon). At one-loop, the transition amplitude $\mathcal{M}(g \to g)$, which is just the gluon two-point function, does have m_f dependence, due to an intermediate quark loop. One can show that the effect of heavy fermion loops is to add the following piece to the effective QCD Lagrangian

$$\delta\mathcal{L} = \frac{-\alpha_s}{24\pi} G^a_{\mu\nu}G^{\mu\nu a} \sum_f \ln\left(\frac{\Lambda^2_{UV}}{m^2_f}\right),\tag{2.58}$$

where Λ_{UV} is the ultraviolet cutoff. Using eq. (2.57), we obtain the following effective Lagrangian governing the Higgs–gluon interaction

$$\mathcal{L}_{\phi^0 gg} = \frac{N_H \alpha_s}{12\pi v} \phi^0 G^a_{\mu\nu}G^{\mu\nu a},\tag{2.59}$$

where N_H is the number of heavy quark flavors. Here, "heavy" means that N_H is the number of quarks heavier than ϕ^0 and the confinement scale of QCD (denoted by Λ_{QCD}).[‡] Remarkably, eq. (2.59) gives precisely the same answer as the triangle diagram calculation resulting in eq. (2.34); namely, the $\phi^0 gg$ matrix element is constant in the limit of $m_q \to \infty$. This technique has also been used to obtain the effective $\phi^0\gamma\gamma$ interaction in the soft Higgs limit [9]. The result matches the triangle diagram calculation in which all particles which can appear in the triangle (charged spin 0, 1/2, and 1 fields) are infinitely massive. In this case, the coefficient of the effective interaction is proportional to the electromagnetic β-function, since the latter depends only on the photon two-point function. Likewise, this method provides an easy way to obtain the large fermion mass radiative correction to the $\phi^0 W^+ W^-$ and $\phi^0 ZZ$ couplings.

Consider now the application of the low energy theorems to the study of Higgs interactions with mesons and baryons at low energy. The mesons and baryons are complicated bound state systems made up of light quarks and gluons. The Higgs bosons can interact with these systems in three distinct ways: (i) interaction with the gluons via eq. (2.59); (ii) direct interactions with the light constituent quarks; and (iii) via a weak interaction process, where the quarks exchange a W or Z boson, and the Higgs interacts with the exchanged vector boson. It is convenient to develop low energy theorems which separate out the interactions via gluons from the direct interactions with fermions and vector bosons as follows. First, we divide up the quarks into "light" ($m_q < \Lambda_{QCD}$) and "heavy" ($m_q > \Lambda_{QCD}, m_{\phi^0}$). The heavy quarks are important in that they are responsible for the $\phi^0 gg$ interaction; hence, we remove the heavy quarks from eq. (2.57). Instead, we derive a new low energy theorem by observing that eq. (2.59) can be combined with the gluon kinetic energy term to obtain

$$\mathcal{L} = \frac{-1}{4g^2_s}(\partial_\mu A^a_\nu - \partial_\nu A^a_\mu - f_{abc}A^b_\mu A^c_\nu)^2 \left(1 - \frac{N_H \alpha_s}{3\pi v}\phi^0\right),\tag{2.60}$$

where $\alpha_s \equiv g^2_s/(4\pi)$, and we have rescaled the gluon field, $A^a_\mu \to g^{-1}_s A^a_\mu$. In the zero momentum limit where ϕ^0 is a constant field, we see that the ϕ^0

[‡]Λ_{QCD} is expected to be of the order of a few hundred MeV. This means that the strange quark is neither heavy nor light. For simplicity, we shall henceforth regard the strange quark as being light.

interactions can be reproduced simply by rescaling α_s. Thus, if we denote the corresponding change by $\alpha_s \rightarrow \alpha_s + \delta\alpha_s$, then to first order,

$$\delta\alpha_s = \frac{N_H \alpha_s^2}{3\pi v} \phi^0 . \tag{2.61}$$

The following low energy theorem is thereby obtained [27]

$$\lim_{p_\phi \rightarrow 0} \mathcal{M}(A \rightarrow B + \phi^0)\big|_{\text{gluons}} = \frac{N_H \alpha_s^2}{3\pi v} \frac{\partial}{\partial \alpha_s} \mathcal{M}(A \rightarrow B) , \tag{2.62}$$

where the subscript "gluons" indicates that we are exhibiting the partial contribution to $\mathcal{M}(A \rightarrow B + \phi^0)$ due to the $\phi^0 gg$ interactions induced by the heavy quark loops. The remaining contributions due to the interactions with the light constituent quarks and weak vector bosons are obtained by deleting the heavy quarks from eq. (2.57)

$$\lim_{p_\phi \rightarrow 0} \mathcal{M}(A \rightarrow B + \phi^0)\big|_{q,W,Z} = \frac{1}{v} \left(\sum_{q=u,d,s} m_q \frac{\partial}{\partial m_q} + \sum_V m_V \frac{\partial}{\partial m_V} \right) \mathcal{M}(A \rightarrow B) \tag{2.63}$$

where we sum over *light* quarks and $V = W$ and Z.

As an example of this formalism, let us work out the Higgs–nucleon coupling using the low energy theorems. We will assume that the dominant mechanism leading to the $\phi^0 NN$ coupling is via the $\phi^0 gg$ coupling; therefore eq. (2.62) is the relevant low energy theorem. (In practice, the strange quark contribution to the proton can also play an important role; see eq. (2.43) and surrounding discussion.) We therefore find that

$$g_{\phi^0 NN} = \frac{N_H \alpha_s^2}{3\pi v} \frac{\partial m_N}{\partial \alpha_s} . \tag{2.64}$$

To proceed, we must evaluate $\partial m_N/\partial \alpha_s$. In QCD, m_N must be proportional to the intrinsic scale of QCD, Λ. This scale is defined as follows [36]. We use the following normalization in the definition of the QCD β-function

$$\mu \frac{\partial \alpha_s}{\partial \mu} = \alpha_s \beta(\alpha_s) , \tag{2.65}$$

with

$$\beta(\alpha_s) = \frac{-b\alpha_s}{2\pi} + \mathcal{O}(\alpha_s^2) , \tag{2.66}$$

where $b \equiv 11 - \frac{2}{3} N_f$, and N_f is the number of quark flavors. Then, Λ is defined as

$$\Lambda = \mu \exp\left\{ -\int \frac{d\alpha_s}{\alpha_s \beta(\alpha_s)} \right\} . \tag{2.67}$$

Note that by using eq. (2.66), $d\Lambda/d\mu = 0$, which implies that Λ is a physical parameter of the theory. It then follows that

$$\frac{\partial \Lambda}{\partial \alpha_s} = \frac{-\Lambda}{\alpha_s \beta(\alpha_s)} .$$ (2.68)

Since m_N is proportional to Λ, m_N satisfies precisely the same equation. In order to apply this equation in eq. (2.64), we must understand the apparent scale dependence of this equation. Here, we must resort to the arguments of §2.1, in which we computed the Higgs–nucleon coupling by employing the trace of the energy-momentum tensor [20]. In that argument, it was noted that Θ^μ_μ was independent of scale, so it was appropriate to perform the analysis in a regime in which the one-loop approximation of the β-function is valid. In this case, we immediately find that

$$g_{\phi^0 NN} \simeq \frac{2N_H m_N}{3bv} ,$$ (2.69)

where b should be computed in a theory where the heavy flavors are decoupled, namely $b = 11 - \frac{2}{3}n_L$, with n_L being the number of light flavors. This result agrees with the analysis given in §2.1. The error we make in the analysis by using the one-loop approximation for the β-function is the neglect of terms of order $\alpha_s(m_c)$, *i.e.* α_s evaluated at the last heavy quark which has been decoupled [35].

We shall now apply the Higgs low energy theorems to the study of low energy interactions of a light Higgs boson and the pseudoscalar mesons. We will employ the chiral Lagrangian approach; our notation is that of ref. 37. To briefly review the approach, consider the chiral Lagrangian which describes $\Delta S = 0$ and $\Delta S = 1$ processes involving the pseudoscalar mesons. Keeping only the terms with the fewest number of derivatives, we have

$$\begin{aligned}
\mathcal{L} = & \tfrac{1}{4}f^2 \mathrm{Tr} \partial^\mu \Sigma \partial_\mu \Sigma^\dagger + \tfrac{1}{2}f^2 \left[\mathrm{Tr} \mu M \Sigma^\dagger + \mathrm{h.c.} \right] \\
& + \tfrac{1}{4}f^2 \left[\lambda \mathrm{Tr} \, h \partial^\mu \Sigma \partial_\mu \Sigma^\dagger + \mathrm{h.c.} \right] \\
& + \tfrac{1}{2}f^2 \left[A \mathrm{Tr} \, h\mu (M\Sigma^\dagger + \Sigma M) + \mathrm{h.c.} \right] \\
& + \tfrac{1}{4}f^2 \left[a T^{ij}_{k\ell} (\Sigma \partial^\mu \Sigma^\dagger)^k_i (\Sigma \partial_\mu \Sigma^\dagger)^\ell_j + \mathrm{h.c.} \right] ,
\end{aligned}$$ (2.70)

where $f \equiv f_\pi = 93$ MeV, $\Sigma \equiv \exp(2i\Pi^a T^a/f)$, with the SU(3) generators normalized to $\mathrm{Tr} T^a T^b = \frac{1}{2}\delta_{ab}$ and $\Pi \equiv \Pi^a T^a$ denoting the pseudoscalar octet

$$\Pi = \frac{1}{\sqrt{2}} \begin{pmatrix} \pi^0/\sqrt{2} + \eta/\sqrt{6} & \pi^+ & K^+ \\ \pi^- & -\pi^0/\sqrt{2} + \eta/\sqrt{6} & K^0 \\ K^- & \overline{K}^0 & -2\eta/\sqrt{6} \end{pmatrix} ,$$ (2.71)

and M is the quark mass matrix. The matrix h appears in a $\Delta S = 1$ term which transforms as $(8,1)$ under $\mathrm{SU}(3)_L \times \mathrm{SU}(3)_R$; it is a pure $\Delta I = 1/2$ operator. The tensor T appears in an operator which transforms as $(27,1)$; it contains both $\Delta I = 1/2$ and $\Delta I = 3/2$ pieces. Specifically,

$$M = \begin{pmatrix} m_u & 0 & 0 \\ 0 & m_d & 0 \\ 0 & 0 & m_s \end{pmatrix}, \qquad h = \begin{pmatrix} 0 & 0 & 0 \\ 0 & 0 & 1 \\ 0 & 0 & 0 \end{pmatrix}, \qquad (2.72)$$

and T is a traceless tensor, symmetric in its upper and lower indices. For example, the nonzero components of T corresponding to $\Delta S = 1$ and $\Delta I = 3/2$ are

$$T_{13}^{12} = T_{13}^{21} = T_{31}^{12} = T_{31}^{21} = -T_{23}^{22} = -T_{32}^{22} = \tfrac{1}{2}. \qquad (2.73)$$

If CP-violating effects are neglected, then the coefficients λ, A and a are real.

The parameter μ is adjusted in order that the meson masses come out correctly. For example, if we focus on the charged meson sector, then the chiral Lagrangian takes the form

$$\mathcal{L} = \partial^\mu \pi^+ \partial_\mu \pi^- + \partial^\mu K^+ \partial_\mu K^- + \tfrac{1}{2}\lambda(\partial^\mu \pi^+ \partial_\mu K^- + \partial^\mu \pi^- \partial_\mu K^+)$$
$$- m_\pi^2 \pi^+ \pi^- - m_K^2 K^+ K^- - m_{K\pi}^2(\pi^+ K^- + \pi^- K^+) + \text{interactions}\,, \qquad (2.74)$$

where

$$m_\pi^2 = \mu(m_u + m_d)$$
$$m_K^2 = \mu(m_u + m_s) \qquad (2.75)$$
$$m_{K\pi}^2 = \tfrac{1}{2}A\mu(m_d + m_s)\,.$$

At this point, we can remove λ and $m_{K\pi}^2$ from the quadratic part of the Lagrangian [eq. (2.74)] by performing wave function renormalization and diagonalizing the pseudoscalar mass matrix, respectively. Wave function renormalization in this context means diagonalization of the pseudoscalar kinetic energy, followed by a rescaling of the fields to obtain canonically normalized terms. To diagonalize the pseudoscalar mass matrix, we perform an SU(3)$_L \times$ SU(3)$_R$ transformation, $\Sigma \to L\Sigma R^\dagger$, and we find

$$\text{Tr}[1 + Ah + A^* h^\dagger]\mu M \Sigma^\dagger \to \text{Tr}\mu M_D \Sigma^\dagger\,, \qquad (2.76)$$

where

$$M_D = L^\dagger[1 + Ah + A^* h^\dagger]MR \qquad (2.77)$$

is the diagonalized mass matrix. Since we shall always work to first order in λ and A, this rotation has no affect on the other (interaction) terms of the Lagrangian. Thus, we can simply set $m_{K\pi} = 0$. On the other hand, when wave function renormalization is performed (as described above), interaction terms are generated which depend on λ. Thus, λ is a physical parameter, which can be measured in the $\Delta S = 1$, $\Delta I = 1/2$ weak interactions of pseudoscalar mesons. Fitting to data, Cohen and Manohar [38] find* $|\lambda| = 3.2 \times 10^{-7}$. They also have fit the parameter a to the measured data on $\Delta I = 3/2$ transitions, and find $|a| = 1.0 \times 10^{-8}$. Thus, for most purposes, the $\Delta I = 3/2$ piece can be neglected.

We now consider coupling the Higgs boson into this system. By the rules of the chiral Lagrangian technique, we consider the most general coupling (with the fewest derivatives) [39]

$$
\begin{aligned}
\Delta \mathcal{L} = {} & \tfrac{1}{4} f^2 c_1 \frac{\phi^0}{v} \mathrm{Tr} \partial^\mu \Sigma \partial_\mu \Sigma^\dagger + \tfrac{1}{2} f^2 c_2 \frac{\phi^0}{v} \left[\mathrm{Tr} \mu M \Sigma^\dagger + \mathrm{h.c.} \right] \\
& + \tfrac{1}{4} f^2 c_3 \frac{\phi^0}{v} \left[\lambda \mathrm{Tr}\, h \partial^\mu \Sigma \partial_\mu \Sigma^\dagger + \mathrm{h.c.} \right] \\
& + \tfrac{1}{2} f^2 c_4 \frac{\phi^0}{v} \left[A \mathrm{Tr}\, h \mu (M \Sigma^\dagger + \Sigma M) + \mathrm{h.c.} \right] \\
& + \tfrac{1}{4} f^2 c_5 \frac{\phi^0}{v} \left[a T_{k\ell}^{ij} (\Sigma \partial^\mu \Sigma^\dagger)_i^k (\Sigma \partial_\mu \Sigma^\dagger)_j^\ell + \mathrm{h.c.} \right] ,
\end{aligned}
\tag{2.78}
$$

where the c_i are strong interaction real parameters which cannot be fixed on the basis of chiral symmetry alone. In fact, these parameters can be determined by applying the Higgs low energy theorems, as we will soon demonstrate. When we consider the full Lagrangian $\mathcal{L} + \Delta \mathcal{L}$ [eqs. (2.70) and (2.78)], we must again perform wave function renormalization and diagonalize the pseudoscalar mass matrix, as described earlier.[†] Working to first order in the small parameters μ/v, λ, A and a, we find

$$
\begin{aligned}
\mathcal{L} = {} & \tfrac{1}{4} f^2 \left(1 + c_1 \frac{\phi^0}{v} \right) \mathrm{Tr} \partial^\mu \Sigma \partial_\mu \Sigma^\dagger + \tfrac{1}{2} f^2 \left(1 + c_2 \frac{\phi^0}{v} \right) \left[\mathrm{Tr} \mu M \Sigma^\dagger + \mathrm{h.c.} \right] \\
& + \tfrac{1}{4} f^2 (c_3 - c_1) \frac{\phi^0}{v} \left[\lambda \mathrm{Tr}\, h \partial^\mu \Sigma \partial_\mu \Sigma^\dagger + \mathrm{h.c.} \right] \\
& + \tfrac{1}{2} f^2 (c_4 - c_2) \frac{\phi^0}{v} \left[A \mathrm{Tr}\, h \mu (M \Sigma^\dagger + \Sigma M) + \mathrm{h.c.} \right] \\
& + \tfrac{1}{4} f^2 c_5 \frac{\phi^0}{v} \left[a T_{k\ell}^{ij} (\Sigma \partial^\mu \Sigma^\dagger)_i^k (\Sigma \partial_\mu \Sigma^\dagger)_j^\ell + \mathrm{h.c.} \right] ,
\end{aligned}
\tag{2.79}
$$

---

*The sign of λ is not directly determined by current experiment. However, since λ is implicitly proportional to $V_{ud}^* V_{us} \equiv \cos \theta_c \sin \theta_c$, the sign of λ is well-defined once we adopt the usual convention where $\sin \theta_c \approx 0.22$ is positive. Then, if one matches the chiral Lagrangian to the $\Delta S = 1$ weak Lagrangian of the full theory at 1 GeV, one finds that λ is *negative* in the vacuum insertion approximation.

[†]It is possible to do calculations with a non-diagonal mass matrix and kinetic energy, as long as one includes Feynman rules for off-diagonal two-point vertices. For example, in $K \to \pi \phi^0$, K-π mixing diagrams would exist where one starts with a $\phi^0 KK$ (or $\phi^0 \pi \pi$) interaction and converts a K to a π (or vice versa) on an external leg via the two-point (off-diagonal) vertex. We prefer to perform the diagonalization as described above, thereby avoiding such diagrams.

where we have written M for the diagonal mass matrix M_D [see eqs. (2.76) and (2.77)]. We see explicitly that the term proportional to A is not completely removed by an $SU(3)_L \times SU(3)_R$ transformation, unlike in the case in which no Higgs field is present. That is, A is a *new* parameter which arises when Higgs boson interactions are incorporated, and thus it cannot be determined from experiment at present.

Before proceeding to apply eq. (2.79) to some processes of interest, we should comment on the generality of the chiral Lagrangian being used here. We have restricted our considerations to terms with at most two derivatives; nevertheless, some terms have been apparently left out of eq. (2.78)

$$(i) \quad \frac{\partial_\mu \phi^0}{v} \mathrm{Tr} \Sigma \partial^\mu \Sigma^\dagger \,,$$

$$(ii) \quad \tfrac{1}{2} \frac{\phi^0}{v} f^2 \left[A' \mathrm{Tr}\, h\mu (M\Sigma^\dagger - \Sigma M) + \text{h.c.} \right] \qquad (2.80)$$

$$(iii) \quad \tfrac{1}{4} f^2 \frac{\partial_\mu \phi^0}{v} \left[B\lambda \mathrm{Tr}\, h\Sigma \partial^\mu \Sigma^\dagger + \text{h.c.} \right] .$$

Term (i) above is actually identically equal to zero, since $\Pi^a \mathrm{Tr}\, T^a = 0$. Terms (ii) and (iii) violate CPS symmetry, [40] which is a discrete symmetry that combines CP and the interchange of s and d quarks. The CPS symmetry is respected (in the chiral symmetry limit) by all quark operators of the full effective $\Delta S = 1$ electroweak Lagrangian (including gluonic corrections), even when CP-violation is taken into account [41]. Since chiral symmetry breaking in the fundamental theory is due to $\sum_q m_q q\bar{q}$, we can still make use of the CPS symmetry to restrict possible terms in the chiral Lagrangian if we formally interchange m_s and m_d when performing the S symmetry operation. For example, an explicit computation of the quadratic terms contained in term (ii) above results in a CPS-even expression multiplied by $m_s - m_d$. Thus, assuming that A' and B are mass independent constants, we can omit terms (ii) and (iii) above from the chiral Lagrangian.

In the discussion above, we noted that if no Higgs field is present, then the term proportional to A can be removed by an $SU(3)_L \times SU(3)_R$ rotation. This is no longer the case when $\Delta \mathcal{L}$ [eq. (2.78)] is included. However, we could remove the A term completely if we perform an appropriate ϕ^0-*dependent* $SU(3)_L \times SU(3)_R$ transformation, $\Sigma \to L\Sigma R^\dagger$, in eq. (2.79). Clearly, L and R will depend on the coordinate (through $\phi^0(x)$); hence the kinetic energy term in eq. (2.79) will not be invariant under this transformation. (Other terms with derivatives are also not invariant, but the extra terms generated are second order in the small parameters and thus negligible.) It is easy to compute L and R; to first order we find

$$
\begin{aligned}
L &\simeq I + \left(1 + \frac{(c_4 - c_2)\phi^0}{v} \right) \frac{m_s^2 + m_d^2}{m_s^2 - m_d^2} \left(Ah - A^* h^\dagger \right) , \\
R &\simeq I + \left(1 + \frac{(c_4 - c_2)\phi^0}{v} \right) \frac{2 m_s m_d}{m_s^2 - m_d^2} \left(Ah - A^* h^\dagger \right) ,
\end{aligned}
\qquad (2.81)
$$

where I is the 3×3 identity matrix and we have assumed that $(m_s + m_d)|A| \ll |m_s - m_d|$. The extra terms in $\mathcal{L}$ which are generated are

$$\mathcal{L}_{\text{extra}} = \frac{1}{4} \frac{\partial_\mu \phi_0}{v}$$
$$\times \left[B_1 \lambda \operatorname{Tr} h(\Sigma \partial^\mu \Sigma^\dagger + \Sigma^\dagger \partial^\mu \Sigma) + B_2 \lambda \operatorname{Tr} h(\Sigma \partial^\mu \Sigma^\dagger - \Sigma^\dagger \partial^\mu \Sigma) + \text{h.c.} \right] \tag{2.82}$$

with[‡]

$$\lambda B_1 = A(c_4 - c_2) \left(\frac{m_s + m_d}{m_s - m_d} \right),$$
$$\lambda B_2 = A(c_4 - c_2) \left(\frac{m_s - m_d}{m_s + m_d} \right). \tag{2.83}$$

The ratio of quark masses can be rewritten in terms of meson masses by using eq. (2.75), with $m_u \approx m_d$. Using the arguments analogous to those below eq. (2.80), it is clear that $\mathcal{L}_{\text{extra}}$ is CPS-even since the coefficients B_1 and B_2 are odd under interchange of m_d and m_s. This is no surprise since these terms have been generated from the CPS-even "A term". If we now expand out the two terms proportional to B_1 and B_2, we find that the former contains only even powers of the pseudoscalar field Π, while the latter contains only odd powers of Π. Thus, if one wishes to parametrize $e.g.$, the amplitude for $K \to \pi \phi^0$, it is a matter of convenience whether one works with an A-term or a B-term; between them there is only one independent parameter.[*]

Finally, we turn to the parameters $c_1, \ldots, c_5$. These parameters cannot be determined by chiral symmetry. However, they can be determined by applying the Higgs low energy theorems [in particular eqs. (2.62) and (2.63)] discussed above [42–44]. To apply these equations, we need to investigate the m_q, m_W and α_s dependence of the terms appearing in eq. (2.70). It is easy to argue that

$$f, \mu \sim \Lambda; \qquad M \sim m_q; \qquad A, \lambda, a \sim \frac{\Lambda^2}{m_W^2}, \tag{2.84}$$

where the α_s dependence of Λ has been given in eq. (2.67). In the case of the dimensionless parameters (A, λ, a), the m_W^{-2} behavior is clear, since these terms arise from the charged current weak interactions. By dimensional analysis, the Λ^2 behavior follows. From eq. (2.63), it follows that all terms in eq. (2.70) which are proportional to the quark mass matrix will be multiplied by $(1 + \phi^0/v)$, and those terms proportional to A, λ or a will be multiplied by $(1 - 2\phi^0/v)$. (It is sufficient to work to leading order in ϕ^0/v.) The latter factor is easily understood as being the result of summing a geometric series of

---

[‡]Here, we have used $\Sigma^\dagger \partial^\mu \Sigma = -(\partial^\mu \Sigma^\dagger)\Sigma$, which follows from taking the divergence of the equation, $\Sigma^\dagger \Sigma = 1$.

[*]It is important to appreciate that B_2 is predicted once B_1 is known [see eq. (2.83)] which is a consequence of the CPS symmetry of the quark operators of the underlying theory.

multiple Higgs emission from the exchanged W. This leads to a factor of $(1 + \phi^0/v)^{-2}$, which is the expected result. Finally, we turn to the consequence of eq. (2.62). We work in the one-loop approximation for the β-function; $i.e.$, we combine eqs. (2.66) and (2.68) to obtain

$$\frac{\partial \Lambda}{\partial \alpha_s} = \frac{2\pi\Lambda}{b\alpha_s^2} . \qquad (2.85)$$

Using the Λ dependence shown in eq. (2.84), we find that a term in eq. (2.70) which behaves like Λ^p will be multiplied by

$$\left(1 + \frac{2pN_H}{3b}\frac{\phi^0}{v}\right) . \qquad (2.86)$$

Combining the results of the low energy theorem, we find

$$
\begin{aligned}
c_1 &= \frac{4N_H}{3b} , \\
c_2 &= 1 + \frac{2N_H}{b} , \\
c_3 &= c_5 = -2 + \frac{8N_H}{3b} , \\
c_4 &= -1 + \frac{10N_H}{3b} .
\end{aligned}
\qquad (2.87)
$$

The values of $c_1, \ldots, c_5$ have also been obtained in refs. 39 and 44 (we have corrected their value of c_4). Our results also disagree slightly with those of ref. 43.

In the above derivation, there is one sleight of hand which needs to be justified. Namely, there is a factor of f hidden in the definition of $\Sigma = \exp(2i\Pi^a T^a/f)$. Nevertheless, even though we stated that $f \sim \Lambda$, we neglected this dependence in Σ when we applied the low energy theorem [eq. (2.62)]. In fact, one can show that this is the correct procedure [45]. First we note that the effective low energy theory is invariant under the combined scaling of dimensionful parameters and the dilatation transformation of the fields [44]. Thus we can write

$$\Theta^\mu_\mu = \partial_\mu s^\mu = -\left(\Lambda\frac{\partial}{\partial\Lambda} + \sum_{q=u,d,s} m_q\frac{\partial}{\partial m_q} + \sum_V m_V\frac{\partial}{\partial m_V}\right)\mathcal{L} , \qquad (2.88)$$

where s^μ is the scale current. However, as shown in ref. 35, the field Σ (and likewise the Π field) do not scale under a dilatation transformation; $i.e.$, they have scale dimension equal to zero. It follows that for the leading term of the chiral Lagrangian, $\mathcal{L} = \frac{1}{4}f^2 \text{Tr}\partial^\mu\Sigma\partial_\mu\Sigma^+$, one can derive $\partial_\mu s^\mu = -2\mathcal{L}$. As a result, to be consistent with eq. (2.88), we must use $\partial\Sigma/\partial\Lambda = 0$ in the low energy theorem, which is the justification for the procedure described above.

Equation (2.88) also provides the link between the Higgs boson low energy theorems [eqs. (2.62) and (2.63)] and eq. (2.45) of the previous section. That is, by comparing eqs. (2.88) and (2.62) [using eqs. (2.65)–(2.68)], we can eliminate $\Lambda\,\partial/\partial\Lambda$ in favor of Θ_μ^μ, thereby obtaining eq. (2.45).

We now illustrate the above formalism by computing the amplitude for $K \to \pi\phi^0$ using the chiral Lagrangian derived above. Before proceeding, it is useful to consider the various mechanisms for $K \to \pi\phi^0$ on the more fundamental quark level. As we shall describe in detail in §3.1, there are two classes of diagrams to consider. First, the direct quark decay $s \to d\phi^0$ generates a Higgs coupling to an effective two-quark operator. Second, there are contributions in which the Higgs boson couples to effective four-quark operators. An example of such a process is $u\bar{s} \to u\bar{d}\phi^0$, where ϕ^0 is emitted from the exchanged W. The process $s \to d\phi^0$ has been computed by several groups, [46–49,33,39] and can be summarized by the effective Lagrangian

$$\mathcal{L}_{sd\phi^0} = \frac{\phi^0}{2v}\left[\zeta m_s\,\bar{d}(1+\gamma_5)s + \zeta^* m_d\bar{s}(1+\gamma_5)d + \text{h.c.}\right]\,, \qquad (2.89)$$

where

$$\zeta = \frac{3\alpha}{16\pi m_W^2\sin^2\theta_W}\sum_i m_i^2 V_{id}^* V_{is}\,, \qquad (2.90)$$

and the sum is taken over up-type quarks (u,c,t). The m_i and V_{ij} are the corresponding quark masses and Kobayashi-Maskawa mixing angles. When we evaluate $\langle\pi|\mathcal{L}_{sd\phi^0}|K\rangle$, the γ_5 pieces give no contribution. Thus, following ref. 39, we may simply make the standard chiral Lagrangian replacement

$$\tfrac{1}{2}\sum_{ij}[N_{ij}\bar{\psi}_i\psi_j + \text{h.c.}] \to -\tfrac{1}{2}f^2\text{Tr}\mu N\Sigma^\dagger + \text{h.c.}\,, \qquad (2.91)$$

to represent the two-quark operator contribution to $K \to \pi\phi^0$. Thus, in the chiral Lagrangian, the effect of the two-quark operator is[†]

$$\mathcal{L}_{2q} = -\tfrac{1}{2}f^2\frac{\phi^0}{v}\left[\zeta\text{Tr}\,h\mu(M\Sigma^\dagger + \Sigma M) + \text{h.c.}\right]\,, \qquad (2.92)$$

where the matrices M and h are defined in eq. (2.72). Note that this term is precisely of the same form as the "A-term" of eq. (2.78).

Next, we consider the effects of the four-quark operators. Here, we will simply invoke the full chiral Lagrangian given in eq. (2.79). For convenience, we will keep separate the contributions of the two-quark and four-quark operators. That is, we add eqs. (2.79) and (2.92), so that A (as well as λ and a) is, by definition, a consequence of the four-quark operators. It is then

---

[†]QCD corrections can be incorporated by using running masses defined at the chiral Lagrangian scale [39].

straightforward to extract the full amplitude for $K \to \pi \phi^0$. The result is

$$
\begin{aligned}
\mathcal{M}(K^- \to \pi^- \phi^0) &= \mathcal{M}^*(K^+ \to \pi^+ \phi^0) \\
&= \frac{-\lambda(m_K^2 + m_\pi^2 - m_{\phi^0}^2)}{2v} \left(1 - \frac{2N_H}{3b}\right) \\
&\quad + \frac{A m_K^2}{v} \left(1 - \frac{2N_H}{3b}\right) + \frac{m_K^2 \zeta}{2v},
\end{aligned}
\tag{2.93}
$$

$$
\mathcal{M}(K_L^0 \to \pi^0 \phi^0) = -\mathrm{Re}\,\mathcal{M}(K^- \to \pi^- \phi^0),
$$

$$
\mathcal{M}(K_S^0 \to \pi^0 \phi^0) = \frac{i m_K^2}{2v} \,\mathrm{Im}\,\left[2A\left(1 - \frac{2N_H}{3b}\right) + \zeta\right].
$$

In deriving these results, we have omitted the effect of the $\Delta I = 3/2$ contribution [the term proportional to a in eq. (2.79)]. We have also neglected the effects of ϵ (which measures the CP-violation in the physical states of the K^0–$\overline{K^0}$ system) and Im λ (which is related to ϵ'/ϵ), as both these effects are numerically small. We shall examine the implications of these results in §3.1. For now, let us simply make the following observation. If we suppose that A is of order λ, then all the terms in eq. (2.93) are of order 10^{-10}, and we would then expect $BR(K \to \pi \phi^0) \sim 10^{-5}$. However, a definite prediction cannot be made without precise knowledge of the parameter A (with some additional uncertainty due to presently unknown t-quark mass and mixing angles).

Even in the absence of knowledge of the value of A, there is some theoretical information which can be extracted. Consider the CP-violating effects in eq. (2.93). We argued above that it was numerically safe to neglect ϵ and Im λ. Nevertheless the CP-violating effects can be significant [50] because Im ζ can be phenomenologically important. The key point to observe is the presence of m_i^2 in the definition of ζ in eq. (2.90). In particular, the top-quark contribution is greatly enhanced in the sum, and this allows Im $\zeta/\mathrm{Re}\,\zeta$ to be of $\mathcal{O}(1)$. The m_t^2 amplitude growth is a feature which is unique to the two-quark operator. Simple dimensional reasoning suggests that the contributions to the amplitude from the four-quark operators can at best approach a constant in the large m_t limit. It follows that the ratio of the imaginary to the real part of the four-quark amplitudes can be no larger than (roughly) Im $V_{td}^* V_{ts}/(V_{ud}^* V_{us})$ which is of order 10^{-3}. Thus, we conclude that Im $A \sim \mathcal{O}(10^{-3})\,\mathrm{Re}\,A$. The consequences of this result for the $K \to \pi \phi^0$ decay rate are discussed in §3.1.

As a final exercise in the use of the formalism described in this section, we shall derive the branching ratio for $\eta' \to \eta \phi^0$. The $\eta \eta' \phi^0$ coupling can be found [42,43] in a similar manner to the $K \pi \phi^0$ coupling. We begin by writing the physical η and η' states in terms of the SU(3)$_{\text{flavor}}$ singlet and octet states,

η_1 and η_8,

$$\eta' = \eta_1 \cos \theta + \eta_8 \sin \theta$$
$$\eta = \eta_8 \cos \theta - \eta_1 \sin \theta . \tag{2.94}$$

The mixing angle θ can be determined from data, which favor [51] a value of $\theta \simeq -20°$. In terms of the $SU(3)_{\text{flavor}}$ states, the mass terms in the Lagrangian are

$$\mathcal{L}_M = -\tfrac{1}{2} m_1^2 \eta_1^2 - \tfrac{1}{2} m_8^2 \eta_8^2 - \mu^2 \eta_1 \eta_8 . \tag{2.95}$$

Diagonalizing the mass matrix we find,

$$\mu^2 = \sin \theta \cos \theta (m_{\eta'}^2 - m_\eta^2)$$
$$m_1^2 = \cos^2 \theta \, m_{\eta'}^2 + \sin^2 \theta \, m_\eta^2 \tag{2.96}$$
$$m_8^2 = \sin^2 \theta \, m_{\eta'}^2 + \cos^2 \theta \, m_\eta^2 .$$

The $\eta \eta' \phi^0$ coupling can now be derived using the low energy theorem techniques described above via

$$\mathcal{L} = \frac{\phi^0}{v} \left[\sum_q m_q \frac{\partial}{\partial m_q} + \frac{N_H \alpha_s^2}{3\pi} \frac{\partial}{\partial \alpha_s} \right] \mathcal{L}_M , \tag{2.97}$$

where N_H is the number of heavy flavors (c, b, t, ...). We note that m_8 and μ depend linearly on the strange quark mass m_s (we neglect m_u and m_d), whereas m_1 does *not* vanish in the chiral limit (*i.e.*, there is no ninth Goldstone boson). Thus, by dimensional analysis we obtain the following Λ dependence for the three mass parameters

$$m_1^2 \sim \Lambda^2$$
$$m_8^2 \sim m_s \Lambda \tag{2.98}$$
$$\mu^2 \sim m_s \Lambda.$$

Using eqs. (2.97), (2.98) and (2.68) we can easily compute the $\eta \eta' \phi^0$ interaction Lagrangian

$$\mathcal{L} = \eta \eta' \frac{\phi^0}{v} \cos \theta \sin \theta (\cos^2 \theta \, m_{\eta'}^2 + \sin^2 \theta \, m_\eta^2) \left(-1 + \frac{2N_H}{3b} \right)$$
$$= \eta \eta' \phi^0 (1.1 \times 10^{-3} \, GeV) \left(1 - \frac{2N_H}{3b} \right) . \tag{2.99}$$

Hence the branching ratio for $\eta' \to \eta \phi^0$ is

$$BR(\eta' \to \eta \phi^0) = 1.2 \times 10^{-4} \left(1 - \frac{2N_H}{3b} \right)^2 B_{\phi^0} \tag{2.100}$$

where we have used the particle data group value of $\Gamma(\eta' \to \text{all}) = 2.1 \times 10^{-4}$ GeV [52] and $B_{\phi^0} = 2p_{\phi^0}/m_{\eta'}$ in the η' rest frame. It is appropriate to use $N_H = 3$, $b = 9$ in eq. (2.100).

In §3.1, we shall employ many of the results in this section in assessing the limits placed by current experiments on a light Higgs boson.

2.3 Decays with a Higgs Boson in the Final State

Quarkonium Couplings and Decays to Higgs

The Higgs boson coupling to heavy quarks is expected to be significant since it is proportional to gm_Q/m_W. This led to the early suggestion by Wilczek [16] that a light Higgs boson could be seen in the decays of quarkonia. In this section we consider the ϕ^0 coupling to the S- and P-wave quarkonium states of a heavy quark. We will use these couplings to find limits on a light Higgs boson from the decay $\Upsilon \to \phi^0\gamma$ in §3.1 and to search for the intermediate mass Higgs boson from the decay $(\overline{Q}Q) \to \phi^0 Z$ in §3.4.

We consider the quarkonium states $\eta(0^{-+})$, $\psi(1^{--})$, χ_J (where $J^{PC} = 0^{++}, 1^{++}$, or 2^{++}), and $\xi(1^{+-})$. The Higgs boson ϕ^0 has $J^{PC} = 0^{++}$. The following quarkonium decays to Higgs bosons are allowed by angular momentum and CP conservation [53]

$$\eta, \psi, \xi, \chi_1, \chi_2 \to Z\phi^0$$
$$\psi, \xi \to \phi^0\gamma \qquad\qquad (2.101)$$
$$\chi_0, \chi_2 \to \phi^0\phi^0 \ .$$

If the quarkonium state is weakly bound (binding energy $\ll m_Q$) then non-relativistic quantum mechanics is appropriate and the decay widths can be calculated in perturbation theory. As an aside, we note [54] that χ_0 has the same quantum numbers as ϕ^0 and they can mix; hence, if $m_{\chi_0} \simeq m_{\phi^0}$ the production of ϕ^0 could be enhanced.

The heavy quark couplings to the Higgs boson and the photon are scalar and vector respectively, so parity is conserved in the decays $(Q\overline{Q}) \to \phi^0\gamma$. Hence only the $C = -1$ states ψ and ξ can decay to $\phi^0\gamma$. The decay width for $\psi \to \phi^0\gamma$ was derived by in ref. 16, while the rate for $\xi \to \phi^0\gamma$ was computed by Kuhn [55],

$$\Gamma(\psi \to \gamma\phi^0) = \frac{\alpha_W \alpha e_Q^2}{2m_W^2}(1 - R_{\phi^0})|R_S(0)|^2$$
$$\Gamma(\xi \to \gamma\phi^0) = \frac{6\alpha_W \alpha e_Q^2}{M^2 M_W^2}(1 - R_{\phi^0})|R_P'(0)|^2, \qquad (2.102)$$

where M is the mass of the $Q\overline{Q}$ quarkonium state ($M \simeq 2m_Q$), e_Q is the electric charge of quark Q in units of e, and

$$\alpha_W = \frac{\alpha}{\sin^2\theta_W}$$
$$R_i = \frac{M_i^2}{M^2} \ . \qquad (2.103)$$

The quark radial wave functions at the origin are related to the derivative of the potential,

$$|R_S(0)|^2 = m_Q \left\langle \frac{dV}{dr} \right\rangle$$

$$|R'_P(0)|^2 = \frac{m_Q}{9} \left\langle \frac{1}{r^2} \frac{dV}{dr} + \frac{4(E-V)}{r^3} \right\rangle, \tag{2.104}$$

where $E = M - 2m_Q$. The results are obviously very sensitive to the choice of potential model. We discuss several typical cases. The Cornell potential [56] is parameterized by

$$V(r) = -\frac{k}{r} + \frac{r}{a^2} \tag{2.105}$$

where a and k are fixed by the charm and upsilon data. For a pure Coulomb potential $a = \infty$ and $k = (4/3)\alpha_s(m_Q^2) \simeq 0.1\text{–}0.2$. A fit to the data gives $k \simeq 1/2$ and $a \simeq 2.3$ GeV^{-1}. An alternate potential which is linearly confining at large distances and asymptotically free at short distances is the Richardson potential, whose Fourier transform is [57]

$$V(Q^2) = -\frac{4}{3} \frac{12\pi}{33 - 2N_{lf}} \frac{1}{Q^2} \frac{1}{\ln(1 + Q^2/\Lambda^2)} \tag{2.106}$$

where N_{lf} is the number of light flavors. A value of $\Lambda \simeq 400$ MeV is needed to reproduce the data. A discussion of the virtues of these potentials is given in ref. 53. In fig. 2.10 we show the square of the S-wave radial wave function at the origin and in fig. 2.11 we show the derivative of the P-wave radial wave function for several potential models.

If $M > m_{\phi^0} + m_Z$, then all states except for χ_0 can decay into $Z\phi^0$. The widths for these decays are enhanced by factors of $(M/m_Z)^2$ and could be sizable [58,53]

$$\Gamma(\eta \to Z\phi^0) = \frac{3\alpha_Z^2 a_Q^2 \beta_{Z\phi^0}^3}{4} \frac{M^2}{m_Z^4} |R_S(0)|^2,$$

$$\Gamma(\psi \to Z\phi^0) = \frac{\alpha_Z^2 v_Q^2 \beta_{Z\phi^0}}{2m_Z^2} \frac{|R_S(0)|^2}{(1 - R_Z)^2(1 - R_Z - R_{\phi^0})^2}$$

$$\times \left\{ \left[(1 - R_Z)^2 - R_{\phi^0}(1 - 3R_Z) \right]^2 \right. \tag{2.107}$$

$$\left. + \tfrac{1}{2}R_Z \left[(1 - R_Z)^2 + R_{\phi^0}(2 - R_{\phi^0}) \right]^2 \right\},$$

$$\Gamma(\xi \to Z\phi^0) = \frac{6\alpha_Z^2 v_Q^2 \beta_{Z\phi^0}^3}{M^2 M_Z^2} \frac{|R'_P(0)|^2}{(1 - R_Z - R_{\phi^0})^2}, \tag{2.108}$$

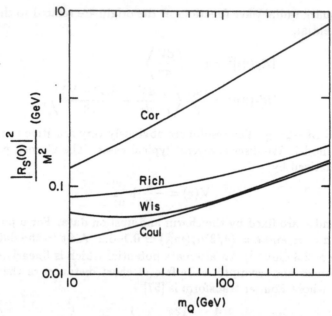

Figure 2.10 The S-wave radial wave at the origin squared, divided by M^2 (where M is the bound state mass) for various quarkonium potential models (Cornell, Richardson, Wisconsin, and Coulomb). This figure is from ref. 53.

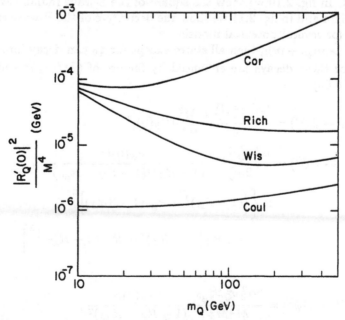

Figure 2.11 The derivative of the P-wave radial wave function at the origin squared, divided by M^4. This figure is from ref. 53.

$$\Gamma(\chi_1 \to Z\phi^0) = \frac{6\alpha_Z^2 a_Q^2 \beta_{Z\phi^0}}{m_Z^4} \frac{|R_P'(0)|^2}{(1 - R_Z - R_{\phi^0})^2}$$

$$\left\{ (1 + R_Z - R_{\phi^0})^2 \left(1 - R_Z + \frac{R_{\phi^0} R_Z}{1 - R_Z} \right)^2 \right.$$

$$+ \tfrac{1}{2} R_Z \left[3(1 - R_Z) + R_{\phi^0} \right.$$

$$\left. \left. + 4 R_{\phi^0} R_Z \left(\frac{1}{1 - R_Z} + \frac{1}{1 - R_Z - R_{\phi^0}} \right) \right]^2 \right\},$$

$$\Gamma(\chi_2 \to Z\phi^0) = \frac{9\alpha_Z^2 a_Q^2 \beta_{Z\phi^0}^5}{5M^2 m_Z^2} \frac{|R_P'(0)|^2}{(1 - R_Z - R_{\phi^0})^4},$$

$$(2.109)$$

where

$$\alpha_Z = \frac{\alpha}{\sin^2 \theta_W \cos^2 \theta_W}$$

$$(2.110)$$

$$\beta_{Z\phi^0} = \sqrt{(1 - R_Z - R_{\phi^0})^2 - 4 R_Z R_{\phi^0}},$$

and v_Q and a_Q are the quark couplings to the Z

$$v_Q = T_Q^{3L}/2 - e_Q \sin^2 \theta_W$$

$$a_Q = T_Q^{3L}/2,$$

$$(2.111)$$

where $T_Q^{3L} = \pm\frac{1}{2}$ is the third component of weak isospin of Q and e_Q is the charge of Q in units of e.

The $\phi^0 \phi^0$ system has $J^{PC} = 0^{++}$, 2^{++}, etc. so angular momentum and CP conservation allow only the decays $\chi_0 \to \phi^0 \phi^0$ and $\chi_2 \to \phi^0 \phi^0$ [53]. The widths are

$$\Gamma(\chi_0 \to \phi^0 \phi^0) = \frac{3\alpha_W^2 \beta_{\phi^0}}{32 m_W^4} \left(\frac{5 - 8 R_{\phi^0}}{(1 - 2 R_{\phi^0})^2} - \frac{9 R_{\phi^0}}{1 - R_{\phi^0}} \right)^2 |R_P'(0)|^2$$

$$(2.112)$$

$$\Gamma(\chi_2 \to \phi^0 \phi^0) = \frac{3\alpha_W^2 \beta_{\phi^0}^5}{80 m_W^4} \frac{1}{(1 - 2 R_{\phi^0})^4} |R_P'(0)|^2$$

where $\beta_{\phi^0} = \sqrt{1 - 4 R_{\phi^0}}$.

For the charm and bottom quarks, it is only the rates into $\phi^0 \gamma$ which are significant since all of the other Higgs-quarkonium couplings have additional factors of $(m_Q/m_W)^2$. If there exist quarks with masses greater than m_W then the quarkonium decays into $\phi^0 \phi^0$ and $\phi^0 Z$ will be enhanced and offer a possible discovery mechanism for the intermediate mass Higgs boson.

Decays of Gauge Bosons to Higgs

A very important class of decays that will be useful in searching for the ϕ^0 are the rare decays of the W and Z in which the ϕ^0 appears as a final state particle.

$W, Z \rightarrow \phi^0 f\bar{f}$

The most important of these modes was first suggested by Bjorken [59],

$$Z \rightarrow \phi^0 \ell^+ \ell^- \tag{2.113}$$

as shown in fig. 2.12.

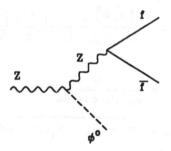

Figure 2.12 Feynman diagram for $Z \rightarrow \phi^0 f\bar{f}$.

More generally, let us consider $Z \rightarrow \phi^0 f\bar{f}$. The differential decay rate is given by

$$\frac{1}{\Gamma(Z \rightarrow f\bar{f})} \frac{d\Gamma(Z \rightarrow \phi^0 f\bar{f})}{dx} = \frac{g^2(12 - 12x + x^2 + 8y^2)\sqrt{x^2 - 4y^2}}{192\pi^2 \cos^2\theta_W (x - y^2)^2} \tag{2.114}$$

where,

$$x = \frac{2E_{\phi^0}}{m_Z}, \qquad y = m_{\phi^0}/m_Z, \tag{2.115}$$

with E_{ϕ^0} equal to the energy of the Higgs boson in the Z rest frame. The kinematical limits on x are $2y \le x \le 1 + y^2$. Integrating over x, we obtain the following expression for the branching ratio

$$\frac{BR(Z \rightarrow \phi^0 f\bar{f})}{BR(Z \rightarrow f\bar{f})} = \frac{g^2}{192\pi^2 \cos^2\theta_W}\left[\frac{3y(y^4 - 8y^2 + 20)}{\sqrt{4 - y^2}} \cos^{-1}\left(\frac{y(3 - y^2)}{2}\right)\right.$$

$$\left. - 3(y^4 - 6y^2 + 4)\ln y - \tfrac{1}{2}(1 - y^2)(2y^4 - 13y^2 + 47)\right]. \tag{2.116}$$

In the above equations, we have neglected the fermion mass (m_f) and the total Z-width (Γ_Z). This accounts for the logarithmic singularity in eq. (2.116) at

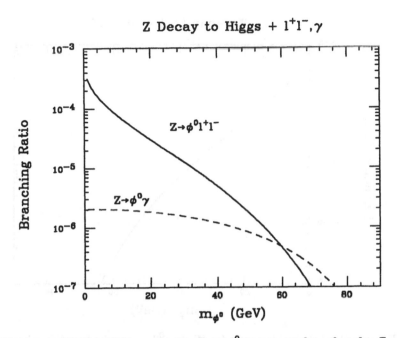

Figure 2.13 Branching ratio for $Z \to \phi^0 \gamma$, compared to that for $Z \to \phi^0 \ell^+ \ell^-$ (for fixed final lepton flavor), as a function of m_{ϕ^0}.

$y = 0$. In practice, for the light fermions ($m_f \ll \Gamma_Z$), the non-zero Z-width smooths out the singularity. Equation (2.116) becomes unreliable for $y \lesssim \Gamma_Z/m_Z$. In this region, we may approximate the $Z \to \phi^0 f\bar{f}$ branching ratio by using a Breit-Wigner for the virtual Z-propagator and setting $m_{\phi^0} = 0$. (For a more exact expression see ref. 60.) The result is

$$\frac{BR(Z \to \phi^0 f\bar{f})}{BR(Z \to f\bar{f})}(m_{\phi^0} = 0) = \frac{g^2}{192\pi^2 \cos^2 \theta_W}\left[\left(6 - \frac{\Gamma_Z^2}{2m_Z^2}\right) \ln\left(\frac{\Gamma_Z^2 + m_Z^2}{\Gamma_Z^2}\right) \right.$$
$$\left. + \frac{12\Gamma_Z}{m_Z} \tan^{-1}\left(\frac{m_Z}{\Gamma_Z}\right) - \frac{23}{2}\right]$$
$$\simeq 10^{-2}.$$

$$\tag{2.117}$$

Thus, we see that Higgs production in Z-decay can never be more than a 1% effect on the total Z-width. We illustrate the rate for $Z \to \phi^0 \ell^+ \ell^-$ in fig. 2.13 (assuming $m_Z < 2m_t$). A final subtlety that should be mentioned would arise if the top quark mass turns out to be near $m_Z/2$. In this case, it is possible that $Z \to t\bar{t}$ would be allowed, but $Z \to \phi^0 t\bar{t}$ would not. In this case, one must correct the ratio of eq. (2.116) to reflect the absence of the latter decay in the numerator.

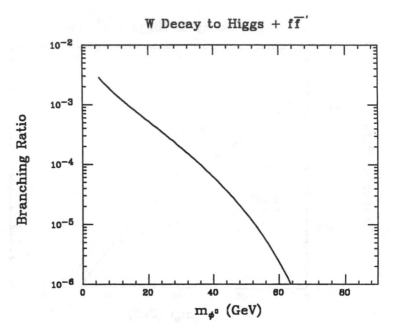

Figure 2.14 Branching ratio for $W \to \phi^0 f\bar{f}'$, summed over fermions, as a function of m_{ϕ^0}. (We assume that the t-quark is too heavy to appear in W decays.)

Of course, in complete analogy with the Z decays we have just discussed, we can consider

$$W \to \phi^0 f\bar{f}' , \qquad (2.118)$$

where (f, f') is a fermionic weak doublet (which couples to the W). The formula for $BR(W \to \phi^0 f\bar{f}')/BR(W \to f\bar{f})$ is simply obtained from eq. (2.116) by removing the $1/\cos^2\theta_W$ factor, and, of course, using $y = m_{\phi^0}/m_W$. We illustrate the total branching ratio for $W \to \phi^0 f\bar{f}'$, summed over all possible fermion final states, in fig. 2.14. (For simplicity, we assume that the t-quark is too heavy to appear in W decays.)

$Z \to \phi^0 \gamma$

A second process which can lead to Higgs production in Z decay is

$$Z \to \phi^0 \gamma \qquad (2.119)$$

whose rate was first computed by Cahn, Chanowitz and Fleishon [61] (see also ref. 13). This decay proceeds via a one loop triangle diagram, where the internal triangle consists of either W-bosons or the quarks and leptons. (Note that there is no tree-level $\phi^0 Z \gamma$ vertex, since at tree level, the photon couples only to charged particles.) Explicit formulae have been given in ref. 13. We find

$$\frac{\Gamma(Z \to \phi^0 \gamma)}{\Gamma(Z \to \mu^+ \mu^-)} = \frac{\alpha^2}{8\pi^2} \left(1 - \frac{m_{\phi^0}^2}{m_Z^2}\right)^3 \frac{|A_F + A_W|^2}{[1 + (1 - 4\sin^2 \theta_W)^2]}, \qquad (2.120)$$

where A_F and A_W [which are explicitly given in eq. (2.22)] are the contributions of the fermion loops and the W-loop, respectively. The contribution of the fermion loops is rather small compared to that of the W-loops. For small m_f/m_W, A_F is proportional to m_f^2/m_W^2 and is negligible. As $m_f \to \infty$, the fermion loop does not decouple, but approaches a constant

$$\lim_{m_f \to \infty} A_F = \sum_f \frac{2 N_c e_f (T_f^{3L} - 2 e_f \sin^2 \theta_W)}{3 \sin \theta_W \cos \theta_W}, \qquad (2.121)$$

where f has charge e_f in units of e, weak isospin T_f^{3L}, and $N_c = 3$ for color triplets and $N_c = 1$ for leptons. For example, the limiting values for the contributions of different types of superheavy fermions to A_F in eq. (2.121) are 0.063, 0.613 and 0.549 for a charged lepton L^-, and up and down type quarks U and D, respectively.

For comparison, we note the following approximate formula for the W-loop contribution (assuming $\sin^2 \theta_W = 0.23$), which is rather accurate (to within a few percent) for moderate values of the Higgs mass [13]

$$A_W = - \left[9.50 + 0.65 \frac{m_{\phi^0}^2}{m_W^2}\right]. \qquad (2.122)$$

Note that the contribution of the W-loops is opposite in sign to that of the fermion loops. However, the contribution of the W-loop is substantially larger than that of the fermion loops. Thus, to a good approximation, we may simply ignore the fermions, and obtain

$$\frac{\Gamma(Z \to \phi^0 \gamma)}{\Gamma(Z \to \mu^+ \mu^-)} = 6.94 \times 10^{-5} \left(1 - \frac{m_{\phi^0}^2}{m_Z^2}\right)^3 \left(1 + 0.07 \frac{m_{\phi^0}^2}{m_W^2}\right)^2. \qquad (2.123)$$

(We have used $\alpha \simeq 1/128$ for the coupling constant appropriate to the Z-mass scale in obtaining the above result.) The numerical value of the branching ratio is shown in fig. 2.13, compared to the branching ratio for $Z \to \phi^0 e^+ e^-$. Note that the rate for $Z \to \phi^0 \gamma$ is larger than the rate for $Z \to \phi^0 e^+ e^-$ only for very large Higgs masses (roughly $m_{\phi^0} \gtrsim 60$ GeV). An important point regarding $Z \to \phi^0 \gamma$ is that non-standard physics can contribute to the loop and possibly enhance the overall rate. One must be careful, since different loops come in with different signs, and it is also possible to suppress the Standard Model rate. But it is very important to be aware that this decay of the Z is a probe of higher mass scales (because it only occurs via a loop).

When several one-loop processes, such as $Z \rightarrow \phi^0 \gamma$, $\phi^0 \rightarrow \gamma\gamma$, and $gg \rightarrow \phi^0$, are measured, very valuable information on heavy particles that couple to the ϕ^0 can be obtained.

2.4   Radiative Corrections (and the "Screening Theorem")

As we shall see, direct observation of the Higgs boson most probably must await new accelerators, and will, in any case, be very challenging. Here we consider indirect effects of the Higgs boson through its contribution to various radiative corrections. We first summarize the limits on Higgs boson masses which can be obtained from precision measurements of the electroweak ρ parameter and $\sin^2 \theta_W$. We then discuss the "screening" theorem, which gives a general understanding of why sensitivity of radiative corrections to the Higgs mass is so small, even in the limit of large m_{ϕ^0}.

Limits on the Higgs Boson Mass From Radiative Corrections

In this section, we discuss numerical limits on the Higgs mass which can be obtained from precision measurements of weak neutral currents and the W and Z masses. As we show below, one-loop radiative corrections involving the Higgs mass grow only as $\ln(m_{\phi^0}/m_W)$ and so are numerically small. The issue is further complicated by the unknown top quark mass which enters into the same radiative corrections.

A comprehensive analysis of neutral current data on ν-hadron, νe, eN, μN, and e^+e^- scattering, atomic parity violation, and the W and Z masses has been performed in refs. 62 and 63. The analysis makes use of the one-loop radiative corrections to all processes [64]. A global fit to existing data gives the result,

$$\sin^2 \theta_W \equiv 1 - \frac{m_W^2}{m_Z^2} = 0.230 \pm 0.0048 , \qquad (2.124)$$

where all statistical, systematic, and theoretical errors have been included. This result assumes three families of quarks and leptons, $m_t < 100$ GeV, and $m_{\phi^0} < 1$ TeV. Since the radiative corrections are sensitive to the definition of $\sin^2 \theta_W$, it is important to note that ref. 62 uses a renormalization scheme in which the definition of $\sin^2 \theta_W$ given in eq. (2.124) is *exact* even when radiative corrections are taken into account. (Note that m_W and m_Z are the *physical* W and Z masses.) This definition of $\sin^2 \theta_W$ makes sense in any SU(2) $\times$ U(1) electroweak model in which all symmetry breaking is due to Higgs doublets.

However, in more general models, one must generalize eq. (2.124) to read

$$\rho \equiv \frac{m_W^2}{m_Z^2 \cos^2 \theta_W} \tag{2.125}$$

That is, we first compute ρ at tree level, and then (fixing ρ) we define $\sin^2 \theta_W$ so that the relation given in eq. (2.125) remains exact when radiative corrections are included.

As a result of the definition of $\sin^2 \theta_W$ above, the effect of radiative corrections is absorbed into the relation between the standard Fermi constant, G_F, of muon decay, the W mass and $\sin \theta_W$:

$$m_W^2 = \frac{\pi \alpha}{\sqrt{2} G_F} \frac{1}{\sin^2 \theta_W (1 - \Delta r)}, \tag{2.126}$$

where Δr is due to the one-loop radiative corrections. As we shall discuss in chapter 4, $\rho \neq 1$ is possible if there are Higgs multiplets with weak isospin greater than 1/2. A value of $\rho \neq 1$ can also be generated by mixing the Z with an additional Z boson. A two parameter fit to the existing data has also been presented in ref. 62 [with the same assumptions stated below eq. (2.124)], with the result

$$\sin^2 \theta_W = 0.229 \pm 0.0064$$
$$\rho = 0.998 \pm 0.0086 \ . \tag{2.127}$$

The sensitivity of $\sin^2 \theta_W$ to the Higgs mass is rather weak. Typically, it produces an uncertainty of 0.002 in the $\sin^2 \theta_W$ value extracted from most reactions other than deep inelastic scattering [62]. For example, if $m_t = 45$ GeV, then the global fit to $\sin^2 \theta_W$ is 0.230 for $m_{\phi^0} = 100$ GeV, and 0.231 for $m_{\phi^0} = 1$ TeV. Clearly, no limit on the Higgs mass can be obtained at present from neutral current experiments. Whether systematic errors can ever be reduced below 0.002 is not clear. The weak sensitivity of one-loop radiative corrections to m_{ϕ^0} can also be demonstrated explicitly by noting that [with the definition of eq. (2.124)] the contribution of a Higgs boson to the Δr parameter of eq. (2.126) is [64]

$$\Delta r_{\phi^0} = \frac{11 g^2}{96 \pi^2} \ln \left(\frac{m_{\phi^0}}{m_W} \right), \tag{2.128}$$

in the limit of large m_{ϕ^0}.

It is often said that the effects of radiative corrections can be studied by examining the ρ-parameter. However, this is not true for ρ as defined in eq. (2.125). Since $\sin^2 \theta_W$ is defined so that eq. (2.125) is exact even when radiative effects are included, it follows that ρ defined above is simply fixed at its tree-level value (e.g., $\rho = 1$ in the Standard Model). Nevertheless, there are other quantities called "the ρ-parameter" which have been defined in the literature which are sensitive to radiative corrections. Although these quantities are all defined to equal 1 at tree level in the Standard Model, they differ at one loop. We list here two common definitions of "the ρ-parameter", which we will call $\hat{\rho}$ and ρ_{NC} to avoid confusion. Unlike ρ of eq. (2.125), these quantities receive radiative corrections of order $g^2 \ln(m_{\phi^0}/m_W)$. The first is

$$\hat{\rho} \equiv \frac{m_W^2}{m_Z^2 \cos^2 \theta_W (m_W)}, \tag{2.129}$$

where $\cos^2 \theta_W (m_W)$ is a running parameter evaluated at an energy scale of m_W (see ref. 65). With this definition, the contribution of a heavy Higgs boson is [66][‡]

$$\hat{\rho} = 1 - \frac{11g^2}{96\pi^2} \tan^2 \theta_W \ln\left(\frac{m_{\phi^0}}{m_W}\right). \tag{2.130}$$

A second quantity which is often called "the ρ-parameter" is extracted from neutral current scattering and we will denote it by ρ_{NC}. The ratio of the tree level contribution to the total cross section for neutral current scattering to that for charged current scattering is defined as

$$R_\nu^0 \equiv \frac{\sigma_0(\nu_\mu N \to \nu_\mu X)}{\sigma_0(\nu_\mu N \to \mu^- N)} \tag{2.131}$$

and R_ν^1 is the corresponding quantity when the one-loop radiative corrections are included. Then $(\rho_{NC})^2$ is defined to be

$$R_\nu^1 = (\rho_{NC})^2 R_\nu^0. \tag{2.132}$$

With this definition,

$$\rho_{NC} = 1 - \frac{3g^2}{32\pi^2} \tan^2 \theta_W \ln\left(\frac{m_{\phi^0}}{m_W}\right) \tag{2.133}$$

in the large Higgs mass limit [64]. It should be emphasized that neither eq. (2.130) nor eq. (2.133) can be directly compared with the fitted value of ρ given in eq. (2.127) since the various ρ parameters are defined differently.

The logarithmic sensitivity of ρ_{NC} to the Higgs mass makes it very difficult to obtain Higgs mass constraints from precision electroweak measurements. Furthermore, the effects of other radiative corrections may be much more significant. For example, if one adds a fermion doublet (U,D) to the Standard Model, then the correction to ρ_{NC} at one loop is [68]

$$\delta\rho_{NC} = \frac{g^2 N_c}{64\pi^2 m_W^2} \left[m_U^2 + m_D^2 - \frac{2m_U^2 m_D^2}{m_U^2 - m_D^2} \ln\left(\frac{m_U^2}{m_D^2}\right) \right] \tag{2.134}$$

where N_c is the number of color degrees of freedom of the fermions.[*] Note that $\delta\rho_{NC} = 0$ if $m_U = m_D$, which is due to the existence of a "custodial"

[‡] Other definitions of $\sin^2 \theta_W (m_W)$ also appear in the literature [e.g., the $\overline{MS}$ definition, $\sin^2 \hat{\theta}_W (m_W)$, used in ref. 67], which lead to slightly different results for the one-loop corrections to $\hat{\rho}$.

[*] If we compute $\delta\hat{\rho}$ in this case, we would obtain $\delta\hat{\rho} = \delta\rho_{NC}+$ additional terms which are (at most) logarithmic in the fermion masses.

SU(2) symmetry in this limit. This symmetry is violated if $m_U \neq m_D$, and affects $\delta\rho_{NC}$ at one loop due to fermion-loop corrections to the gauge boson propagators. The quadratic sensitivity of $\delta\rho_{NC}$ to new heavy fermion masses (assuming $m_U \neq m_D$) is to be contrasted with the much weaker logarithmic behavior (for large $m_{\phi 0}$) exhibited in eq. (2.133). Due to the experimental bounds deduced in refs. 62 and 63, one can already place interesting limits on the magnitude of the top-quark mass or other new fermions. For example, using eq (2.134) we see that $\delta\rho_{NC} < 0.01$ implies that $m_t < 180$ GeV.

The "Screening" Theorem

Veltman [69] was the first to consider radiative corrections due to a large Higgs boson mass. For a large Higgs mass, the quartic scalar coupling (which is proportional to $m_{\phi 0}^2$) becomes large and the theory is strongly interacting. One might then expect that radiative corrections will be large as $m_{\phi 0}$ grows. However, Veltman concluded that at one loop there are only $g^2 \ln(m_{\phi 0}/m_W)$ corrections to physical processes, while at two or more loops, there are corrections of order $g^4(m_{\phi 0}^2/m_W^2)$. Radiative corrections which are dependent on the Higgs mass are thus of the form,

$$g^2 \left[\ln\left(\frac{m_{\phi 0}}{m_W}\right) + g^2 \left(\frac{m_{\phi 0}^2}{m_W^2}\right) + ... \right]. \tag{2.135}$$

For large Higgs masses, the theory is strongly coupled, but the effects of the strong interactions are screened by an extra factor of g^2. This is the so-called "screening theorem" and explains why low energy observables are relatively insensitive to the Higgs mass. A deeper understanding of the source of this screening was developed in ref. 70.

Radiative corrections to the Weinberg Salam model in the limit $m_{\phi 0} \to \infty$ have been extensively studied [71,72]. When $m_{\phi 0} \to \infty$, the resulting theory is a non-linear σ model coupled to an SU(2)$_L \times$ U(1) Yang-Mills theory. Since the theory is non-renormalizable, there are physical effects which depend on the cutoff, which is taken to be $m_{\phi 0}$.

The scalar fields of the theory may be written as

$$M(x) = \sigma(x) + i\vec{\tau} \cdot \vec{\pi}(x), \tag{2.136}$$

where the Higgs field is $\sigma(x)$, $\vec{\pi}(x)$ is a triplet of would be Goldstone bosons, and $\vec{\tau}$ are the Pauli matrices. The most general renormalizable scalar potential consistent with the SU(2)$_L \times$ U(1) gauge symmetry is

$$V(M, M^\dagger) = \frac{\lambda}{4} \left[\frac{1}{2} Tr(MM^\dagger) - f^2 \right]^2, \tag{2.137}$$

where $\lambda = m_{\phi 0}^2/2f^2$. The scalar interactions are then

$$\mathcal{L} = \frac{1}{4} Tr \partial_\mu M \partial^\mu M^\dagger - V(M, M^\dagger). \tag{2.138}$$

$\mathcal{L}$ is invariant under a global $SU(2)_L \times SU(2)_R$ transformation

$$M(x) \to e^{i\vec{\ell}_L \cdot \vec{\tau}/2} M(x) e^{-i\vec{\ell}_R \cdot \vec{\tau}/2} . \tag{2.139}$$

When the $SU(2)_L \times U(1)$ symmetry is gauged, the scalar sector loses this chiral symmetry. In the limit $\theta_W = 0$, however, the global chiral symmetry is restored. It is this accidental $SU(2)_L \times SU(2)_R$ symmetry of the potential which is responsible for the tree-level relation $\rho = m_W^2 / (m_Z^2 \cos^2 \theta_W) = 1$.

As $m_{\phi^0} \to \infty$, the theory becomes the non-linear σ model, with the constraint

$$MM^\dagger = f^2 = \frac{2m_W^2}{g^2} . \tag{2.140}$$

The gauge theory coupled to the non-linear σ model is formally equivalent to a Yang-Mills theory with a vector boson mass added by hand. The theory is non-renormalizable, but may be made finite to any order in perturbation theory by adding counterterms to the tree level non-linear Lagrangian. (The counterterms must be invariant under $SU(2)_L \times U(1)$ since this symmetry is unbroken by the constraint of eq. (2.140).) The cutoff dependence of the new counterterms is calculable, since they must cancel the divergences which occur when m_{ϕ^0} is removed from the Weinberg-Salam model. The Higgs boson mass is then identified with the cutoff. In this manner, the radiative corrections to various physical parameters may be calculated in the limit $m_{\phi^0} \to \infty$. At one loop, the greatest sensitivity to m_{ϕ^0} is logarithmic [71,72], as is clearly seen in the previous section.[†]

2.5   Higgs Mass Bounds: Vacuum Stability and Triviality

As repeatedly emphasized, the most important unknown parameter of the Standard Model is the Higgs boson mass, m_{ϕ^0}. In this section we give a brief overview of the lower and upper bounds on the mass of a Higgs boson deriving from considerations of vacuum stability and triviality, respectively. Other reviews of these subjects can be found in refs. 74 and 75, with the former containing reprints of many of the original papers on the subject. We focus on the minimal Standard Model case of a single Higgs doublet field. To

---

[†]The absence of a quadratic sensitivity to $m_{\phi^0}^2$ at one loop can also be attributed to the accidental $SU(2)_L \times SU(2)_R$ symmetry of the tree-level Higgs potential. For example, Toussaint [73] has constructed a model with two Higgs doublets which does not have this symmetry and found that ρ_{NC} does indeed receive corrections proportional to $m_{\phi^0}^2$ at one loop.

fix our definitions we define the scalar potential as

$$V(\phi) = -\mu^2 \phi^\dagger \phi + \lambda(\phi^\dagger \phi)^2. \tag{2.141}$$

Our convention for the vacuum expectation v is $\langle \phi \rangle = v/\sqrt{2}$, so that $v = 2m_W/g \approx 246$ GeV. At the tree level, λ and m_{ϕ^0} are related by

$$m_{\phi^0}^2 = 2\lambda v^2, \tag{2.142}$$

or, equivalently,

$$\lambda = \frac{g^2 m_{\phi^0}^2}{8 m_W^2}. \tag{2.143}$$

Both the questions of vacuum stability and triviality are conveniently discussed using the renormalization group equations at one-loop[‡] (though this can only be viewed as a perturbative guide in the case of triviality). We retain only the top quark Yukawa coupling for which we simplify our notation by defining $h \equiv g_t = \sqrt{2} m_t/v$. The renormalization group equations for the gauge couplings are

$$\begin{aligned}
\frac{d}{dt} g_s^2 &= -\frac{(33 - 4N_g)}{48\pi^2} g_s^4, \\
\frac{d}{dt} g^2 &= -\frac{(22 - 4N_g - \frac{1}{2}N_d)}{48\pi^2} g^4, \\
\frac{d}{dt} g'^2 &= \frac{(\frac{20}{3} N_g + \frac{1}{2}N_d)}{48\pi^2} g'^4,
\end{aligned} \tag{2.144}$$

where N_g is the number of generations (we shall always take $N_g = 3$), N_d is the number of complex scalar doublets (which we take here to be $N_d = 1$), and $t = \ln(Q^2/Q_0^2)$. Note that these equations are independent of both λ and h. The top quark Yukawa coupling evolves according to

$$\frac{d}{dt} h = \frac{1}{16\pi^2} \left[\frac{9}{4} h^3 - 4g_s^2 h - \frac{9}{8} g^2 h - \frac{17}{24} g'^2 h \right]. \tag{2.145}$$

Note that the evolution of h does not depend on λ. Finally, for λ we have

$$\frac{d}{dt} \lambda = \frac{1}{16\pi^2} \left\{ 12\lambda^2 + 6\lambda h^2 - 3h^4 - \frac{3}{2}\lambda(3g^2 + g'^2) + \frac{3}{16} \left[2g^4 + (g^2 + g'^2)^2 \right] \right\}. \tag{2.146}$$

[‡]See ref. 76 for a convenient summary of the one-loop equations; their conventions are, however, not quite the same as ours.

A Lower Bound on the Higgs Mass

The Linde-Weinberg lower bound [77] on the Higgs mass is derived by considering the evolution equation for λ and the limit where the Higgs mass is very small (corresponding to small λ). Neglecting terms containing λ we obtain

$$\frac{d}{dt}\lambda = \beta_\lambda, \tag{2.147}$$

where

$$\beta_\lambda = \frac{1}{16\pi^2}\left[-3h^4 + \frac{3}{16}\left(2g^4 + (g^2 + g'^2)^2\right)\right]. \tag{2.148}$$

Approximating β_λ by a constant and integrating up to scales Q of order ϕ itself produces the logarithmic form

$$\lambda(\phi) = \lambda(Q_0) + \beta_\lambda \ln(\phi^2/Q_0^2), \tag{2.149}$$

where by convention we have defined $\phi^2 \equiv \phi^\dagger\phi$. As noted in ref. 78, the $\lambda(\phi)$ so defined can be used to specify the one-loop effective potential

$$V \simeq -\mu^2\phi^2 + \lambda(\phi)\phi^4 = -\mu^2\phi^2 + \lambda\phi^4 + \beta_\lambda\phi^4\ln(\phi^2/Q_0^2). \tag{2.150}$$

The vacuum expectation value can be determined by the requirement

$$\left.\frac{\partial V}{\partial\phi}\right|_{\phi=v/\sqrt{2}} = 0, \tag{2.151}$$

yielding

$$-\mu^2 + \lambda v^2 + \beta_\lambda v^2[\ln(v^2/2Q_0^2) + \tfrac{1}{2}] = 0. \tag{2.152}$$

The mass of the Higgs particle is given by

$$m_{\phi^0}^2 = \tfrac{1}{2}\left.\frac{\partial^2 V}{\partial\phi^2}\right|_{\phi=v/\sqrt{2}} = 2v^2\left[\lambda + \beta_\lambda\left(\ln(v^2/2Q_0^2) + \tfrac{3}{2}\right)\right], \tag{2.153}$$

where we have used eq. (2.152) to eliminate μ^2. We can rewrite this expression in terms of the value of the potential at the minimum

$$V(\phi)|_{\phi=v/\sqrt{2}} = -\tfrac{1}{2}\mu^2 v^2 + \tfrac{1}{4}\lambda v^4 + \tfrac{1}{4}\beta_\lambda v^4 \ln(v^2/2Q_0^2). \tag{2.154}$$

Substituting for μ^2 [eq. (2.152)], we see that we can express m_{ϕ^0} as follows

$$m_{\phi^0}^2 = \frac{-8}{v^2}V\left(v/\sqrt{2}\right) + \beta_\lambda v^2. \tag{2.155}$$

Since $V(0) = 0$ when there is no symmetry breaking, the requirement that the symmetry breaking vacuum is preferred, i.e. $V(v/\sqrt{2}) < V(0)$, immediately implies that

$$m_{\phi^0}^2 > \beta_\lambda v^2 = \frac{3}{16\pi^2 v^2}\left[2m_W^4 + m_Z^4 - 4m_t^4\right], \tag{2.156}$$

where we have made use of the standard formulae for W, Z and t masses. From this equation we find that if all fermions are light then the lower bound on m_{ϕ^0} is roughly 7 GeV (for $\alpha = 1/128$, $m_W = 80$ GeV and $\sin^2 \theta_W = 0.23$); but if the top-quark mass is chosen near 78 GeV then the expression in eq. (2.156) vanishes and there is no lower bound on m_{ϕ^0} coming from this argument.

For m_t values such that the right-hand side of eq. (2.156) is negative, different types of considerations may impose a lower bound on m_{ϕ^0}. In particular, if β_λ is negative the potential of eq. (2.150) would be unbounded from below in the limit of $\phi \to \infty$. Naively, this appears to exclude m_t values above about 78 GeV. However, such an argument is clearly too simple. First, since perturbation theory will break down as $\phi \to \infty$, the one-loop approximation is not valid for ϕ values that are too large. Thus, in ref. 79 the constraint that $V(\phi_1) > V(v/\sqrt{2})$ was imposed, where $\phi = \phi_1$ was an estimate of the point at which perturbation theory breaks down. This led to bounds on the minimum m_{ϕ^0} as a function of m_t, indicating that m_{ϕ^0} near zero is only consistent if $m_t \sim 78$ GeV (and not just $\gtrsim 78$ GeV). Additional work along these lines can be found in ref. 80. One might attempt to improve upon the analysis by incorporating in β_λ the terms in eq. (2.146) proportional to λ^2 and λh^2, which have been neglected in our discussion so far. However, as first emphasized in ref. 76, a more complete approach must incorporate the full set of renormalization group equations, including the running of all the couplings constants as a function of field strength, ϕ. This more complete treatment will be discussed shortly.

The Linde-Weinberg bound is the minimum Higgs mass corresponding to a stable electroweak symmetry breaking vacuum. A related question is: what is the minimum Higgs mass taking into account the history of the early universe, in which the universe evolves from a symmetric phase to an electroweak symmetry breaking phase? We will shortly demonstrate that the minimum Higgs mass consistent with the standard hot Big Bang model of cosmology is given (roughly) by the Coleman-Weinberg mass [78], m_{CW} which is a factor of $\sqrt{2}$ larger than the Linde-Weinberg bound. Thus, we turn first to a brief look at the Coleman-Weinberg mechanism.

A Special Value for the Higgs Mass: the Coleman-Weinberg Mechanism

There is one "special" value for the Higgs mass which corresponds to the assumption that the tree-level Higgs mass parameter μ^2 *vanishes*. Here, we must be careful to state that there is no symmetry principle which can guarantee $\mu^2 = 0$. (One might be tempted to invoke scale symmetry, but this symmetry is anomalous and is violated by quantum corrections.) Thus, the choice of $\mu^2 = 0$ is just as unnatural as any other choice, and from this point of view, $\mu^2 = 0$ is not special. Nevertheless, this particular choice leads to the interesting phenomenon of spontaneous symmetry breaking via radiative corrections, known as the Coleman-Weinberg mechanism, first studied in ref. 78. If we set $\mu^2 = 0$ in eq. (2.141), then there is no spontaneous symmetry breaking at tree level. We can include the effects of radiative corrections by computing the effective potential in the one-loop approximation. It is sufficient to examine the effective potential as a function of $\phi_c \equiv \sqrt{2}\,\mathrm{Re}\,\phi^0$

$$V_{\mathrm{eff}}(\phi_c) = \tfrac{1}{4}\lambda\phi_c^4 + B\phi_c^4 \ln(\phi_c^2/M^2)\,, \qquad (2.157)$$

where

$$B = \frac{1}{64\pi^2 v^4} \sum_i C_i (2J_i + 1)(-1)^{2J_i} m_i^4\,, \qquad (2.158)$$

and the sum i is taken over the vector bosons and fermions of mass m_i, spin J_i and counting factor C_i (which counts color and electric charge: $C_i = 1, 2, 2$ and 6 for the Z-boson, W-boson, charged lepton and quark, respectively). The mass M is an arbitrary scale (subtraction point) which is used to define the coupling λ. It is straightforward to check that the potential given in eq. (2.157) possesses a global minimum at a nonzero value of ϕ_c; thus, the symmetry is spontaneously broken as a result of including radiative corrections. We denote $\langle\phi_c\rangle = v$, and compute the Higgs mass using

$$m_{CW}^2 = \left.\frac{d^2 V_{\mathrm{eff}}}{d\phi_c^2}\right|_{\phi_c = v} = 8Bv^2\,, \qquad (2.159)$$

where B is given by eq. (2.158). This is the famous result of Coleman and Weinberg [78]. Given the values of the vector boson and fermion masses, m_{CW} can be computed. If all the fermion masses are negligible, we find

$$m_{CW}^2 = \frac{3\alpha m_W^2}{4\pi \sin^2\theta_W}(1 + \tfrac{1}{2}\sec^4\theta_W) \simeq (10\ \mathrm{GeV})^2\,, \qquad (2.160)$$

using the same values of α, m_W, and $\sin^2\theta_W$ as above. A complete two-loop calculation gives a somewhat larger value, $m_{CW} = 10.4 \pm 0.3$ GeV [81]. On the other hand, m_{CW} is decreased when the fermion masses become appreciable. In particular, for $m_t > 78$ GeV, one would find that m_{CW}^2 is negative, which implies that the Coleman-Weinberg mechanism no longer results in the spontaneous breakdown of the electroweak symmetry.

From eqs. (2.156) and (2.159), it follows that

$$m_{CW} = \sqrt{2}m_{LW}\,, \qquad (2.161)$$

where m_{LW} is the lower Higgs mass bound obtained by Linde and Weinberg discussed above. It is instructive to see how the Linde-Weinberg bound follows from the present analysis. We have seen that even when $\mu^2 = 0$, the electroweak symmetry is spontaneously broken when the one-loop radiative corrections are taken into account. Consider now the effect of turning on a small *positive* tree-level Higgs squared mass, $m^2 \equiv -\mu^2 > 0$. It is easy to show that there is still a symmetry-breaking minimum to the potential when we add $\tfrac{1}{2}m^2\phi_c^2$ to V_{eff} given in eq. (2.157); the corresponding Higgs mass is now

$$m_{\phi 0}^2 = m_{CW}^2 - 2m^2 . \tag{2.162}$$

We now demand that the symmetry-breaking minimum be a global one, *i.e.*, $V_{\text{eff}}(v) < V(0)$. This provides us with an *upper* bound for m^2, which when substituted in eq. (2.162) results in the Linde-Weinberg bound [77]

$$m_{LW}^2 = 4Bv^2 = \tfrac{1}{2}m_{CW}^2 . \tag{2.163}$$

That is, the requirement that the symmetry-breaking minimum be a global minimum implies that $m_{\phi 0} > m_{LW}$. This is precisely the result obtained in the previous subsection.

Although the Linde-Weinberg bound is the minimum Higgs mass corresponding to a stable electroweak symmetry breaking vacuum, it is not necessarily the relevant bound when we take into account how the electroweak symmetry breaking vacuum was reached in the early universe. We will briefly discuss this question in the next subsection. The upshot of the arguments presented there is that the true lower bound for the Higgs boson in the minimal model is approximately the Coleman-Weinberg mass, m_{CW}, rather than the Linde-Weinberg bound, m_{LW}. This conclusion is based on the standard hot big bang scenario and an analysis of the phase transition which converts the universe from the unbroken to the broken electroweak symmetry phase.

Cosmology and Light Higgs Boson Mass Bounds

Let us consider the implications of cosmology for the Linde-Weinberg bound. When the evolution of the early universe is taken into account, one finds that, even for small m_t, the $m_{\phi 0} \gtrsim 7$ GeV bound need not be satisfied. As remarked above, this bound follows from the requirement that $V(\phi = v/\sqrt{2}) < V(0)$, *i.e.* that the SU(2) × U(1) violating (asymmetric) vacuum is a global minimum. Suppose that the universe was initially in the asymmetric state $\langle \phi \rangle = v/\sqrt{2}$, but that this vacuum state is *not* a global minimum, but possesses a higher energy density than the SU(2) × U(1) preserving (symmetric) minimum. This need not be in conflict with experiment (which indicates that we presently live in an asymmetric vacuum), if the lifetime of the asymmetric vacuum state is much longer than the age of the universe. The asymmetric vacuum state would be metastable, and would decay (in cataclysmic fashion) to the symmetric state at some time in the (hopefully) distant future. In such a scenario, $V(v/\sqrt{2}) > V(0)$ which would imply that $m_{\phi 0} < m_{LW}$. As one modifies the parameters of the Higgs potential so that the physical Higgs mass (computed from the radiatively corrected potential) is lowered below m_{LW}, the lifetime of the metastable asymmetric vacuum state *decreases*. By demanding that the asymmetric vacuum not decay during the present lifetime of the universe, we obtain a new lower bound on the Higgs mass.*

---

*In this picture, the experimental determination of the Higgs mass could provide crucial information as to the future fate of our universe!

This possibility of relaxing the Linde-Weinberg bound was suggested by Frampton [82]. Soon after Linde [83] reanalyzed and corrected some of the assumptions of the calculation. The decay of the vacuum state proceeds by the spontaneous generation of bubbles with $\langle \phi \rangle = 0$ inside them. This is a tunneling process. Coleman [84] has shown that to calculate the tunneling probability we need the $O(4)$ symmetric solution of

$$\left(\frac{\partial^2}{\partial \tau^2} + \vec{\nabla}^2 \right) \phi = \frac{dV}{d\phi} \tag{2.164}$$

with the boundary condition $\phi = v/\sqrt{2}$ at $t^2 + \mathbf{x}^2 \to \infty$. The number of such tunnelings which have occurred in the past is

$$N = F e^{-S_4} \tag{2.165}$$

where S_4 is the action, corresponding to the $O(4)$ symmetric solution of eq. (2.164). F is proportional to the space-time volume of the backward light cone and is approximately $F \sim 10^{170}$. The universe remains in the asymmetric vacuum when $N < 1$. This is the condition that Linde used to determine the new lower limit for the Higgs boson mass, which must be computed numerically by solving the tunneling problem. A few years later, Steinhardt [85] repeated Linde's calculation and employed more recent values for the weak interaction parameters. The result of this calculation is that $m_{\phi 0} \gtrsim 450$ MeV in order that the metastable asymmetric vacuum be sufficiently long-lived to explain our present existence.

The calculation described above implicitly assumes that the universe was initially in the asymmetric vacuum. However, this assumption is difficult to arrange in most models of the early universe. In the standard big bang picture, the universe is initially very hot, $T \gg v$, and the usual Higgs boson potential must be changed to include an additional field dependent term, V_T,

$$V_T = \frac{g'^2 + 3g^2}{32} T^2 \phi^2 + \mathcal{O}(T\phi^3) \quad . \tag{2.166}$$

For $T \gg v$ there is clearly no minimum at $\phi = v/\sqrt{2}$ and so the hot universe begins in the $SU(2) \times U(1)$ symmetric phase. As the universe cools, the transition to the spontaneously broken phase can occur.

In ref. 86, it was shown that there is a range of temperature $T_{C1} < T < T_C$, for which the asymmetric minimum is lower than the symmetric minimum, but there is a finite barrier between the two vacua. In this regime, the universe (which is still in the symmetric phase) is metastable, and the transition proceeds by the formation and growth of bubbles and is first order.[†] To reproduce the observed universe, one must demand that the lifetime of the metastable state be shorter than the age of the universe. For $m_{\phi 0} > m_{CW}$, this restriction imposes no constraints, since for $T \leq T_{C1}$, the barrier between

---

[†] For the large Higgs masses, $T_{C1} \simeq T_C$, and the phase transition is effectively second order.

the two vacua disappears and the transition is rapidly completed. At $m_{\phi 0} = m_{CW}$, $T_{C1} = 0$, and for $m_{LW} \lesssim m_{\phi 0} < m_{CW}$, T_{C1} does not exist. That is, the latter mass region corresponds to one in which there is a finite barrier between the symmetric vacuum and the true asymmetric vacuum all the way down to zero temperature. In ref. 83, Linde argued that for $m_{LW} \lesssim m_{\phi 0} \lesssim m_{CW}$, the metastable vacuum would never decay to the true asymmetric vacuum during the lifetime of the universe, and therefore such Higgs masses are ruled out.[‡] Guth and Weinberg reached a similar conclusion [87] by requiring that the phase transition not generate too much entropy. If $m_{LW} \lesssim m_{\phi 0} < m_{CW}$, the phase transition, if completed, would have only occurred after extreme supercooling, resulting in the generation of too much entropy (which would dilute the baryon number of the universe way below the observed value in any presently known model of baryogenesis). One additional effect, suggested by Witten [88], involves the chiral symmetry phase transition which would also break SU(2) × U(1). It turns out that for values of the Higgs mass very near m_{CW}, this effect can drive the transition to the asymmetric minimum thereby halting the period of supercooling and reducing the entropy produced in the transition. Flores and Sher [89] have summarized the results of these calculations in a figure, reproduced here as fig. 2.15. The upshot of these considerations is that the Higgs mass lower bound is approximately $m_{\phi 0} \gtrsim m_{CW}$, where $m_{CW} \approx 10$ GeV if $m_t^4/m_W^4 \ll 1$. A full discussion and summary of the subtleties of these arguments has been given in a review by Sher [75].

We have noted above that $m_{CW} \approx 10$ GeV only if $m_t^4/m_W^4 \ll 1$. For $m_t \approx m_W$, m_{CW} is near zero, thereby removing all Higgs mass lower bounds. Furthermore, even for $m_t^4 \ll m_W^4$, we have seen that the Higgs mass lower bound is substantially weakened if slightly unconventional assumptions are made about the evolution of the early universe, as discussed above. For these reasons, we feel it is most prudent in the phenomenological analyses presented in chapter 3 to disregard the lower bounds on $m_{\phi 0}$ discussed in this and the preceding subsections. In the future, when data about Higgs physics exists, perhaps it will teach us about these other matters.

An Upper Bound on the Higgs Mass

Let us begin our discussion by considering the pure ϕ^4 theory. Neglecting all couplings other than λ in eq. (2.146) we obtain a solution of the form

$$\frac{1}{\lambda(v)} - \frac{1}{\lambda(Q)} = \frac{3}{4\pi^2} \ln(Q^2/v^2), \qquad (2.167)$$

[‡] Higgs masses less than m_{LW} are also ruled out since in this case the symmetric vacuum is the true minimum at all T.

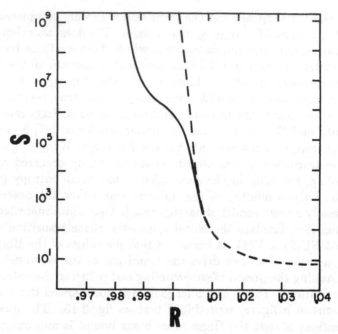

Figure 2.15 Entropy production, S, generated in the transition from the symmetric phase to the spontaneously broken phase. $R = m_{\phi 0}/m_{CW}$, where $m_{CW} \approx 10$ GeV if $m_t^4 \ll m_W^4$. The dashed line is the entropy produced if tunnelling alone drives the transition. The full line takes chiral symmetry breaking into account. In order to be consistent with the presently observed baryon number of the universe, in the context of all currently known mechanisms of baryogenesis, we must demand that $S < 10^6$. This implies that $m_{\phi 0} \geq m_{CW}$. This figure is taken from ref. 89.

where we have chosen to evolve from some large scale Q down to v.* The term "triviality" [90] refers to the fact that the coupling $\lambda(v)$ at low energy is driven to 0 (implying a trivial non-interacting theory) if we are allowed to take $Q \to \infty$ keeping $\lambda(Q) > 0$ (as required for stability of the theory). An alternative view of this same result is obtained by rewriting eq. (2.167) in the form

---

* One may wonder why we choose to evolve down to v, rather than, say, $v/\sqrt{2}$ or some other quantity of order v. The difference between these choices is non-leading in a leading-log approximation. Since we only work to one-loop accuracy in the renormalization group equations, our analysis cannot really distinguish between these choices of scale. The extent to which a change in scale affects numerical results reflects the uncertainty in the calculation.

$$\lambda(Q) = \frac{\lambda(v)}{1 - \frac{3\lambda(v)}{4\pi^2}\ln\left(\frac{Q^2}{v^2}\right)}, \tag{2.168}$$

showing that regardless of how small $\lambda(v)$ is, $\lambda(Q)$ will eventually blow up at some large Q (called the Landau pole), *if* the one-loop β function remains an adequate description of the theory at large λ. However, it is apparent that one cannot reliably use the one-loop equations when the coupling constant $\lambda(Q)$ becomes too large. Above a certain point higher-order or non-perturbative corrections to the beta function would have to be incorporated. For example, non-perturbative corrections can yield a fixed point where $\beta(\lambda)$ vanishes, thereby preventing growth of $\lambda(Q)$ beyond the fixed point location. Thus, even though the one-loop evolution equation for λ indicates a problem for the pure ϕ^4 theory if the theory applies at arbitrarily large scales, non-perturbative techniques are required to ascertain whether this problem actually exists.

Of course, as one considers larger and larger scales, it becomes more and more likely that new physics, beyond the Standard Model, will enter the picture. If we imagine that the Standard Model is embedded in a more complete theory at energy scale Λ, then Λ acts as a cutoff for our effective Standard Model theory. In this case, Λ will presumably be smaller than the Planck mass $m_{Pl} \simeq 10^{19}$ GeV. It might be set, for instance, by the scale of grand unification or of supersymmetry breaking. As we will see in §4.2, the latter type of model is an example of a theory in which it is natural to have a light Higgs boson because of the presence of a higher symmetry that cures the fine-tuning and hierarchy problems. However, in new physics models without such a higher symmetry, it is not *a priori* obvious that a large value for the Higgs mass leads to an inconsistency.

For sufficiently small $\lambda(v)$ [corresponding to a small Higgs mass by eq. (2.142)], the weak logarithmic dependence of $\lambda(Q)$ on the scale Q could allow $\lambda(\Lambda)$ to still be in the perturbative regime, so that the one-loop equation remains valid over the complete range of energy scales from v to Λ. We will characterize this possibility by saying that $\Lambda < \Lambda_{NP}$, where we have introduced Λ_{NP} to denote the energy scale at which the one-loop renormalization group equations break down, and higher order (or nonperturbative) effects become important. That is, $\Lambda < \Lambda_{NP}$ means that there is no "window" where it is necessary to use non-perturbative techniques (in particular, lattice approaches) to ascertain the relation between Λ and the maximum allowed Higgs mass [91]. On the other hand there is the alternative possibility that Λ is significantly larger than Λ_{NP}, in which case there is a window between Λ_{NP} and Λ where non-perturbative techniques would have to be employed and where the Higgs mass might also lie.

In order to illustrate the physics involved in assuming that $\Lambda < \Lambda_{NP}$, let us regard the one-loop β function as being exact for *all* λ. It is then amusing to ask what the maximum Higgs mass is if we take $\Lambda = m_{Pl}$. In determining an upper bound for $m_{\phi 0}^2$, we see from eq. (2.142) that we are interested in the maximum value of λ. However, we must specify at what scale λ should be evaluated. For the moment, we choose to use $\lambda = \lambda(v)$, $i.e.$ to evaluate λ at the scale of electroweak symmetry breaking. (The ambiguity involved in choosing the scale at which to evaluate λ will be addressed shortly.) It is clear from eq. (2.167) that the largest value of $\lambda(v)$ corresponds to $\lambda(\Lambda) = \infty$. For this choice we find $m_{\phi^0} \simeq 140$ GeV. (We note that this result is very insensitive to the precise choice of $\lambda(\Lambda)$; for example, $\lambda(\Lambda) = 1$, which is well into the perturbative domain, would yield nearly the same numerical result.) This type of bound has been obtained by a number of authors, beginning with the work of ref. 76 (for a review of this and related bounds, see ref. 92). Further work in this area appears in refs. 93 and 94.

So far, we have limited the discussion to a pure ϕ^4 theory. It is clear that a similar analysis can be made for the full electroweak model. Since the gauge couplings are small, the only new feature of the calculation is the fact that a heavy fermion mass (which we will take to be the t-quark) can somewhat alter the results. Following the logic of the previous paragraph, let us require that the theory never approach a region where λ or h (the top-quark Yukawa coupling) are large. In practice, this is indistinguishable from the requirement that $\lambda(m_{Pl})$ and $h(m_{Pl})$ be finite. In this case, the one-loop renormalization group equations (2.144)–(2.146) are an entirely adequate description of the theory. We now proceed to explore their implications.

If m_t ($i.e.$, h) is small, then we should expect to reproduce our previous Higgs upper bound, namely $m_{\phi^0} \lesssim 140$ GeV.[†] Thus, we examine the implications of large m_t. Following ref. 76, we note that if g_s^2 is taken to be constant, then eq. (2.145) implies that the top-quark Yukawa coupling would develop a singularity for large Q if $h^2(v) \gtrsim (16/9)g_s^2(v)$ corresponding to

$$m_t^2 > \frac{128\pi\alpha_s m_W^2}{9g^2} = (259 \text{ GeV})^2 \left(\frac{\alpha_s}{0.1}\right). \qquad (2.169)$$

More generally, there will be a maximum m_t value beyond which a consistent solution is no longer possible. Next, we observe that if the Q^2 dependence of h, g, and g' is neglected in eq. (2.146), then the right-hand side is quadratic in λ with roots $\lambda_\pm$. For $4m_t^4 > 2m_W^4 + m_Z^4$, one can show that $\lambda_- < 0 < \lambda_+$. For any initial value $\lambda(v)$ in the region $0 \leq \lambda(v) \leq \lambda_+$ the right-hand side of eq. (2.146) is negative and $\lambda(Q)$ is eventually driven to negative values as Q increases. Since the large field behavior of the effective potential is determined by $\lambda(\phi)$ [see eq. (2.150)], a negative value of $\lambda(Q)$ for Q larger than some critical value, Q_c, would imply that the potential is unstable (unless new physics enters below Q_c to alter the effective potential). The requirement of a stable renormalization-group-improved potential for all $Q < Q_c$ imposes a Q_c-dependent lower bound on the Higgs mass. As a crude estimate, we can

---

[†]Corrections due to the inclusion of the gauge couplings tend to raise the bound by about 10 to 20 GeV. This is a consequence of the negative coefficient in front of the $\lambda(3g^2 + g'^2)$ term in eq. (2.146).

impose $\lambda(v) > \lambda_+$, which is a sufficient (even for $Q_c \to \infty$) but not a necessary condition for stability. The resulting Higgs mass bound is

$$m_{\phi 0}^2 > m_W^2 + \tfrac{1}{2}m_Z^2 - m_t^2 + \sqrt{(m_W^2 + \tfrac{1}{2}m_Z^2 - m_t^2)^2 + 4m_t^4 - 2m_W^4 - m_Z^4}\,.$$

(2.170)

This bound is useful only for rather large m_t (and m_ϕ). Finally, we note that at sufficiently large m_t, the lower and upper bounds on $m_{\phi 0}$ tend to coalesce. This occurs because eq. (2.146) implies very rapid variation for $\lambda(Q)$ once h is large.

A detailed analysis consisting of a full numerical treatment of the coupled equations was first presented in ref. 76. Cabibbo et al. selected the minimal grand unified SU(5) model as a prototype for a model in which there is no new physics between the electroweak scale and the grand unification scale M_U. (In their model, they employed a rather large value for the grand unification mass, $M_U = 8.4 \times 10^{15}$ GeV, with corresponding electroweak angle $\sin^2 \theta_W \simeq 0.2$.) The assumption that the Higgs self coupling is finite at all scales below M_U leads to a bound on the Higgs mass as described above. (By using M_U instead of m_{Pl}, the resulting bound on $m_{\phi 0}$ will be slightly larger.) The results of ref. 76 are shown in fig. 2.16. Even more restrictive bounds are obtained with additional assumptions. For instance, in ref. 93 the additional assumption that one should approach a fixed point in the ratio of coupling constants $z \equiv h^2/g_1^2$ is shown to lead to an upper bound on $m_{\phi 0}$ of order 125 GeV for small m_t. In any case, it is clear that, under the assumptions of no new physics and perturbative unification at some large scale, small values of $m_{\phi 0}$ are preferred. It should be emphasized that this result does not solve the general problem of naturalness in a theory with elementary scalars. Indeed, the existence of a light Higgs boson means that the renormalization group scaling is consistent. But renormalization group scaling is derived from the logarithmic divergences of the unrenormalized theory and is completely oblivious of the quadratic divergences. The "unnatural" hierarchy between the Higgs mass and the Planck scale remains unexplained.

A lower limit on $m_{\phi 0}$ as a function of m_t, for $m_t \gtrsim m_W$, using a full one-loop renormalization group analysis of the renormalization-group-improved potential, has been given in ref. 95. Recently, the calculation of this bound has been refined in ref. 96 by solving the renormalization group equations to two-loop order. The results differ numerically somewhat from those given in fig. 2.16, but the qualitative features of the lower bound curve are much the same. Thus, given the Linde-Weinberg bound for $m_t \lesssim 80$ GeV and this second stability bound for $m_t \gtrsim 80$ GeV, it seems that one must conclude that the SM Higgs boson can only have mass $\lesssim 5$ GeV if m_t lies very near 80 GeV. Consequently, in discussing experimental constraints on such a very light ϕ^0 we shall generally adopt this value for the top-quark mass.

Before proceeding, it is worth mentioning that this same type of renormalization group analysis can be extended to the case of more than one Higgs doublet in the SM context. In particular, the SM with two Higgs doublets

Bounds on m_{ϕ^0} as a function of m_t

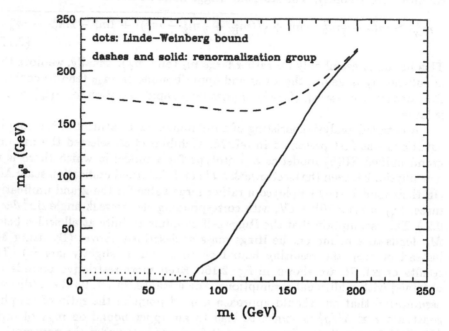

Figure 2.16 Upper and lower bounds on m_{ϕ^0} as a function of m_t, coming from the requirement of a perturbative theory at all energy scales from v to M_U. This figure is taken from ref. 76, where M_U is taken to be the grand unification scale $M_U \simeq 10^{16}$ GeV and $\sin^2\theta_W \simeq 0.2$. Three generations of quarks and leptons have been assumed.

was analyzed in ref. 97. As in ref. 93 one requires that the ratios of Higgs quartic couplings to gauge couplings remain finite at m_{Pl}. Then, when both the vacuum expectation values v_1 and v_2 of the neutral members of the two Higgs doublets are non-zero, an upper bound on the lightest neutral Higgs boson mass of about 85 GeV is obtained.

We now would like to broaden our considerations by assuming that new physics does enter at an energy scale Λ which is below m_{Pl}. To be definite, we again take $\lambda(\Lambda) = \infty$ and use the evaluation of λ at the electroweak scale v to determine the Higgs mass. The result for the maximum Higgs mass squared is

$$(m_{\phi^0}^{\max})^2 = \frac{8\pi^2 v^2}{3\ln(\Lambda^2/v^2)}. \tag{2.171}$$

In fig. 2.17 we plot the resulting value of $m_{\phi^0}^{\max}$ as a function of Λ. Note that as $\Lambda \to \infty$ we find that $m_{\phi^0}^{\max}$ quickly asymptotes to fairly low values.

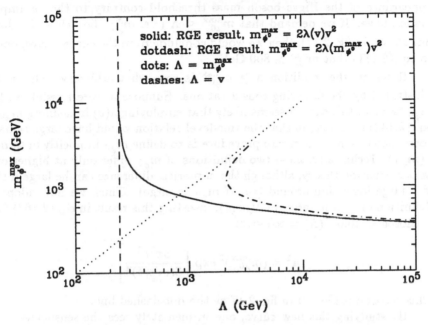

Figure 2.17 The maximum value of $m_{\phi0}$ as a function of the largest scale Λ for which we may apply the pure ϕ^4 renormalization group equation, assuming $\lambda(\Lambda) = \infty$.

Of course, we also see from fig. 2.17 that the lower the scale at which there is new physics the larger the maximum $m_{\phi0}$ value can be. What is the lowest value of Λ that we might allow? There is no definite criterion for this. One extreme is to presume that new physics could set in at scales of order v, in which case the phenomenology and detection of the Higgs boson would clearly be influenced and, indeed, preceded by the new physics itself. In this case, the question of the value of the Higgs boson mass may not be a meaningful one (*e.g.*, in technicolor models, there may not be any obvious candidate which would play the role of the conventional Higgs boson). Although this possibility is quite interesting, it is not the concern of this section. Here, we wish to investigate the implications of an alternative picture: namely, we assume the existence of an energy scale in which a (well-defined) effective field theory exists whose particle spectrum is precisely that of the Standard Model. That is, suppose that one can deduce that there are no thresholds for new physics beyond the Standard Model up to some energy scale Λ. Is there a maximum allowed value, $m_{\phi0}^{max}$, based on the requirement of theoretical consistency? In the context of our previous discussion, this means that we must demand that $m_{\phi^0}^{\max} \lesssim \mathcal{O}(\Lambda)$; otherwise new physics will enter before the appearance of the Higgs boson mass threshold contrary to the assumption made above. If we demand that $m_{\phi^0}^{\max} \leq \Lambda$ (*i.e.* $m_{\phi^0}^{\max}$ lie below the dotted line shown in fig. 2.17), then we would conclude from the curve corresponding to eq. (2.171) that $m_{\phi^0}^{\max} \lesssim 800$ GeV.

However, the condition $\Lambda \geq m_{\phi^0}^{\max}$ is somewhat arbitrary. This is best illustrated by the following considerations. Suppose it were possible to have $m_{\phi^0} \gg v$. In this case, it seems likely that employing $\lambda(v)$ in defining m_{ϕ^0} via eq. (2.142) is unwise, in that the tree-level relation would have large radiative corrections. A more accurate procedure is to define m_{ϕ^0} implicitly in terms of $\lambda(m_{\phi^0})$. Technically, these two definitions of m_{ϕ^0} differ only at higher order in perturbation theory, although the numerical difference can be large if there is a large logarithm around (*i.e.* if m_{ϕ^0} is large). Thus, we now adopt the definition of m_{ϕ^0} in terms of $\lambda(m_{\phi^0})$; inserting this result in eq. (2.167) (with λ chosen so that $\lambda(\Lambda) = \infty$) gives

$$\Lambda^2 = (m_{\phi^0}^{\max})^2 \exp\left[\frac{8\pi^2 v^2}{3(m_{\phi^0}^{\max})^2}\right]. \tag{2.172}$$

This relation is plotted in fig. 2.17 as the dot-dashed line.

By studying this new curve, one immediately sees the sensitivity to the choice of the condition $\Lambda \gtrsim m_{\phi^0}^{\max}$. If one were to simply require $\Lambda \geq m_{\phi^0}^{\max}$, one finds no bound on the Higgs mass; in fact the vanishing of $\lambda(v)$ as $\Lambda \to \infty$ appears to be disconnected from the value of $m_{\phi^0}^{\max}$. However, the domain in which the solid and dot-dashed lines of fig. 2.17 significantly diverge is precisely the regime where the neglected higher-order perturbative effects are non-negligible. Furthermore, we know that an effective theory as defined by integrating out the heavy degrees of freedom is well defined only at scales sufficiently below the heavy mass thresholds so that power corrections inversely proportional to the heavy masses are small. These arguments suggest that we can be confident in our analysis only if we require $\Lambda \geq a m_{\phi^0}^{\max}$, for some number a sufficiently larger than 1. For example, if we demand that $\Lambda \geq 2 m_{\phi^0}^{\max}$ and use the dot-dashed line of fig. 2.17, we would conclude that $m_{\phi^0}^{\max} \sim 800$ GeV, which is (roughly) the same result we obtained earlier. For $m_{\phi^0}^{\max} \lesssim \Lambda \lesssim a m_{\phi^0}^{\max}$, the question of the existence of a Higgs boson is intricately tied to the nature of the new physics. This still leaves open the possibility that somehow one can arrange a situation in which a Higgs-like object appears just below the mass threshold for new physics thereby evading the bound obtained above.

However, both curves in fig. 2.17 enter a region for λ that is non-perturbative when $m_{\phi^0}^{\max} \gtrsim 800$ GeV. Thus, we would like to use an alternative approach to the question of what happens if $\Lambda > \Lambda_{NP}$ — that is the one-loop approximation to the beta function breaks down before the scale Λ is reached. In the Standard Model, this scale can be estimated by noting that the tree-level unitarity limit for the Higgs quartic coupling is [98] $\lambda < 12\pi/5$, which would imply that $m_{\phi^0} \lesssim 750$ GeV. This suggests that one can never really deduce a rigorous upper bound for the Higgs mass using a perturbative argument, and that non-perturbative methods (such as lattice calculations) are required. We now turn to a brief discussion of those calculations and their implications for the value of $m_{\phi^0}^{max}$.

Non-perturbative modifications have been the subject of much recent investigation; this area has been reviewed in refs. 92, 99 and 74 and, more briefly, in refs. 100–103. Especially numerous have been various attacks on the problem using lattice techniques [106–108] which are based on a proposal first put forth by Dashen and Neuberger [94]. The hope is that one can go beyond the perturbative domain and, in the context of the electroweak theory, determine an absolute upper bound on the Higgs mass. In the case of a relatively light top quark, it is possible that all gauge and Yukawa couplings are sufficiently small that results from a pure ϕ^4 investigation would be sufficient. All such investigations provide strong evidence confirming the triviality of the (lattice) ϕ^4 theory. In addition, there is good numerical agreement among the Monte Carlo [104–107] and the analytic strong coupling [108] techniques. These studies also tend to confirm the general feature that a large value for m_{ϕ^0} requires new physics at a fairly modest energy scale Λ, and that the crossover point between new physics and m_{ϕ^0} is near 1 TeV. In particular, Higgs mass upper limits around 650 GeV have been claimed on the basis of Monte Carlo simulations of lattice ϕ^4 scalar field theory [104–107]. Part of the impact of the lattice investigations has been to confirm that the perturbative approach is reliable over a remarkably large range of energy scales and matches rather smoothly onto the strong coupling λ regime.

Nevertheless, one must inject a note of caution here. The lattice ϕ^4 models studied above are still somewhat crude approximations to the Higgs sector of the Standard Model. Although the statements made above are almost certainly qualitatively correct, the above conclusions have not yet become unambiguously quantitative. As an example, consider the fact that the lattice approaches rely on the ability to detect non-scaling effects at separations corresponding to a significant number of lattice spacings. Thus, there is sensitivity to finite size effects on the lattice and to the exact lattice scheme employed. For instance, it seems [101,102] that there could be lattice artifact effects associated with the non-Lorentz-invariant nature of a finite lattice that would obscure the non-scaling effects associated with triviality. Thus, it is not yet clear whether there is a window of energies between Λ_{NP} (defined earlier as the scale where non-perturbative effects enter) and the new physics scale Λ where perturbative techniques are not meaningful, but a strongly interacting ϕ^4 theory is still an appropriate description of the physics of the effective theory. The analysis of ref. 104 suggests that such a window does not exist, but only if their lattice calculations turn out to be free of the ambiguities described above.

Certainly, lattice techniques based on the effective theory alone are not relevant to the situation where new physics occurs below the scale of m_{ϕ^0}. However, a more delicate issue is the question of what happens if $m_{\phi^0} \sim \Lambda$ at large Λ substantially above the scale Λ_{NP} at which perturbation theory breaks down. Lattice investigations are very restricted in the effective Λ that can be probed, being only fully reliable when $m_{\phi^0}/\Lambda \lesssim 0.2$, though perhaps one can trust results even for $m_{\phi^0}/\Lambda \lesssim 0.5$. When $\Lambda \sim m_{\phi^0}$, effects due to the size of the lattice spacing will be dominant; additional terms, that affect this domain on the lattice but do not affect the continuum limit, can be added to the Lagrangian. In other words, the lattice computations cannot be used to explore the domain where the Compton wave length of the Higgs is comparable to the lattice cutoff. Furthermore, new physics will enter at the scale Λ, and so it is very hard to make precise statements about the scalar spectrum in this situation.

We believe that it is premature to argue against the possibility of a strongly interacting WW sector and/or scalar particles in the mass region above 1 TeV. The lattice calculations do suggest that such features are incompatible with a theory based solely on the Standard Model. However, for $\Lambda \sim m_{\phi^0}$, new physics enters in the vicinity of Λ and precise predictions are harder to come by. Consequently, experimentalists should not be discouraged from searching for Higgs bosons over the full mass range accessible to a given machine.

2.6   Unitarity Constraints on the Higgs Sector

The possibility of a strongly interacting WW sector and/or scalar particles in the mass region above 1 TeV is an interesting alternative to the weakly coupled Higgs sector [69,109–111]. Such a possibility presents a formidable theoretical challenge, since perturbation theory is not generally reliable in describing the consequences of a strongly interacting Higgs sector. Nevertheless, there are theoretical techniques that can be used to provide important constraints on the theory. In this section, we will discuss the implications of unitarity and briefly describe the Equivalence Theorem and its uses. It has often been claimed that the requirements of unitarity impose an upper bound on the value of the Higgs mass. However, a more accurate statement would be that the unitarity bound indicates the mass scale at which the Standard Model must be superseded by new nor non-perturbative physics of some kind (e.g., resonance formation in a strongly interacting WW sector). Even in this case where normal perturbative techniques must fail, low energy theorems can be derived which give accurate predictions for amplitudes of electroweak processes at energies sufficiently below the unitarity bound.

The connection between unitarity and an upper bound on the Higgs mass was described in great detail in a classic paper by Lee, Quigg and Thacker [109]. (This idea was first exploited in ref. 112. Related work was

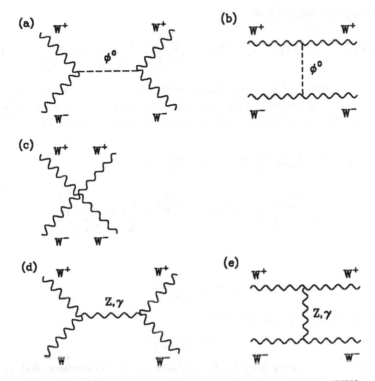

Figure 2.18 Complete set of gauge invariant diagrams for WW scattering.

also carried out in refs. 110, 111 and 113.) We illustrate the salient ideas by considering WW elastic scattering. Suppose we compute the scattering amplitude for $W^+W^- \rightarrow W^+W^-$, using Standard Model Feynman rules, but omitting all diagrams involving the Higgs boson. Let s be the center-of-mass energy squared for the process. Then, we find that the amplitude for this process grows with s, in violation of unitarity. This, of course, is no surprise, since the theory without the Higgs boson is not renormalizable. It was shown in ref. 114 that when the full content of the gauge theory is employed including the Higgs boson (see the diagrams of fig. 2.18), then all tree-level amplitudes approach a constant value (or else vanish) at infinite energy. Since we can remove the Higgs boson from the spectrum by taking $m_{\phi^0} \rightarrow \infty$, it follows that tree-level unitarity must eventually break down for some large value of m_{ϕ^0}.

Let us illustrate this by explicitly evaluating the amplitude for WW elastic scattering. It is easy to show that the potential for bad high energy behavior occurs when the external W-bosons are longitudinal. [To see this, note that in the large energy limit we may approximate the polarization vector of W_L by: $\epsilon_L^\mu(p) \simeq p^\mu/m_W$.] The amplitude for $W_L^+ W_L^- \rightarrow W_L^+ W_L^-$ in the limit where $s, m_{\phi^0}^2 \gg m_W^2, m_Z^2$ is[‡]

$$A(W_L^+ W_L^- \to W_L^+ W_L^-) = -\sqrt{2}\, G_F m_{\phi^0}^2 \left[\frac{s}{s - m_{\phi^0}^2} + \frac{t}{t - m_{\phi^0}^2} \right]. \quad (2.173)$$

As advertised above, the amplitude grows linearly in s if we take $m_{\phi^0} \to \infty$. However, even at finite m_{ϕ^0}, this amplitude may violate unitarity. To see this, we compute the contribution to the $J = 0$ partial wave from eq. (2.173)

$$
a_0 = \frac{1}{16\pi s} \int\limits_{-s}^{0} A(W_L^+ W_L^- \to W_L^+ W_L^-)\, dt
$$

$$
= -\frac{G_F m_{\phi^0}^2}{8\pi\sqrt{2}} \left[2 + \frac{m_{\phi^0}^2}{s - m_{\phi^0}^2} - \frac{m_{\phi^0}^2}{s} \ln\left(1 + \frac{s}{m_{\phi^0}^2} \right) \right]. \quad (2.174)
$$

For $s \gg m_{\phi^0}^2$,

$$a_0 = -\frac{G_F m_{\phi^0}^2}{4\pi\sqrt{2}}. \quad (2.175)$$

Partial wave unitarity requires that

$$|a_J|^2 \leq |\mathrm{Im}\, a_J|, \quad (2.176)$$

for the Jth partial wave amplitude. Equation (2.176) implies that $|a_J| \leq 1$, which is the condition applied in refs. 109, 113, and 110. However, we may derive a second condition, which will be even more useful for our purposes, as recently emphasized in ref. 98. It is easy to see that eq. (2.176) implies that

$$(\mathrm{Re}\, a_J)^2 \leq |\mathrm{Im}\, a_J|(1 - |\mathrm{Im}\, a_J|). \quad (2.177)$$

Since the right hand side of eq. (2.177) is bounded by 1/4, we find the restriction

$$|\mathrm{Re}\, a_J| \leq \tfrac{1}{2}. \quad (2.178)$$

It turns out that $J = 0$ leads to the strictest bounds, so we will not discuss the higher partial waves any further. Applying eq. (2.178) to eq. (2.175) implies that $m_{\phi^0} < 850$ GeV. Lee, Quigg and Thacker showed in ref. 109 that the most stringent unitarity bound is obtained by considering the implications of partial wave unitarity for the full coupled-channel system involving W, Z and ϕ^0. Namely, one computes the $J = 0$ partial-wave amplitude matrix for the normalized states $W_L^+ W_L^-$, $\sqrt{1/2}Z_L Z_L$, $Z_L \phi^0$ and $\sqrt{1/2}\phi^0\phi^0$. (The factor of $\sqrt{1/2}$ is required for correctly normalized amplitudes involving identical

---

[‡]We neglect the width of the ϕ^0 since it is not relevant for the simple unitarity arguments presented here, as long as we choose s to lie outside the Higgs resonance region. The effects of a large Γ_{ϕ^0} (for the case of large m_{ϕ^0}) will be briefly discussed at the end of this section.

particles.) The most restrictive bound is derived from the largest eigenvalue of the amplitude matrix, and the end result is

$$m_{\phi 0}^2 \leq \frac{4\pi\sqrt{2}}{3G_F} \simeq (700 \text{ GeV})^2 \,. \tag{2.179}$$

It is interesting to note that this bound is rather close to the Higgs mass upper limits obtained in lattice models discussed in the previous section.

How should we interpret this bound? Formally, all we have shown is that the *tree-level* amplitude violates unitarity (at sufficiently large energy) if the Higgs mass violates the bound of eq. (2.179). Certainly, in a consistent theory the full amplitude cannot violate unitarity. However, it is not possible to determine whether unitarity is restored in the Standard Model simply by including higher order corrections to the tree-level amplitude. This is because in the large Higgs mass regime, the Higgs self-coupling is strong, implying that perturbation theory is no longer reliable. Thus, the mass bound of eq. (2.179) is best thought of as the largest Higgs mass for which perturbation theory can be reliable for all s. Attempts have been made to include the leading contributions to the one-loop radiative corrections to the unitarity bound that are enhanced in the large $m_{\phi 0}$ limit [115]. When these one-loop corrections are included, it is no longer possible to obtain a bound on the Higgs mass which is independent of energy. However, for all s, these corrections *increase* the $J = 0$ partial wave amplitude, and hence do not help to restore unitarity.

Conceptually, it is perhaps more useful to consider the unitarity constraints in a different limit where m_W^2, $m_Z^2 \ll s \ll m_{\phi 0}^2$. In this case, if s becomes too large, we also find that unitarity of the tree-level amplitudes is violated. We shall denote by s_c the maximum s allowed by tree-level unitarity in the limit where $m_{\phi 0} \to \infty$. Then $\sqrt{s_c}$ can be interpreted as an energy scale, above which new physics beyond the Standard Model (*e.g.*, a strongly interacting gauge boson sector) must enter. Using the explicit example of $W_L^+ W_L^-$ elastic scattering considered above, we find, for $s \ll m_{\phi 0}^2$, that the $J = 0$ partial wave is

$$a_0 = \frac{G_F s}{16\pi\sqrt{2}} \,, \tag{2.180}$$

showing that for large enough s, unitarity is violated. As above, the most stringent bound is obtained by doing the full coupled-channel computation. Here, the analysis is simpler since the Higgs channels do not contribute in the limit of $m_{\phi 0} \to \infty$. The best bound emerges from the isospin zero channel $\sqrt{1/6}\,(2W_L^+ W_L^- + Z_L Z_L)$, and is

$$s \leq s_c \equiv \frac{4\pi\sqrt{2}}{G_F} \lesssim (1.2 \text{ TeV})^2 \,. \tag{2.181}$$

[This differs from the result of ref. 110 because we use the more restrictive unitarity constraint given in eq. (2.178).] The significance of this latter bound

goes beyond the perturbative calculation which was employed in its derivation. As emphasized in ref. 110, even in the limit of a strongly interacting
Higgs sector, one can obtain rigorous low-energy theorems (valid for $s \ll m_{\phi 0}^2$)
involving the Goldstone bosons which depend solely on the group theoretical
properties of the electroweak symmetry breaking. These can be reinterpreted
as theorems involving the longitudinal gauge bosons by using the Equivalence
Theorem. For example, in a theory where the electroweak symmetry break-
ing is such that $\rho \equiv m_W^2 / (m_Z^2 \cos^2 \theta_W) = 1$, one finds that eq. (2.180) is a
correct low-energy theorem* *to all orders in the Higgs self couplings* λ, and
to leading order in the electroweak coupling g^2. The low-energy theorem is
valid so long as $s \ll \min[M_{SB}^2, (4\pi v)^2]$, where M_{SB} is the characteristic scale
of spontaneous symmetry breaking [110] (*e.g.*, $M_{SB} \sim m_{\phi 0}$ in the Standard
Model).

The Equivalence Theorem permits one to compute the interactions of the
longitudinal gauge bosons, in the limit of $s \gg m_W^2$, by employing an effective theory of interacting scalars. The theorem states that for a scattering
process involving external longitudinally polarized W's and Z's, the S-matrix
element can be calculated, to $\mathcal{O}(m_W / \sqrt{s})$, by replacing the external gauge
bosons with the corresponding Goldstone bosons of the R_ξ gauge [116]. Since
calculations in the effective theory involve only scalars as external particles,
they are extremely simple to perform. The Equivalence Theorem is clearly
true for tree-level S-matrix elements [109], but the question of its validity
when radiative corrections are included is somewhat subtle [117,118]. Following ref. 118, we distinguish two limits. Consider first the limit of $g \to 0$ at
fixed vacuum expectation value v. In this limit, we apply the Equivalence
Theorem for $m_W^2 \ll m_{\phi 0}^2$, s, which implies that $g^2 \ll \lambda$ (where λ is the Higgs
self coupling). One can show that in this limit, the Equivalence Theorem
holds to all orders in λ and at leading order in g^2, provided one chooses to
renormalize the Goldstone boson wave function at a momentum scale small
compared to $m_{\phi 0}$. The $\mathcal{O}(g^2)$ corrections violate the theorem, but this is not
a problem, since perturbation theory in g is known to be reliable. On the
other hand, in the limit where m_W^2, $m_{\phi 0}^2 \ll s$, with the W and Higgs mass of
the same order, we see $g^2 \sim \lambda$, and there is no natural renormalization scheme
in which the Equivalence Theorem is satisfied beyond tree level.[†]

Another way to view the Equivalence Theorem is as follows. In the limit,
$m_{\phi 0}^2 \gg m_W^2$, interactions of enhanced electroweak strength, $\mathcal{O}(G_F m_{\phi 0}^2)$, arise
only from diagrams in which the *internal* particles are also Goldstone bosons
or the Higgs boson. Hence, for many purposes, it is sufficient to consider an
effective theory of scalars in which the interactions of the Goldstone bosons

---

*In theories with $\rho \neq 1$, the low-energy theorem is modified simply by multi-
plying the right hand side of eq. (2.180) by $4 - 3/\rho$ [113,111].

[†]Since no physical observable fixes the Goldstone boson wave function renormalization constant, one can define this constant such that the Equivalence Theorem
is true to all orders in λ and g. In practice, such a definition is not very useful [118].

and the Higgs scalar are summarized by the following Lagrangian

$$\mathcal{L} = -\lambda \left(G^+ G^- + \tfrac{1}{2} G^0 G^0 + \frac{\phi^2}{2} + v_0 \phi + \frac{v_0^2}{2} - \frac{\mu_0^2}{2\lambda} \right)^2 \qquad (2.182)$$

where $G^\pm$ and G^0 are the Goldstone bosons, ϕ is the physical Higgs scalar field, v_0 is the tree-level Higgs vacuum expectation value which gives rise to spontaneous symmetry breaking, and $\lambda = G_F m_{\phi 0}^2 / \sqrt{2}$ is the bare coupling of the $\lambda \phi^4$ theory. (The last two terms of eq. (2.182), which cancel at tree level, yield a tadpole counterterm which is fixed at each order in perturbation theory in such a way that the physical Higgs field has zero vacuum expectation value.) From eq. (2.182), it is immediately apparent that large $m_{\phi 0}$ corresponds to strong interactions between the longitudinal gauge bosons and the Higgs. It is often quite convenient to work in the Landau gauge, where the Goldstone bosons are massless and there is no G–W mixing. The Feynman rules for this effective theory are given in ref. 119. The scattering amplitude for $W_L^+ W_L^- \rightarrow W_L^+ W_L^-$ [see eq. (2.173)], can be easily derived using the effective Lagrangian of eq. (2.182).

It is instructive to compare the computation based on the exact $WW \rightarrow WW$ amplitude computed from the graphs of fig. 2.18 [see eq. 3.89] with one based on using the Equivalence Theorem (in which the W's are replaced by their Goldstone boson equivalents). At large enough $M_{WW} \equiv \sqrt{s}$ the two calculations should, in principle, agree. However, a subtlety arises in the case of large $m_{\phi 0}$ due to the large size of $\Gamma_{\phi 0}$. As shown in ref. 120, if one includes the effect of the Higgs width by making the usual replacement $s - m_{\phi 0}^2 \rightarrow s - m_{\phi 0}^2 + i\, m_{\phi 0} \Gamma_{\phi 0}$ in the Higgs propagator, then the two calculations will give different results. For example, for Higgs masses above 1 TeV, the calculation based on eq. (2.182) gives a substantially larger cross section than given by the exact (tree-level) WW scattering amplitude, when M_{WW} is below $m_{\phi 0}$. This discrepancy arises due to the fact that the inclusion of a resonance width in the Higgs propagator is equivalent to summing only a partial set of diagrams to all orders in λ, whereas for $g^2 \ll \lambda$, the Equivalence Theorem holds order by order in λ (as noted in the discussion above). In any case, for Higgs masses large enough for this discrepancy to matter, it is clear that perturbation theory (in λ) is unreliable, and only the rigorous low-energy theorem (for appropriately restricted values of M_{WW} as indicated earlier) is trustworthy.

REFERENCES

1. J. Ellis, M.K. Gaillard, and D.V. Nanopoulos, *Nucl. Phys.* **B106** (1976) 292.
2. J. Fleischer and F. Jegerlehner, *Phys. Rev.* **D23** (1981) 2001.
3. S. Dawson and S. Willenbrock, *Phys. Lett.* **211B** (1988) 200.
4. E. Braaten and J.P. Leveille, *Phys. Rev.* **D22** (1980) 715.
5. N. Sakai, *Phys. Rev.* **D22** (1980) 2220.
6. T. Inami and T. Kubota, *Nucl. Phys.* **B179** (1981) 171.
7. B.W. Lee, C. Quigg and G.B. Thacker, *Phys. Rev. Lett.* **38** (1977) 883; *Phys. Rev.* **D16** (1977) 1519.
8. W.J. Marciano and S.S.D. Willenbrock, *Phys. Rev.* **D37** (1988) 2509.
9. A.I. Vainshtein, M.B. Voloshin, V.I. Zakharov, and M.S. Shifman, *Sov. J. Nucl. Phys.*, **30** (1979) 711.
10. L.B. Okun, *Leptons and Quarks* (North-Holland, Amsterdam, 1982).
11. M.B. Gavela, G. Girardi, C. Malleville, and P. Sorba, *Nucl. Phys.* **B193** (1981) 257.
12. H. Leutwyler and M. Shifman, *Phys. Lett.* **221B** (1989) 384.
13. L. Bergstrom and G. Hulth, *Nucl. Phys.* **B259** (1985) 137.
14. J.F. Gunion, G.L. Kane, and J. Wudka, *Nucl. Phys.* **B299** (1988) 231.
15. W.-Y. Keung and W.J. Marciano, *Phys. Rev.* **D30** (1984) 248.
16. F. Wilczek, *Phys. Rev. Lett.* **39** (1977) 1304.
17. J. Ellis, M.K. Gaillard, D.V. Nanopoulos and C.T. Sachrajda, *Phys. Lett.* **83B** (1979) 339.
18. T.G. Rizzo, *Phys. Rev.* **D22** (1980) 178.
19. H.M. Georgi, S.L. Glashow, M.E. Machacek, and D.V. Nanopoulos, *Phys. Rev. Lett.* **40** (1978) 692.
20. M.A. Shifman, A.I. Vainshtein, and V.I. Zakharov, *Phys. Lett.* **78B** (1978) 443.
21. A.I. Vainshtein, V.I. Zakharov, and M.A. Shifman, *Sov. Phys. Usp.* **23** (1980) 429.
22. R. Crewther, *Phys. Rev. Lett.* **28** (1972) 1421; M. Chanowitz and J. Ellis, *Phys. Lett.* **40B** (1972) 397; *Phys. Rev.* **D7** (1973) 2490; J. Collins, L. Duncan and S. Joglekar, *Phys. Rev.* **D16** (1977) 438.
23. D.B. Kaplan and A. Manohar, *Nucl. Phys.* **B310** (1988) 527.
24. J. Donoghue, in *Proceedings of the Second International Conference on πN Physics*, edited by W.R. Gibbs and B.M.K. Nefkens, LA-11184-C (Los Alamos National Lab, 1987).
25. T.P. Cheng, *Phys. Rev.* **D38** (1988) 2869.
26. H.Y. Cheng, *Phys. Lett.* **219B** (1989) 347.
27. M. Voloshin, *Sov. J. Nucl. Phys.* **44** (1986) 478.
28. M. Voloshin and V. Zakharov, *Phys. Rev. Lett.* **45** (1980) 688.
29. V.A. Novikov and M.A. Shifman, *Z. Phys.* **C8** (1981) 43.
30. S. Raby and G. West, *Phys. Rev.* **D38** (1988) 3488.
31. T. Appelquist and H.D. Politzer, *Phys. Rev. Lett.* **34** (1975) 43.

32. T.N. Truong and R.S. Willey, *Phys. Rev.* **D40** (1989) 3635.

33. B. Grinstein, L. Hall, and L. Randall, *Phys. Lett.* **211B** (1988) 363.

34. We thank David Kaplan and Ann Nelson for discussions on this point.

35. R.S. Chivukula, A. Cohen, H. Georgi, B. Grinstein, and A.V. Manohar, *Ann. Phys.* **192** (1989) 93.

36. E. de Rafael, in *Quantum Chromodynamics*, Lecture Notes in Physics No. 118, ed. by J.L. Alonso and R. Tarrach (Springer-Verlag, Berlin, 1980), p. 1.

37. H. Georgi, *Weak Interactions and Modern Particle Theory* (Benjamin-Cummings, New York, 1984).

38. A. Cohen and A.V. Manohar, *Phys. Lett.* **143B** (1984) 481.

39. R.S. Chivukula and A. V. Manohar, *Phys. Lett.* **207B** (1988) 86 [E: **217B** (1989) 568].

40. C. Bernard, T. Draper, A. Soni, H.D. Politzer and M.B. Wise, *Phys. Rev.* **D32** (1985) 2343.

41. J. Flynn and L. Randall, *Nucl. Phys.* **B326** (1989) 31.

42. M.B. Voloshin, *Sov. J. Nucl. Phys.* **45** (1987) 122.

43. R. Ruskov, *Phys. Lett.* **187B** (1987) 165.

44. R. Chivukula, A. Cohen, H. Georgi and A.V. Manohar, *Phys. Lett.* **222B** (1989) 258.

45. A.V. Manohar, private communication.

46. R. Willey and H. Yu, *Phys. Rev.* **D26** (1982) 3287.

47. R. Willey, *Phys. Lett.* **173B** (1986) 480.

48. B. Grzadkowski and P. Krawczyk, *Z. Phys.* **C18** (1984) 43.

49. F. Botella and C. Lim, *Phys. Rev. Lett.* **56** (1986) 1651.

50. H.-Y. Cheng and H.-L. Yu, *Phys. Rev.* **D40** (1989) 2980.

51. J.F. Donoghue, B.R. Holstein, and Y-C. R. Lin, *Phys. Rev. Lett.* **55** (1985) 2766; F.J. Gilman and R. Kauffman, *Phys. Rev.* **D36** (1987) 2761; A. Seiden, H.F.-W. Sadrozinski, and H.E. Haber, *Phys. Rev.* **D38** (1988) 824.

52. G.P. Yost *et al.* (Particle Data Group), *Phys. Lett.* **204B** (1988) 1.

53. V. Barger, *et al.*, *Phys. Rev.* **D35** (1987) 3366.

54. H.E. Haber, G.L. Kane, and T. Sterling, *Nucl. Phys.* **B161** (1979) 493.

55. J. Kuhn, *Acta Phys. Pol.* **B12** (1981) 347.

56. E. Eichten, K. Gottfried, T. Kinoshita, K.D. Lane, and T.M. Yan, *Phys. Rev.* **D17** (1979) 3090; *ibid.* **21** (1980) 203.

57. J. Richardson, Phys. Lett. **82B** (1979) 272; W. Buchmuller and S.-H. Tye, *Phys. Rev.* **D24** (1981) 132.

58. J. Kaplan and J. Kuhn, *Phys. Lett.* **78B** (1978) 252; B. Guberina, *et al.*, *Nucl. Phys.* **B174** (1981) 317.

59. J. Bjorken, in *Proceedings of the 1976 SLAC Summer Institute on Particle Physics*, ed. by M.C. Zipf (SLAC Report 198, 1977) p. 1.

60. F.A. Berends and R. Kleiss, *Nucl. Phys.* **B260** (1985) 32.

61. R.N. Cahn, M.S. Chanowitz and N. Fleishon, *Phys. Lett.* **82B** (1979) 113.

62. U. Amaldi, A. Bohm, L.S. Durkin, P. Langacker, A. K. Mann, W.J. Marciano, *Phys. Rev.* **D36** (1987) 1385.
63. G. Costa, J. Ellis, G.L. Fogli, D.V. Nanopoulos and F. Zwirner, *Nucl. Phys.* **B297** (1988) 244.
64. A. Sirlin, *Phys. Rev.* **D22** (1980) 972; **D29** (1984) 89; W. Marciano and A. Sirlin, *Phys. Rev.* **D22** (1980) 2695; **D29** (1984) 75; **D29** (1984) 945 [E: **D31** (1985) 213]; *Nucl. Phys.* **B189** (1981) 442; S. Sarantakos, A. Sirlin and W.J. Marciano *Nucl. Phys.* **B217** (1983) 84; B. Lynn and J.F. Wheater, editors, *Proceedings of the Trieste Workshop on Radiative Corrections in* $SU(2)_L \times U(1)$, 6–8 June 1983, (World Scientific, Singapore, 1984).
65. W.J. Marciano, *Phys. Rev.* **D20** (1979) 274.
66. See footnote 33 in ref. 65.
67. W.J. Marciano and A. Sirlin, *Phys. Rev. Lett.* **46** (1981) 163; S. Sarantakos, A. Sirlin and W.J. Marciano *Nucl. Phys.* **B217** (1983) 84.
68. M. Veltman, *Nucl. Phys.* **B123** (1977) 89; M. Chanowitz, M. Furman and I. Hinchliffe, *Phys. Lett.* **78B** (1978) 285; *Nucl. Phys.* **B153** (1979) 402.
69. M. Veltman, *Acta Phys. Pol.* **B8** (1977) 475.
70. M. B. Einhorn and J. Wudka, *Phys. Rev.* **D39** (1989) 2758.
71. T. Appelquist and R. Shankar, *Nucl. Phys.* **B22** (1979) 317; T. Appelquist and C. Bernard, *Phys. Rev.* **D22** (1980) 200.
72. A. Longhitano, *Phys. Rev.* **D22** (1980) 1166; *Nucl. Phys.* **B188** (1981) 118.
73. D. Toussaint *Phys. Rev.* **D18** (1978) 1626.
74. *The Standard Model Higgs Boson*, edited by M.B. Einhorn, (North-Holland, Amsterdam, 1991).
75. M. Sher, *Phys. Rep.* **179** (1989) 273.
76. N. Cabibbo, L. Maiani, G. Parisi, R. Petronzio, *Nucl. Phys.* **B158** (1979) 295.
77. A.D. Linde, *JETP Lett.* **23** (1976) 64; *Phys. Lett.* **62B** (1976) 435; S. Weinberg, *Phys. Rev. Lett.* **36** (1976) 294.
78. S. Coleman and E. Weinberg, *Phys. Rev.* **D7** (1973) 1888.
79. H.D. Politzer and S. Wolfram, *Phys. Lett.* **82B** (1978) 242 (E: **83B** (1979) 421).
80. E. Gross and E. Duchovni, *Phys. Rev.* **D38** (1988) 2308.
81. K.T. Mahanthappa and M. Sher, *Phys. Rev.* **D22** (1980) 1711.
82. P. Frampton, *Phys. Rev. Lett.* **37** (1976) 1378.
83. A.D. Linde, *Phys. Lett.* **70B** (1977) 306; **92B** (1980) 119.
84. S. Coleman, *Phys. Rev.* **D15** (1977) 2929; C. Callan and S. Coleman, *Phys. Rev.* **D16** (1977) 1762.
85. P.J. Steinhardt, *Phys. Lett.* **97B** (1980) 147.
86. D.A Kirzhnits and A.D. Linde, *Ann. of Phys.* **101** (1976) 195; A.D. Linde, *Rep. Prog. Phys.* **42** (1979) 389.
87. A.H. Guth and E.J. Weinberg, *Phys. Rev. Lett.* **45** (1980) 1131.

88. E. Witten, *Nucl. Phys.* **B177** (1981) 477.

89. R.A. Flores, and M. Sher, *Ann. Phys. (NY)* **148** (1983) 95.

90. K.G. Wilson, *Phys. Rev.* **B4** (1971) 3184; K.G. Wilson and J. Kogut, *Phys. Rep.* **12** (1974) 75.

91. H. Neuberger, *Nucl. Phys. B (Proc. Suppl.)* **9** (1989) 495.

92. D. Callaway, *Phys. Rep.* **167** (1988) 241.

93. M.A.B. Beg, C. Panagiotakopoulos, and A. Sirlin, *Phys. Rev. Lett.* **52** (1984) 883.

94. R. Dashen and H. Neuberger, *Phys. Rev. Lett.* **50** (1983) 1897.

95. M.J. Duncan, R. Philippe, and M. Sher, *Phys. Lett.* **153B** (1985) 165.

96. M. Lindner, M. Sher and H.W. Zaglauer, *Phys. Lett.* **228B** (1989) 139.

97. K.S. Babu and E. Ma, *Phys. Rev.* **D31** (1985) 2861; A. Bovier and D. Wyler, *Phys. Lett.* **154B** (1985) 43.

98. See, for example, M. Luscher and P. Weisz, *Phys. Lett.* **212B** (1988) 472.

99. J. Fröhlich, in *Proceedings of the XXIV International Conference on High Energy Physics*, Munich, 1988, edited by R. Koffhaus and J.H. Kühn (Springer-Verlag, Berlin, 1989) p. 219.

100. C. Wetterich, in *Superstrings, Unified Theories, and Cosmology, 1987*, the ICTP Series in Theoretical Physics, vol. 4, edited by G. Furlan *et al.* (World Scientific, Singapore, 1988) p. 403.

101. M. B. Einhorn and D. N. Williams, *Phys. Lett.* **211B** (1988) 457.

102. H. Neuberger, in *Proceedings of the Lattice Higgs Workshop*, Tallahassee, FL, May 16–18, 1988, edited by B. Berg *et al.* (World Scientific, Singapore, 1988) p. 197.

103. I. Montvay, in *INFN Eloisatron Workshop Working Group Report*, Erice, Italy, CCSEM Report EL-88/1 (1988) p. 34.

104. J. Kuti, L. Lin, and Y. Shen, *Phys. Rev. Lett.* **61** (1988) 678; *Nucl. Phys. B (Proc. Suppl.)* **4** (1988) 397; in *Proceedings of the Lattice Higgs Workshop*, edited by B. Berg *et al.* (World Scientific, Singapore, 1988) p. 140, 186; Y. Shen, J. Kuti, L. Lin, and S. Meyer, *ibid.*, p. 216.

105. K. Bitar and G. Bhanot, in *Proceedings of the Lattice Higgs Workshop*, edited by B. Berg *et al.* (World Scientific, Singapore, 1988) p. 21.

106. J. Jersák, in *Higgs Particle(s): Physics Issues and Experimental Searches in High Energy Collisions*, edited by A. Ali, (Plenum Press, New York, 1989) p. 39.

107. H.G. Evertz *et al.*, *Nucl. Phys.* **B285** (1987) 590; A. Hasenfratz *et al.*, *Phys. Lett.* **B199** (1987) 531; A. Hasenfratz *et al.*, *Nucl. Phys.* **B317** (1989) 81.

108. M. Luscher and P. Weisz, *Nucl. Phys.* **B290** (1987) 25; **B295** (1988) 65; **318** (1989) 705.

109. B.W. Lee, C. Quigg and G.B. Thacker, *Phys. Rev. Lett.* **38** (1977) 883; *Phys. Rev.* **D16** (1977) 1519.

110. M. Chanowitz and M.K. Gaillard, *Phys. Lett.* **142B** (1984) 85; *Nucl. Phys.* **B261** (1985) 379.

111. M. Chanowitz, M. Golden and H. Georgi, *Phys. Rev. Lett.* **57** (1986) 2344; *Phys. Rev.* **D36** (1987) 1490.

112. D.A. Dicus and V.S. Mathur, *Phys. Rev.* **D7** (1973) 3111.

113. H.A. Weldon, *Phys. Rev.* **D30** (1984) 1547; *Phys. Lett.* **146B** (1984) 59.

114. J.S. Bell, *Nucl. Phys.* **B60** (1973) 427; C.H. Llewellyn Smith, *Phys. Lett.* **46B** (1973) 233; J.M. Cornwall, D.N. Levin, and G. Tiktopoulos, *Phys. Rev. Lett.* **30** (1973) 1268, and *Phys. Rev.* **D10** (1974) 1145.

115. S. Dawson and S. Willenbrock, *Phys. Rev. Lett.* **62** (1989) 1232; *Phys. Rev.* **D40** (1989) 2880.

116. J. Cornwall *et al.*, in ref. 114; C. Vayonakis, *Lett. Nuovo Cim.* **17** (1976) 383; M. Chanowitz and M. Gaillard, in ref. 110; and G.J. Gounaris, R. Kogerler, and H. Neufeld, *Phys. Rev.* **D34** (1986) 3257.

117. Y.P. Yao and C.P. Yuan, *Phys. Rev.* **D38** (1988) 2237.

118. J. Bagger and C. Schmidt, *Phys. Rev.* **D41** (1990) 264.

119. W.J. Marciano and S.S.D. Willenbrock, *Phys. Rev.* **D37** (1988) 2509.

120. C.P. Yuan, University of Michigan thesis (1988); C.P. Yuan and G.L. Kane, *Phys. Rev.* **D40** (1989) 2231.

Chapter 3

How to Search for the Minimal Higgs Boson

In this chapter we shall focus on techniques for searching for the minimal Higgs boson, ϕ^0, of the Standard Model. Since, as we have discussed, its mass is unconstrained by the perturbative theory, we shall consider which experiments and accelerators are capable of searching for the ϕ^0 as a function of its mass m_{ϕ^0}. Much of what is said about the ϕ^0 in the region $m_{\phi^0} \lesssim 100$ GeV will also apply to the lightest neutral scalar of a supersymmetric theory, or even to a composite Higgs boson. Some experimental information exists at very low mass, but for m_{ϕ^0} above about 5 GeV, one must turn to SLC, LEP and future accelerators.

As an aid to the reader, we present four figures which summarize the content of the sections to follow. In fig. 3.1 we summarize the existing limits on m_{ϕ^0} arising from various nuclear experiments, rare kaon decays, and rare B-meson decays. The possible range over which Upsilon decays can place a limit on m_{ϕ^0}, is indicated, assuming that the present data sample is roughly doubled. Also shown are the approximate maximum m_{ϕ^0} values that can be probed at a Z factory with 100 pb^{-1} of integrated luminosity in the two main discovery modes. The minimum value of m_{ϕ^0} that can be probed depends upon experimental details, but potentially the Z factories could reach down into the $m_{\phi^0} \sim 10$ MeV range. In fig. 3.2 we survey future e^+e^- colliders of various energies and luminosities with regard to their ability to detect the ϕ^0, ranging from a possible intermediate energy linear collider designed specifically to probe up to $m_{\phi^0} \sim 2m_W$, to a super high energy machine that could reach ϕ^0 masses as high as 1 TeV and perhaps above. Also indicated are the typical production modes in which the search would be conducted, and the required luminosity and corresponding numbers of events. In fig. 3.3 we survey the range of m_{ϕ^0} over which various signatures are amenable to detection at the SSC. At a hadron collider, the best means for discovering the ϕ^0 is strongly dependent upon its mass. In the Intermediate Mass domain, with m_{ϕ^0} below $2m_W$, one must employ rare ϕ^0 decay modes or $\phi^0 + W^{\pm}$ associated production in order to overcome backgrounds; this is a difficult

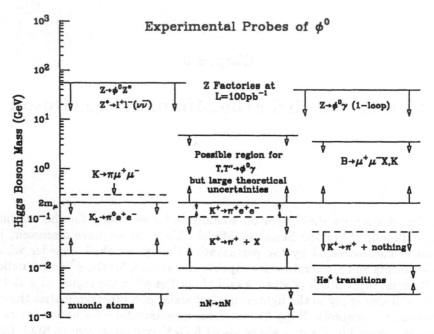

Figure 3.1 Summary of existing limits on m_{ϕ^0} and expected sensitivity in various modes at SLC and LEP-I.

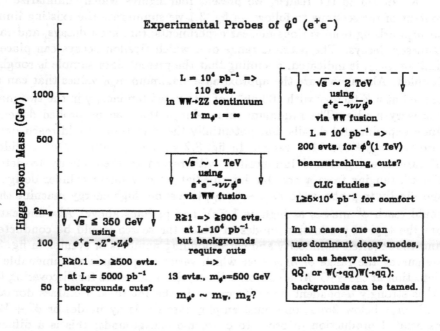

Figure 3.2 Experimental reach in m_{ϕ^0} and probable detection modes at future e^+e^- colliders of various energies and luminosities.

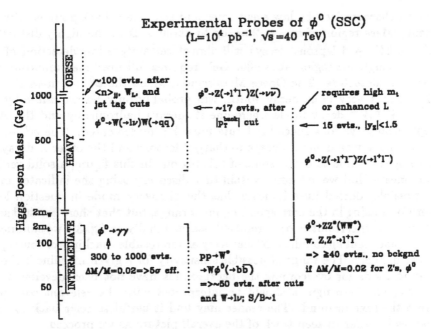

Figure 3.3 Experimental reach in m_{ϕ^0} and probable detection modes at the SSC.

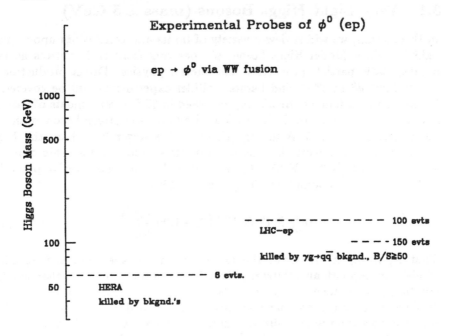

Figure 3.4 Experimental reach in m_{ϕ^0} for ep colliders.

region characterized by low event rates and significant backgrounds. For the Heavy Mass region of $m_{\phi^0} > 2m_Z$ up to about 1 TeV, the highly distinctive $\phi^0 \rightarrow ZZ \rightarrow 4$ leptons decays will almost certainly allow detection of the ϕ^0, although the region above 800 GeV requires full machine luminosity and excellent detectors. The Obese Mass region, where the ϕ^0 has mass above 1 TeV and a very large decay width, will probably only be accessible by the $\phi^0 \rightarrow W^+W^-$ decay mode where one W decays hadronically and the other leptonically; various specialized cuts must be implemented. The $\phi^0 \rightarrow ZZ$ decay mode where one Z decays to charged leptons and the other Z decays to $\nu\bar{\nu}$ might be able to probe some of this region. In this figure, a solid vertical line means that we are quite certain that discovery using the indicated mode is possible; dotted lines indicate that the discovery mode in question looks very promising in the corresponding mass range, but that there are still some remaining uncertainties or unresolved issues. Finally, in fig. 3.4 we summarize the situation at ep colliders; ϕ^0 discovery at foreseeable machines of this type is quite unlikely. The range of possible sensitivity for a given machine is shown, but methods for overcoming the backgrounds have not been developed. Of course, all these figures should be interpreted with the relevant comments from the text in mind. The reader may find it useful to refer back to these figures in order to keep track of the overall picture as we proceed.

3.1 Very Light Higgs Bosons (mass $\lesssim$ 5 GeV)

In this section, we will review a variety of limits and constraints upon a very light Standard Model Higgs boson, ϕ^0, deriving from such sources as rare decays, static particle properties, and nuclear physics. Direct production of a very light ϕ^0 in e^+e^- and hadron collider experiments will be covered in the section that follows. As already reviewed in §2.5, requirements of vacuum stability can be used to derive various limits on the Higgs boson mass. In particular, Linde, and Weinberg [1], derived a lower bound on the Higgs boson mass by requiring that the asymmetric vacuum be the state of lowest energy, $i.e.$, $V(v/\sqrt{2}) - V(0) < 0$, where V is the one-loop effective potential. For $m_t \lesssim m_W$ this bound implies [see eq. (2.156)]

$$m_{\phi^0}^2 \gtrsim (7 \text{ GeV})^2 \left(1 - 1.09\frac{m_t^4}{m_W^4}\right). \tag{3.1}$$

That is, if $m_t \simeq 78$ GeV, the lower bound vanishes and the Standard Model can support an arbitrarily light Higgs boson. On the other hand, if $(m_t/m_W)^4 \ll 1$, then a very light SM Higgs boson is not possible. For $m_t > 78$ GeV, this bound no longer applies. Instead, a different lower bound on m_{ϕ^0} emerges from renormalization group analyses. As described in §2.5 (see $e.g.$, fig. 2.16), these analyses imply that if m_t is significantly larger than m_W then, again, it is not possible to have a light ϕ^0. Thus, in our examination of

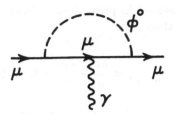

Figure 3.5 The contribution of the standard model neutral Higgs boson, ϕ^0, to the anomalous magnetic moment of the muon.

those experimental constraints on a very light SM Higgs boson that are sensitive to the top quark mass, we shall assume that $m_t \approx m_W$.* Note that this assumption is not inconsistent with current t-quark mass limits, discussed in §1.5.

Limits from the Muon Anomalous Magnetic Moment

The incredible agreement between theory and experiment in the measurement of the muon anomalous magnetic moment puts stringent limits on new physics. The experimental value of $a_\mu \equiv (\frac{g-2}{2})_\mu$ is [2],

$$a_\mu^{exp} = (1 \quad 165 \quad 924.0 \pm 8.5) \times 10^{-9} \qquad (3.2)$$

while the Standard Model prediction is [3,4],

$$a_\mu^{SM} = (1 \quad 165 \quad 920.2 \pm 2.0) \times 10^{-9}. \qquad (3.3)$$

Any new contribution to $(g-2)_\mu$ is then constrained at the 95% confidence level by

$$-13 \times 10^{-9} < \Delta a_\mu < 21 \times 10^{-9}. \qquad (3.4)$$

The neutral Higgs boson of the standard Weinberg-Salam model, ϕ^0, contributes to a_μ by the diagram of fig. 3.5. The result was first calculated by Jackiw and Weinberg [3],

$$\Delta a_\mu \big|_{\text{fig. 3.5}} = \frac{G_F m_\mu^4}{4\pi^2 \sqrt{2}} \int_0^1 \frac{y^2(2-y)}{m_\mu^2 y^2 + m_{\phi^0}^2(1-y)} dy. \qquad (3.5)$$

If $m_{\phi^0} \gg m_\mu$, then the Higgs contribution to the muon anomalous magnetic moment is smaller than that from diagrams containing W^+, W^- and Z gauge bosons and can be neglected, (the contribution from weak bosons is 2 ×

---

*In models with an extended Higgs sector, the Linde-Weinberg bound does not normally apply to the lightest physical Higgs boson (see §4.1). Thus, in such models, a very light Higgs boson is *not* inconsistent with other values of m_t.

10^{-9} and cannot be observed in present experiments). If the Higgs boson is very light, $m_{\phi^0} \ll m_\mu$, then

$$\Delta a_\mu|_{\text{fig. 3.5}} = \frac{3m_\mu^2 G_F}{8\pi^2\sqrt{2}} = 3.4 \times 10^{-9}, \tag{3.6}$$

of the same order of magnitude as the W^+, W^- and Z contributions.* We see that present experiments do not constrain a light Higgs boson mass. A proposed Brookhaven experiment which will improve the current limit on a_μ by a factor of 20 [5] may, however, be capable of putting a limit on a light ($m_{\phi^0} \ll m_\mu$) Higgs boson. However, by then other limits on such a light ϕ^0 will exist.

The precision with which $(g-2)_\mu$ is measured can be used to restrict the parameters of physics beyond the Standard Model. The contribution of a typical non-standard neutral Higgs boson to Δa_μ has been computed in a general theory by Leveille [6] for a Lagrangian,

$$\mathcal{L}_{\text{int}} = \overline{\mu}(C_S + C_P\gamma_5)Fh. \tag{3.7}$$

where F is any charge -1 fermion. The contribution to Δa_μ comes from the diagram of fig. 3.6 and is,

$$\Delta a_\mu|_{\text{fig. 3.6}} = \frac{m_\mu^2}{8\pi^2} \int_0^1 x^2\, dx \frac{\{C_S^2(1-x+m_F/m_\mu) + C_P^2(1-x-m_F/m_\mu)\}}{m_\mu^2 x^2 + (m_F^2 - m_\mu^2)x + m_h^2(1-x)}. \tag{3.8}$$

Using this formula, it is straightforward to obtain the contribution of Higgs bosons to $(g-2)_\mu$ in any new theory. Note that for $m_F \gtrsim m_\mu$ a cancellation between scalar and pseudoscalar couplings is possible.

The simplest extension of the standard model is the addition of extra Higgs SU(2) doublets [7]. The charged Higgs bosons would contribute to $(g-2)_\mu$ by the diagram of fig. 3.7. This contribution is small because it is suppressed by a factor of $(m_\mu/m_{H^\pm})^2$ relative to eq. (3.6). Grifols and Pascual [8] find

$$\Delta a_\mu|_{\text{fig. 3.7}} = \frac{G_F m_\mu^2}{48\sqrt{2}\pi^2}\kappa^2\left(\frac{m_\mu}{m_{H^\pm}}\right)^2, \tag{3.9}$$

where $\kappa < 1$ is a factor which accounts for possible mixings between the Higgs doublets.

---

*From eq. (3.6) we see that the contribution of a Higgs boson to $(g-2)_e$ will be suppressed by an additional factor of $(m_e/m_\mu)^2$ and thus will be of order 10^{-13}, which is unmeasurable.

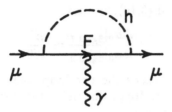

Figure 3.6 The contribution to the anomalous magnetic moment of the muon from a non-standard Higgs boson, h, which couples with the Lagrangian, $\mathcal{L}_{\text{int}} = \overline{\mu}(C_S + C_P\gamma_5)Fh$.

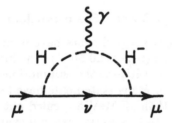

Figure 3.7 The contributions of charged scalars to the anomalous magnetic moment of the muon.

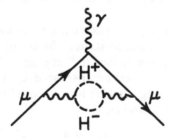

Figure 3.8 Two-loop contribution to the muon anomalous magnetic moment from a charged Higgs boson pair.

The two-loop contribution to $(g-2)_\mu$ from charged Higgs bosons (see fig. 3.8) is not suppressed and is [8]

$$\Delta a_\mu|_{\text{fig. 3.8}} = \frac{1}{180}\left(\frac{\alpha}{\pi}\right)^2\left(\frac{m_\mu}{m_{H^\pm}}\right)^2 + \mathcal{O}\left[\left(\frac{m_\mu}{m_{H^\pm}}\right)^4\ln\left(\frac{m_{H^\pm}}{m_\mu}\right)\right]. \quad (3.10)$$

Comparing eqs. (3.9) and (3.10), we find

$$\frac{\Delta a_\mu|_{\text{fig. 3.7}}}{\Delta a_\mu|_{\text{fig. 3.8}}} \simeq \frac{\kappa^2}{170} \tag{3.11}$$

and so for light Higgs bosons the two-loop contribution of fig. 3.8 can be significant. For N SU(2) Higgs doublets, the contribution of the charged Higgs bosons is approximately

$$\Delta a_\mu|_{\text{fig. 3.8}} \sim (N-1)\ 3 \times 10^{-8} \left(\frac{m_\mu}{m_{H\pm}}\right)^2. \tag{3.12}$$

For $m_{H\pm} = 300$ MeV, $\Delta a_\mu|_{\text{fig. 3.8}} \sim 3 \times 10^{-9}(N-1)$.

Limits on Light Higgs Bosons From Nuclear Physics

A variety of nuclear physics experiments have been used to place limits on a very light SM Higgs boson ($m_{\phi^0} \lesssim 20$ MeV). The limits derived from these experiments are extremely sensitive to the assumptions made about the nucleon–Higgs coupling (see §2.1). Many statements have been made in the literature that Higgs bosons less than 15 MeV are ruled out by nuclear physics limits on long range forces [9]. In this section we will examine these limits in detail and emphasize the assumptions involved in each limit.

Limits for $m_{\phi^0} < 2m_e$ are difficult to obtain since in this region the Higgs decays exclusively to $\gamma\gamma$. There are good limits from electron–neutron and electron–deuteron scattering and from measurements of the transition energies in muonic atoms.

The best limits on Higgs bosons with a mass less than 1 MeV come from X-ray transition rates in muonic atoms. Beltrami $et\ al.$ [10] have measured the $3d_{5/2} - 2p_{3/2}$ X-ray transition in ^{24}Mg and ^{28}Si. They looked for any muon–nucleon isoscalar interaction mediated by a neutral scalar or vector boson. The presence of a light scalar boson would induce a Yukawa potential for the muon–nucleon interaction,

$$V(r) = \frac{-g_{\phi^0 NN} g_{\phi^0 \mu\mu}}{4\pi} A \frac{e^{-rm_{\phi^0}}}{r}, \tag{3.13}$$

where A is the atomic mass number and $g_{\phi^0 NN}$ and $g_{\phi^0 \mu\mu}$ are the Higgs couplings to nucleons and muons, respectively. The corresponding change in the transition wavelength due to this induced Yukawa potential is

$$\frac{\delta\lambda}{\lambda} = -\frac{g_{\phi^0 NN} g_{\phi^0 \mu\mu}}{10\pi\alpha Z} [9f(2) - 4f(3)], \tag{3.14}$$

where Z is the nuclear charge, $f(n) = (1 + nm_{\phi^0}/2\alpha Z m_\mu)^{-2n}$, and n is the principal quantum number of the muonic state. For small m_{ϕ^0}, $f(n) \to 1$ and the shift in the wavelength is independent of the scalar mass. In this limit, Beltrami $et\ al.$ [10], find the result

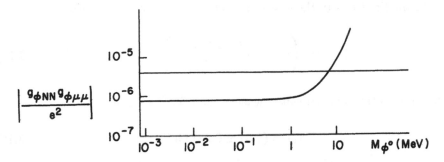

Figure 3.9 Limits on light SM Higgs bosons from the measurement of the $3d_{5/2} - 2p_{3/2}$ transition rates in muonic atoms of ref. 10. The region above the curved line is excluded and the straight line is the prediction of the SM with $\eta = 0.3$.

$$|g_{\phi^0 NN} g_{\phi^0 \mu\mu}| < 7 \times 10^{-8} \quad \text{for } m_{\phi^0} < 1 \text{ MeV}. \tag{3.15}$$

With the SM coupling for $g_{\phi^0 \mu\mu}$ and

$$g_{\phi^0 NN} = \frac{g M_N}{2 m_W} \eta \tag{3.16}$$

we find from eq. (3.15) that this experiment requires $\eta < 0.04$ for $m_{\phi^0} <$ 1 MeV. If one believes that $\eta > 0.3$, (see §2.1), then these experiments exclude Higgs bosons which are lighter than about 8 MeV. The results of ref. 10 are shown in fig. 3.9, where the region above the curved line is excluded. The prediction using SM couplings and $\eta = 0.3$ is also shown. (For $\eta = 0.1$, this experiment excludes Higgs masses less than about 3 MeV.)

An alternate limit on very light Higgs bosons can be found from electron–neutron and electron–deuteron scattering. This experiment is more sensitive to the value of η than the limits from muonic atoms, but it is instructive to analyze it. Adler *et al.* [11] have noted that the presence of a scalar field will modify the electron–neutron Coulomb interaction,

$$V_{eN} \rightarrow -i \left(\frac{e^2 G_e^N(t)}{t} - \frac{g_{\phi^0 ee} g_{\phi^0 NN}}{t - m_{\phi^0}^2} \right), \tag{3.17}$$

where $G_e^N(t) = -a_N t$ is the neutron charge form factor, t is the usual Mandelstam variable, and $g_{\phi^0 ee}$ and $g_{\phi^0 NN}$ are the Higgs couplings to the electron and the neutron, respectively.

In the limit $t \to 0$, the electron–neutron interaction becomes

$$V_{eN} = i \left(e^2 a_N - \frac{g_{\phi^0 ee} g_{\phi^0 NN}}{m_{\phi^0}^2} \right) \equiv i e^2 a_{\text{eff}}. \qquad (3.18)$$

Thermal neutron experiments measure the effective slope

$$a_{\text{eff}} = (0.51 \pm 0.02) \text{ GeV}^{-2}. \qquad (3.19)$$

Electron–deuteron scattering experiments, however, are performed at large values of t $(0.01 < t < 0.1 \text{ GeV}^2)$ and so the scalar exchange term of eq. (3.17) can be ignored. These experiments then measure the true value of a_N and give the limit [12]

$$\frac{a_N - a_{\text{eff}}}{a_{\text{eff}}} \leq 0.5 , \qquad (3.20)$$

which translates into the limit on m_{ϕ^0},

$$m_{\phi^0}^2 > \frac{\eta m_e m_N}{2 m_W^2 \sin^2 \theta_W} \frac{1}{a_{\text{eff}}}. \qquad (3.21)$$

This gives,

$$m_{\phi^0} \geq (0.5 \text{ MeV}) \sqrt{\eta}. \qquad (3.22)$$

Consequently, the region of Higgs masses that is excluded by this experiment depends strongly on the exact magnitude of $g_{\phi^0 NN}$. However, it is clear that under any assumptions, a massless ϕ^0 is not allowed.

In the mass region $2m_e < m_{\phi^0} < 20$ MeV both low energy neutron–nucleon scattering and certain $0^+ \to 0^+$ nuclear transitions would see the effect of a light Higgs boson. Barbieri and Ericson [13] have looked at the angular effects of a scalar exchange interaction on neutron–nucleon scattering. They consider a potential of the form

$$V_{nN}(r) = \frac{-A g_{\phi^0 NN}^2}{4\pi} \frac{e^{-rm_{\phi^0}}}{r} , \qquad (3.23)$$

where A is the atomic number of the nucleus and $g_{\phi^0 NN}$ is the Higgs-nucleon coupling constant. For $m_{\phi^0} \to 0$, large effects could be obtained since V_{nN} looks like a Coulomb interaction, which is not normally expected between the neutron and the nucleus. Experiments with neutrons and energies E in the range 1 KeV $< E <$ 26 KeV produce the limit

$$m_{\phi^0}(\text{MeV}) > \left(\frac{g_{\phi^0 NN}^2}{4\pi} \frac{10^{11}}{3.4} \right)^{1/4} \tag{3.24}$$

$$> 13\sqrt{\eta} \quad \text{MeV}.$$

For $\eta > 0.3$, this gives $m_{\phi^0} > 7$ MeV. Unfortunately, this experiment was only sensitive to $m_{\phi^0} \geq 1$ MeV.*

Finally, a search has been made for the $0^+ \rightarrow 0^+$ transitions [14]

$$
\begin{aligned}
{}^{16}\text{O}(6.05 \text{ MeV}) &\rightarrow (\text{ground state}) + \phi^0 \\
{}^{4}\text{He}(20.1 \text{ MeV}) &\rightarrow (\text{ground state}) + \phi^0.
\end{aligned}
\tag{3.25}
$$

These experiments then looked for the decay of the Higgs boson to $e^+ e^-$. As in all the nuclear physics experiments, the limit obtained here is extremely sensitive to the value of $g_{\phi^0 NN}$. The experimental analysis used a mass dependent value for $g_{\phi^0 NN}$,†

$$\frac{g_{\phi^0 NN}^2}{4\pi} = 1.8 \times 10^{-8} e^{-.26 m_{\phi^0}/\text{MeV}}. \tag{3.26}$$

In terms of $g_{\phi^0 NN}$ as defined in eq. (3.16), this corresponds to $\eta > 0.3$ for $m_{\phi^0} < 7$ MeV. These experiments claim to exclude light scalars with SM couplings and masses between $2m_e$ and 18 MeV. However, the Helium experiment of Kohler et al. has been reanalyzed by two other groups [16,17], who conclude that it was incorrectly interpreted and, in fact, provides no bounds on the Higgs boson mass. (Using the ^{16}O experiment alone implies $2m_e \lesssim m_{\phi^0} \lesssim 5.8$ MeV; this bound remains sound.) Freedman et al. [17] have performed a new experiment involving the excited ^{4}He decay. They conclude that Higgs masses $2.8 \lesssim m_{\phi^0} \lesssim 11.5$ MeV are excluded for $\eta \simeq 0.3$.

Numerous other mechanisms for searching for a Higgs boson with a mass less than 20 MeV have been proposed. As far as we are aware, none of them are capable of improving the limits listed above.

Combining all of the limits from nuclear physics we find that the exact limits depend on the assumed value of the Higgs-nucleon coupling and hence on η. However, for $\eta > 0.3$, Higgs bosons with all masses less than 11.5 MeV are excluded by the measurements of the X-ray transitions in muonic atoms and by the $0^+ \rightarrow 0^+$ transition experiments in ^{4}He.

---

*Barbieri and Ericson claim that for $m_{\phi^0} < 1$ MeV, the measured angular dependence differs violently from the prediction derived from the potential of eq. (3.23), but they cite no experiment sensitive to this region.

†At the time this experiment was performed, there was a discrepancy between the level splittings in muonic Helium and the predictions of QED [15]. This discrepancy could be explained by the existence of a light scalar boson with a mass dependent coupling. This discrepancy subsequently vanished.

A Direct Production Limit

In the previous subsections, the limits obtained on a light Higgs boson have relied to a greater or lesser degree on our theoretical understanding of the physics of some secondary system. Here we summarize the result from a direct production limit experiment [18]. The technique employed is to look for electron bremsstrahlung of a Higgs boson in a nuclear field. The rate of production is completely determined by the (known) $\phi^0 e^+ e^-$ coupling. A beam dump experimental set up is used; on the far side of the beam dump is a decay region followed by detectors sensitive to the $e^+ e^-$ decay mode of the ϕ^0. In other words, one is searching for the process $eN \rightarrow e\phi^0 N$, with $\phi^0 \rightarrow e^+ e^-$. The range of m_{ϕ^0} values to which the experiment is sensitive is determined by the cross section (which falls off as $1/m_{\phi^0}^2$), and by the size and location of the beam dump (because of the rapid variation of the Higgs lifetime as a function of m_{ϕ^0}). The particular experimental arrangement employed had maximum sensitivity to the ϕ^0 at $m_{\phi^0} \simeq 30$ MeV, decreasing for m_{ϕ^0} values on either side because of the changing Higgs lifetime. For small values of m_{ϕ^0} the lifetime of the ϕ^0 becomes too long for the $\phi^0 \rightarrow e^+ e^-$ decay to have occurred before the ϕ^0 has passed through the decay region. For large m_{ϕ^0}, the lifetime of the ϕ^0 is too short, and most decays occur inside the beam dump. The result is that the range $1.2 \leq m_{\phi^0} \leq 52$ MeV is excluded in the case of the SM Higgs boson.[†]

Higgs Bosons in Pion Decays

The rate for Higgs boson production in charged pion decay, where the Higgs boson is emitted from the virtual W can be easily obtained using the chiral Lagrangian techniques of §2.2. The result is [19,20]

$$\frac{\Gamma(\pi^+ \rightarrow e^+ \nu_e \phi^0)}{\Gamma(\pi^+ \rightarrow \mu^+ \nu_\mu)} = \frac{\sqrt{2} G_F m_\pi^4 f(x)}{96\pi^2 m_\mu^2 \left(1 - m_\mu^2/m_\pi^2\right)^2} \left(1 - \frac{2N_H}{3b}\right)^2$$

$$= 1.9 \times 10^{-9} f(x) \,, \tag{3.27}$$

where $f(x) \equiv (1 - 8x + x^2)(1 - x^2) - 12x^2 \ln(x)$ with $x \equiv m_{\phi^0}^2/m_\pi^2$, and we have taken $N_H = 3$ and $b = 11 - (2/3)n_L$ with $n_L = 3$. The N_H-dependent piece of eq. (3.27) arises from the inclusion of graphs in which the Higgs couples to the gluons in the pion via N_H heavy-quark intermediate loops.

The SINDRUM Collaboration [21] has recently searched for decays of the type $\pi^+ \rightarrow e^+ \nu_e e^+ e^-$. A Higgs boson in the mass range $2m_e < m_{\phi^0} \lesssim m_\pi$ would yield events of this type, since it would decay primarily to an $e^+ e^-$

[†]Taking appropriate account of alterations in Higgs lifetime and $e^+ e^-$ coupling, limits are also placed on light non-Standard-Model scalar and pseudoscalar Higgs bosons.

pair. In ref. 21, a 90% confidence level upper limit on $BR(\pi^+ \rightarrow e^+ \nu_e \phi^0)$ is obtained as a function of m_{ϕ^0}. This limit is roughly a factor of 3 below the theoretical prediction quoted above[‡] over the range of Higgs masses from 10 MeV to 100 MeV, which clearly excludes a SM Higgs boson in this mass range.

A closely related decay is $K^- \rightarrow \phi^0 \mu^- \bar{\nu}$. This can be computed in a similar manner [19,20], and also has branching ratio $\mathcal{O}(10^{-9})$. To date, no experiment has measured this decay with sufficient sensitivity that limits on a SM Higgs can be obtained.

Higgs Bosons in η' Decays

The branching ratio for this decay was derived in §2.2, as an example of the application of low energy theorems to Higgs couplings. The result appears in eq. 2.98. The experiment of ref. 22 has searched for the decay $\eta' \rightarrow \eta \phi^0 \rightarrow \eta \mu^+ \mu^-$. This experiment is sensitive to 250 MeV $\lesssim m_{\phi^0} \lesssim$ 400 MeV. The theoretical estimate for this branching ratio depends crucially on $BR(\phi^0 \rightarrow \mu^+ \mu^-)$. Using $N_H = 3$ and $b = 9$ in eq. 2.98, we obtain the following theoretical prediction

$$BR(\eta' \rightarrow \eta\phi^0 \rightarrow \eta\mu^+\mu^-) \simeq 7.2 \times 10^{-5} B_{\phi^0} BR(\phi^0 \rightarrow \mu^+\mu^-), \qquad (3.28)$$

where $B_{\phi^0} \equiv 2p_{\phi^0}/m_{\eta'} < 0.53$ in the kinematical range of interest. This should be compared to the quoted (90% confidence level) experimental upper limit of 1.5×10^{-5}. Clearly, there is a range of ϕ^0 masses, 250 MeV$\lesssim m_{\phi^0} \lesssim 2m_\pi$, which is ruled out by the data. The actual Higgs mass upper limit which can be ruled out depends sensitively on $BR(\phi^0 \rightarrow \mu^+\mu^-)$ [23]. For example, for $2m_\pi \lesssim m_{\phi^0} \lesssim m_{\eta'} - m_\eta$, ref. 23 claims that $BR(\phi^0 \rightarrow \mu^+\mu^-) \gtrsim 0.07$, whereas ref. 24 obtains $BR(\phi^0 \rightarrow \mu^+\mu^-) \gtrsim 0.2$. An improvement by about a factor of 10 of the experimental upper limit is needed to definitively rule out Higgs bosons over most of the above mass range.

Higgs Bosons in Kaon Decays

The decays $K^\pm \rightarrow \pi^\pm \ell^+ \ell^-$ and $K^0_L \rightarrow \pi^0 \ell^+ \ell^-$ can be used to place limits on a Higgs boson with a mass less than $m_K - m_\pi$, which could contribute through the decay sequence $K \rightarrow \pi\phi^0$, $\phi^0 \rightarrow \ell^+\ell^-$. Similarly, $K \rightarrow \pi\gamma\gamma$ limits could perhaps place limits on the Higgs boson in the m_{ϕ^0} range where the ϕ^0 decays significantly to two photons [25]. Experimental limits are significant [25], and will be quoted below. In addition, dedicated searches have been made for the decay $K^+ \rightarrow \pi^+\phi^0, \phi^0 \rightarrow e^+e^-$ [26-28] , and $K^+ \rightarrow \pi^+\phi^0, \phi^0 \rightarrow \mu^+\mu^-$ [29] Hence, if the rate for $K \rightarrow \pi\phi^0$ can be reliably calculated, we may be able

---

[‡]In ref. 21 they do not include the $(1 - 2N_H/3b)^2$ factor of eq. (3.27) in comparing theory to experiment; this only affects the region above 100 MeV where the theoretical prediction is of the same order as the experimental upper limit.

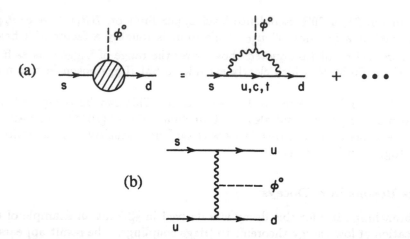

Figure 3.10 a) Two quark operators which contribute to the transition $K \to \pi\phi^0$. b) Four quark operators which contribute to the transition $K \to \pi\phi^0$.

to rule out the existence of a very light Higgs boson. However, there is some controversy in the literature over the theoretical reliability of the calculation of $BR(K \to \pi\phi^0)$. We will discuss some of the issues involved, and examine to what extent the data can constrain the existence of a light Higgs boson.

There are two classes of diagrams which contribute to a quark model calculation of $K^{\pm} \to \pi^{\pm}\phi^0$. The first class consists of spectator diagrams corresponding to the two-quark operators which contribute to the transition $s \to d\phi^0$ (fig. 3.10a). The second class consists of non-spectator diagrams corresponding to four-quark operators (fig. 3.10b). The calculation of the amplitude for $s \to d\phi^0$ is straightforward and has been checked by several groups [30-35],

$$\mathcal{L}_{sd\phi^0} = \frac{G_F^{3/2}}{2^{1/4}} \frac{3}{16\pi^2} \sum_i m_i^2 V_{id}^* V_{is} \left[m_s \bar{d}(1 + \gamma_5)s + m_d \bar{d}(1 - \gamma_5)s \right] \phi^0 + \text{h.c.},$$

(3.29)

where V_{ij} are the elements of the Kobayashi-Maskawa matrix and m_i are the corresponding quark masses. [Notice that all contributions of the form $\ln(m_i^2/m_W^2)$ cancel among the diagrams of fig. 3.10a.] In deriving eq. (3.29), the assumption has been made that $m_W, m_t \gg m_s, m_d$. However, no approximation has been made on the ratio m_t/m_W.

The two-quark operator of eq. (3.29) can be related to the divergence of the difference between the vector and axial-vector currents using the Dirac equation,

$$m_s \bar{d}(1 + \gamma_5)s = i\partial^\mu \bar{d}\gamma_\mu(1 - \gamma_5)s + \mathcal{O}(m_d)$$
$$= i\partial^\mu(V_\mu^{6+i7} - A_\mu^{6+i7}),$$

(3.30)

where the superscripts identify the SU(3) transformation properties of the currents. To obtain the amplitude for $K \to \pi\phi^0$, we must compute the matrix elements of these currents between an on-shell kaon and pion state. The axial vector current does not contribute (due to its quantum numbers), and the matrix element of $i\partial^\mu V_\mu^{6+i7}$ can be related to a form factor of $K_{\ell 3}$ decay as follows. Using [36]

$$\langle \pi^0(p')|i\partial^\mu V_\mu^{4+i5}|K^-(p)\rangle = f_+(0)(m_K^2 - m_\pi^2) + \mathcal{O}(q^2),$$

(3.31)

where $q \equiv p - p'$ and $f_+(0) \simeq 1/\sqrt{2}$ as measured in $K^- \to \pi^0 \ell^- \bar{\nu}$, we perform an isospin rotation which gives

$$\langle \pi^-|m_s\bar{d}(1 + \gamma_5)s|K^-\rangle \simeq \langle \pi^-|i\partial^\mu V_\mu^{6+i7}|K^-\rangle \simeq m_K^2,$$

(3.32)

where we have dropped the term proportional to the pion mass to be consistent with the approximation $m_s \gg m_d$ made in eq. (3.30). Combining the results of eqs. (3.29) and (3.32), we obtain the contribution of the two-quark operator to the matrix element for $K^- \to \pi^- \phi^0$

$$\mathcal{M}_1(K^- \to \pi^- \phi^0) \simeq \frac{G_F^{3/2}}{2^{1/4}} \frac{3}{16\pi^2} m_K^2 \sum_i m_i^2 V_{id}^* V_{is}.$$

(3.33)

Note that eq. (3.33) agrees with the expression obtained in §2.2 [see eqs. (2.89)–(2.93)] using the chiral Lagrangian technique.

The non-spectator diagram of fig. 3.10b ($s + u \to d + u + \phi^0$) gives rise to the (low-energy) effective four-fermion-Higgs interaction

$$\mathcal{L} = \frac{2G_F^{3/2}}{2^{1/4}} V_{ud}^* V_{us} \bar{d}\gamma^\mu(1 - \gamma_5)u \bar{u}\gamma_\mu(1 - \gamma_5)s\phi^0 + \text{h.c.}$$

(3.34)

Gluonic corrections introduce additional four-quark operators [37]. (For a recent review of the effective Hamiltonian for $\Delta S = 1$ weak non-leptonic processes, see ref. 38.) One could compute the matrix elements of the four-quark operators between an on-shell kaon and pion state using the vacuum insertion approximation. However, this method is known not to be very reliable; for example, it does not reproduce in detail the observed $\Delta I = 1/2$ rule. Instead, we prefer to use the chiral Lagrangian technique described in §2.2. The $\Delta I = 1/2$ rule is parametrized by a parameter λ whose magnitude (but not its sign) is determined by the data. In addition, there is also a $\Delta I = 3/2$ contribution which is parametrized by a parameter a. However, we know from experiment that $|a| \ll |\lambda|$, so we will henceforth neglect effects due to the $\Delta I = 3/2$ contribution. The contribution of the non-spectator diagrams to $K^- \to \pi^- \phi^0$ was obtained in eq. (2.90). In addition to λ, we saw in §2.2 that the amplitude also depends on a new parameter A. For $N_H = 3$ heavy quarks and $b =$

$11 - 2n_L = 9$ (for $n_L = 3$ light quarks), we obtain

$$\mathcal{M}_2(K^- \to \pi^- \phi^0) = G_F^{1/2} 2^{1/4} \left[\frac{7 A m_K^2}{9} - \frac{7\lambda(m_K^2 + m_\pi^2 - m_{\phi^0}^2)}{18} \right]. \quad (3.35)$$

It should be noted that the parameters λ and A are of order G_F, so that $\mathcal{M}_2$ is of order $G_F^{3/2}$ (and is thus of the same order as $\mathcal{M}_1$). By fitting to K decays, the authors of ref. 39 determined that $|\lambda| = 3.2 \times 10^{-7}$. The parameter A is expected to be roughly of the same order as λ, but its precise value is not known at the present time. In principle, A and λ are complex. However we have argued in §2.2 that Im $\lambda \ll$ Re λ and Im $A \ll$ Re A, so we can neglect these imaginary parts in the subsequent analysis.

The amplitudes $\mathcal{M}_1$ and $\mathcal{M}_2$ must be combined coherently in the $K \to \pi\phi^0$ amplitude. It is here that much of the controversy in the literature has been generated, since the relative sign between the two amplitudes is physically important. First, let us assume that it is proper to simply add $\mathcal{M}_1$ and $\mathcal{M}_2$. Then, we must discuss the signs of three quantities: $\sum_i m_i^2 V_{id}^* V_{is}$, which occurs in $\mathcal{M}_1$, and A and λ, which occur in $\mathcal{M}_2$. In the usual convention which we adopt here, where $V_{ud}^* V_{us} = \cos\theta_c \sin\theta_c$ is positive, we see that $V_{cd}^* V_{cs}$ and $V_{td}^* V_{ts}$ are negative [40] (ignoring for the moment the imaginary parts of these quantities). The relative signs of A and λ are not determined *a priori*. (Note that λ implicitly depends on $V_{ud}^* V_{us}$, so that its sign relative to $\mathcal{M}_1$ is well-defined.) However, if one matches the chiral Lagrangian to the $\Delta S = 1$ weak Lagrangian in the full theory at 1 GeV, one finds that λ is *negative* in the vacuum insertion approximation. That is, neglecting A in eq. (3.35), $\mathcal{M}_2$ would be positive. This result has also been found in refs. 31 and 20; there is no disagreement on this point. On the other hand, $\mathcal{M}_1$ is dominated by the c- and t-quark contributions, and is therefore negative. Thus, adding the two amplitudes would lead to destructive interference (if A were zero) resulting in a smaller predicted value for $BR(K \to \pi\phi^0)$ than naively expected. A similar result was claimed by Pham and Sutherland [41], who attempted to incorporate non-perturbative effects into the calculation of $K \to \pi\phi^0$, by using vector meson dominance and SU(3) × SU(3) chiral symmetry to calculate $\langle K|\mathcal{H}_{\text{weak}}|\pi\rangle$. They found strong cancellation between the perturbative and non-perturbative contributions.

In ref. 31, Willey argued that the correct procedure is to combine $\mathcal{M}_1$ and $\mathcal{M}_2$ with a relative minus sign. This would lead to constructive interference (if $A = 0$) and a larger value for $BR(K \to \pi\phi^0)$. He presented two arguments to support his claim. First, recall that in the QED calculation of $e^+e^- \to e^+e^-$, one inserts an extra minus sign between the annihilation graph and the t-channel exchange graph, due to Fermi statistics. Willey argued that in the present context, the non-spectator diagrams (fig. 3.10b) resemble the annihilation graph and the spectator diagrams (fig. 3.10a) resemble the exchange graph. Thus one must insert an extra minus sign between $\mathcal{M}_1$ and $\mathcal{M}_2$. Willey supported this heuristic argument with a detailed examination of the relative sign in a Bethe-Saltpeter approach (in which the quarks are bound into the observed meson states). Once again, he concluded that an extra relative minus sign between the spectator and non-spectator amplitudes is required by Fermi statistics. However, these arguments appear to contradict the result of the chiral Lagrangian technique described in §2.2. This approach predicts a well-defined relative sign between $\mathcal{M}_1$ and $\mathcal{M}_2$ which implies destructive interference between the two amplitudes when $A = 0$, as noted above. Furthermore, once the quark fields have been replaced with the meson fields, there should be no extra minus signs generated by Fermi statistics. It is hard to see how the chiral Lagrangian approach can produce an incorrect relative sign.

Admittedly, it would be useful to understand the relation between the two approaches for the determination of the relative sign. However, once we decide to work in the chiral Lagrangian framework, there is a more crucial theoretical uncertainty to deal with, namely the existence of a new parameter A which appears in eq. (3.35). Clearly, without any knowledge of the value of A (other than its order of magnitude), the sign of $\mathcal{M}_2$ can be positive or negative. Hence the controversy over the relative sign between $\mathcal{M}_1$ and $\mathcal{M}_2$ is one of academic interest only, since without the explicit knowledge of the numerical value of A, there is no definite theoretical prediction for $BR(K \to \pi\phi^0)$. However, it was recently pointed out in ref. 20 that there is a prediction for a *lower* limit for $BR(K^{\pm} \to \pi^{\pm}\phi^0)$ due to the fact that the amplitude for $K^{\pm} \to \pi^{\pm}\phi^0$ is complex. This suggestion would make sense only if the dominant source of the imaginary part is due to the heavy quark mixing angles which appear in $\mathcal{M}_1$. In fact, we believe that this is correct, due to an argument given in §2.2, which we summarize here. The two-quark operator is special in that the amplitude contains an explicit factor of the quark mass squared [see eq. (3.33)]. The t-quark contribution in the sum is greatly enhanced, which permits Im $\mathcal{M}_1/$Re $\mathcal{M}_1$ to be of $\mathcal{O}(1)$. On the other hand, simple dimensional analysis suggests that the contributions to the amplitude from the four-quark operators can at best approach a constant in the large m_t limit. It follows that the ratio of the imaginary to the real part of the four-quark amplitudes can be no larger than (roughly) Im $V_{td}^* V_{ts}/(V_{ud}^* V_{us})$ which is of order 10^{-3}. Thus,

$$\text{Im } \mathcal{M}_2 \sim \mathcal{O}(10^{-3}) \text{ Re } \mathcal{M}_2. \qquad (3.36)$$

Since we expect the real parts of $\mathcal{M}_1$ and $\mathcal{M}_2$ to be roughly of the same order of magnitude, we conclude that Im $\mathcal{M}_1$ should be the dominant source of the imaginary part of the total amplitude.*

We therefore obtain a lower limit by setting the real part of the total amplitude to zero (which would happen for some choice of the parameter A).

---

*The above argument implies that Im $\lambda \ll$ Re λ and Im $A \ll$ Re A. We can estimate that Im $\lambda \sim 10^{-4}$ Re λ based on the experimentally measured value of ϵ'/ϵ.

It is convenient to use the Wolfenstein parametrization [42] of the Kobayashi-Maskawa matrix, where

$$V_{ud}^* V_{us} \simeq -V_{cd}^* V_{cs} \simeq \sin \theta_c \cos \theta_c$$
$$V_{td}^* V_{ts} \simeq -A_w^2 \sin^5 \theta_c (1 - \rho + i\eta). \tag{3.37}$$

Experimentally, $\sin \theta_c \simeq 0.22$, the parameter A_w is close to unity ($A_w = 1.05 \pm 0.17$ according to ref. 43), and $\rho \lesssim 0$ (based on the observed B-$\overline{B}$ mixing). In addition, the measurement of ϵ'/ϵ at CERN [44] may be used to determine a value of η. Values in the range $0.1 \lesssim \eta \lesssim 0.6$ have been obtained in the literature [20,45].[†] In the analysis below, we shall take a modest value, $\eta = 0.2$. Thus, taking the imaginary part of the amplitude as determined from eq. (3.33), we get

$$|\text{Im} \mathcal{M}(K^\pm \to \pi^\pm \phi^0)| \gtrsim \frac{G_F^{3/2}}{2^{1/4}} \frac{3 m_K^2 m_t^2 \eta A_w^2 \sin^5 \theta_c}{16\pi^2} \gtrsim 7.9 \times 10^{-11} \text{ GeV}, \tag{3.38}$$

where we have taken $m_t \approx 80$ GeV, as required in order to have a very light Standard Model Higgs boson. Using $\Gamma(K \to \pi \phi^0) = B_{\phi^0} |\mathcal{M}|^2 /(16\pi m_K)$, where $B_{\phi^0} = 2p_{\phi^0}/m_K$, and normalizing to the total decay rate, $\Gamma(K^\pm) = 5.32 \times 10^{-17}$ GeV, we find

$$BR(K^\pm \to \pi^\pm \phi^0) \gtrsim 4.3 \times 10^{-6} B_{\phi^0}. \tag{3.39}$$

Let us next turn to the decay $K_L^0 \to \pi^0 \phi^0$. The one important change in the analysis results from the fact that the matrix element for this process is real.[‡] In particular, as derived in §2.2, we have

$$\mathcal{M}_1(K_L^0 \to \pi^0 + \phi^0) = -\frac{G_F^{3/2}}{2^{1/4}} \frac{3}{16\pi^2} m_K^2 \sum_i m_i^2 \text{Re}(V_{id}^* V_{is}),$$
$$\mathcal{M}_2(K_L^0 \to \pi^0 + \phi^0) = -\text{Re } \mathcal{M}_2(K^- \to \pi^- \phi^0). \tag{3.40}$$

Thus, we cannot obtain such a definitive theoretical bound for $K_L^0 \to \pi^0 \phi^0$ without the explicit knowledge of A, since (unlike in the case above) there exists a particular value of A for which this amplitude vanishes. Nevertheless, we can examine the branching ratio as a function of the parameter A in $\mathcal{M}_2$. We do this by considering several values for A (which is presumably of order λ) and computing the decay width as a function of $m_t^2 |\text{Re } V_{td}^* V_{ts}|$.

[†] This line of reasoning may be less secure, given the most recent measurement of ϵ'/ϵ at the Tevatron [46] which is consistent with zero and differs from the CERN result by two standard deviations.

[‡] $\mathcal{M}(K_L^0 \to \pi^0 \phi^0)$ is purely real in the limit where $\epsilon \ll \arg V_{td}^* V_{ts}$; see eq. (2.93) and the discussion following. Since $\epsilon \simeq 2 \times 10^{-3}$, the numbers quoted above suggest that this is a good approximation. For instance, $\text{Im} \mathcal{M}(K_L^0 \to \pi^0 \phi^0) = \pm \text{Re} \epsilon \text{ Im} \mathcal{M}(K^\pm \to \pi^\pm \phi^0)$, yielding a negligible contribution to the $K_L^0 \to \pi^0 \phi^0$ amplitude squared.

The resulting width is normalized with respect to the total width $\Gamma(K_L^0) = 1.27 \times 10^{-17}$ GeV. According to the discussion above, for $m_t \approx 80$ GeV, we expect $m_t^2|\mathrm{Re}\, V_{td}^* V_{ts}| \gtrsim 2.5$ GeV^2, with large uncertainty. Thus, we plot in fig. 3.11, $BR(K_L^0 \to \pi^0\phi^0)$ vs. $m_t^2|\mathrm{Re}\, V_{td}^* V_{ts}|$ for three different values of A. We see that only for a very narrow range of $m_t^2|\mathrm{Re}\, V_{td}^* V_{ts}|$, and only for positive values of A does the branching ratio for $K_L^0 \to \pi^0\phi^0$ ever fall below 10^{-5}. Therefore, barring a very unlikely conspiracy of parameters, we obtain a predicted branching ratio which satisfies

$$BR(K_L^0 \to \pi^0\phi^0) \gtrsim 10^{-5} B_{\phi^0}. \tag{3.41}$$

By a similar computation, we can calculate the amplitude for the CP-violating decay $K_S^0 \to \pi^0\phi^0$. If we can neglect Im A and Im λ as suggested above, (and neglecting small CP-violating effects in the kaon state mixing), then $\mathcal{M}_2$ can be ignored. $\mathcal{M}_1$ is modified by replacing Re $V_{td}^* V_{ts}$ with Im $V_{td}^* V_{ts}$ in eq. (3.40). Using the numbers quoted above, we find

$$BR(K_S^0 \to \pi^0\phi^0) \simeq 10^{-8} B_{\phi^0}. \tag{3.42}$$

Much of the above uncertainty can be eliminated if a reliable calculation of A becomes available. In this regard, we would like to cite a recent computation by Leutwyler and Shifman [47] in which the parameter A is estimated using the vacuum saturation approximation in the evaluation of the relevant hadronic matrix elements. They find a rather small value for A, roughly $A \sim -0.1\lambda$ (with about a factor of two uncertainty). Thus, numerically, the kaon decay amplitudes are well approximated by simply neglecting A! This leads to predictions for $BR(K \to \pi\phi^0)$ somewhat larger than the estimates quoted above. It should also be noted that as the lower limits on the t-quark mass become larger, the relative importance of $\mathcal{M}_1$ increases, in which case the uncertainty in A (and λ) becomes moot.

We now turn to a discussion of the extent to which currently available experimental data can be used to constrain the existence of a light Higgs boson.* Potentially relevant experimental data are of two types: first, there are limits on the branching ratio for K decay to a variety of final states to which production of a ϕ^0 might contribute—these typically must be examined critically in order to determine the extent to which they lead to limitations on the Higgs boson; and secondly there have been a number of direct searches for Higgs bosons in many of the same final states. We begin by reviewing data of the first type. Available limits are the following [†]

L1. $BR(K_L^0 \to \pi^0\mu^+\mu^-) < 1.2 \times 10^{-6}$ [50];

L2. $BR(K_L^0 \to \pi^0 e^+ e^-) < 2.3 \times 10^{-6}$ [50], with recent limits of $< 3.2 \times 10^{-7}$ [51] and $< 4.2 \times 10^{-8}$ [52];

---

*Some of these issues have also been considered in refs. 48, 49 and 20.

[†]We quote only those that are potentially useful, given the branching ratios of eqs. (3.39) and (3.41).

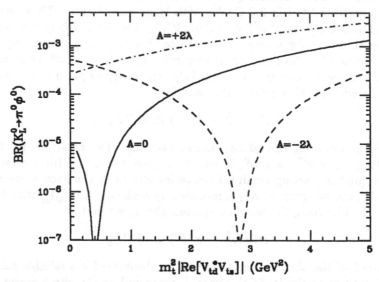

Figure 3.11 The $K_L^0 \rightarrow \pi^0\phi^0$ branching ratio as a function of m_t^2 $\times|\text{Re}\,V_{td}^*V_{ts}|$, assuming $m_{\phi^0} \ll m_K$. The three curves shown are: $A = 0$ (solid), $A = 2\lambda$ (dotdashed), and $A = -2\lambda$ (dashes), where $\lambda = -3.2 \times 10^{-7}$.

L3. $BR(K^+ \rightarrow \pi^+e^+e^-) = (2.7 \pm 0.5) \times 10^{-7}$ [53], and $BR(K^+ \rightarrow \pi^+e^+e^-) < 2.7 \times 10^{-7}$ [54];

L4. $BR(K^+ \rightarrow \pi^+\mu^+\mu^-) < 2.4 \times 10^{-6}$ [55], and $BR(K^+ \rightarrow \pi^+\mu^+\mu^-) < 2.3 \times 10^{-7}$ [29];

L5. $BR(K^+ \rightarrow \pi^+\gamma\gamma) < 8 \times 10^{-6}$ [25], and $BR(K^+ \rightarrow \pi^+\gamma\gamma) < 1.4 \times 10^{-6}$ [56]; and

L6. $BR(K^+ \rightarrow \pi^+ + nothing) < 3.8 \times 10^{-8}$ [57], and $BR(K^+ \rightarrow \pi^+ + nothing) < 3.4 \times 10^{-8}$ [58].

Some discussion concerning the above limits is necessary. In particular, to ascertain the implications of the above limits for the SM Higgs, we must know the branching ratios for $\phi^0 \rightarrow \gamma\gamma$, $\phi^0 \rightarrow e^+e^-$ and $\phi^0 \rightarrow \mu^+\mu^-$. We must also consider the expected lifetimes for the ϕ^0 when these various decay modes are dominant. Finally, we must examine the extent to which the quoted limits are restricted in applicability to the ϕ^0 search by limitations coming from experimental acceptance, cuts and backgrounds. For example, the lepton pair final state limits have typically assumed a phase space distribution over a certain invariant mass range, while the $\gamma\gamma$ and *nothing* final state limits can only be applied to Higgs boson decays after accounting for the Higgs boson lifetime.

First, consider the limits L2 and L3 for the e^+e^- final state. There is a question as to how far down in $M(e^+e^-)$ they can be applied. Unfortunately, the recent very strong limit of ref. 51 does not apply for $M(e^+e^-) < 2m_\mu$ (the region of interest for ϕ^0 decays). In the case of ref. 51 one finds that their branching ratio limit deteriorates very rapidly as $M(e^+e^-)$ decreases, rising to 1.2×10^{-6} at their lowest quoted point of $M(e^+e^-) = 240$ MeV. A similar deterioration occurs in the case of ref. 50. L. Littenberg, has analyzed this question for that experiment [59]. The branching ratio limit falls very steeply as $M(e^+e^-)$ increases, passing the nominal limit 2.3×10^{-6} at $M(e^+e^-) \sim$ 250 MeV and reaching 1×10^{-6} at $M(e^+e^-) \sim 275$ MeV. But in the region $M(e^+e^-) < 200$ MeV the branching ratio limit is much worse, $\lesssim 10^{-5}$, with very large error bars.[‡]

The limit on $K_L \to \pi e^+e^-$ from the FNAL experiment E731 [52] is also of dubious utility because of the cuts made in obtaining the explicit acceptance plots as a function of $M(e^+e^-)$ that are presented [60]. While these show good acceptance over the region $50 \ MeV \lesssim M(e^+e^-) \lesssim 350$ MeV, in particular covering the region of $M(e^+e^-) \sim m_\pi$ which presents a problem for the other experiments, their current method of analysis is such that the e^+e^- would have to emerge within roughly 1 m of the initial $K_L^0 \to \pi^0$ decay vertex. Since their initial K_L^0 has momentum of ~ 60 GeV on average, the ϕ^0 emerging in $K_L^0 \to \pi^0\phi^0$ would be quite energetic. For $m_{\phi^0} \leq 2m_\mu$, where the decay $\phi^0 \to e^+e^-$ is dominant, the ϕ^0 will have a long lifetime and its path length generally exceeds 1 m. For instance, for $m_{\phi^0} = 140$ MeV fig. 2.9 yields $c\tau \sim$ 0.8 cm; to first approximation the π and ϕ^0 will share the K_L^0 momentum in proportion to their masses, yielding a boost factor of order 220 and a path length for the ϕ^0 of order 1.6 m. Even for $m_{\phi^0} \sim 200$ MeV, the path length is only just below 1 m. Thus, the E731 group plans to reanalyze their data in order to assess the restrictions it places on a light Higgs [61].

Turning to the other experiments, we note that explicit branching ratio limits as a function of $M(e^+e^-)$ are not available for the $K^+ \to \pi^+e^+e^-$ experiments of ref. 53 and 54. Reference 53 quotes a lower limit on accepted $M(e^+e^-)$ of 140 MeV. Reference 54 quotes a lower limit of 50 MeV. In analogy with the two analyses discussed above, we expect that the actual lower limits in the region $M(e^+e^-) < 2m_\mu$ are significantly worse than those quoted above based on a phase space distribution of $M(e^+e^-)$. In the case of the $\mu^+\mu^-$ final states, there is again uncertainty as to how the branching ratio limits depend on $M(\mu^+\mu^-)$; some deterioration as $M(\mu^+\mu^-)$ approaches $2m_\mu$ (the lower limit of interest for ϕ^0 decays) is probable, but hopefully not severe.

For $m_\phi^0 < 2m_e$ the $\phi^0 \to \gamma\gamma$ mode dominates. However, the ϕ^0 lifetime is so long (see fig. 2.9) that it would have escaped the apparatus used to establish the experimental limit L5, which is not sufficiently strong to have been useful in any case. Fortunately, because of the long ϕ^0 lifetime the limit L6 becomes

---

[‡]In fact the difference in the limits quoted in refs. 50 and 51 appears to be largely due to a difference in the range of $M(e^+e^-)$ included in quoting the limit.

relevant and clearly rules out $m_{\phi^0} < 2m_e$, given the predicted branching ratio lower bound of eq. (3.39).

For $2m_e < m_{\phi^0} < 2m_\mu$, the ϕ^0 decays to e^+e^- and $\gamma\gamma$, with the branching ratio for the e^+e^- mode being close to 1 over a large part of this mass range.* If we adopt the conservative lower limit of eq. (3.41) away from the cancellation dips in fig. 3.11 we have $BR(K_L^0 \to \pi^0\phi^0) \times BR(\phi^0 \to e^+e^-) \gtrsim 0.8 \times 10^{-5}$. Unfortunately, as we have discussed, the $K_L^0 \to \pi^0 e^+ e^-$ limit from E731 is not applicable. Further, the upper limits on this decay for $M(e^+e^-) < 200$ MeV ($BR \leq 10^{-5}$) from the other experiments do not allow us to rule out any m_{ϕ^0} values below $2m_\mu$ on the basis of this decay if we are anywhere near the cancellation dips of fig. 3.11. Turning to charged kaon decays, the nominal limits on $BR(K^+ \to \pi^+ e^+ e^-)$ are a factor of more than 30 below that expected [eq. (3.39)] from $K^+ \to \pi^+\phi^0(\to e^+e^-)$. But, as discussed above, we cannot be certain that there are any useful limits for $M(e^+e^-) < 2m_\mu$. Thus, a conservative conclusion (away from the cancellation dips of fig. 3.11) is that the $K \to \pi e^+ e^-$ experiments that are not explicitly dedicated to searching for the ϕ^0 rule out only a very small range of Higgs masses, 200 MeV $\lesssim m_{\phi^0} \lesssim 2m_\mu$.

However, there is still the possibility that the $K^+ \to \pi^+ + nothing$ limit L6 in our list can be used to rule out the ϕ^0 for some range of m_{ϕ^0} above $2m_e$. The idea is that the ϕ^0 has a sufficiently long lifetime, even when the $\phi^0 \to e^+e^-$ channel is allowed, that some of the decays will have been invisible to the detector. Since the detector of ref. 57 is much smaller than that of BNL-787 [58], only the former experiment will be considered in the following discussion. Consider, for example, $m_{\phi^0} = 50$ MeV; the lifetime of such a ϕ^0 is 8.5×10^{-11} sec and the time dilation factor in the K rest frame is ~ 4.6, implying a decay path length of order 0.1 m. The fiducial dimension of the apparatus in the experiment of ref. 57 was about 0.37 m. A fraction $e^{-3.7} \sim 0.03$ of the ϕ^0 decays would have occurred outside the apparatus. Thus, the theoretical prediction of eq. (3.39) would imply an effective branching ratio for $K^+ \to \pi^+\phi^0$, $\phi^0 \to nothing$ of $\gtrsim 1.3 \times 10^{-7}$, which is clearly excluded by limit L6. Even for $m_{\phi^0} = 60$ MeV, where the ϕ^0 lifetime is 6.5×10^{-11}, the decay path length is of order 0.072 m, and the effective branching ratio for $K^+ \to \pi^+\phi^0$, $\phi^0 \to nothing$ computed using eq. (3.39) is $\sim 2.3 \times 10^{-8}$, only slightly below the limit of ref. 57. To determine exactly how high in m_{ϕ^0} this limit continues to be useful requires a more thorough understanding of the

---

*This result for the e^+e^- mode branching ratio employs the formulae for the $\gamma\gamma$ width given in eq. (2.15), taking $m_t = 80$ GeV and including all SM particles in the loop graphs. If constituent quark masses for the light quarks are employed (as we have done in fig. 2.5), then $BR(\phi^0 \to \gamma\gamma)$ increases with m_{ϕ^0} (in the interval $2m_e < m_{\phi^0} < 2m_\mu$), reaching a value of 0.14 just below the two-muon threshold. On the other hand, if very small current quark masses are chosen ($m_{u,d} \sim 10$ MeV and $m_s \sim 150$ MeV), then $\Gamma(\phi^0 \to \gamma\gamma)$ remains small ($\lesssim 10^{-2}$) throughout this mass range.

apparatus. At our suggestion, T. Shinkawa from the collaboration of ref. 57 has had a closer look at the above considerations. He has arrived at essentially the same conclusions [62]. In particular, he has supplied the precise average apparatus size used in the computations above, and has checked that the range of the detected pion has a high probability of being in the background-free region above 36 g/cm^2. For instance, combining all efficiencies he finds that the 90% confidence level upper limit for $BR(K^+ \to \pi^+\phi^0)$ is 1.3×10^{-6} for $m_{\phi^0} = 50$ MeV, 6.3×10^{-6} for $m_{\phi^0} = 60$ MeV, and 4.6×10^{-5} for $m_{\phi^0} = 65$ MeV, indicating that masses up to $m_{\phi^0} \sim 55$ MeV are probably excluded. He has also pointed out that the limit (L5 above) from their ref. 56 on $K^+ \to \pi^+\gamma\gamma$ can be reinterpreted to yield limits on $K^+ \to \pi^+\phi^0$, where the ϕ^0 decays either to $\gamma\gamma$ or to e^+e^- inside their lead glass detector. According to Shinkawa's analysis, in order for the ϕ^0 decay to be observed, the decay must take place after 5 cm and before 2 radiation lengths short of the 37 cm lead glass average size, i.e. before 32 cm. Including also the probability that the π^+ be in the background-free range region above 36 g/cm^2, Shinkawa finds 90% confidence level upper limits on $BR(K^+ \to \pi^+\phi^0)$ of 15, 4.5, 2.5, 5.1, and 5.9 $\times 10^{-6}$ at m_{ϕ^0} values of 10, 20, 50, 75, and 80 MeV. Comparing to eq. (3.39) we conclude that m_{ϕ^0} values between about 25 and 40 MeV would be ruled out on this basis. The range of m_{ϕ^0} values excluded in this way would broaden to include the range from 20 to 80 MeV if there is any significant real part to the charged K decay amplitude. The analysis of data from the BNL-787 experiment at Brookhaven yields even stronger results [58]. Taking the Higgs lifetime into account, the BNL-787 group interprets their search for $K^+ \to \pi^+ + nothing$ as a limit on $K^+ \to \pi^+\phi^0$. They conclude that for $m_{\phi^0} = 0$, $BR(K^+ \to \pi^+\phi^0) < 6.4 \times 10^{-9}$; the limit becomes less severe as the Higgs mass is increased [58].

For $2m_\mu < m_{\phi^0} < 2m_\pi$, the ϕ^0 decays almost entirely to $\mu^+\mu^-$, and we believe that the limits L1 and L4 definitively rule out this region of m_{ϕ^0}. For $m_{\phi^0} > 2m_\pi$, and employing the benchmark branching ratio of eq. (3.41), we require $B_{\phi^0} \times BR(\phi^0 \to \mu^+\mu^-) \gtrsim 0.15$ in order that the prediction for $BR(K_L^0 \to \pi^0\phi^0)$ be clearly larger than the limit L1 for the $\pi^0\mu^+\mu^-$ final state. Since the kinematical factor B_{ϕ^0} is significantly smaller than 1 ($B_{\phi^0} = 0.53$ for $m_{\phi^0} \sim 2m_\pi$), $BR(\phi^0 \to \mu^+\mu^-)$ is only marginally large enough (see §2.1 and fig. 2.8) even in the absence of $\pi\pi$ decay mode enhancements. Thus, any amount of enhancement in the $\phi^0 \to \pi\pi$ decay width (whether the moderate amount found in ref. 24 or the very large enhancement found in ref. 23) would make it impossible to eliminate any $m_{\phi^0} > 2m_\pi$ on the basis of the limit L1. However, note that if $BR(K_L^0 \to \pi^0\phi^0) > 10^{-4}$ (which may be possible according to fig. 3.11), then $B_{\phi^0} \times BR(\phi^0 \to \mu^+\mu^-)$ values as small 0.015 would still allow exclusion of a ϕ^0 using the limit L1. Severe suppression from B_{ϕ^0} only sets in very near threshold; for $m_{\phi^0} < 340$ MeV we find $B_{\phi^0} > 0.2$ and by $m_{\phi^0} = 2m_\pi$ $B_{\phi^0} > 0.5$. Then, for example, $BR(\phi^0 \to \mu^+\mu^-) > 0.075$ at $m_{\phi^0} = 340$ MeV, would be sufficient to violate the bound L1. Such a branching ratio is consistent with the moderate $\pi\pi$ enhancements predicted in ref. 24, but not with the larger enhancements predicted in ref. 23.

In contrast, the $K^+ \to \pi^+\mu^+\mu^-$ limit of ref. 55 (the first L4 limit) cannot be used to exclude a Higgs boson with $m_{\phi^0} > 2m_\pi$. Combining this limit with the lower bound of eq. (3.39), we find that a ϕ^0 is excluded only if $BR(\phi^0 \to \mu^+\mu^-)B_{\phi^0} \gtrsim 0.5$, which is inconsistent with the values of B_{ϕ^0} quoted above. However, as noted above, the lower bound of eq. (3.39) on $BR(K^+ \to \pi^+\phi^0)$ would be a substantial underestimate away from zeroes of the real part of the amplitude, and may allow for some restrictions on m_{ϕ^0}. The second L4 limit [29] was obtained as a byproduct of a direct Higgs search $K^+ \to \pi^+\phi^0$, $\phi^0 \to \mu^+\mu^-$ and will be discussed below.

We now turn to the dedicated Higgs search analyses that have been performed for K decays. We shall see that these eliminate all the gaps left by the restrictions obtained in the analysis of limits L1 through L6. Let us focus first on the dedicated searches for a Higgs boson that have been performed for $K^+ \to \pi^+\phi^0$ in refs. 26–28, and 58. The first experiment places a limit on a light Higgs decaying to e^+e^- pairs which is dependent on its lifetime (see fig. 3.12). To calibrate this figure, we note that a Higgs boson of mass 1.8, 10, 50, 70, 100 and 200 MeV has a lifetime of approximately 4×10^{-9}, 4.3×10^{-10}, 8.5×10^{-11}, 6×10^{-11}, 4.3×10^{-11} and 2×10^{-11} seconds, respectively. From the available curves, we estimate that the limit of Baker *et al.* [26] falls below the prediction of eq. (3.39) at $m_{\phi^0} \gtrsim 60$ MeV. In addition, we have learned [63] that the published analysis only applies for $m_{\phi^0} < 100$ MeV. Furthermore, the most recent data from the Brookhaven experiment BNL-777 [27] gives a limit

$$BR(K^+ \to \pi^+\phi^0) \times BR(\phi^0 \to e^+e^-) < 10^{-8}, \qquad (3.43)$$

for all $M(e^+e^-)$ above 120 MeV. Thus, these experiments [26,27] rule out Higgs masses in the range $60 < m_{\phi^0} < 100$ MeV and 120 MeV $< m_{\phi^0} < 2m_\mu$, respectively.

The second direct search K^+-decay experiment relies on measuring the decay spectrum of the π^+ and does not rely on detecting the ϕ^0 decay products. It is thus potentially applicable for any m_{ϕ^0} kinematically accessible in K^+ decays. From the results of ref. 28 (which are summarized in their fig. 2) one sees that there is no sensitivity to the mass range $110 < m_{\phi^0} < 150$ MeV due to the large rate for $K^+ \to \pi^+\pi^0$. Above 150 MeV their limit is above 10^{-5}, which is too high to be useful for the theoretical lower bound branching ratio of eq. (3.39), but below 100 MeV their limit is below 4×10^{-6} and $m_{\phi^0} < 100$ MeV is ruled out (down to their lowest plotted point of about 10 MeV).

A third direct search has been carried out at Brookhaven by the BNL-787 experiment. The results of this search were recently reported in ref. 29

$$BR(K^+ \to \pi^+\phi^0) \times BR(\phi^0 \to \mu^+\mu^-) < 1.5 \times 10^{-7}, \qquad (3.44)$$

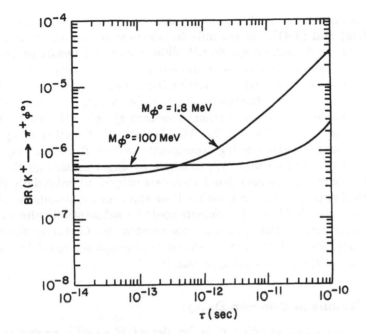

Figure 3.12 The experimental limit on the kaon decay width to a light Higgs boson as a function of its lifetime from the mode $K^+ \to \pi^+\phi^0$, $\phi^0 \to e^+e^-$ (from ref. 26).

for $2m_\mu < m_{\phi^0} < 320$ MeV at 90% CL. Combining the lower bound on $BR(K^+ \to \pi^+\phi^0)$ of eq. (3.39) with $B_{\phi^0} \sim 0.34$ at $m_{\phi^0} = 320$ MeV, we see that $BR(\phi^0 \to \mu^+\mu^-) \gtrsim 0.1$ is required in order to exclude a ϕ^0 of this mass. Since m_{ϕ^0} values below 320 MeV yield even larger values for B_{ϕ^0} than obtained at $m_{\phi^0} = 320$ MeV, we conclude that the BNL-787 results exclude m_{ϕ^0} values throughout the range quoted above, provided that $BR(\phi^0 \to \mu^+\mu^-)$ is not unduly suppressed.

The final direct Higgs search in K decays was performed by the CERN NA-31 experiment on K_L^0 decays. They have obtained a 90% confidence level upper limit of [64]

$$BR(K_L^0 \to \pi^0\phi^0) \times BR(\phi^0 \to e^+e^-) < 2 \times 10^{-8} \qquad (3.45)$$

throughout the range $20 \lesssim m_{\phi^0} \lesssim 350$ MeV, except for the region of m_{ϕ^0} within 15 MeV on either side of m_π where their limit for the above branching ratio product is $< 3 \times 10^{-8}$. Using the results of eq. (3.41) and fig. 3.11, we see that this experiment conclusively eliminates all m_{ϕ^0} values in the range 20 MeV $< m_{\phi^0} < 2m_\mu$, where $BR(\phi^0 \to e^+e^-)$ is near 1.

In summary, even if we adopt the conservative branching ratios of eqs. (3.39) and (3.41), we are able to conclude from current experimental information that kaon decays do not allow a standard model Higgs boson in the mass region $m_{\phi^0} \lesssim 2m_\mu$, with the strongest information coming from the limits of refs. 27, 57 and 64. In addition the region $2m_\mu < m_{\phi^0} < 320$ MeV is excluded on the basis of the results of ref. 29 (with additional support in some mass regions from a variety of other experiments) unless the $\pi\pi$ enhancement is as large as predicted in ref. 23. We also note that further analysis of the data from many of the existing experiments on rare K decays may yield additional restrictions on m_{ϕ^0}. We conclude by again cautioning that most of the restrictions on m_{ϕ^0} considered above are subject to potentially significant theoretical uncertainties. In addition, if we allow for the possibility of a fourth generation, the CKM matrix elements could be substantially altered, thereby invalidating some of the more marginal constraints. Certainly, the improved limits from rare K decay experiments now in progress should further reduce the impact of the theoretical uncertainties.

Higgs Bosons in B-meson Decays

Here we may consider either inclusive decays $B \to \phi^0 X$ or exclusive decay channels such as $B \to \phi^0 K$. Let us focus first on the former. An important advantage of B decays over K decays is that, in the B meson case, the two-quark operator term is completely dominant because of the much larger CKM matrix elements occurring in this term ($|V_{tb}| \gg |V_{td}|$). In addition, for B meson decays the quark spectator approach is expected to be valid; in particular, uncertainties from non-perturbative terms such as those discussed in connection with the chiral Lagrangian approach to K decays should be quite small. We can obtain the $b \to \phi^0 s$ branching ratio directly from a formula like that of eq. (3.29), with appropriate substitutions. The most reliable way of predicting the rate for the decay $B \to \phi^0 X$ is then to compute the ratio $\Gamma(B \to \phi^0 X)/\Gamma(B \to e\nu X)$. This is because hadronic operators which occur in the quark level transitions $b \to s\phi^0$ and $b \to ce\nu$ are closely related. Thus uncertainties in the matrix element due to strong interaction effects will be minimized in the ratio. A convenient form for the relative widths is[†]

$$\frac{\Gamma(B \to \phi^0 X)}{\Gamma(B \to e\nu X)} = \frac{27g^2}{256\pi^2} \left(\frac{m_t^4}{m_b^2 m_W^2}\right) \frac{\left(1 - \frac{m_{\phi^0}^2}{m_b^2}\right)^2}{f(m_c/m_b)} \left|\frac{V_{ts}^* V_{tb}}{V_{cb}}\right|^2. \tag{3.46}$$

In the above $f(m_c/m_b)$ is the phase space factor for semileptonic B decay. If we take $m_b = 4.5$ GeV, then we find $f(m_c/m_b) = 0.51$, and we may rewrite

---

[†]Equation (3.45) was first derived in ref. 30 and subsequently quoted in refs. 32, 65 and 66 with the omission of one power of $1 - m_{\phi^0}^2/m_b^2$. This omission was corrected in ref. 34.

eq. (3.46) in the form

$$\frac{\Gamma(B \to \phi^0 X)}{\Gamma(B \to e\nu X)} = 2.95 \left(\frac{m_t}{m_W}\right)^4 \left(1 - \frac{m_{\phi^0}^2}{m_b^2}\right)^2 \left|\frac{V_{ts}^* V_{tb}}{V_{cb}}\right|^2 . \tag{3.47}$$

If we now employ the experimental branching ratio for $B \to e\nu X$ of 0.123, we find

$$BR(B \to \phi^0 X) = 0.36 \left(\frac{m_t}{m_W}\right)^4 \left(1 - \frac{m_{\phi^0}^2}{m_b^2}\right)^2 \left|\frac{V_{ts}^* V_{tb}}{V_{cb}}\right|^2 . \tag{3.48}$$

If there are only three families, unitarity of the CKM matrix implies that $|V_{tb}| \simeq 1$ and $|V_{ts}| \simeq |V_{cb}|$. In addition, as mentioned below eq. (3.1), in order that to have a very light Higgs boson in the Standard Model we must require $m_t \approx m_W$.

Since the ϕ^0 will, in part, decay to lepton pairs, it is useful to consider the implications of available experimental limits on the branching ratio for $B \to \ell^+ \ell^- X$, with $\ell = e$ or μ. Because we feel that the existing discussions are confusing we go into some detail. We shall consider separately various m_{ϕ^0} regions, characterized by different dominant decay modes, and discuss the available experimental data appropriate to each m_{ϕ^0} range.

For $m_{\phi^0} < 2m_\mu$, the ϕ^0 decays to $e^+ e^-$ and $\gamma\gamma$, with the branching ratio for the $e^+ e^-$ mode being nearly 1. The MARK II Collaboration at PEP has searched for $B \to \phi^0 X$ followed by $\phi^0 \to e^+ e^-$ and have found no evidence for a Higgs boson in the range 70 MeV $< \phi^0 <$ 210 MeV [67]. Turning next to the region $2m_\mu < m_{\phi^0} < 2m_\pi$, we first note that the decay $\phi^0 \to \mu^+ \mu^-$ has branching ratio near 1. The most directly relevant experimental information derives from the TASSO collaboration [68] who obtain a limit of

$$BR(B \to \mu^+ \mu^- X) < 0.02 . \tag{3.49}$$

This limit assumes that the muon momentum is greater than 1.2 GeV. Since the muons coming from the ϕ^0 decay would be quite energetic, the probability that they would pass the momentum cut is high. Combining this with the branching ratio of eq. (3.48) and $BR(\phi^0 \to \mu^+ \mu^-) \sim 1$ (in the mass region under discussion), we obtain a clear conflict with the experimental limit of eq. (3.49).

In the region $m_{\phi^0} > 2m_\pi$ we are again faced with the various uncertainties regarding the relative branching ratio for $\phi^0 \to \pi\pi$ compared to that for $\phi^0 \to \mu^+ \mu^-$ discussed in chapter 2. The major uncertainty is due to the possibility of large final state interactions which enhance the $\pi\pi$ over the $\mu^+ \mu^-$ final state. As an example, if we neglect final state interactions, the results of fig. 2.8 show that $BR(\phi^0 \to \mu^+ \mu^-) \sim 0.4 - 0.2$ for 300 MeV $\lesssim m_{\phi^0} < 2m_K$. However, including $\pi\pi$ enhancements can reduce the value of $BR(\phi^0 \to \mu^+ \mu^-)$. For example, ref. 23 obtains $BR(\phi^0 \to \mu^+ \mu^-) \sim 5 \times 10^{-2}$ at $m_{\phi^0} \sim 300$ MeV falling to $\sim 10^{-2}$ at $m_{\phi^0} \sim 800$ MeV. In contrast, ref. 24 finds $BR(\phi^0 \to$

$\mu^+\mu^-) > 0.25$ in this same range. Above 800 MeV, the $\phi^0 \to \mu^+\mu^-$ branching ratio precipitously drops in the vicinity of the $f_0(975)$ 0^{++} scalar resonance peak. For $m_{\phi^0} > 2m_K$ final states involving kaons become the dominant Higgs decay modes, and presumably there is some resonant enhancement in the 1–2 GeV region. However, for $m_{\phi^0} \gtrsim 2$ GeV we expect the spectator model to be a reasonable approximation for Higgs boson decay, implying a branching ratio to $\mu^+\mu^-$ that is 1–2 $\times 10^{-2}$ up to $m_{\phi^0} \simeq 2m_\tau$ (see fig. 2.8). Assuming perfect acceptance, the TASSO branching ratio limit of eq. (3.49) combined with eq. (3.48) implies that one can only rule out a Higgs boson with $BR(\phi^0 \to \mu^+\mu^-) \gtrsim 5.6 \times 10^{-2}$. In light of the above discussion, the TASSO limits may rule out the existence of a light Higgs boson with $2m_\pi \lesssim m_{\phi^0} \lesssim 800$ MeV, if $BR(\phi^0 \to \mu^+\mu^-)$ is not too suppressed by $\pi\pi$ decay mode enhancements.

For $m_{\phi^0} > 500$ MeV, various limits from the CLEO collaboration become relevant [69,70]. Combining the best limits from refs. 69 and 70, we find

$$BR(B \to \phi^0 X) \times BR(\phi^0 \to \mu^+\mu^-)$$
$$< \begin{cases} 8 \times 10^{-3}, & 0.5 \text{ GeV} < m_{\phi^0} < 1 \text{ GeV}, \\ 1 - 8 \times 10^{-4}, & 1 \text{ GeV} < m_{\phi^0} < 3 \text{ GeV}, \\ 4 - 6 \times 10^{-5}, & 3.2 \text{ GeV} < m_{\phi^0} < 2m_\tau. \end{cases} \quad (3.50)$$

These branching ratio limits (given at 90% confidence level) have been read off from a set of curves which are quite jagged due to the small number of candidate events over the relevant invariant mass regions. We note that a clear signal is seen in $B \to \mu^+\mu^- X$ for $M(\mu^+\mu^-) \simeq 3.1$ GeV, which is interpreted as being due to $B \to J/\psi + X$. Outside the J/ψ region, the branching ratio limits can be interpreted as limits for $B \to \phi^0 X$. From eq. (3.50) we see that in the range between 500 MeV and 1 GeV a Higgs boson can be ruled out if the $\mu^+\mu^-$ branching ratio is larger than 2×10^{-2}. From the discussion of the $\phi^0 \to \mu^+\mu^-$ branching ratio given above, we see that some fraction of the lower part of this Higgs mass range may be ruled out. Clearly, a definitive statement depends sensitively on the magnitude of the $\pi\pi$ mode enhancement. For the mass range 1 GeV$< m_{\phi^0} < 2m_\tau$, the above limits of eq. (3.50) combined with eq. (3.48) imply that a Higgs is ruled out so long as $BR(\phi^0 \to \mu^+\mu^-) > 2.5 \times 10^{-3}$ (we exclude the J/ψ region). We have seen that the $\mu^+\mu^-$ branching ratio is larger than this value throughout this mass region, except (perhaps) near 1 GeV, where $\pi\pi$ resonant effects are large. (Even in the J/ψ region the data is close to ruling out a Standard Model Higgs.)

As pointed out in ref. 34, complementary data on $BR(B \to \mu^+\mu^- X)$, applicable over some of this same mass range, is available from the ARGUS collaboration [71], as a by-product of their measurement of $B \to \psi X$ at the $\Upsilon(4S)$. Their overall sample contained $\gtrsim 170,000$ B's. In selecting their e^+e^- and $\mu^+\mu^-$ events, they impose cuts of $p(\ell^+\ell^-) < 2$ GeV and $p(\ell) > 0.9$ GeV as well as an acoplanarity cut. The efficiency for observing $B \to \psi(\to \ell^+\ell^-)X$ with these cuts is of order 50%. It is not possible to be certain how this efficiency varies with $\ell^+\ell^-$ invariant mass, but it is clear that as $M(\ell^+\ell^-)$ decreases, $p(\ell^+\ell^-)$ will increase on average and the first of the above cuts will eliminate more and more events. Since the B's are produced pretty much at rest, if the spectator system X is treated as massless then once $M(\ell^+\ell^-) \lesssim$ 2.5 GeV we would find $p(\ell^+\ell^-) > 2$ GeV. However, in the observed $B \to \psi X$ decays, the momentum spectrum of the ψ is quite soft, implying that the spectator system X is fairly massive on average. Indeed, their plotted $M(\ell^+\ell^-)$ spectra extend all the way down to $M(\ell^+\ell^-) = 1.5$ GeV. Thus, it may not be unreasonable to assume that their acceptance for $B \to \phi^0(\to \mu^+\mu^-)X$ events remains substantial (say > 0.20) over the entire plotted mass range, from 1.5 GeV up to the point where phase space suppression becomes severe. This upper boundary is also quite sensitive to the average mass of the X system produced in $B \to \phi^0 X$ decays, and also cannot be determined with precision. However, it seems apparent from the mass spectra that it probably extends up to the upper limit of interest here, namely $2m_\tau$. For any $M(\ell^+\ell^-)$ value with efficiency greater than 20%, and using eq. (3.48), we estimate that the 170,000 B's would produce $\gtrsim 10^4 \times BR(\phi^0 \to \mu^+\mu^-)$ events in one of the experimental 50 MeV wide bins of the $\mu^+\mu^-$ spectrum. Outside the J/ψ region the experimental distributions contain at most 10 to 15 events in each such bin. Since resonant enhancements are unlikely to yield $BR(\phi^0 \to \mu^+\mu^-) < 10^{-3}$ in the mass region of interest, we conclude that the ARGUS experiment does indeed rule out a significant, but imprecise, range of m_{ϕ^0} masses between about 1.5 GeV and $2m_\tau$. We do not feel that the lower boundary of this mass range can be stated with the type of precision given in ref. 34.

For $m_{\phi^0} > 2m_\tau, 2m_c$, the dominant decays for the ϕ^0 are $\phi^0 \to \tau^+\tau^-, c\bar{c}$, for which there are no reliable limits. In summary, the most important B inclusive decay limits are as follows.

1. There are no constraints for $m_{\phi^0} \lesssim 70$ MeV.
2. Higgs masses in the range 70 MeV $< m_{\phi^0} \lesssim 2m_\mu$ are ruled out by the MARK II limit.
3. Higgs masses in the range $2m_\mu < m_{\phi^0} < 2m_\pi$ are ruled out by the TASSO limit. This conclusion requires only a 5% efficiency for a Higgs decay event, relative to the $\mu^+\mu^-$ acceptance assumed in their analysis.
4. It may be possible to rule out Higgs masses between 500 MeV and about 900 MeV if the $\phi^0 \to \pi\pi$ enhancement does not unduly suppress the $\phi^0 \to \mu^+\mu^-$ branching ratio. Higgs masses in the vicinity of the $f_0(975)$ scalar meson resonance cannot be ruled out due to the strong final state $\pi\pi$ interaction.
5. Higgs masses between about 1.2 GeV and 3 GeV are ruled out based on experimental results from CLEO and ARGUS.
6. For $3.0 < m_{\phi^0} < 3.2$ GeV the decay $B \to \psi X$ is an important Standard Model background for Higgs detection. Thus, even the recent CLEO data cannot definitively rule out a Higgs in this mass range.

7. Higgs masses between 3.2 GeV and $2m_\tau$ are ruled out by the ARGUS and CLEO data.

8. There are no constraints for $m_{\phi^0} > 2m_\tau$.

Finally, we remind the reader that all the above conclusions apply only for three generations as implicitly assumed in obtaining eq. (3.48), and for $m_t \approx m_W$ as required to have a very light Standard Model Higgs.

Let us now consider the exclusive channel $B \to \phi^0 K$, which is quite analogous to $K \to \pi \phi^0$ with the exception that in the B decay case the contribution from the two-quark transition operator should be much larger than that from the four-quark operator. The width for the B decay case [65,66,34] is best computed using the ratio

$$r_K \equiv \frac{BR(B \to \phi^0 K)}{BR(B \to \phi^0 X)} \tag{3.51}$$

and the result for the inclusive branching ratio of eq. (3.48). The ratio r_K has been computed in a non-relativistic approximation in ref. 66, but, as pointed out in ref. 34, this approximation cannot really be justified since there is considerable sensitivity to the component of the K-meson wave function with large relative quark momentum. For small m_{ϕ^0} values, ref. 66 estimates that $r_K \sim 0.08$, rising to roughly $r_K \sim 0.2$ by $m_{\phi^0} \sim 3$ GeV. Reference 34 obtains similar results at small m_{ϕ^0} for one choice of wave function, but also finds that r_K could be as much as a factor of 7 to 10 smaller for reasonable alternative choices for the wave function parameters.

As we have seen, in the mass range of interest (below $m_B - m_K$), the ϕ^0 decays primarily to $\mu^+\mu^-$, $\pi\pi$, and to KK and multi-meson final states for larger m_{ϕ^0}. Published experimental upper limits are available from CLEO [72] $BR(B^0 \to K^0\mu^+\mu^-) < 4.5 \times 10^{-4}$ and $BR(B^+ \to K^+\mu^+\mu^-) < 3.2 \times 10^{-4}$. Given the uncertainties due to experimental acceptance and theoretical uncertainties, these experimental limits are not strong enough to rule out any range of m_{ϕ^0}.

However, more stringent limits from CLEO have recently appeared [70]. For charged B decays (summed over charge conjugate final states) they find

(i) $BR(B \to \mu^+\mu^- K) \lesssim 4 - 6 \times 10^{-5}$ for $2m_\mu < M_{\mu^+\mu^-} < 3.6$ GeV;

(ii) $BR(B \to \pi^+\pi^- K) \lesssim 2.5 - 20 \times 10^{-5}$ for $2m_\pi < M_{\pi^+\pi^-} < 3.6$ GeV; and

(iii) $BR(B \to K^+K^- K) \lesssim 2.5 - 20 \times 10^{-5}$ for $2m_K < M_{K^+K^-} < 3.6$ GeV.

In specific mass regions, the limits are often near the minimum of the stated ranges. However, in numerous mass bins, the limits peak sharply at values near the maximum limits indicated above. These mass bins correspond to bins where candidate events have been observed. These results can be converted to limits on $BR(B \to \phi^0 K)$ as follows. For $2m_\mu < m_{\phi^0} < 1$ GeV we know that $BR(\phi^0 \to \mu^+\mu^-) + BR(\phi^0 \to \pi^+\pi^-) + BR(\phi^0 \to \pi^0\pi^0) = 1$; using (i) and (ii) (and isospin conservation) gives

$$BR(B^- \rightarrow \phi^0 K^-) \lesssim 10^{-4}. \qquad (3.52)$$

Using eq. (3.48) and the conservative $r_K = 0.01$ choice yields a result which is about 36 times larger, and ϕ^0 masses in the above range would be ruled out. For m_{ϕ^0} above $2m_K$ ref. 70 uses the experimentally measured ratio of $\sigma(e^+e^- \rightarrow K^+K^-)/\sigma(e^+e^- \rightarrow \text{all})$ to estimate $BR(\phi^0 \rightarrow K^+K^-)$ in order to convert their limit (iii) above to a limit on $B \rightarrow \phi^0 K$. Using these estimates, and assuming that $r_K \gtrsim 0.01$, we find that the above limit translates to $BR(B \rightarrow \phi^0 X) \lesssim 0.06$ for $2m_K < m_{\phi^0} < 2$ GeV, $\lesssim 0.3$ for 2 $GeV < m_{\phi^0} <$ 2.5 GeV— thereby eliminating $2m_K < m_{\phi^0} < 2.5$ GeV. The less conservative exclusive to inclusive ratio assumed in ref. 70, based on refs. 66 and 65, allows exclusion of Higgs masses in the full range $2m_K < m_{\phi^0} < 2m_\tau$. A slightly different approach is to use eq. (3.48) to convert the exclusive limits (i)–(iii) into the following

$$BR(\phi^0 \rightarrow \mu^+\mu^-) < 1 - 3 \times 10^{-4}/r_K$$
$$BR(\phi^0 \rightarrow \pi^+\pi^-) < 1 - 6 \times 10^{-4}/r_K \qquad (3.53)$$
$$BR(\phi^0 \rightarrow K^+K^-) < 1 - 6 \times 10^{-4}/r_K ,$$

in the relevant mass ranges. Clearly multi-body final states would have to be the dominant component of ϕ^0 decays in order that *all* of these two-body modes have branching ratio below the $\sim$1–6% upper limit obtained for $r_K =$ .01. These numbers can be compared to the estimates from ref. 70 for the K^+K^- component of ϕ^0 decays which, for example, falls below 3% only for $m_{\phi^0} \gtrsim 2$ GeV. A close examination of the limits as a function of KK mass would imply that $2m_K \lesssim m_{\phi^0} \lesssim 2.5$ GeV is excluded.

Finally, it is worth remarking on the influence of the ϕ^0 on the rare $B_s \rightarrow \tau^+\tau^-$ decays. In ref. 73 it is noted that the ϕ^0 contributes as an s-channel decay diagram, using the $b\bar{s}\phi^0$ coupling intrinsic to the $b \rightarrow s\phi^0$ decays. And, of course, the ϕ^0 couples significantly to $\tau^+\tau^-$. Obviously, one obtains a very large effect if the B_s and ϕ^0 masses are degenerate, but even for m_{B_s} away from the resonance pole, there is sensitivity. For instance, for $m_t = 200$ GeV, 10^6 B_s's would probe m_{ϕ^0} between 3 and 7 GeV, but if $m_t = 100$ GeV then sensitivity is confined to a very narrow range of m_{ϕ^0} near the B_s mass. The decay mode $B_s \rightarrow \mu^+\mu^-$ would, of course, be cleaner, but the $\phi^0\mu^+\mu^-$ coupling is much smaller, and, for instance, it takes 10^8 B_s's to probe m_{ϕ^0} between 4 and 6.5 GeV for $m_t = 200$ GeV.

Let us now summarize the existing limits on the ϕ^0 from rare B decays. Altogether, we see that B decays come close to ruling out m_{ϕ^0} in the entire range 70 MeV $< m_{\phi^0} < 2m_\tau$. Data for $B \rightarrow e^+e^-X$ rules out 70 MeV $< m_{\phi^0} < 2m_\mu$, and data for $B \rightarrow \mu^+\mu^-X$ can reliably rule out $2m_\mu < m_{\phi^0} < 2m_\pi$ and 1.5 GeV $\lesssim m_{\phi^0} < 2m_\tau$, excluding a small window around $m_{\phi^0} \approx m_\psi$. Exclusive charmless B decays, discussed above, can rule out $2m_\mu <$

$m_{\phi^0} \lesssim 2.5$ GeV, if the model-dependent parameter r_K is larger than 0.01. While the region $m_{\phi^0} > 2m_\tau$ up to $m_B - m_K$ is kinematically accessible via rare B decays, it has been shown in ref. 74 that limits in this region are only possible if one adopts optimistic theoretical assumptions and/or quark masses larger than the $m_t \simeq 79$ GeV value required for a light SM ϕ^0.

Higgs Bosons in Upsilon Decays

As first suggested by Wilczek [75], the Higgs boson can be searched for in the radiative decays of heavy vector mesons, V, such as the ψ and Υ. The decay rate for $V \to \phi^0 \gamma$ has been given in chapter 2 and to lowest order is,

$$R_0 \equiv \frac{\Gamma_0(V(1^{--}) \to \phi^0\gamma)}{\Gamma_0(V(1^{--}) \to \mu^+\mu^-)} = \frac{G_F m_Q^2}{\sqrt{2}\pi\alpha} \left(1 - \frac{m_{\phi^0}^2}{M^2} \right) \qquad (3.54)$$

for a 1^{--} bound state. m_Q is the heavy quark mass, $M \simeq 2m_Q$ is the mass of the bound state.

The QCD radiative corrections to the rate for $V \to \phi^0 \gamma$ have been calculated in refs. 76 and 77, and are found to be large

$$\Gamma(V \to \phi^0\gamma) = \Gamma_0(V \to \phi^0\gamma) \left[1 - \frac{\alpha_s C_F}{\pi} a_H(z) + \mathcal{O}(\alpha_s^2) \right] , \qquad (3.55)$$

where $C_F = 4/3$ and $z = 1 - m_{\phi^0}^2/M^2$. For $m_{\phi^0} \ll M$,

$$a_H(1) = 7 + 6\ln(2) - \frac{\pi^2}{8} \qquad (3.56)$$

and for $m_{\phi^0} \simeq M$,

$$a_H(0) = 1 + \frac{4}{3}\pi \frac{1}{\sqrt{1 - m_{\phi^0}^2/M^2}} . \qquad (3.57)$$

For completeness, we also display the one-loop QCD corrected decay rate for $V \to A^0\gamma$, where A^0 is a pseudoscalar Higgs boson which arises in non-minimal models [77,78]

$$\Gamma(V \to A^0\gamma) = \Gamma_0(V \to A^0\gamma) \left[1 - \frac{\alpha_s C_F}{\pi} a_P(z) + \mathcal{O}(\alpha_s^2) \right] . \qquad (3.58)$$

A plot of the functions $a_H(z)$ and $a_P(z)$ is shown in fig. 3.13. For $m_{\phi^0} \ll M$, the radiative corrections result in an 84% decrease in the absolute rate for the scalar case and a 56% decrease for a pseudoscalar for $\alpha_s = 0.2$. The large size of these corrections indicates that higher orders in perturbation theory may be significant. We can also determine the one-loop QCD corrections to the ratio $R \equiv \Gamma(V \to \phi^0\gamma)/\Gamma(V \to \mu^+\mu^-)$ by including corrections to the

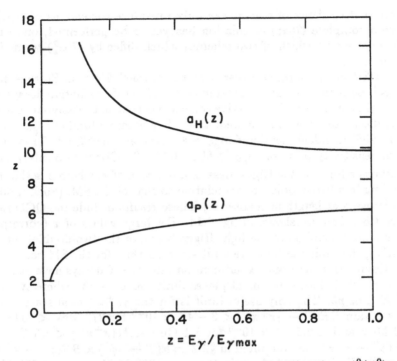

Figure 3.13 Radiative corrections to the rate for $V(1^{--}) \rightarrow \phi^0(A^0)\gamma$, where $\Gamma[V \rightarrow \phi^0(A^0)\gamma] = \Gamma_0[V \rightarrow \phi^0(A^0)\gamma][1 - (\alpha_s C_F)/\pi\; a_{H(P)}(0)]$, from ref. 77.

denominator. One finds [79]

$$\Gamma(V \rightarrow \mu^+\mu^-) = \Gamma_0(V \rightarrow \mu^+\mu^-)\left[1 - \frac{4\alpha_s C_F}{\pi}\right], \qquad (3.59)$$

which is a 34% reduction for $\alpha_s = 0.2$. Therefore, the one-loop QCD corrections to the ratio R will be somewhat less than the large radiative corrections to the numerator alone. Unfortunately, there is some ambiguity as to how one determines the numerical effect of the QCD corrections to R. The standard procedure is to expand out the denominator, thereby obtaining a series in powers of α_s for R. We find that the order α_s QCD corrections reduce R_0 by 50% [eq. (3.54)] in the scalar case and by 22% in the pseudoscalar case (for $\alpha_s = 0.2$). On the other hand, if we simply compute $(R_0 - R)/R_0$, we obtain reductions of 76% and 33%, respectively. The numerical difference between these two approaches is another indication of the large uncertainty due to order α_s^2 effects not yet computed. A related ambiguity also arises over the question of scheme dependence of these corrections [80]. In fact, eqs. (3.55) and (3.58) are manifestly independent of the subtraction scheme;

an alternative choice of scheme can only introduce corrections of $\mathcal{O}(\alpha_s^2)$. Because a complete $\mathcal{O}(\alpha_s^2)$ calculation has yet to be performed, one cannot *a priori* determine which of two schemes which differ by $\mathcal{O}(\alpha_s^2)$ terms is more appropriate.

CUSB has performed a search at the Cornell Electron Storage Ring for Higgs bosons in $\sim 8 \times 10^5 \Upsilon(1S)$ and $\sim 6 \times 10^5 \Upsilon(3S)$ radiative decays [81,82]. When they include QCD radiative corrections (but not relativistic corrections to be discussed below) in the prediction, they obtain a limit of $m_{\phi^0} > 5.4$ GeV at the 90% confidence level and $m_{\phi^0} > 4.8$ GeV at the 95% CL. They currently claim sensitivity only for $m_{\phi^0} \gtrsim 600$ MeV [83]. CUSB also interprets their results as a limit on the Higgs mass as a function of x, where x is the Higgs-$b\bar{b}$ coupling in a two doublet model relative to that of the SM. (Such models will be discussed at length in chapter 4.) These results include the QCD radiative corrections and are shown in fig. 3.14. For large values of x (corresponding to enhanced coupling of the light Higgs boson of the two-doublet model to b quarks), the limit can be quite a bit stronger than for the SM case.

Results from a second experiment on radiative Υ decays have also recently appeared [84]. This experiment places limits on $\Upsilon \to \gamma X$, where $X \to \pi^+\pi^-$, K^+K^-, or $p\bar{p}$. The only useful limit is for the $\gamma\pi^+\pi^-$ final state, for which they obtain $BR(\Upsilon \to \gamma\pi^+\pi^-) < 3 - 4.5 \times 10^{-5}$ for 270 MeV $< M(\pi^+\pi^-) \lesssim$ 670 MeV and $< 2.5 - 3 \times 10^{-5}$ for 1.1 GeV $< M(\pi^+\pi^-) < 3.5$ GeV. Using $BR(\phi^0 \to \pi^+\pi^-)$ as computed in §2.2, $BR(\Upsilon \to \gamma\phi^0) \times BR(\phi^0 \to \pi^+\pi^-)$ exceeds the limits quoted above in the range 290 MeV $< m_{\phi^0} <$ 570 MeV; $\pi\pi$ mode enhancements allow one to exclude an even larger range—270 MeV $< m_{\phi^0} <$ 670 MeV. Above $m_{\phi^0} = 2m_K$, the $\pi\pi$ modes will be rapidly overshadowed by the $K\bar{K}$ modes. The experimental limits from ref. 84 are somewhat weaker for these modes, and precise estimates for ϕ^0 decays to $K\bar{K}$ in the region $1 < M(K^+K^-) < 3.5$ GeV (where the experimental data exists) are not currently available. Thus, restrictions on m_{ϕ^0} from the data of ref. 84 in the region $m_{\phi^0} > 1$ GeV are not currently possible. In the region from $\sim$ 670 MeV to ~ 1 GeV there is a large ρ resonance contribution to the $\pi\pi$ final state that yields branching ratios far above those expected from ϕ^0 contributions.

Unfortunately, the above conclusions are beset with additional theoretical uncertainties. In addition to the QCD radiative corrections discussed above, there may also be large bound state and relativistic corrections. First, we consider bound state corrections, which were first examined in ref. 85. For $m_{\phi^0} \ll M_\Upsilon$, the photon emitted in the decay $\Upsilon \to \phi^0\gamma$ is hard and the perturbative calculation is valid. If, however, $m_{\phi^0} \sim M_\Upsilon$, then the photon is soft and bound state effects must be considered. In this case, after emitting the soft γ, the propagating quark is still strongly affected by the potential. A key ingredient to the calculation of these bound state effects is that of gauge invariance: the same $b\bar{b}$ interaction must be used to calculate both the $\Upsilon b\bar{b}$ and $\phi^0 b\bar{b}$ vertices. (This may be the source of the problem with the calculation of ref. 86, whose results disagree with all other published calculations [87].) The importance of gauge invariance has been recently stressed in a detailed

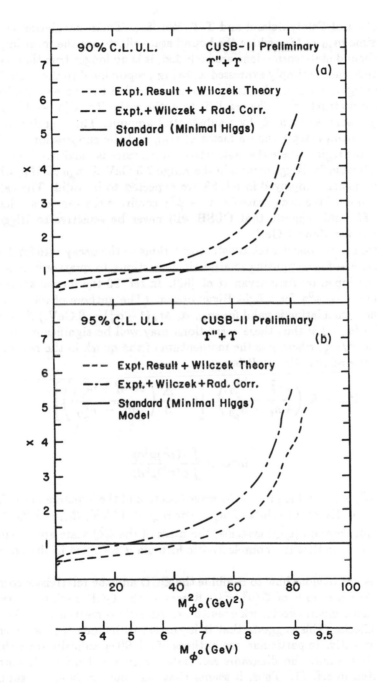

Figure 3.14 Limits on the Higgs mass at the 90% and 95% confidence levels from CUSB, plotted as a function of x, the ratio of the Higgs-$b\bar{b}$ coupling in the two Higgs doublet model to the $\phi^0 b\bar{b}$ coupling.

calculation of Faldt, Osland and T.T. Wu [88]. Their calculation is valid in the regime $m_{\phi^0} \sim M_{\Upsilon}$, where the bound state effects are the most important. When bound state corrections are included, it is no longer true that the decay amplitude can be simply expressed as being proportional to the wave function at the origin. Further, the decay rate is controlled by two amplitudes which interfere destructively. The result is that the ratio $R \equiv \Gamma(\Upsilon \to \phi^0 \gamma)/\Gamma(\Upsilon \to \mu^+\mu^-)$ is smaller than R_0 by a substantial amount. The overall suppression factor is always larger than a factor of two, and the suppression can be very large in the region where the destructive interference is maximal. These results are applicable for Higgs masses in the range 7.5 GeV $\lesssim m_{\phi^0} \lesssim M_{\Upsilon}$, where the approximations employed in ref. 88 are expected to be valid. The calculated suppression of the decay rate for $\Upsilon \to \phi^0 \gamma$ confirms the previous calculations of ref. 85, and suggests that CUSB will never be sensitive to Higgs boson masses much above 7 GeV.

Second, we consider relativistic corrections to the decay rate for $\Upsilon \to \phi^0 \gamma$. A formalism for computing relativistic corrections to quarkonium decay has been developed by Aznauryan *et al.* [89]. In ref. 90 the formalism is used to calculate $\Upsilon \to \phi^0 \gamma$ for a light Higgs boson. (The authors of ref. 90 estimate that their calculation is valid for $m_{\phi^0} \ll M_{\Upsilon}(1 - \alpha_s^2) \sim 8$ GeV.) Note that for the Υ, $v/c \sim 0.3$; thus these corrections may well be significant. To leading order in p^2/m_b^2, where p is the momentum of the quark in the rest system of the $b\bar{b}$, the relativistic result is

$$R_{\text{rel}} = R_0 \left(\frac{M_{\Upsilon}^2 - m_{\phi^0}^2}{4m_b^2 - m_{\phi^0}^2} \right)^2 \left[1 - \frac{\Delta}{3} \left(\frac{36m_b^2 + m_{\phi^0}^2}{4m_b^2 - m_{\phi^0}^2} \right) \right], \tag{3.60}$$

where

$$m_b^2 \Delta \equiv \frac{\int \phi(p^2) p^4 dp}{\int \phi(p^2) p^2 dp}, \tag{3.61}$$

and $\phi(p^2)$ is the radial part of the wave function of the b-quarks in the Υ. Note that $R_{\text{rel}} < R_0$ for all values of m_{ϕ^0}. For $m_{\phi^0} < 4$ GeV, $R_{\text{rel}} \sim \frac{1}{2} R_0$. For very heavy quarkonium ($Q\bar{Q}$) systems, the mass of the $Q\bar{Q}$ state approaches $2m_Q$ and $\Delta \to 0$, so that the nonrelativistic formula is regained in the appropriate limit.

It is not obvious how to combine the QCD and the relativistic corrections since they are both of $\mathcal{O}(v^2/c^2)$. However, the QCD radiative corrections reflect hard gluon effects, whereas the relativistic corrections are due to soft gluon effects. This suggests that it is correct to include the two corrections independently. In particular, the authors of ref. 90 specifically state that their calculation omits the diagrams associated with the hard QCD corrections computed in ref. 77. Thus, it seems that one must include the suppression from relativistic effects on top of the QCD corrected prediction. It would then follow that no limit on m_{ϕ^0} can be found from present CUSB data. Clearly, additional theoretical work is needed before a firm limit on the Higgs boson

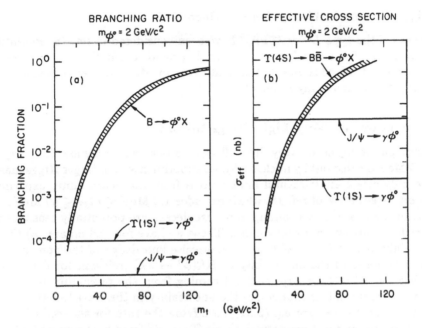

Figure 3.15 A comparison of the rate for $B \to \phi^0 X$ with that for $\Upsilon \to \phi^0 \gamma$ for a fixed $m_{\phi^0} = 2$ GeV, from ref. 66. (QCD, but not relativistic, corrections have been included in the Υ decay rate.)

mass can be extracted from the decay $\Upsilon \to \phi^0 \gamma$. In particular, the relativistic corrections of Aznauryan *et al.* [90] need to be confirmed by an independent calculation.

While the corrections discussed above are clearly the most significant ones, there may also be non-negligible corrections from the electroweak sector. The subset of such corrections coming from the Higgs sector itself have been computed in ref. 91. For a general $Q\overline{Q}\ {}^3S_1$ state, the corrections are $\lesssim 2\%$ for $m_{\phi^0}/m_Q \gtrsim 0.5$, reaching 6% for $m_{\phi^0} \lesssim 0.1 m_Q$. If one were in a multi-doublet model with enhanced couplings of a neutral Higgs to the $Q\overline{Q}$, then these corrections could be substantially larger.

To summarize, limits obtained from Υ decays are sensitive to theoretical uncertainties, including those associated with higher order QCD corrections and relativistic corrections. With our present understanding of the nature of these corrections, no limit can be ascertained from current data. It is important to continue to refine the experimental searches, both in Υ decays and in B meson decays. In particular, for top masses larger than about 40 GeV, the rate for $B \to \phi^0 X$ is certainly larger than that for $\Upsilon \to \phi^0 \gamma$ (see fig. 3.15) and so both search techniques are important.

Higgs Bosons in Heavy Lepton Decays

Decays of the type $L^+ \to \ell^+ \nu \bar{\nu} \phi^0$ have been considered in ref. 48. Potentially both $L = \mu$, $\ell = e$ and $L = \tau$, $\ell = \mu$ could provide useful limits. However, the discussion of ref. 48 shows that current experimental limits are not sufficiently strong to eliminate any range of m_{ϕ^0}.

Conclusions on Very Light Higgs Bosons

From our analysis of very light Higgs bosons, we conclude that $m_{\phi^0} <$ 11.5 MeV is ruled out by nuclear physics experiments. For larger Higgs masses, the only direct experimental limit derives from the beam dump experiment on $eN \to ee^+e^- N$ of ref. 18, which excludes 1.2 MeV $\leq m_{\phi^0} \leq$ 50 MeV. The bounds on the ϕ^0 mass coming from rare decays are potentially sensitive to a variety of theoretical uncertainties. Those we have focused on are: a) the uncertainty in $BR(K \to \pi \phi^0)$ due to imprecise knowledge of the non-spectator contributions; b) the uncertainty in $BR(\phi^0 \to \pi\pi)$ [relevant for determining $BR(\phi^0 \to \mu^+\mu^-)$ for $m_{\phi^0} > 2m_\pi$] coming from the possibility of resonant enhancements in this channel; c) the uncertainty in the wave function dependent parameter r_K [see eq. (3.51)] specifying the rate for the exclusive $B \to K\phi^0$ decays; and d) the uncertainties in $\Upsilon \to \gamma\phi^0$ from higher order QCD corrections and relativistic wave function corrections. Below we summarize the Higgs mass limits that we have discussed. In dealing with the uncertainties described above, we shall proceed as follows. We assume that the imprecisely known parameters (such as $m_t^2 V_{td}^* V_{ts}$ and the A parameter) are not fine-tuned in such a way as to make the $K \to \pi\phi^0$ amplitude artificially small. We adopt a conservatively small choice for r_K. We assume that resonant enhancement of the $\phi^0 \to \pi\pi$ decays is smaller than a factor of 10 in the mass regions of interest. We then have the following results

1. 10 MeV $< m_{\phi^0} \lesssim$ 100 MeV is ruled out by SINDRUM data on $\pi^+ \to e^+ \nu e^+ e^-$;

2. $\sim$ 20 MeV $< m_{\phi^0} \lesssim 2m_\mu$ is excluded by preliminary NA-31 results on $K_L^0 \to \pi^0 e^+ e^-$, with little sensitivity to theoretical uncertainties;

3. $m_{\phi^0} \lesssim$ 55 MeV is excluded by $K^+ \to \pi^+ +$ nothing and $K^+ \to \pi^+ + \gamma\gamma$ limits from refs. 57 and 56, respectively;

4. $\sim$ 60 MeV $< m_{\phi^0} \lesssim$ 100 MeV is excluded by the $K^+ \to \pi^+ e^+ e^-$ decay experiment of ref. 26;

5. $\sim$ 120 MeV $\lesssim m_{\phi^0} \lesssim 2m_\mu$ is strongly excluded by the $K^+ \to \pi^+ e^+ e^-$ decay results of experiment BNL-777 [27];

6. $2m_\mu < m_{\phi^0} <$ 320 MeV is excluded by measurements of $K \to \pi\mu^+\mu^-$, in particular by the analysis of BNL-787 in ref. 29, unless $BR(\phi^0 \to \mu^+\mu^-)$ is strongly suppressed due to a large enhancement of the $\phi^0 \to \pi\pi$ decay mode;

7. 10 MeV $< m_{\phi^0} \lesssim$ 100 MeV is excluded by limits on $K^+ \to \pi^+ + X$;

8. 70 MeV $< m_{\phi^0} < 2m_\mu$ is ruled out by MARK II data on $B \to e^+e^- X$;

9. $2m_\mu < m_{\phi^0} < 2m_\pi$ is ruled out by TASSO data on $B \to \mu^+\mu^- X$;

10. $2m_\pi < m_{\phi^0} < 800$ MeV is ruled out by TASSO data if $BR(\phi^0 \to \mu^+\mu^-) \gtrsim 0.1$, assuming that the acceptance for Higgs decay events is greater than 60%.

11. 500 MeV $\lesssim m_{\phi^0} \lesssim 900$ MeV is ruled out by CLEO data on $B \to \mu^+\mu^- X$ unless $BR(\phi^0 \to \mu^+\mu^-)$ is strongly suppressed;

12. 1.2 GeV $\lesssim m_{\phi^0} \lesssim 3$ GeV and 3.2 GeV $< m_{\phi^0} < 2m_\tau$ are ruled out by the same CLEO data;

13. 1.5 GeV $< m_{\phi^0} < 2m_\tau$, with a hole near $m_{\phi^0} \sim m_\psi$, is ruled out by ARGUS data as a by-product of their $B \to \psi X$ measurement, for any m_{ϕ^0} such that $BR(\phi^0 \to \mu^+\mu^-) > 10^{-3}$;

14. $2m_\mu < m_{\phi^0} < 2.5$ GeV is ruled out by $B \to K\mu^+\mu^-, \pi^+\pi^-, K^+K^-$ CLEO results;

15. no limit from $\Upsilon \to \phi^0\gamma$ can be deduced at this time, due to substantial theoretical uncertainties in the QCD and relativistic corrections to the non-relativistic prediction.

Modifications to the above statements for different theoretical assumptions have been given in the preceding sections. However, our general conclusion is that there is no mass region below about $2m_\tau$ where a light Standard Model Higgs boson might still be allowed. This conclusion should be progressively sharpened as the Brookhaven experiments on rare K decays continue to accumulate data, provided they analyze their data with a possible Higgs boson signal and its possible backgrounds in mind. Also, analysis of the E-731 experimental results [52] as a direct Higgs search may provide results complementary to those available in preliminary form from NA-31. Extension of limits on m_{ϕ^0} to the mass region above $2m_\tau$ may possibly come from Υ decays, but substantial theoretical uncertainties prohibit such an extension at this time. Extension of limits from rare B decays to the region $m_{\phi^0} \gtrsim 2m_\tau$, while kinematically possible, requires having confidence in r_K values larger than 0.01 and/or top quark masses larger than about 80 GeV.

The extent to which the above conclusions apply also to very light non-standard Higgs bosons, such as that which is always present in a low energy supersymmetric theory, is model dependent. Various modifications are possible and will be discussed in more detail in chapter 4.

Finally, we note for the reader's convenience that direct production of a very light Higgs boson in e^+e^- or hadron collisions will be covered in the next section at the same time that we discuss the m_{ϕ^0} region above 5 GeV.

3.2   Light Higgs Boson ($5 \lesssim m_{\phi^0} \lesssim 85$ GeV)

p$\bar{\text{p}}$, pp, and ep Colliders

In discussing the possible discovery of Higgs bosons with mass below 85 GeV, let us first quickly mention the hadron–hadron and ep colliders which are presently (or about to become) available. At a hadron–hadron collider, the most relevant process for Higgs production (in the mass range considered in this section) is via two-gluon annihilation [92]. We will have much to say about this process in later sections. The inclusive Higgs boson cross section due to this mechanism was given in eq. (2.30). For now, we present in fig. 3.16 the cross section for inclusive Higgs production in $p\bar{p}$ annihilation, for Higgs masses below 200 GeV. The produced Higgs decays dominantly into hadronic jets (most probably into $b\bar{b}$ pairs). We imagine that data samples of 15 pb^{-1} will be accumulated at the Tevatron in the next 2 to 3 years. The problem here is twofold. First is the rather small number of total Higgs bosons produced. Second, and more problematical, is the difficulty in isolating a Higgs signal. The QCD background due to continuum $b\bar{b}$ pairs is so large as to make the detection of the Higgs boson in its dominant decay mode impossible. One might consider looking for rarer Higgs boson decays [see the branching ratios of fig. 2.5]. However, now one quickly sees that the paucity of initial Higgs bosons in the original sample (which would survive the necessary cuts required to isolate a signal) makes such a proposal impractical. Thus, it is likely that Higgs boson detection at a hadron collider is possible only if the integrated luminosities are extremely large and the corresponding energies are also large (in order to guarantee a sufficiently large cross section). Even under such conditions, it is a challenge to experimentally isolate an inclusively produced Higgs boson whose dominant decays are into quark pairs. This will be discussed in great detail in §3.4.

Thus, it is worthwhile considering Higgs production mechanisms at hadron colliders which provide some form of associated trigger. One example is $W^* \rightarrow W\phi^0$. The leptonic decay of the W would provide a trigger and the ϕ^0 would be very likely to decay to $b\bar{b}$ allowing possible discrimination against backgrounds coming from $W+$ jets events via vertex tagging the b's. This technique might prove useful at the SSC, see §3.4, but yields very few events at the Tevatron and $Sp\bar{p}S$.

Of course, if the Higgs is light enough, one could also try to make use of real W (or Z) decays followed by, for example, $W \rightarrow \phi^0 W^* \rightarrow \phi^0 f\bar{f}'$. These decays were considered in §2.2. If we take cross sections for W production at the $Sp\bar{p}S$ ($\sqrt{s} = 0.63$ TeV) and the Tevatron ($\sqrt{s} = 1.8$ TeV) to be roughly 6 nb and 23 nb,[‡] respectively, then including the branching ratio of fig. 2.14 yields the effective cross sections of fig. 3.17. The effective cross section from this source is clearly less than that from gg fusion. At the lowest mass plotted

---

[‡]These are obtained from the measured values of $\sigma(W) \times BR(W \rightarrow e\nu)$ [93] at these colliders by assuming $BR(W \rightarrow e\nu) = 0.1$.

Cross Sections for $p\bar{p} \to \phi^0 + X$

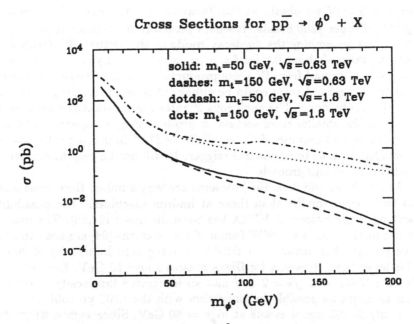

solid: m_t=50 GeV, $\sqrt{s}$=0.63 TeV
dashes: m_t=150 GeV, $\sqrt{s}$=0.63 TeV
dotdash: m_t=50 GeV, $\sqrt{s}$=1.8 TeV
dots: m_t=150 GeV, $\sqrt{s}$=1.8 TeV

Figure 3.16 Cross section for $p\bar{p} \to \phi^0 + X$ as a function of m_{ϕ^0}. Results are given at $\sqrt{s} = 630$ GeV and 1.8 TeV, for top quark masses of $m_t = 50$ and 150 GeV.

Higgs Cross Section from W Decays

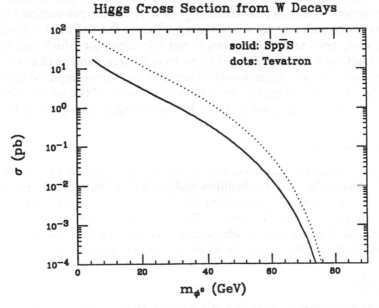

solid: $Sp\bar{p}S$
dots: Tevatron

Figure 3.17 Effective ϕ^0 production cross section from $W \to \phi^0 f\bar{f}'$ at the $Sp\bar{p}S$ and the Tevatron, summed over all $f\bar{f}'$ channels.

of $m_{\phi^0} = 5$ GeV we obtain at the Tevatron, with integrated luminosity of 15 pb^{-1}, of order 1200 events. It would probably be impossible to trigger on these unless we employ the $\ell\nu$ decay mode of the virtual W^* (with a total branching ratio of about 0.2 for $\ell = e, \mu$), which would yield 240 events. The principal background from $W+$ jets production would be large. However, the signal to background ratio could be enhanced if vertex tagging of the $\phi^0 \to b\bar{b}$ decays can be done. The analogous Z rates are substantially smaller, both because of the smaller cross section to produce Z as compared to W at the two colliders, and because of the small $Z^* \to \ell^+\ell^-$ branching ratio that must be included for a leptonic channel trigger. It will not be easy to extract these signals from the backgrounds.

As for the ep colliders, the problems are very similar. Here cross sections tend to be even smaller than those at hadron machines. The possibility of observing Higgs bosons at HERA has been discussed [94,95]. The main production mechanism is via WW fusion. Cross sections [94] are not expected to be very large. For instance, at HERA an integrated luminosity of 200 pb^{-1} yields fewer than 10 events for Higgs masses above 10 GeV. Even at a futuristic ep collider with $\sqrt{s} = 2$ TeV and an integrated luminosity of 1000 pb^{-1} (such as might be possible in connection with the LHC pp collider), one obtains only $\lesssim 400$ signal events at $m_{\phi^0} = 80$ GeV. Since such a Higgs decays to heavy fermions, the relevant background is $\gamma g \to Q\overline{Q}$. Studies indicate that the background level is overwhelming; see, for example, the summary of ref. 95.

Thus, at the current generation of hadron colliders and ep colliders some Higgs bosons would be produced if $m_{\phi^0} \lesssim m_Z$, but the production rates do not appear to be sufficient to isolate the signal given the anticipated backgrounds. Of course, one can imagine exceptional situations for which this conclusion might not hold. For example, if there were a large number of heavy families, the gg fusion cross section would be much larger than computed for the three-family case. Or, if there are extra W bosons, some of the 1-loop initiated rare decay mode branching ratios could be much larger than given in fig. 2.5.

e^+e^- Colliders

The most dramatic improvements on the Higgs mass limits will certainly be made at the e^+e^- collider facilities SLC and LEP [96]. If there exists a Higgs boson with mass below (roughly) 80–85 GeV, then it will be here that the Higgs boson will be discovered. There is one major unknown parameter, namely the top-quark mass, which may play an important role in the discovery of the Higgs. Current mass limits on m_t are summarized in §1.5.

$\Theta \to \phi^0\gamma$

We have already discussed Higgs production in quarkonium decay in the previous section. In the case of toponium decay, some obvious advantages become apparent. First, the rate is proportional to m_t^2 and is thus greatly enhanced compared with Υ decay. Second, the uncertainties due to relativistic corrections are far less severe than they are for Υ, and we are safe in neglecting them here. Finally, whereas the perturbative first-order QCD corrections to $\Gamma(\Upsilon \rightarrow \phi^0\gamma)$ reduce the tree-level result by about 85%, the corresponding reduction in $\Gamma(\Theta \rightarrow \phi^0\gamma)$ is only about 50%, due to the smaller value of the running coupling constant. Thus, we return to the Wilczek process, whose rate is given by

$$\frac{\Gamma(\Theta \rightarrow \phi^0\gamma)}{\Gamma(\Theta \rightarrow \gamma^* \rightarrow \mu^+\mu^-)} = \frac{m_\Theta^2}{8m_W^2 \sin^2\theta_W}\left(1 - \frac{m_{\phi^0}^2}{m_\Theta^2}\right), \qquad (3.62)$$

where Θ is the lowest lying $^3S_1$ $t\bar{t}$ ground state, and $m_\Theta \simeq 2m_t$. [In order to write down a formula which normalizes $\Gamma(\Theta \rightarrow \phi^0\gamma)$ to the *full* $\Gamma(\Theta \rightarrow \mu^+\mu^-)$, one must include the Z-contribution to the latter decay.] If we include the perturbative QCD correction to both numerator and denominator, the ratio of branching ratios is reduced by about 30%. (For a more detailed discussion of the numerics involved, see refs. 97-99.) In fig. 3.18, we exhibit the branching ratio for $\Theta \rightarrow \phi^0\gamma$. Higgs masses up to roughly $0.8m_\Theta$ should present no problem for detection. The possibility of discovering the Higgs boson in toponium decay at LEP has been described in great detail in ref. 100. We will summarize the main points here and refer the reader to the relevant literature. In ref. 100, toponium masses between 70 GeV and 110 GeV are considered. For the moment, let us remove from consideration the possibility that the Θ is roughly degenerate with the Z. This particular case will require special attention. From fig. 3.18, it is clear that we can expect $\phi^0\gamma$ branching ratios near 1%, except when m_Θ is in the vicinity of m_Z. In fact, for m_Θ near m_Z (but $|m_\Theta - m_Z| \gtrsim 2\Gamma_Z$, so that Θ–Z interference can be neglected), the cross section for Θ production increases, so that $\sigma(e^+e^- \rightarrow \Theta)BR(\Theta \rightarrow \phi^0\gamma)$ falls rather slowly as m_t increases in this regime. Nevertheless, detection of the Higgs boson from the decay of toponium whose mass is near m_Z presents additional problems due to an increase in the background associated with Z production. All of these issues are discussed at great length in ref. 100. Based on Monte-Carlo simulations presented there, the authors of ref. 100 conclude that running at LEP for one year at a luminosity of 10^{31} cm^{-2} sec^{-1} will allow a mass limit for the Higgs of about 0.7–$0.9m_\Theta$.

In principle, if $m_t > 55$ GeV, higher mass toponia may be of use for Higgs searches at LEP-II. In fact, two problems limit the utility of superheavy toponia. First, the rate for toponium production decreases with energy, so for toponia with masses above around 140 GeV, there is insufficient rate to observe the Higgs [100]. Second, for m_{ϕ^0} in the vicinity of m_Z (say, within 10 GeV), background due to the process $e^+e^- \rightarrow Z\gamma$ makes the signal-to-background too small for the Higgs to be detected. Thus, for toponia between 110 and 140 GeV, the largest Higgs mass which could be detected remains

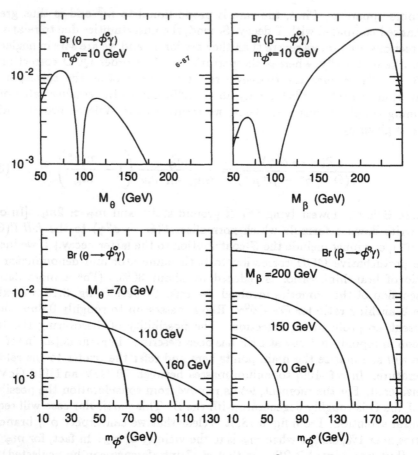

Figure 3.18 Branching ratios for an $n = 1$, $^3S_1$ quarkonium state to decay into a Higgs boson plus a photon. Θ denotes a $t\bar{t}$ bound state, and β denotes a $b'\bar{b}'$ bound state, where b' is a hypothetical fourth generation quark of charge $-e/3$. H denotes the Higgs boson of the Standard Model (ϕ^0). The above figure is taken from ref. 99, and is based on calculations which include $\mathcal{O}(\alpha_s)$ perturbative QCD corrections.

rather constant at around 85 GeV. For very large m_t ($m_t \gtrsim 125$ GeV), the t-quark decays too rapidly for toponium to form [101].

Finally, let us return to the case where $m_\Theta \simeq m_Z$. Two effects make this region problematical. First, due to Θ–Z mixing effects, the prediction of the rate for the production of $\phi^0\gamma$ depends on the beam energy resolution. Hall *et al.* [102] show that for a beam resolution of $W \lesssim \Gamma_Z$,

$$\frac{\sigma(e^+e^- \to \Theta, Z \to \phi^0\gamma)}{\sigma(e^+e^- \to Z \to \phi^0\gamma)} = 1 + \sqrt{\frac{\pi}{8}\frac{\Gamma_Z}{W}}\frac{\Gamma(\Theta \to \phi^0\gamma)}{\Gamma(Z \to \phi^0\gamma)}\frac{(m_\Theta^2 - m_Z^2)^2 + m_Z^2\Gamma^2}{m_Z^2\Gamma^2},$$

$$(3.63)$$

and for $W \gtrsim \Gamma_Z$,

$$\frac{\sigma(e^+e^- \to \Theta, Z \to \phi^0\gamma)}{\sigma(e^+e^- \to Z \to \phi^0\gamma)} = 1 + \frac{\Gamma(\Theta \to \phi^0\Gamma)}{\Gamma(Z \to \phi^0\gamma)}. \tag{3.64}$$

The decay $Z \to \phi^0\gamma$ will be discussed below. The numbers are such that the enhancement for the production of $\phi^0\gamma$ in the limit of large W due to Z–Θ mixing is only a factor of 1.8. For $W < \Gamma_Z$, the enhancement factor can be larger, $e.g.$ a factor of 29 for $W = 50$ MeV (which is not an unrealistic possibility). In the latter case, detection of the Higgs might be possible (it would correspond to about 6×10^{-5} of all Z decays). One possible troublesome background would be due to $Z \to q\bar{q}\gamma$; ref. 100 estimates that after cuts, the signal-to-background would be somewhat below 1. More analysis is needed to fully explore this small region of parameter space.

$Z \to \phi^0 Z^*$

It is entirely possible that no toponium factory will ever exist due to an inconveniently placed top-quark mass. Nevertheless, the e^+e^- collider facilities at SLC and LEP will be able to discover (or set significant mass limits) on the Higgs boson due to the $\phi^0 ZZ$ coupling in the theory. In particular, these colliders will be high statistics Z factories, and therefore will be sensitive to rare Z decays. The rare decays of interest to us here are the decays of the Z into final states which include the Higgs boson. The most important of these modes is $Z \to \phi^0 f\bar{f}$ which was discussed in §2.3. In fig. 3.19, we display the branching ratio for $Z \to \phi^0 + X$ for various choices X, based on eq. (2.114).

Experimentally, the Higgs boson can be detected without observing the specific final-state decay products of the Higgs boson. One simply measures the four-momenta of the outgoing fermions, and uses energy-momentum conservation to obtain the Higgs boson four-vector. In addition, the invariant mass distribution of the outgoing fermion pair ($M_{f\bar{f}}$) is strongly peaked at very large values of the invariant mass [103]; roughly, $M_{f\bar{f}} \simeq r(m_Z - m_{\phi^0})$, where r ranges from 0.96 at $m_{\phi^0} = 0.1 m_Z$ to 0.73 at $m_{\phi^0} = 0.9 m_Z$. This result is obtained from eq. (2.114) by using $M_{f\bar{f}}^2 = m_{\phi^0}^2 + m_Z^2(1 - x)$. Physically, this result is easily understood: as $M_{f\bar{f}}$ approaches its maximum value allowed by kinematics, the virtual Z boson is as close as possible to its mass shell.

At the SLC and LEP, the initial Higgs search will consist of looking for electron or muon pairs recoiling against the Higgs boson. The process $Z \to \phi^0 \ell^+\ell^-$ (where $\ell = e$ or μ) is expected to be rather clean. First, let us consider low-mass Higgs bosons, with $m_{\phi^0} < 10$ GeV. This range of mass has been the least studied, partly due to the assumption that this range of masses would be accessible in Υ decays. The dominant background is due to $e^+e^- \to \ell^+\ell^- q\bar{q}$. In ref. 104, it is claimed that an appropriate set of cuts

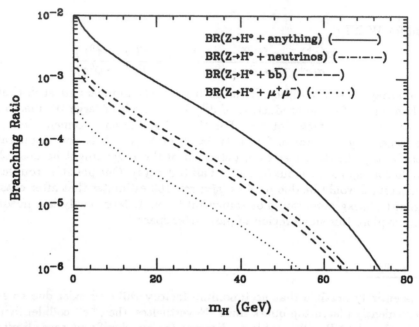

Figure 3.19 Branching ratios for Higgs production in Z decay are given, $BR(Z \rightarrow \phi^0 + X)$, for various possible final states X. The solid curve depicts X summed over all possible final state quark and lepton pairs (with $t\bar{t}$ omitted, assuming that $m_t > m_Z/2$), the dashed-dot curve depicts $X = \nu\bar{\nu}$ summed over three neutrino generations, the dashed curve depicts $X = b\bar{b}$, and the dotted curve depicts $X = \mu^+\mu^-$ (or any other charged lepton pair).

can be devised such that the background is negligible compared to the signal (at least for Higgs masses above 1 GeV). The discussion in ref. 104 is rather sketchy, and we believe that more work is needed to establish a *minimum* Higgs mass that can be detected by observing the recoiling $\ell^+\ell^-$ pair at SLC and LEP. In particular, the resolution with which one is able to measure the $\ell^+\ell^-$ invariant mass and energy becomes increasingly demanding below about 10 GeV. Certainly, it would be very valuable to be able to search for the ϕ^0 as far down in m_{ϕ^0} as possible; *e.g.*, limits in the region $2m_\mu$ and below would be free of theoretical ambiguities in interpretation, unlike the rare decays discussed in §3.1. For Higgs masses above 10 GeV, the analysis of ref. 97 demonstrates that there is no significant background to the $\phi^0\ell^+\ell^-$ signal. Thus, it is basically the branching ratio for $Z \rightarrow \phi^0\ell^+\ell^-$ which determines the upper limit for the Higgs mass which can be detected via this decay mode at SLC and LEP. In particular, ref. 97 concludes that one needs at least 2×10^5 Z's to observe Higgs bosons of mass 10 GeV (or less), 10^6 Z's to observe

Higgs bosons with masses up to 30 GeV, and 10^7 Z's to observe Higgs bosons with masses up to 50 GeV. These results are quoted for $\ell = e$; one can roughly double the statistics by also searching for $\ell = \mu$.

In order to have maximal sensitivity to Higgs masses below 10 GeV, one will have to examine the specific ϕ^0 decay products in $Z \rightarrow \phi^0 f\bar{f}$. Different Higgs mass ranges imply different signatures (as in the case of B decays discussed in §3.1), and a combination of various signatures will have to be examined to study the entire Higgs mass range below 10 GeV. To give one example, for $m_{\phi^0} \lesssim 800$ MeV, the decay mode $\phi^0 \rightarrow \mu^+\mu^-$ may have a significant branching ratio (see the discussion in §2.1) and would provide an excellent signature for a ϕ^0 produced in the process $Z \rightarrow \phi^0 Z^*(\rightarrow f\bar{f})$. The feasibility of this mode has been explored in ref. 105. For such light Higgs bosons, $BR(Z \rightarrow \phi^0 f\bar{f})/BR(Z \rightarrow f\bar{f}) \sim 10^{-2}$ and $BR(\phi^0 \rightarrow \mu^+\mu^-) \gtrsim 10^{-2}$ (including the effects of some $\pi\pi$ mode enhancement, which reduces the Higgs branching ratio into $\mu^+\mu^-$, as shown in refs. 23 and 24). Reference 105 demonstrates that reasonable cuts may be made that exclude all backgrounds, while retaining acceptance efficiency of $\sim 20\%$ for signal events. Thus, 10^6 Z's would produce more than 20 events of this type.

One may be able to improve the Higgs search by examining choices other than $\ell^+\ell^-$ for the final state fermion pair coming from the Z^*. This has the advantage of increasing the raw event rate. The disadvantage is that the backgrounds are larger. First, consider $Z \rightarrow \phi^0 \nu\bar{\nu}$. Summing over three generations of neutrinos, we immediately observe that

$$\frac{\sum_\nu BR(Z \rightarrow \phi^0 \nu\bar{\nu})}{BR(Z \rightarrow \phi^0 e^+ e^-)} \simeq 6. \tag{3.65}$$

Generally speaking, the signal would be characterized by substantial missing energy accompanied by hadronic jets from the Higgs decay. Consider first the case of $m_{\phi^0} > 2m_b$ so that the dominant decay of the Higgs is $\phi^0 \rightarrow b\bar{b}$. Most of the time, such events would consist of two distinct hadronic jets (and missing energy). The major background is due to ordinary hadronic events in which energy is lost for a variety of possible reasons, e.g., undetected neutrals, mismeasurement, lack of detector acceptance, holes in the detector, etc. The first realistic studies of this process appeared in ref. 106 (relevant for the MARK-II detector at SLC) and in ref. 97 (relevant for LEP). Reference 97 claims a somewhat better signal-to-background ratio, due to somewhat different cuts and assumptions regarding the detector. It is clear that one must work hard to eliminate the background, and a better understanding of the realistic backgrounds in a given detector will be necessary. Due to the higher branching ratio into the $\phi^0 \nu\bar{\nu}$ final state, it seems possible to observe Higgs bosons with masses up to 20 GeV, with as few as 10^5 Z's (a result better than the one obtained in the $\phi^0 e^+ e^-$ analysis mentioned above). For 10^6 Z's, Higgs masses up to 30 GeV should be observable; higher masses may be harder to detect due to the decrease in signal rate and mass resolution. For

Higgs bosons with mass below 10 GeV, the analysis must be repeated, taking into account the appropriate final states arising from the Higgs decay. In ref. 106, it is shown that for Higgs masses in the range of 2–3 GeV, the signal rate and backgrounds are roughly comparable.

This same missing energy channel has been reexamined in ref. 107 with similar conclusions. Somewhat improved cuts were employed that led to essentially zero background for $m_{\phi^0} \lesssim 15$ GeV. Based on their analysis, ref. 107 concludes that as few as 1–2×10^4 events will be sufficient to either limit or observe the ϕ^0 in the mass range $5 \lesssim m_{\phi^0} \lesssim 15$ GeV. They agree with earlier references that for $m_{\phi^0} \gtrsim 20$ GeV the $Z^* \rightarrow \ell^+ \ell^-$ channel is superior. They are currently studying the region $m_{\phi^0} \lesssim 5$ GeV, for which different cuts must be employed. Thus, still more analysis will be required to determine the precise limits on the Higgs mass that can be observed in the $\phi^0 \nu \bar{\nu}$ channel.

Finally, one might attempt to examine the possibility of observing the Higgs via the decay $Z \rightarrow \phi^0 q \bar{q}$. Summing over five flavors would give a branching ratio roughly twenty times that of the $\phi^0 e^+ e^-$ final state. Of course, the major problem here is a large multijet QCD background, which according to ref. 108 would lead to a signal-to-background ratio below 1%. Of the purely hadronic final states, the one channel which could provide a plausible strategy for Higgs detection is the $\phi^0 b \bar{b}$ final state, where $\phi^0 \rightarrow b \bar{b}$. The resulting final state, $b \bar{b} b \bar{b}$, would be identified using a vertex detector. Ref. 108 concludes that signal-to-backgrounds of $1 : 1$ may be achievable in this channel (given sufficient luminosity).

In summary, it is clear that the best Higgs mass limits will be obtained by combining a variety of techniques which are sensitive both to the Higgs decay products and to the various possible final state $f \bar{f}$ pairs coming from the Z^*. The ultimate Higgs mass limits can be estimated from fig. 3.19.

$Z \rightarrow \phi^0 \gamma$

As already discussed in §2.2, the process $Z \rightarrow \phi^0 \gamma$ can also lead to ϕ^0 production in Z decays. The numerical value of the branching ratio was shown in fig. 2.13. Note that even with a sample of 10^7 Z's, one ends up with at most 20 events (i.e., without the phase space factor included). In addition, as remarked earlier, the rate for $Z \rightarrow \phi^0 \gamma$ is larger than the rate for $Z \rightarrow \phi^0 e^+ e^-$ only for $m_{\phi^0} \gtrsim 60$ GeV. For such large Higgs masses, the signal is too small to be observed at SLC or LEP. However, it should be noted that the decay rate can be enhanced (or suppressed) by contributions from new particles (beyond the Standard Model) appearing in the loop which mediates the decay. Thus, a measurement of this rate can provide a sensitive probe of new physics.

$e^+ e^- \rightarrow \phi^0 Z$

To improve on the Higgs mass bounds further, one will need to increase the center-of-mass energy of the $e^+ e^-$ collider. This is anticipated as being the

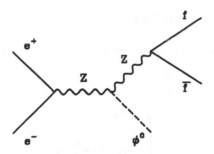

Figure 3.20 Feynman diagram for $e^+e^- \to \phi^0 Z \to \phi^0 f\bar{f}$.

next step for LEP (which will be denoted here by LEP-II), where the eventual goal is a collider energy at least larger than $2m_W$, and ideally larger than $2m_Z$. To be optimistic, let us assume that a maximum energy of $\sqrt{s} = 200$ GeV will be achievable. The process relevant for Higgs detection is now

$$e^+e^- \to \phi^0 Z \to \phi^0 f\bar{f}, \qquad (3.66)$$

which occurs via the diagram shown in fig. 3.20. This figure differs from fig. 2.12 only as to which Z is real and which Z is virtual. For energies above m_Z, there are two possibilities. For values of $\sqrt{s}$ between m_Z and $m_Z + m_{\phi^0}$, both Z's are virtual, whereas for $\sqrt{s} > m_{\phi^0} + m_Z$, the final state Z is real. We first discuss the region of $\sqrt{s} > m_{\phi^0} + m_Z$. In this region, we are interested in $e^+e^- \to \phi^0 Z$, in which a physical Z is produced. The cross section is given by [109]

$$\sigma(e^+e^- \to \phi^0 Z) = \frac{\pi\alpha^2 \lambda^{1/2}[\lambda + 12sm_Z^2][1 + (1 - 4\sin^2\theta_W)^2]}{192s^2 \sin^4\theta_W \cos^4\theta_W (s - m_Z^2)^2} \qquad (3.67)$$

where

$$\lambda = (s - m_{\phi^0}^2 - m_Z^2)^2 - 4m_{\phi^0}^2 m_Z^2 \qquad (3.68)$$

is the usual kinematic factor [with $k = \lambda^{1/2}/(2\sqrt{s})$ equal to the center-of-mass momentum of the produced Z]. The cross section for $e^+e^- \to \phi^0 Z$ versus $\sqrt{s}$ for various values of the Higgs mass is shown in fig. 3.21. Note that the cross section peaks at a center-of-mass energy given by $\sqrt{s} \simeq m_Z + \sqrt{2}m_{\phi^0}$. A calculation of the distribution of final state fermions in eq. (3.66) is given in ref. 110. Electroweak radiative corrections to $e^+e^- \to \phi^0 Z$ have been computed in ref. 111.

The detection of $e^+e^- \to \phi^0 Z$ has many features in common with the process $Z \to \phi^0 f\bar{f}$ examined above. A good review of the relevant issues can be found in refs. 97 and 98, while detailed evaluations of the primary $Z^*, \gamma^* \to \phi^0 f\bar{f}$ backgrounds can be found in refs. 112 and 113. Another recent examination of the relevant issues can be found in ref. 114. Here we will simply summarize the salient points. There are two basic methods for

Associated $Z+\phi^0$ Production in e^+e^- Collisions

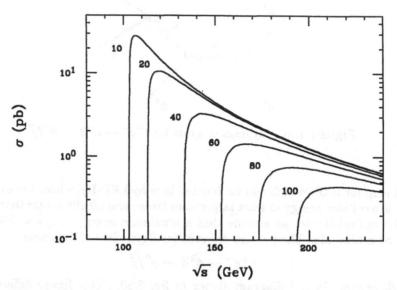

Figure 3.21 Total cross section for $e^+e^- \to \phi^0 Z$ as a function of $\sqrt{s}$ for the fixed m_{ϕ^0} values indicated by the numbers (in GeV) beside each line.

detection of the Higgs in $\phi^0 Z$ production. In the first method, one tags the outgoing Z and reconstructs the Higgs mass, without studying the Higgs decay products. In the second method, one uses all the information in the event to identify both the Z and the Higgs. In the latter case, the analysis will depend on whether the dominant decay of the Higgs is into $b\bar{b}$ or $t\bar{t}$.

As before, there are three classes of processes which must be treated separately

$$
\begin{aligned}
e^+e^- &\to \phi^0 Z, \qquad Z \to \ell^+\ell^- \quad (\ell = e, \mu), \\
e^+e^- &\to \phi^0 Z, \qquad Z \to \nu\bar{\nu}, \\
e^+e^- &\to \phi^0 Z, \qquad Z \to q\bar{q}.
\end{aligned}
\tag{3.69}
$$

In the first case, where $Z \to e^+e^-$ or $\mu^+\mu^-$, one simply identifies the Z by requiring that the invariant mass of the lepton pair be consistent with the Z interpretation. Some additional minor cuts can be made to reduce any QCD background (say from heavy quarks which decay semileptonically). The invariant mass recoiling against the Z can be plotted. If a Higgs signal exists, it will show up as a distinctive peak (typical e^+e^- mass resolutions achieved in Monte-Carlo simulations range between 5 and 10 GeV). There is virtually no background to the signal, unless the Higgs mass is near the Z mass, in which case one must somehow subtract out events due to $e^+e^- \to ZZ$. The only disadvantage of this approach is that demanding that the Z decay into electron or muon pairs immediately eliminates 94% of the original $\phi^0 Z$ events. In order to estimate the ultimate discovery limit of the Higgs boson at LEP-II, we will assume a very high machine luminosity of 10^{32} cm^{-2} sec^{-1}. In a "typical" year consisting of 10^7 sec, we will assume that 1000 pb^{-1} of data have been collected. Suppose we demand a total of 40 $\phi^0 Z$ events, where $Z \to e^+e^-$ or $\mu^+\mu^-$, (above a negligible background) in order to claim discovery. Then, we would be sensitive to Higgs bosons cross sections of $\sigma(e^+e^- \to \phi^0 Z) \gtrsim 0.7$ pb. This would correspond to a maximum Higgs mass of about 80 GeV, assuming that the center-of-mass energy of LEP-II is in the range 180–200 GeV. On the other hand, if $\sqrt{s} = 160$ GeV, then Higgs masses above 60 GeV would be inaccessible, simply due to the kinematic suppression of the cross section near threshold.

Let us briefly turn to the other Z-decay channels. For $Z \to \nu\bar{\nu}$, the larger Z branching ratio provides us with three times the number of events with electron or muon pairs. This has been studied using a detailed Monte-Carlo simulation in refs. 97 and 115. The relevant signature consists of events with hadronic jets (from the Higgs decay) and missing energy. By requiring that the missing mass be large (in the vicinity of m_Z), one can substantially reduce the QCD and various other backgrounds. One would then plot the invariant mass of all the hadrons in the events which survive the missing mass cut. The authors of ref. 115 introduced a more refined invariant mass variable. They assume that the visible hadronic energy and momentum each are a constant fraction α of the true Higgs energy and momentum. They then use the beam energy constraint and impose the kinematics of the assumed $e^+e^- \to \phi^0 Z$, $Z \to \nu\bar{\nu}$ process. This allows them to solve for the parameter α. Using this solution, they can compute the "actual" Higgs mass. This method can significantly improve the mass resolution for the Higgs. The conclusion of their analysis is that with 500 pb^{-1} of data, Higgs masses up to 80 GeV can be observed in this channel. In ref. 97, additional cuts are imposed by requiring a minimum missing *transverse* momentum of 10 GeV, which helps to eliminate two-photon induced backgrounds. They find that the mass resolution of the Higgs is somewhat less (perhaps a factor of 2) and the background is not as negligible as it was in the previous case. Nevertheless, a clear Higgs signal can be extracted in this channel.

Finally, consider the case where the Z decays into quark pairs, *i.e.*, the Z decays into hadronic jets. Now, one has to perform a rather sophisticated four-jet analysis, and deal with a significant QCD background. For Higgs masses larger than, say, 40 GeV, the backgrounds are tractable. This is unlike the case of $Z \to \phi^0 q\bar{q}$, where the backgrounds completely swamped the signal (with the possible exception of the $b\bar{b}b\bar{b}$ final state). In the present case, the larger value of the Higgs mass and the fact that the produced Z is on-shell substantially help to control the backgrounds, and allow for the observation of a signal. Typical signal-to-background ratios of 1 : 1 have been achieved in Monte-Carlo simulations described in ref. 97. The results presented in ref. 115, are somewhat more encouraging, with the conclusion that 500 pb^{-1} of data may be sufficient to observe Higgs bosons with masses as large as 70 GeV in the four-jet final state. Nevertheless, the backgrounds must be fully understood before one is convinced that such a technique is truly viable. Even if one does not discover the Higgs by this technique, it seems likely that such methods could be used to confirm its existence, and, more importantly, to measure its couplings, having established it in the other channels described above.

To conclude, given 1000 pb^{-1} of data at LEP-II, with $\sqrt{s} > 2m_Z$, Higgs bosons with masses up to around 80 GeV should be observable. To push to any higher masses at LEP-II seems rather difficult; especially since one would be approaching the particularly difficult region of $m_{\phi^0} \approx m_Z$. We can see of no way of dealing with the ZZ background without going to a machine of still higher luminosity and energy.

Other Production Mechanisms

For completeness, we end this section with some odds and ends. First, as mentioned above, the Higgs boson could be produced via a mechanism discussed above, in which *both* Z's are virtual, *i.e.*, continuum production $e^+e^- \rightarrow \phi^0 f \bar{f}$ [116]. Not surprisingly, the cross section for this process has a dip in the region where both Z's are virtual.

Exact analytic formulae for the cross section for this process can be found in ref. 117; the corresponding figure is shown in fig. 3.22. (See ref. 114 for further discussion of this process.) Two regions are of particular interest. First, at low $\sqrt{s}$ near the TRISTAN upper energy of 60 GeV, we see that the ϕ^0 cross section from this process (with $f = \mu$ in order to have a good trigger) is at most $\sim 10^{-3}$ pb. Given that maximum yearly luminosities per experiment are not likely (even after all upgrades) to significantly exceed 100 pb^{-1}, we see that even a very light SM Higgs is not likely to be detected at such a machine. Another region of machine energy with decreased sensitivity to the ϕ^0 is that of $m_Z \lesssim \sqrt{s} \lesssim m_{\phi^0} + m_Z$, which lies between the regions of real Z decay and real $\phi^0 Z$ production discussed above. Generally, the cross sections are too small to be observable, except for light mass Higgs bosons (which would have been discovered at SLC and LEP), or near the kinematic regime where one of the Z's is nearly on-shell. In addition, new diagrams involving ZZ and WW fusion also contribute for final states involving e^+e^- or $\nu_e\bar{\nu}_e$ pairs. These diagrams will be discussed in much more detail in §3.4. As another possibility, one can examine double Higgs production (*e.g.*, due to the $\phi^0\phi^0 ZZ$ vertex). Cross sections for this process [118] seem too small to be observable at SLC and LEP. Finally, we mention the process of ϕ^0 bremsstrahlung from a $b\bar{b}$ pair produced via the usual annihilation process. (The final state is actually the same as that in $e^+e^- \rightarrow Z^* \rightarrow Z^*(\rightarrow b\bar{b}) + \phi^0$. Since the diagrams for the two processes

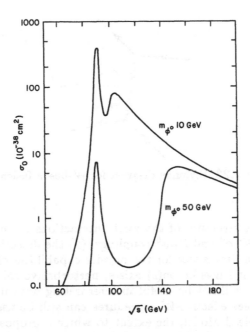

Figure 3.22 Cross section for $e^+e^- \rightarrow \mu^+\mu^-\phi^0$ as a function of $\sqrt{s}$, for $m_{\phi^0} = 10$ and 50 GeV. Parameter choices are $m_Z = 90$ GeV, $\Gamma_Z = 2.5$ GeV, and $\sin^2\theta_W = 0.22$, from ref. 117.

form separate gauge invariant sets, and there is very little interference, the processes can be discussed separately to a good approximation.) This process could conceivably have been of interest for energies below that of SLC and LEP. However, for instance, at $\sqrt{s} = 60$ GeV we have computed this process and found that the cross section for a very light ϕ^0 is $\sim 2 \times 10^{-3}$ pb, dropping rapidly to below 10^{-4} pb by $m_{\phi^0} = 10$ GeV.

3.3   The Effective-W Approximation

While there are many mechanisms that contribute to the production of the SM Higgs boson, the fusion of two vector bosons is of particular importance. Moreover, when the ϕ^0 mass is sufficiently large that its decays to VV pairs dominate, it becomes possible to study the scattering of (nearly) two on-shell-vector bosons, $VV \rightarrow VV$, where the ϕ^0 can occur as an s-channel resonance or as a t- or u-channel exchange. Note that by gauge invariance, one must (in principle) include all $VV \rightarrow VV$ diagrams, not just those involving the ϕ^0-exchange. This area has been the subject of intense theoretical study [119–125]. The subject is of particular interest in the case of very massive Higgs bosons, $m_{\phi^0} \simeq 1$ TeV. The reason for this is three-fold: (1) massive Higgs

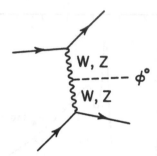

Figure 3.23 Feynman diagram for W-boson fusion processes.

bosons probe the structure of the weak interactions because they are sensitive to the $W^+W^-\phi^0$ and $ZZ\phi^0$ couplings; (2) the detection of very massive Higgs bosons is a litmus test for the physics capabilities of new accelerators; and (3) if no light Higgs boson(s) exists, perturbative calculations involving WW production will not be useful in interpreting data if WW interactions are strong, whereas effective-W procedures can still be used. Given the success of the Standard Model, the extent to which a proposed accelerator can address the physics of the Higgs sector is clearly a crucial test of its scientific justification.

In this section we study the production of Higgs bosons through the "W-fusion" mechanism shown in fig. 3.23. We begin by discussing the effective-W approximation [119–122], which is a reliable technique for calculating the cross section, using the amplitude of fig. 3.23. The effective-W approximation consists of treating the W and Z bosons as on-shell partons in the proton. In a subsequent subsection, we discuss the production and subsequent decay of the Higgs boson in the effective-W approximation for processes involving W (or Z) bosons in the intermediate state. There we will emphasize the importance of WW scattering processes. We will then be in a position to proceed with a detailed discussion of the phenomenology of heavy Higgs bosons at both high energy e^+e^- and hadron–hadron colliders.

The Effective-W Technique

At high energies, the parton sea of the proton contains not only the familiar quarks and gluons, but also the W- and Z-gauge bosons of the SU(2) $\times$ U(1) model. By considering these bosons as partons, calculations involving gauge bosons in the intermediate states can be considerably simplified. The treatment of the W and Z bosons as constituents of the proton is analogous to the effective photon method [126] in which the "effective number of photons" in an electron is calculated. We consider processes of the type

$$
\begin{aligned}
f_1 + f_2 &\to \quad W^+ + W^- + X \quad \to X' + X, \\
f_1 + f_2 &\to \quad\quad Z + Z + X \quad\quad \to X' + X,
\end{aligned}
\tag{3.70}
$$

in which the incident fermions, f_1 and f_2, radiate W or Z bosons which then annihilate to form X'. (Here we have in mind $X' = \phi^0$). The cross sections for such processes are suppressed by at least $(\alpha/\sin^2 \theta_W)$ relative to processes involving a single gauge boson, such as

$$f_1 + f_2 \to W + X',$$
$$f_1 + f_2 \to Z + X'. \tag{3.71}$$

The two-W (or two-Z) process will obviously be completely negligible at low energies. At high energies, $\sqrt{\hat{s}} \gg m_W, m_Z$, the situation is radically different. The protons can radiate transverse W's and Z's so easily that cross sections involving transverse W's are enhanced by factors of $(\ln \hat{s}/m_V^2)^2$. Processes involving longitudinal W's which are scattered into small angles are also enhanced. This is important since at high energies heavy particles often couple mainly to the longitudinal bosons. For example, the $V_L V_L \phi^0$ coupling is proportional to $m_{\phi 0}^2/m_W$ which leads to a sizable enhancement for heavy Higgs bosons.

Here we derive the probability distributions of transverse and longitudinal gauge bosons in the proton at energies much greater than the gauge boson masses. Our derivation of the distribution functions follows closely the calculation of Brodsky, Kinoshita, and Terazawa [127] for the distribution of photons in an electron. Some care is needed, however, because the massive vector bosons, unlike the photon, have both longitudinal as well as transverse degrees of freedom.

We begin by defining orthogonal polarization tensors for a gauge boson V with 4-momentum $k = (k_0, 0, 0, |\vec{k}|)$ and mass m_V,

$$\epsilon_\pm = \frac{1}{\sqrt{2}}(0, 1, \pm i, 0)$$
$$\epsilon_L = \frac{1}{m_V}(|\vec{k}|, 0, 0, k_0). \tag{3.72}$$

For $k_0 \gg m_V$, the longitudinal polarization tensor can be written as,

$$\epsilon_L \simeq \frac{k}{m_V} + \frac{m_V}{2k_0}(-1, 0, 0, 1) . \tag{3.73}$$

For vector boson couplings to massless quarks, the first term in ϵ_L gives zero and so the longitudinal polarization tensor is suppressed by a factor of m_V/k_0 relative to $\epsilon_\pm$.

In order to treat the $W^\pm$ and Z bosons as partons, we consider them as on-shell physical bosons. We work in unitary gauge and everywhere neglect light quark masses. We make the approximation that the partons have zero transverse momentum, which ensures that the longitudinal and transverse projections of the W and Z partons are uniquely specified. (It is possible

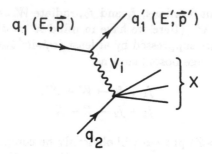

Figure 3.24 Amplitude for a polarized vector boson V_i to scatter from a quark into the final state X.

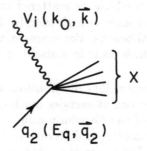

Figure 3.25 Quark–quark scattering by vector boson exchange.

to retain the p_T distribution of the vector bosons in the W and Z structure functions [128].)

The amplitude of a polarized vector boson V_i to scatter from a quark q_2 into the final state X (see fig. 3.24) is

$$iA_i(V_i + q_2 \to X) = \epsilon_i \cdot \mathcal{M}\sqrt{E_q}, \qquad i = \pm, L, \qquad (3.74)$$

where E_q is the quark energy and we have replaced the $V_\mu - q_2 - X$ vertex by an effective coupling $\mathcal{M}_\mu \sqrt{E_q}$. Averaging over the quark spin and working to leading order in $m_V^2/\hat{s}$

$$d\sigma(V_i + q_2 \to X) = \frac{1}{8k_0}|\epsilon_i \cdot \mathcal{M}|^2 d\Gamma, \qquad (3.75)$$

where the Lorentz-invariant phase space of the state X is $d\Gamma$, and k_0 is the vector boson energy in the laboratory frame of the quark. (Note that our derivation is frame dependent.) Now consider the two body scattering process of fig. 3.25, which yields the amplitude

$$iA_i(q_1 + q_2 \rightarrow q'_1 + X) = \bar{u}(p') \not\epsilon_i (C_V + C_A\gamma_5)u(p)\epsilon_i \cdot M \frac{\sqrt{E_q}}{(k^2 - m_V^2)}, \quad (3.76)$$

where k is the momentum of the V and

$$C_V = -C_A = \frac{g}{2\sqrt{2}} \qquad \text{for } W\text{'s}, \quad (3.77)$$

and

$$C_V = \frac{g}{\cos\theta_W} \left(\frac{T_{q_1}^{3L}}{2} - e_{q_1} \sin^2\theta_W \right) \qquad \text{for } Z\text{'s.} \quad (3.78)$$

$$C_A = -\frac{g}{\cos\theta_W} \frac{T_{q_1}^{3L}}{2}$$

Here $T_{q_1}^{3L}(e_{q_1})$ are the weak isospin (electric charge) of q_1 and $g^2/4\pi = \alpha/\sin^2\theta_W$. The spin-averaged total cross section is then

$$\sigma_i(q_1 + q_2 \rightarrow q'_1 + X) = \frac{1}{32EE_q} \int \frac{d^3p'}{(2\pi)^3} \frac{|A_i(q_1 + q_2 \rightarrow q'_1 + X)|^2}{E'} d\Gamma. \quad (3.79)$$

The effective-W approximation consists of replacing the current $|\epsilon_i \cdot M|$ and the phase space $d\Gamma$ by their values obtained when $k^2 \rightarrow m_V^2$ and the V is emitted in the forward direction ($\theta' \rightarrow 0$). In order to interpret V as a parton within a quark, we define the distribution function $f_{q/V_i}(x)$ such that,

$$\sigma_i(q_1 + q_2 \rightarrow q'_1 + X) = \int_{m_V/E}^{1} dx f_{q/V_i}(x)\sigma_i(V_i + q_2 \rightarrow X). \quad (3.80)$$

The distribution functions for the different polarizations can now be found from eqs. (3.76), (3.79) and (3.80),

$$f_{q/V_\pm} = \frac{1}{16\pi^2 x} \ln\left(\frac{4E^2}{m_V^2}\right) \left[(C_V \mp C_A)^2 + (1-x)^2(C_V \pm C_A)^2\right]$$

$$f_{q/V_L} = \frac{C_V^2 + C_A^2}{4\pi^2 x}(1-x). \quad (3.81)$$

It is instructive to consider the longitudinal structure function further. Because of the factor $m_V/\sqrt{\hat{s}}$ in the longitudinal polarization tensor, eq. (3.73), the matrix element squared, $|A_L(q_1 + q_2 \rightarrow q'_1 + X)|^2$, is proportional to $m_V^2/\hat{s}$. The integral over the final state phase space, d^3p', of eq. (3.79) thus becomes,

$$f_{q/V_i} \simeq \frac{m_V^2}{\hat{s}} \int \frac{\theta' \, d\theta'}{(\theta'^2 + 2m_V^2/\hat{s})^2}$$

$$\simeq 1 \, .$$

(3.82)

It is clearly the scattering into small angles which dominates the longitudinal structure function.

This derivation of the structure functions, although straightforward, leaves open the question of whether or not there can be interference between the longitudinal and transverse polarization states. By calculating in the rest frame of the massive final state X, Lindfors [129] has shown that, to leading order in m_V^2/E^2, there is complete factorization of the cross section,

$$\sigma = \sum_{\lambda=L,\pm} f_{q/V_\lambda} \sigma^\lambda \, .$$

(3.83)

Similar results were obtained in ref. 130.

The boson–boson luminosity in a two-quark system is then

$$\left. \frac{d\mathcal{L}}{d\tau} \right|_{qq/VV} = \int_\tau^1 f_{q/V}(x) f_{q/V}\left(\frac{\tau}{x}\right) \frac{dx}{x} \, ,$$

(3.84)

where $\tau = m_X^2/\hat{s}$, and, to reemphasize, $\hat{s}$ is the qq scattering subprocess energy. For $m_V^2 \ll \hat{s}$, the quark–quark luminosity integral [eq. (3.84)] can be performed exactly, yielding

$$\left. \frac{d\mathcal{L}}{d\tau} \right|_{qq/V_T V_T} = \left(\frac{C_V^2 + C_A^2}{8\pi^2}\right)^2 \frac{1}{\tau} \ln\left(\frac{\hat{s}}{m_V^2}\right)^2 [(2+\tau)^2 \ln(1/\tau) - 2(1-\tau)(3+\tau)]$$

$$\left. \frac{d\mathcal{L}}{d\tau} \right|_{qq/V_L V_L} = \left(\frac{C_V^2 + C_A^2}{4\pi^2}\right)^2 \frac{1}{\tau} [(1+\tau) \ln(1/\tau) + 2(\tau - 1)] \, ,$$

(3.85)

where we have averaged over the two transverse polarizations. For some weak processes the two transverse polarizations of eq. (3.81) interact differently [131], since left-handed and right-handed W's have different electroweak interactions. Thus one should be careful in using this formalism for computations of weak processes. The quark–quark system can be regarded as a source of polarized W and Z beams.

The luminosity of polarized vector bosons in a proton–proton system is

$$\left.\frac{d\mathcal{L}}{d\tau}\right|_{pp/V_iV_i} = \sum_{ij} \int_\tau^1 \frac{d\tau'}{\tau'} \int_{\tau'}^1 \frac{dx}{x} f_i(x) f_j\left(\frac{\tau'}{x}\right) \left.\frac{d\mathcal{L}}{d\xi}\right|_{qq/VV}, \tag{3.86}$$

where $\xi = \tau/\tau'$ and $f_i(x)$ are the quark structure functions. Our luminosities are defined such that a hadronic cross section is given by

$$\sigma_{pp \to VV \to X'}(s) = \int_{\tau_{min}}^1 d\tau \left.\frac{d\mathcal{L}}{d\tau}\right|_{pp/VV} \sigma_{VV \to X'}(\tau s). \tag{3.87}$$

Results for the luminosities of transversely and longitudinally polarized W's and Z's are shown in figs. 3.26 and 3.27 for hadronic energies of $\sqrt{s}=$ 20 TeV and 40 TeV. The transverse luminosities are significantly larger than the longitudinal. Luminosities for W^+W^-, $W^\pm Z$, $W^\pm\gamma$, and $Z\gamma$ for pp, $p\bar{p}$, e^+e^- and ep collisions have also been calculated [132,133].

Many improvements to the effective-W approximation have been suggested. It is possible to calculate the $\mathcal{O}(m_V^2/\hat{s})$ corrections to the structure functions in several frameworks [122,129,130,134]. Beyond the leading order in $m_V^2/\hat{s}$, the distribution functions are process dependent. Since all of the calculations in the literature use slightly different definitions of the structure functions, they all obtain different results for the $m_V^2/\hat{s}$ corrections. These corrections are however uniformly small for $m_{\phi^0} > 500$ GeV. We will not discuss them here since our focus is Higgs production where the effective-W approximation is extremely useful as a physical means of understanding the magnitude of the cross section, even if an exact perturbative calculation can also be performed.

The accuracy of the effective-W approximation has been tested in heavy quark production [135,136], heavy lepton production [136] and Higgs production [123,137–142]. In all of these cases cross sections have been computed exactly in the parton model and have been compared with those derived using the effective-W approximation. In all cases the two calculations agree within a factor of two. In the case of Higgs production, the agreement between the exact and the effective-W calculation is approximately 20% for $m_{\phi^0} \simeq 1$ TeV and a factor of 2 for $m_{\phi^0} \simeq 300$ GeV. In the next section we discuss heavy Higgs production in the effective-W approximation in detail.

Higgs Production in the Effective-W Approximation

The most important use of the effective-W approximation is the calculation of Higgs production from resonant W^+W^- (or ZZ) scattering. In the effective-W approximation,

$$\sigma_{\text{eff}} = \frac{16\pi^2}{m_{\phi^0}^3} \Gamma(\phi^0 \to W^+W^-) \tau \left.\frac{d\mathcal{L}}{d\tau}\right|_{pp/WW}. \tag{3.88}$$

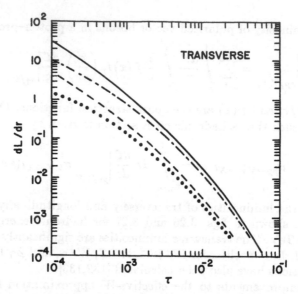

Figure 3.26 Luminosities of transversely polarized gauge bosons in a proton–proton system. The solid and dot-dashed lines are the $W_T^+ W_T^-$ luminosities at 40 and 20 TeV, respectively; the dashed and dotted lines are the $Z_T Z_T$ luminosities at these same respective energies.

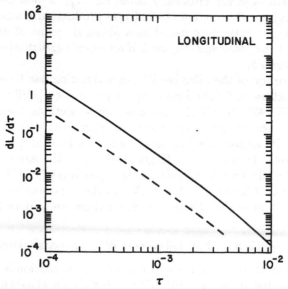

Figure 3.27 Luminosities of longitudinally polarized gauge bosons in a proton–proton system. The solid and dashed lines are the $W_L^+ W_L^-$ and $Z_L Z_L$ luminosities, respectively. $\sqrt{s} = 40$ TeV.

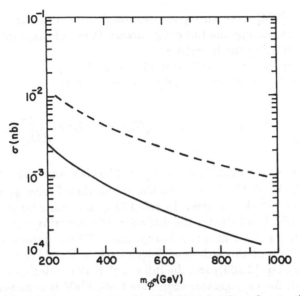

Figure 3.28 Total cross section for Higgs production, due to the fusion of longitudinally polarized vector bosons, in the effective-W approximation for pp collisions at $\sqrt{s} = 20$ TeV (solid) and $\sqrt{s} = 40$ TeV (dashed).

For heavy Higgs bosons ($m_{\phi^0} \simeq 1$ TeV) the longitudinal component dominates and so we have dropped the transverse contribution. At lower masses, ($m_{\phi^0} \simeq 300$ GeV), or away from the Higgs mass peak region, the transverse contribution becomes significant. Figure 3.28 shows the total cross section for Higgs production at $\sqrt{s} = 20$ and 40 TeV due to the fusion of longitudinally polarized vector bosons, in the effective-W approximation.

As the Higgs boson mass approaches 1 TeV, the decay width into W and Z pairs becomes extremely large,

$$\Gamma(\phi^0 \to W^+ W^-) = \frac{\alpha}{16 \sin^2 \theta_W} \frac{m_{\phi^0}^3}{m_W^2}. \qquad (3.89)$$

In this case, the Higgs boson must be treated as an intermediate state in the scattering process $WW \to WW$. At intermediate energies, all of the diagrams of fig. 2.18 must be included in order to obtain an accurate, gauge invariant answer. Duncan *et al.* [124] have investigated this process and find:

$$\mathcal{A}(W_L^+ W_L^- \to W_L^+ W_L^-) = \frac{-ig^2}{2} \left[\frac{1}{\cos^2 \theta_W} (1 + \frac{\hat{s}}{\hat{t}} + \frac{\hat{t}}{\hat{s}}) + \frac{m_{\phi^0}^2}{m_W^2} - i \frac{m_{\phi^0} \Gamma_{\phi^0}}{m_W^2} \right]. \qquad (3.90)$$

We can now use the above equation and the effective-W approximation to find the rate for W-pair production from virtual WW scattering. Chanowitz and Gaillard [120] have also given results for the full theory. Some older calculations do not use the full gauge invariant set of diagrams, so some care is necessary in using the literature.

The invariant mass distribution of the W^+W^- pair arising from the various WW scattering diagrams in fig. 2.18 is given by

$$\frac{d\sigma}{dM_{WW}^2} = \sum_{ij} \int dy \int_{\tau_{min}}^{e^{-2|y|}} d\tau f_i(\sqrt{\tau}e^y) f_j(\sqrt{\tau}e^{-y}) \frac{d\hat{\sigma}}{dM_{WW}^2}(\hat{s}) . \qquad (3.91)$$

In fig. 3.29 we show $d\sigma/dM_{WW}$ for $m_{\phi^0} = 500$ GeV and 1 TeV for pp collisions at $\sqrt{s} = 40$ TeV. W^+W^- pairs can also be produced from $q\bar{q}$ scattering. The rate for $q\bar{q} \to W^+W^-$ is given in ref. 143. This contribution is also shown in fig. 3.29, which illustrates the difficulty of extracting the signal for Higgs production from the continuum background. These figures show separately the contribution of the s-channel Higgs exchange [which is what would be obtained from eq. (2.166)] and the full set of WW scattering diagrams. As we have stressed, for m_{ϕ^0} greater than about 800 GeV it is necessary to include the complete amplitude for WW scattering to obtain accurate results [124].

Of course, one cannot ignore the possibility that m_{ϕ^0} is substantially larger than 1 TeV. In this case, the meaning of the Higgs boson as a particle is lost, and one must study WW scattering *per se*. In this scenario, once M_{WW} is above ~ 0.8 TeV the scattering amplitude for the longitudinal W modes begins to approach its unitarity limit, and WW scattering itself becomes of substantial interest. The perturbation theory result undoubtedly is no longer correct beyond a certain point in M_{WW}, but can still be used as a benchmark against which to compare results. Of course, if the ϕ^0 is light then the perturbative result at large M_{WW} should apply, and comparison with experiment provides an important cross check of the theoretical structure of the Standard Model. However, in using the effective-W approximation, combined with the perturbative result for the on-shell WW scattering amplitudes, some caution is appropriate. It was shown in refs. 139 and 140 that WW continuum scattering is sensitive to the Coulomb pole in the t-channel, and that on-shell WW computations are only an accurate reflection of the full $f_1f_2 \to f_3f_4WW$ subprocess if one is careful to make cuts on the outgoing W's (such as requiring them to have small rapidities and significant energy) such that the singularity of the Coulomb pole is not approached. This has also been studied in refs. 138 and 144. When the effective Higgs mass is extremely large, other approaches to WW scattering, based on low-energy theorems and other non-perturbative procedures, become useful at large M_{WW} [145]. Overall, if $m_{\phi^0} \gg 1$ TeV measurement of the $M_{WW} \sim 1$ TeV region should reveal a wealth of new non-perturbative physics.

It is very important to understand that if there is no physical Higgs scalar with mass below about 0.8 TeV, then the effective-W method becomes the only way to either extract information about VV scattering from pp or e^+e^-

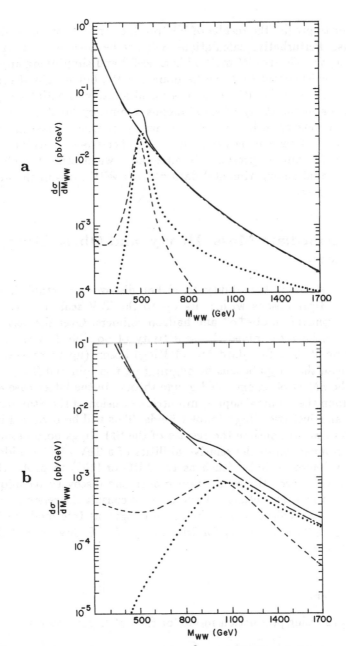

Figure 3.29 $d\sigma/dM_{WW}$ for (a) $m_\phi^0 = 500$ GeV and (b) $m_{\phi 0} = 1$ TeV, for pp collisions at $\sqrt{s} = 40$ TeV, with $|y_W| < 2.5$. The dot-dashed line is the background from $q\bar{q} \rightarrow W^+W^-$. The dashed line is the contribution from vector boson scattering, and the solid line is the total of these two contributions. The dotted line is the contribution from the s-channel Higgs pole only, fig. 2.18. This figure is from ref. 138.

collisions, or to predict the results of VV production from theories about VV interactions. Perturbative calculations will not be adequate. If light Higgs bosons exist, the effective-W method is a useful and simplifying approach to perturbative calculations that can be done exactly, but if light Higgs bosons do not exist the effective-W method is an absolutely essential technique to make progress in the study of spontaneous symmetry breaking. Of course, it would be necessary to be careful to make appropriate cuts—those needed to obtain good agreement between exact and effective-W calculations in the perturbative domain are presumably adequate—when using the effective-W technique to analyze experimental data at high WW mass in the absence of a light Higgs boson.

3.4 Intermediate Mass, Heavy, and Obese Higgs Bosons

As we have stressed, the ability to either discover or establish the non-existence of Higgs bosons with mass up to the TeV scale is a crucial goal for the next generation of e^+e^- and hadron colliders. Over the last few years many authors have examined the sensitivity of possible future colliders to both the one-doublet Standard Model Higgs boson (up to masses of order 1 TeV) and to the Higgs bosons belonging to a non-minimal Higgs representation of the SM or of an extended gauge theory. In the latter case emphasis has been upon the minimal supersymmetric extension of the Standard Model which contains just two Higgs doublet fields. This will be reviewed in a later section. Here we will outline the results of the SM Higgs boson studies, giving a first comparison of the relative abilities of a TeV e^+e^- collider and of a multi-TeV hadron collider (such as the LHC or SSC) to probe the Higgs sector. We will survey the production modes and discovery techniques, and present conclusions; for details the reader must consult the referenced studies.

It will be convenient to use the terminology: a) Intermediate Mass for $m_Z/2 \lesssim m_{\phi^0} \lesssim 2m_Z$; b) Heavy for $2m_Z \lesssim m_{\phi^0} \lesssim 700$ GeV; and c) Obese for $m_{\phi^0} \gtrsim 700$ GeV.

e^+e^- Colliders

The most promising production mode for the SM Higgs boson is

$$e^+e^- \rightarrow \nu\bar{\nu} + \phi^0 \tag{3.92}$$

via W^+W^- fusion [116,123,146,147,141]. An alternative production mode is

$$e^+e^- \rightarrow Z^* \rightarrow Z + \phi^0, \tag{3.93}$$

which has been discussed in §3.2 [see eqs. (3.66)–(3.69)]. Both production modes rely on the VV couplings of the ϕ^0.* To get an idea of the neutral Higgs production cross sections we give in figs. 3.30a and 3.30b, taken from ref. 141, the cross sections for SM Higgs production from the reactions (3.92) and (3.93). These graphs make it clear that the W^+W^- fusion reaction (3.92) rapidly becomes dominant at fixed m_{ϕ^0} as $\sqrt{s}$ increases. Before applying any cuts, reaction (3.93) overtakes reaction (3.92) at roughly

$$\sqrt{s} = 400 \text{ GeV} + 0.6 m_{\phi^0}, \tag{3.94}$$

for m_{ϕ^0} between 50 and 500 GeV. However, for ϕ^0 masses in the intermediate-mass domain ($m_{\phi^0} < 2m_W$) lower energy machines still yield a sizeable ($\sigma \gtrsim 0.1\sigma_{pt}$) cross section.

Of course, various types of new physics would lead to enhanced or distinctive new ϕ^0 production mechanisms. For instance, if a fourth generation of quarks exists, the associated new superheavy quarkonium states could be produced and a variety of new Higgs signals, as discussed at the end of chapter 2, would be available. Similarly, a new Z' would be likely to have decays of the type $Z' \rightarrow \phi^0 + Z$.

For a ϕ^0 in the intermediate mass range, relatively low machine energies are probably fully adequate. Figure 3.30 and eq. (3.94) indicate that the production mode of eq. (3.93), i.e., $e^+e^- \rightarrow Z\phi^0$, is the most useful. As already noted, this was studied at LEP energies in refs. 97 and 98. However, somewhat higher energies must be considered to fully cover the intermediate mass region; a machine energy of about 350 GeV would be ideal. Many of the same considerations that emerged in the LEP studies still apply. In addition, theoretical studies of the important backgrounds for the higher ϕ^0 masses of the intermediate mass domain have appeared in ref. 112 and, more recently, ref. 113. In these references, the ϕ^0 is observed through a mass peak in the mass of the system recoiling against the observed Z. The Z might be observed in its purely leptonic $\ell^+\ell^-$ decay modes or in $q\bar{q}$ decay modes. For example, both studies conclude that for most $m_{\phi^0} < 2m_W$ the backgrounds from $Z^*, \gamma^* \rightarrow Zq\bar{q}$ will be small and event rates adequate. For instance [112], suppose the final Z is detected in its $\ell^+\ell^-$ decay modes, $\ell = e$ or μ. The raw event rate at an integrated luminosity of 10^4 pb^{-1} will be $\gtrsim 40$, and the only difficult region of Higgs mass is $m_{\phi^0} \simeq m_Z$ where the $e^+e^- \rightarrow ZZ$ background becomes comparable to the signal. For such m_{ϕ^0} the authors of ref. 112 conclude that some triggering on the final state ($t\bar{t}$ or $b\bar{b}$) of the Higgs decay would be required. A more detailed study of the intermediate mass region was performed by a group of experimentalists for a machine energy of $\sqrt{s} = 400$ GeV as part of the 1988 Snowmass workshop effort [148]. Their conclusions are very similar. They examine the ability to isolate the $e^+e^- \rightarrow$

---

*In certain extended Higgs models the VV couplings to some of the neutral Higgs bosons can be absent or highly suppressed, thereby making them extremely difficult to produce at an e^+e^- collider.

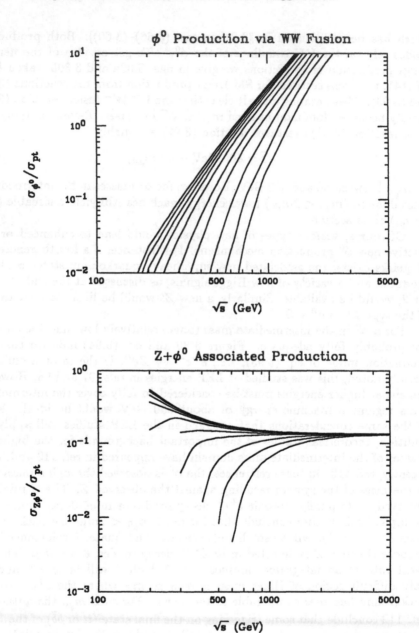

Figure 3.30 Cross sections for ϕ^0 production deriving from reactions (3.92) and (3.93), respectively. We normalize relative to the point cross section for $e^+e^- \rightarrow \mu^+\mu^-$, σ_{pt}. The different lines correspond to m_{ϕ^0} values of 10, 30, 50, 70, 100, 150, 200, 300, 400, and 500 GeV. Cross section curves always decrease in magnitude with increasing m_{ϕ^0}.

$Z\phi^0$ signal in all the decay modes of the Z ($\nu\bar{\nu}$, $\ell^+\ell^-$, and $q\bar{q}$). They conclude that a ϕ^0 with mass up to about 200 GeV should be detectable in any one of these modes for the above-mentioned integrated luminosity 10 fb^{-1}. They employ different cuts *etc.* for each decay mode, and, as in the earlier studies, find that the mass region of $m_{\phi^0} \simeq m_Z$ will require the use of flavor tagging.

Let us now turn to higher machine energies. The results stated here derive largely from the work of the SLAC study group [149,150] which has investigated the physics sensitivity of an e^+e^- linear collider with energy $\sqrt{s}$ of order 600 GeV to 1 TeV. Parallel work has been performed by the European community and their results appear in ref. 151. They focused on a higher energy e^+e^- collider with $\sqrt{s} = 2$ TeV, dubbed CLIC. Of course, numerous theoretical efforts have preceded these works and will be referenced where appropriate, but for complete references the reader should refer to refs. 149–151.

The conclusions of the SLAC study group are relatively straightforward to summarize. They consider machine energies up to 1 TeV and integrated luminosity of order 30 fb^{-1}. Aside from the region $m_{\phi^0} \sim m_W$, they find that if the SM ϕ^0 has mass less than or of order 50% of the total machine energy then backgrounds are not likely to be a problem and discovery will be possible. Similarly optimistic conclusions were reached by the European community in the studies of ref. 151, with estimated mass reach in m_{ϕ^0} extending to $m_{\phi^0} \sim 0.5\sqrt{s}$ at $\sqrt{s} = 2$ TeV. While no serious challenges to these conclusions have appeared, further study of backgrounds and radiative corrections has been continuing. We present some of the details of currently available results in the following.

At a high energy e^+e^- supercollider, the dominant production mechanism for the SM Higgs boson is $e^+e^- \rightarrow \nu_e\bar{\nu}_e\phi^0$. The dominant decay mode of ϕ^0 depends on its mass and couplings; possible channels are W^+W^-, ZZ, $t\bar{t}$ and $b\bar{b}$. One important background to the signal will thus come from continuum production of the same final states. There are two sources of such production. In the first, the particle pairs are produced by direct e^+e^- annihilation (*e.g.*, $e^+e^- \rightarrow q\bar{q}$ when $m_{\phi^0} < 2m_W$ and $e^+e^- \rightarrow W^+W^-$ when $m_{\phi^0} > 2m_W$) and will have the full energy of the initiating e^+e^-. Thus, it might be imagined that this background would not be a problem when looking for a Higgs boson with mass less than $\sqrt{s}$. However, very high energy e^+e^- colliders exhibit the phenomenon of "beamstrahlung" in which the crossing of the beams can cause significant radiation prior to the actual e^+e^- collision. Thus the initiating e^+e^- need not have full machine energy, and a continuum of lower energy particle pairs results. The extent of this low mass tail depends very much on machine design parameters. The SLAC Study Group chose to look at fairly conservative designs for which this tail is significant. For their choices, a mass peak in the particle pair spectra coming from a Higgs with mass above about $0.5\sqrt{s}$ had low statistical significance, for anticipated luminosities. Below that, cuts on the acoplanarity of the observed quark or W jets are very effective in eliminating these types of backgrounds; the Higgs production process (3.92) produces large acoplanarity because of the unobserved transverse momentum carried away by the neutrinos (typically of order m_W). The CLIC studies chose a much more ideal machine design with small beamstrahlung tail, and found that at $\sqrt{s} = 2$ TeV a $m_{\phi^0} = 1$ TeV Higgs signal would not have any serious background from this source.

The second type of background derives from processes such as $\gamma\gamma$ collisions [141,146] leading to $e^+e^- \to e^+e^-W^+W^-$ (initially studied in refs. 146, 147, and 141, and later in refs. 149–151) and $e^+e^- \to e^+e^-t\bar{t}$ [152,153], and γW collisions leading to $e^+e^- \to e^\pm W^\mp \nu_e$ [148–150, 153,154] and $e^+e^- \to e^\pm W^\mp Z\nu_e$ (refs. 151 and the erratum of ref. 150). The $e^+e^- \to e^+e^-W^+W^-$ process is relevant when the Higgs mass is larger than $2m_W$; it is a background to the production process (3.92) when the final e^+ and e^- are not seen because they disappear at small forward or backward angles, or have very low energy. The $e^+e^- \to t\bar{t}$ background is potentially important for $m_{\phi^0} < 2m_W$, if $2m_t < m_{\phi^0}$ (given current bounds on m_t in the SM, this is a relatively improbable scenario). All such backgrounds are enhanced above the levels computed in the naive effective-photon approximation by the presence of the beamstrahlung photons [149,153]. These $\gamma\gamma$, and other such backgrounds, are most effectively eliminated by requiring that the production process be characterized by substantial missing transverse momentum [141,152,151,149]. In practice, the studies of ref. 149 have implemented such a cut by requiring that the missing momentum of the event have a certain minimum angle with respect to the beam direction. In the CLIC studies, sensitivity to the highest Higgs mass considered ($m_{\phi^0} = 1$ TeV) requires eliminating even the large transverse momentum tail from the $\gamma\gamma$ processes by having a forward angle tagger and vetoing events in which an electron or positron tag occurs.

The $e^+e^- \to e^\pm W^\mp \nu_e$ reaction is a background to $e^+e^- \to \nu_e \bar{\nu}_e \phi^0$ when the Higgs mass is in the vicinity of m_W. It is a serious background in this case since the cross section is much larger than the Higgs cross section and it cannot be eliminated by transverse momentum or acoplanarity cuts; the unobserved neutrino typically produces a missing transverse momentum of order m_W, similar to that found for the Higgs from reaction (3.92). Thus, observation of Higgs bosons between about 50 GeV and 150 GeV appears to require cuts which trigger on the heavy quark decays of the Higgs [149]. The extent to which such cuts can be implemented was studied in some detail in ref. 148. Their results from b tagging cuts show that the range over which Higgs observation will be difficult is narrowed to 65 GeV $\lesssim m_{\phi^0} \lesssim 120$ GeV, if hadronic calorimetry resolution of $\sigma_E/E \sim 0.35/\sqrt{E} + 0.02$ is assumed.

The $e^+e^- \to e^\pm W^\mp Z\nu_e$ process turns out to be the most difficult-to-eliminate background to the detection of a heavy Higgs boson which decays to W^+W^- and ZZ pairs. This background is reduced, but not eliminated, by performing cuts requiring large transverse momentum for the vector boson pair, non-forward production of the vector bosons, and (less importantly) by vetoing against a forward or backward e^+ or e^-. (These cuts essentially eliminate the $e^+e^- \to W^+W^-e^+e^-$ background.) The CLIC studies showed that these cuts allow observation of the ϕ^0 for m_{ϕ^0} up to 0.8 TeV for an integrated luminosity of 10 fb^{-1} at a machine energy of $\sqrt{s} = 2$ TeV, using the hadronic decay modes of the vector bosons. To go to $m_{\phi^0} \sim 1$ TeV requires a larger integrated luminosity of ~ 50 fb^{-1}. One can even use the $\phi^0 \to Z(\to \ell^+\ell^-)Z(\to$ hadrons) decay mode of the ϕ^0 if 100 fb^{-1} of integrated luminosity are available. The $e^+e^- \to e^{\pm}W^{\mp}Z\nu_e$ background was not considered in the initial studies of ref. 149, but, as noted in the erratum, this background is not as significant for $m_{\phi^0} \le 0.5$ TeV (the mass range considered in ref. 149) because the width of the Higgs scales as $m_{\phi^0}^3$, leading to a considerably narrower signal at these smaller masses. This background has been estimated (see ref. 148) to be 10% of the signal for $m_{\phi^0} = 300$ GeV and 20% for $m_{\phi^0} = 500$ GeV, neglecting beamstrahlung. This increases to $\sim$40% when a typical beamstrahlung spectrum is included (before the forward e veto, which only reduces the level of background by another 20% or so).

To provide a flavor for the overall results we present one graph from ref. 149 in fig. 3.31 showing signal and background event rates as a function of reconstructed mass in the W^+W^- channel for the case of $\sqrt{s} = 1$ TeV and Standard Model Higgs boson with mass of $m_{\phi^0} = 300$ GeV or 500 GeV. The beamstrahlung parameter δ_B (see ref. 149) was taken to be 0.26. Various cuts on jet mass, acoplanarity, and missing transverse momentum were made, as motivated above and fully specified in ref. 149. The figure shows that the background from $e^+e^- \to W^+W^-$ collisions at less than full machine energy (due to beamstrahlung) has been effectively eliminated. (The $\gamma W \to WZ$ background was not included in this figure; see the discussion above.) However, the very large integrated luminosity assumed above may well be overly optimistic. At CLIC a similarly optimistic integrated luminosity of 50 fb^{-1} was required, in addition to the various cuts reviewed above designed to eliminate the $\gamma\gamma$ and reduce the γW induced backgrounds, in order to obtain a clear signal for $m_{\phi^0} = 1$ TeV (see fig. 3 on page 64 of ref. 151). At an integrated luminosity of 10 fb^{-1} and without forward electron anti-tagging, the Higgs signal at $\sqrt{s} = 2$ TeV becomes marginal for $m_{\phi^0} = 800$ GeV.

Preliminary studies of the production mode of eq. (3.92) have also been performed at the lower center of mass energy of $\sqrt{s} = 0.6$ TeV in ref. 149. For an integrated luminosity of 10 fb^{-1} and with $\delta_B = 0.26$, the Higgs signal becomes marginal above about $m_{\phi^0} = 200$ GeV. Again, the region near m_W is quite problematical.

We have attempted to crudely summarize these conclusions in a tabular format in table 4.2, which appears in §4.3 following the discussion of two-Higgs doublet and supersymmetric models in chapter 4. A SM Higgs boson that can be discovered by an e^+e^- collider with energy $\lesssim 0.5$–0.6 TeV is assigned a weight of 2 in this table. As the machine becomes less likely to be built on a reasonable time scale this weight is reduced, with our best guesses shown. Note that the results of this section assume full coupling of the Higgs boson to WW for the production; in supersymmetric theories and other extensions

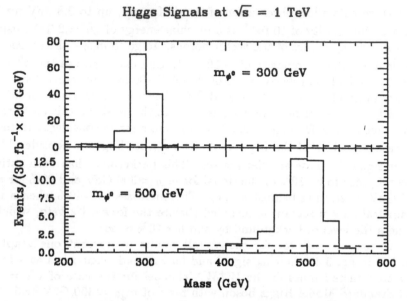

Figure 3.31 Events per 3×10^7 sec ($\sim$ 3 years) at $\mathcal{L} = 10^{33}$ cm^{-2}sec^{-1}, in the W^+W^- channel at $\sqrt{s} = 1$ TeV as a function of the reconstructed W^+W^- mass. (It is important to note that the W^+ and W^- are reconstructed using all possible decay channels and not just those containing leptons.) The dashes show the background (excluding the $e^+e^- \to e^\pm W^\mp Z\nu_e$ process); it is dominated by $e^+e^- \to W^+W^-$ at reduced energy due to beamstrahlung. The beamstrahlung parameter, defined in ref. 149, is a conservative 0.26. Signals for 300 GeV and 500 GeV Higgs signals are shown by the solid histograms.

of the Standard Model this coupling will be reduced or absent for some of the neutral scalar bosons.

As a final remark, let us conclude this section by emphasizing one point. Since there are a number of theoretical reasons to suppose that a fairly light neutral Higgs boson with large VV couplings exists, and since hadron machines may have difficulty with the mass region $m_{\phi^0} \lesssim 2m_Z$, particularly if $m_{\phi^0} > 2m_t$, (as we shall discuss in the next section) it is very desirable to build on a reasonable time scale an e^+e^- collider that could discover such a Higgs boson. If a high energy e^+e^- machine proves sufficiently technologically difficult that a long time scale would be necessary, we conclude that it would be very important to construct an intermediate collider with $\sqrt{s} \sim$ 350–400 GeV, and small δ_B, if only to cover the very crucial $m_{\phi^0} \lesssim 2m_Z$ mass range.

Hadron Colliders

In this section we shall explore the sensitivity of a high energy hadron collider to the Higgs sector of the SM. To begin with we shall focus on the Superconducting Super Collider (SSC), with $\sqrt{s} = 40$ TeV and integrated yearly luminosity of 10^4pb^{-1}. There have been a large number of workshop studies that have explored this area. These include the Snowmass studies, in particular those in 1986 [155,156] and 1988 [157–162], the 1987 Madison workshop (see the summary in ref. 163), the 1987 Berkeley workshop [164–166], and the University of California at Davis workshop [167]. Parallel work for the LHC (with $\sqrt{s} = 17$ TeV) has been performed in Europe and is summarized in the reports of ref. 151 and ref. 168, where the latter focuses on the high luminosity 5×10^5 pb^{-1} option required for good Higgs sensitivity over a substantial range of $m_{\phi 0}$. Some initial studies have also been performed for the super high energy hadron collider, named the Eloisatron, with $\sqrt{s} = 200$ TeV and the same target integrated luminosity as the SSC of 10^4 pb^{-1} per year [169]. Finally, several relevant theoretical studies have appeared recently and will be referenced where appropriate. It is, of course, not possible to give a detailed survey of all the results obtained in these efforts. We will simply highlight the important conclusions and give illustrative examples.

Generally speaking, at a hadron collider the environment for finding a Higgs boson is much more difficult than at an e^+e^- collider. Backgrounds to channels in which the Higgs boson decays to quark pairs, or to vector bosons which in turn decay to quark-pairs, are quite large. Inevitably, it is necessary to use some form of electromagnetic or leptonic trigger in order to reduce these backgrounds. The best strategy is strongly dependent upon both the type and mass of the Higgs boson, and upon the top quark mass (which influences both production and decay rates for any Higgs). However, the production rates are generally very substantial and derive mainly from a combination of

$$gg \rightarrow \phi^0, \tag{3.95}$$

and

$$W^+W^-, ZZ \rightarrow \phi^0. \tag{3.96}$$

The production rate deriving from

$$gg \rightarrow t\bar{t}\phi^0 \tag{3.97}$$

is generally a small fraction of the total Higgs production rate [170].[†]

[†] One should also include the effects of $t\bar{t} \rightarrow \phi^0$ and $gt \rightarrow \phi^0 t$. These effects only become important as $m_{\phi 0}$ is taken much larger than m_t. However, to avoid double counting one must be careful to remove the parts of these amplitudes which are implicitly included in the $gg \rightarrow t\bar{t}\phi^0$ process. A completely consistent calculation of this kind has been carried out in ref. 171, following the techniques developed in ref. 172. The result of this calculation is to increase the $gg \rightarrow t\bar{t}\phi^0$ contribution by less than a factor of 2 for $m_{\phi 0} < 1$ TeV. Thus, according to the results displayed in fig. 3.32, this contribution remains insignificant.

The cross sections for these three processes as a function of m_{ϕ^0} are illustrated for the SSC in fig. 3.32, taken from ref. 173, for two extreme choices of the top quark mass. Note that if m_t is large, the gg fusion reaction can dominate over VV fusion (summed over $V = W^{\pm}, Z$) all the way out to $m_{\phi^0} = 1$ TeV. A comparison of Higgs cross sections from gg fusion (for a moderate choice of $m_t = 80$ GeV) and VV fusion at the LHC, SSC, and Eloisatron is given in fig. 3.33, taken from ref. 174.[‡] This figure illustrates the very important advantage of having a machine with the highest possible energy. It is also instructive to compare the VV cross sections of fig. 3.33 with the VV cross sections presented in fig. 3.28. The results obtained for the LHC and SSC in fig. 3.28 are about a factor of two smaller than those obtained in fig. 3.33. The difference is primarily[*] due to the fact that in fig. 3.28 the structure functions were evolved to scale m_W^2, whereas in fig. 3.33 the evolution scale was set at $\hat{s}$, where $\hat{s}$ is the center-of-mass energy-squared of the quark–quark system initiating the vector-boson fusion process. Higher order computations would be required to determine the actual evolution scale that would yield the most accurate answer. Thus this factor of two discrepancy is a rough guide as to the current theoretical uncertainty in the size of these cross sections.

Higher order QCD corrections to these computations have been given some consideration. In ref. 176 the real gluon emission correction to $gg \rightarrow \phi^0$ has been computed, with the finding that it leads to a fairly broad p_T spectrum for the produced Higgs. In ref. 171 corrections to the process of eq. (3.97) due to evolution effects for the incoming t quarks have been included. However, these corrections are not required for a general study of the experimental observability of the ϕ^0 at a hadron collider, to which we now turn.

Intermediate Mass $(m_Z/2 \lesssim m_{\phi^0} \lesssim 2m_Z)$

By our definition, an intermediate mass Higgs boson is one that is too heavy to be detected by SLC or LEP-I $(m_{\phi^0} \gtrsim 0.5m_Z)$ but too light to have vector-boson decay modes $(m_{\phi^0} \lesssim 2m_Z)$. (We shall use the notation of the Standard Model Higgs boson, but the reader should keep in mind that our remarks apply to any generic Higgs boson with Standard-Model-like couplings to VV and fermion-antifermion channels.) In this mass range we have seen that its primary decays will be to $b\bar{b}$ pairs (if $m_{\phi^0} \leq 2m_t$) or to $t\bar{t}$ pairs (if $m_{\phi^0} > 2m_t$). It is important to note that in this mass range the ϕ^0 is an extremely narrow resonance; even for m_{ϕ^0} near $2m_Z$ its width is

---

[‡]As for fig. 3.32, these cross sections are computed in the on-shell approximation; however, there are minor differences between the procedures of ref. 173 and 174, as outlined in ref. 174.

[*]The calculation which produced fig. 3.28 also neglected the effects of the transversely polarized vector bosons (which are included in fig. 3.33). However, such an omission has a negligible effect on the observable Higgs cross section as shown in ref. 175.

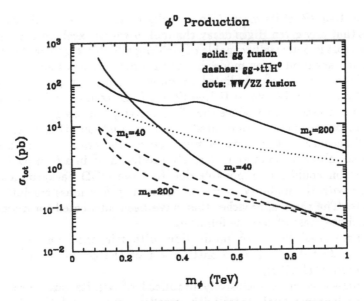

Figure 3.32 Cross sections for ϕ^0 production at the SSC deriving from reactions (3.95), (3.96), and (3.97) are given as a function of the Higgs mass for two extreme values of the top quark mass, $m_t = 40$ and $m_t = 200$ GeV. From ref. 173.

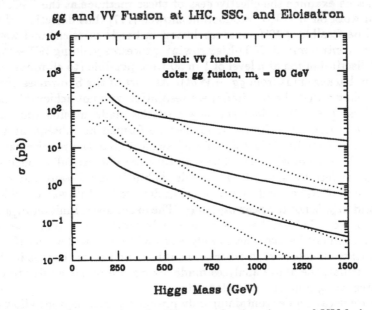

Figure 3.33 Cross sections for Higgs production via gg and VV fusion at the LHC, SSC, and Eloisatron (in order of increasing $\sqrt{s}$ and cross section), computed as described in ref. 174.

still less than 2% of its mass. (Since backgrounds are smoothly varying, this implies that any given Higgs decay channel in which excellent mass resolution is achievable is potentially capable of yielding a highly significant signal.) The two important production modes in the intermediate mass domain are gluon–gluon fusion, $gg \to \phi^0$, mediated by quark triangle graphs, and associated production with a W, $q'\bar{q} \to W^* \to W + \phi^0$. Of course, WW, ZZ fusion and $t\bar{t}$ fusion processes also produce the ϕ^0 inclusively but are significantly smaller in this mass range [173]. Associated $Z + \phi^0$ production is less interesting than $W + \phi^0$ because backgrounds are much larger and cross sections are smaller.

Direct observation of an inclusively produced ϕ^0 in a $t\bar{t}$ or $b\bar{b}$ decay mode at a hadron collider is impossible due to large QCD backgrounds, in combination with the inability to achieve extremely fine mass resolution in such channels. The three approaches that have been advocated for discovering an intermediate mass ϕ^0 are the following

1. associated $W^\pm + \phi^0$ production with triggering on the $W^\pm$ via its leptonic decays and reconstruction of the ϕ^0 in the $t\bar{t}$ or $b\bar{b}$ decay modes [112,177];

2. detection of an inclusively produced ϕ^0 via its rare purely photonic or leptonic final states with excellent mass resolution (to eliminate backgrounds) [112,178]; and

3. detection of ϕ^0 via its rare decay mode $\phi^0 \to \tau^+\tau^-$ either with or without an associated Z or g trigger at high p_T [112,179,180].

Let us examine the effectiveness of these methods at the SSC. The theoretical work on $W^\pm + \phi^0$ production showed [177] that detection of the Higgs would be feasible *if* 10% mass resolution in the $b\bar{b}$ or $t\bar{t}$ channel were achievable (to control mixed QCD-Electroweak processes yielding $W^\pm + b\bar{b}(t\bar{t})$), and *if* b/t discrimination at a level of 1 in 50 were possible (to eliminate misidentification backgrounds from $gg \to Wbt$). As part of the Snowmass 1984 studies it was shown that the required mass resolution was not achievable in the case of $t\bar{t}$ decay [181]. Neutrinos and slow particles made invisible too much of the top-quark momenta. However, studies performed at subsequent workshops and summer studies [182,165,167,159] suggest that the requirements can be met in the case of $\phi^0 \to b\bar{b}$. These studies are not yet fully conclusive (*e.g.*, additional detector simulation is required) but it would seem that an intermediate mass ϕ^0 has a good chance of being detected in this way provided $m_{\phi^0} < 2m_t$ and m_{ϕ^0} is not too near m_Z, m_W. The exact lower limit on m_{ϕ^0} for which this mode can be used has not been clearly established. While detecting the $b\bar{b}$ mode will not be easy, these analyses should be viewed as specifying the requirements on detectors in order to be able to find scalar bosons in this mode. In particular, the above analyses made strong use of the ability to trigger on b quarks using impact parameter cuts.

Rare decay modes containing only photons and/or leptons allow for excellent mass resolution, provided the detector design is appropriate [183]. Three possible such modes are

$$\phi^0 \to \mu^+\mu^-, \qquad \phi^0 \to \gamma\gamma, \qquad \phi^0 \to Z(\to \ell^+\ell^-)Z^*(\to \ell^+\ell^-). \qquad (3.98)$$

Backgrounds to these modes derive from $q\bar{q} \rightarrow \mu^+\mu^-$, $q\bar{q}, gg \rightarrow \gamma\gamma$, and $q\bar{q}, gg \rightarrow Z(\rightarrow \ell^+\ell^-)\ell^+\ell^-$, respectively, where in the $Z\ell^+\ell^-$ case the second $\ell^+\ell^-$ pair comes from a virtual Z^* or γ^* (at the Feynman graph level). The latter two modes are probably only relevant for a Higgs boson with Standard Model couplings, since large values for both the $\gamma\gamma$ branching ratio and the ZZ^* branching ratio depend on a substantial VV coupling. (See chapter 4 for more details on the $\gamma\gamma$ mode case.)

The ϕ^0 branching ratio to $\mu^+\mu^-$ is plotted in fig. 2.5, where we see that it is of order 10^{-4} so long as $m_{\phi^0} < 2m_t$ and we are not too near the WW decay threshold. To illustrate the comparison between signal and background we quote some results for $m_{\phi^0} = 20, 50$, and 150 GeV, assuming $m_t = 90$ GeV. The $\phi^0 \rightarrow \mu^+\mu^-$ branching ratio is 1.7×10^{-4}, 1.2×10^{-4} and 0.5×10^{-4} at these m_{ϕ^0} values, respectively. The ϕ^0 cross sections at the SSC for these m_{ϕ^0} values are 2,500 pb, 300 pb, and 120 pb, respectively. We can improve signal to background ratio by making a cut on the angle with respect to the beam axis of the outgoing $\mu^+\mu^-$ pair in the ϕ^0 rest frame of $|\cos\theta^*| < 0.5$, reducing the ϕ^0 signal by a factor of two. Thus for an integrated luminosity of 10^4 pb^{-1} we obtain 2125, 180, and 30 events, respectively, for $\phi^0 \rightarrow \mu^+\mu^-$. The $q\bar{q} \rightarrow \mu^+\mu^-$ background at the SSC is obtained from the standard γ and Z s-channel diagrams, after performing the above $\cos\theta^*$ cut and integrating over a 2% mass bin centered about m_{ϕ^0}. We find event rates of 70,000, 37,000, and 5500, respectively. When m_{ϕ^0} is in the vicinity of m_Z the signal to background ratio is even worse. Thus, we conclude that this rare decay mode is probably not useful. Fortunately, the situation is not so bad for the other two rare decays of eq. (3.98).

The $\gamma\gamma$ mode is useful for $m_{\phi^0} < 2m_t$, down to a lower limit that depends on m_t and machine luminosity. The upper limit derives from the fact that the branching ratio to $\gamma\gamma$ simply becomes too small for $m_{\phi^0} > 2m_t$ as illustrated in fig. 2.5. The lower limit derives from the fact that the $q\bar{q}, gg \rightarrow \gamma\gamma$ background increases with decreasing $\gamma\gamma$ mass while the $\phi^0 \rightarrow \gamma\gamma$ event rate decreases with decreasing mass in the relevant mass region. The boundaries in m_{ϕ^0}-m_t space within which this mode can be used to yield a 4σ effect in a 2% mass bin are given in fig. 3.34, from ref. 178. In ref. 178 only an estimate of the $gg \rightarrow \gamma\gamma$ background was included; this has since been computed [179,184] giving results consistent with the guess made in ref. 178, though slightly larger. Note also that QCD radiative corrections to the $\phi^0 \rightarrow b\bar{b}$ width reduce its value (see fig. 2.3), thereby increasing the rare branching ratios, perhaps by as much as a factor of 2.

The lower boundary of fig. 3.34 is probably too optimistic [158] since the background is so large that a 4σ effect in a 2% mass bin is very hard to see on a $\gamma\gamma$ invariant mass plot. In fact, a 0.5% mass resolution is needed to see a Higgs bump in the $\gamma\gamma$ spectrum for $m_{\phi^0} = 120$ GeV. In addition, there are a few remaining issues that require further detailed investigation. In particular, when a pair of quark jets is produced there is some probability that each jet will appear to be an isolated photon in the calorimeter. This

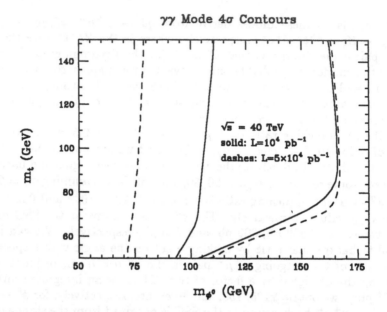

Figure 3.34 We give the boundaries in m_t–m_{ϕ^0} space which separate the region in which the $\gamma\gamma$ mode yields a $> 4\sigma$ effect from that where it does not. The two different boundaries correspond to integrated luminosities of $L = 10^4$ pb^{-1} and 5×10^4 pb^{-1}. A 2% mass bin is assumed for the $\gamma\gamma$ final state, and a cut on the angle with respect to the beam of the outgoing photons in their center of mass of $|\cos\theta_\gamma| < 0.5$ is made. The $q\bar{q} \to \gamma\gamma$ background is doubled to account for $gg \to \gamma\gamma$.

happens, for instance, if a π^0 carries most of the jet momentum. Since the jet pair cross section for pair masses in the range of interest is very large, the detector design must be optimized so as to discriminate between a real photon and a jet mimicking a photon to a level of one part in 10^4. It is also important to discriminate against the direct-γ plus jet cross section [185]. Again, the same factor of one part in 10^4 for discriminating the single jet from a second real photon would suffice. This has been examined in ref. 165, but preliminary results are encouraging. Further, the UA2 and CDF groups have reported results [186] comparing data on large-p_T photon production with QCD predictions. Thus, these groups have developed techniques for preventing jet fluctuations, radiated photons, *etc.*, from obscuring the true isolated photon signal. This amounts to an experimental confirmation that such backgrounds will not hide $\phi^0 \to \gamma\gamma$.

A potentially important property of the $\gamma\gamma$ mode to notice is that it is the only decay that a Higgs boson has that a Z boson does not share. Eventually it may be necessary to measure $h \to \gamma\gamma$ to untangle the physics of the Higgs sector. In addition, as we have pointed out in chapter 2, the rate for $\phi^0 \to$

$\gamma\gamma$ (and most probably for an arbitrary h) is very sensitive to extensions of the gauge theory [9,178] and may be a valuable input to understanding, even if it turns out not to be the discovery technique.

Regarding the $Z\gamma$ mode, this was studied in ref. 178 with the conclusion that it could not be used for discovery of the ϕ^0. This is because there is a large background from continuum $q\bar{q} \to Z\gamma$ production. (The $gg \to Z\gamma$ background has also been evaluated [187] and increases the continuum background by 10–30%.) The continuum background is substantially larger compared to the signal than in the $\gamma\gamma$ final state case just analyzed in detail.

The ZZ^* mode has a large branching ratio as m_{ϕ^0} approaches the upper part of the intermediate mass region, as seen in fig. 2.5. In the analysis that follows, we assume that both the Z and Z^* decay to $\ell^+\ell^-$ ($\ell = e$ or μ). We shall analyze the Higgs signal and background for a 2% bin in the ZZ^* invariant mass. Let us focus on the $\ell^+\ell^-$ pair coming from the Z^*. It turns out that the background in the $Z\ell^+\ell^-$ channel coming from $q\bar{q} \to ZZ^*$ Feynman graphs is quite small and that the main background derives from $q\bar{q} \to Z\gamma^*$ Feynman graphs (of course, interference must also be included). These latter graphs are strongly peaked at small $\ell^+\ell^-$ mass, whereas the $\phi^0 \to ZZ^*$ signal peaks at large $\ell^+\ell^-$ mass. Thus a simple cut on $M(\ell^+\ell^-)$, the exact magnitude depending on the m_{ϕ^0} mass of interest, is sufficient to entirely remove this background while retaining 90% of the signal. This mode is thus rate limited, and we adopt the procedure of requiring that there be 40 events in the uncut signal, where, of course, we demand that both the Z and Z^* decay to e^+e^- or $\mu^+\mu^-$. This number is chosen conservatively so as to allow for the reduction in the number of events deriving from the $M(\ell^+\ell^-)$ cuts and realistic detector acceptance for the final state leptons. The resulting boundaries in m_{ϕ^0}-m_t space within which this mode can be employed are shown in fig. 3.35, from ref. 178. Another potential background not considered above can arise from $gg \to Zb\bar{b}$, where $Z \to \ell^+\ell^-$ and both the b and $\bar{b}$ decay semileptonically [160]. Presumably, this background can be removed by requiring lepton isolation and the absence of isolated b-vertices.

A Higgs decay mode that is complementary to the ZZ^* channel is the WW^* mode, where both the W and W^* are observed in a $\ell\nu$ final state. Even though a Higgs mass peak cannot be directly reconstructed, it has been shown in ref. 188 that a Jacobian peak survives above the background from production of real continuum WW pairs over some of the same mass region that can be covered by the ZZ^* mode.

The $\tau^+\tau^-$ decay mode of the ϕ^0 has been studied by D. Atwood as part of the effort of ref. 165. His results suggest that this mode is rather marginal for the techniques in which one triggers on an associated Z or g and looks for the ϕ^0 using the main τ decay modes to single charged particles. The mass resolution achievable, using the kinematics of the high transverse momentum Z or g trigger, is simply not adequate to obtain a clear signal given the event rates. However, there may be some hope for seeing an inclusively produced ϕ^0 decaying to $\tau^+\tau^-$ if both the τ's are reconstructed in high multiplicity

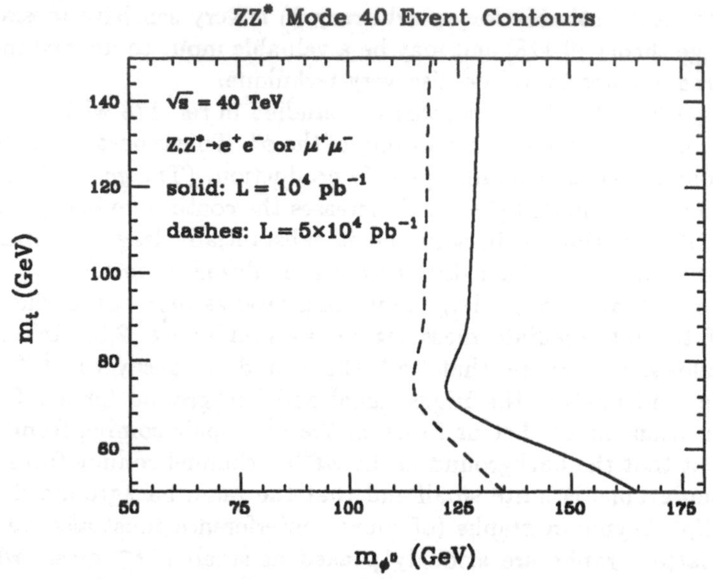

ZZ* Mode 40 Event Contours

Figure 3.35 We present the boundary curves for the ZZ^* mode which separate the region where the number of Higgs decay events is > 40 from that where it is < 40. Both the Z and Z^* are required to decay to e^+e^- or $\mu^+\mu^-$. No cuts are imposed. We give boundaries for $L = 10^4$ pb^{-1} and $L = 5 \times 10^4$ pb^{-1}.

decay modes. This technique is certainly restricted to $m_{\phi^0} < 2m_t$. Currently, it does not appear that any of the $\tau^+\tau^-$ mode techniques can be brought to the same level as the $\gamma\gamma$ and ZZ^* modes. The $\tau^+\tau^-$ channel might, however, provide confirming evidence for a signal found in one of the latter two modes.

Combining the $W + \phi^0(\to b\bar{b})$, $\phi^0 \to \gamma\gamma$, and $\phi^0 \to ZZ^*$ techniques, we see that a very large section of the $m_{\phi^0}-m_t$ parameter space plane can be covered. Only the ZZ^* mode is usable for $m_{\phi^0} > 2m_t$, but if we use the lower bound of $m_t \geq 77$ GeV (see §1.5), then there is no uncovered region. Note from figs. 3.34 and 3.35 that the coverage is significantly increased if these modes could be used at 5 times the canonical SSC luminosity. This is not obviously impossible for purely photonic/leptonic modes, and is an option that should be thoroughly explored. The impact of a high luminosity interaction region on discovery of other types of new physics, which inevitably have some purely photonic/leptonic signals, is also substantial [189].

We also note that any decrease in $\sqrt{s}$ or luminosity would have a disastrous effect on our ability to discover a SM Higgs boson in this intermediate mass region. In particular, the relevant backgrounds tend to decrease less rapidly than the signal with decreasing $\sqrt{s}$, so that the lower signal rate at lower $\sqrt{s}$ would quickly fall below an observable level. As a result, for instance, the LHC would probably have very little chance of finding a SM Higgs in the intermediate region, unless they could run the machine and detectors at a luminosity very much higher than 10^{33}cm^{-2}sec^{-1} for the $\gamma\gamma$, $b\bar{b}$, and ZZ^* modes.

Finally, since coverage of the intermediate mass region using the "conventional" techniques described above is difficult, it is perhaps worth mentioning that various types of new physics could rescue the situation. We mention only one example: the impact of a spin-zero bound state (η_v) of a very heavy fourth generation bottom quark (v) [190]. For $m_{\eta_v} > 250$ GeV $+ m_{\phi^0}$ the dominant decay mode of the η_v will be to $Z + \phi^0$, so long as mixing angles for $v \rightarrow qW$ are smaller than or of order 0.01 (so that charged W decays be adequately suppressed). The production rate via gg fusion of the η_v can still be substantial. For $m_{\phi^0} > 2m_t$, the final state will be $Zt\bar{t}$ and the primary background from $gg \rightarrow Zt\bar{t}$ has been shown [191] to be manageable.

Heavy and Obese Higgs

The strategy for detecting a SM Higgs with mass larger than $2m_Z$ is dependent upon the precise mass, on the luminosity, and on the detector hermeticity. It also may depend upon the dominant production mechanism, which depends on the Higgs mass and the $\sqrt{s}$ of the collider. For example, the production of a Higgs boson at the SSC that has $m_{\phi^0} \lesssim 350$ GeV is still dominated by gg fusion. However, depending on m_t, at higher m_{ϕ^0} vector-boson fusion processes tend to become dominant, as shown in fig. 3.32. An important point to note for later reference is that vector-boson fusion takes place via the subprocess $q\bar{q} \rightarrow q'\bar{q}' + \phi^0$, so that there are spectator quark jets present in the final state in addition to the Higgs. In either case, the total cross section for Higgs production is substantial, so we turn to the question of Higgs signatures. The cleanest Higgs signal (sometimes called the "gold-plated mode") is $\phi^0 \rightarrow ZZ$, where each Z decays to $\ell^+\ell^-$ ($\ell = e$ or μ). This is the unique signature where both Z's (and hence the ϕ^0) can be reconstructed with no ambiguity. Using $BR(\phi^0 \rightarrow ZZ) = 1/3$ and $BR(Z \rightarrow \ell^+\ell^-) = 3.4\%$, it follows that $\sum_{\ell_1,\ell_2} BR(\phi^0 \rightarrow \ell_1^+\ell_1^-\ell_2^+\ell_2^-) = 1.5 \times 10^{-3}$, summed over $\ell_1, \ell_2 = e, \mu$. Even with such a small branching fraction, there is enough cross section for Higgs production at the SSC such that the decay mode $\phi^0 \rightarrow ZZ \rightarrow \ell^+\ell^-\ell^+\ell^-$, with $\ell = e, \mu$, will reveal a Higgs signal if $m_{\phi^0} \lesssim 600$ GeV at an integrated luminosity of 10^4 pb^{-1}. To illustrate this point we give in fig. 3.36 the number of events in this mode as a function of m_{ZZ} in the case of a 600 GeV Higgs, assuming $m_t = 40$ GeV. A small but clear Higgs bump emerges on the tail of the ZZ continuum pair background. For smaller Higgs masses the Higgs bump becomes extremely clear.

To go beyond $m_{\phi^0} \sim 600$ GeV via bump-hunting in the four charged lepton channel at a moderate top mass like $m_t = 40$ GeV requires a high luminosity interaction region, as suggested in ref. 189. This idea has been

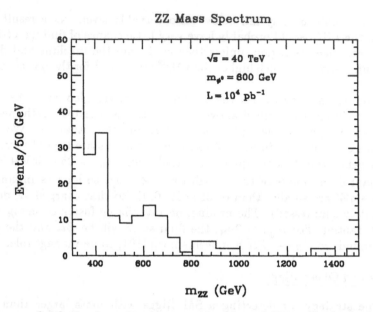

Figure 3.36 Number of events as a function of ZZ mass obtained in one year's running $(10^4 \mathrm{pb}^{-1})$ at the SSC (assuming $m_t = 40$ GeV). Both continuum ZZ production and $\phi^0 \to ZZ$ processes are included. The two Z's are detected in e^+e^- or $\mu^+\mu^-$ modes. This graph is from ref. 164. An ideal detector was assumed, and both Z's are required to have rapidity less than 1.5. The Higgs mass is 600 GeV.

pursued in ref. 192 by F. Paige. He proposes to trigger on two very energetic electromagnetic showers and two very energetic muons $(e^+e^-\mu^+\mu^-)$ with a mild hadronic activity veto and isolation requirements imposed. At the SSC and for an integrated luminosity of $10^5 \mathrm{pb}^{-1}$ the preliminary results indicate that the usefulness of the $e^+e^-\mu^+\mu^-$ mode could be extended up to $m_{\phi^0} \sim$ 800 GeV. A preliminary graph of a typical m_{ZZ} spectrum is given in fig. 3.37, for the case of $m_{\phi^0} = 800$ GeV. It includes detector simulation and the triggering procedure mentioned above. Note that since the Higgs is very broad at this mass, a clear bump is not guaranteed. It will be important to know the normalization and shape of the ZZ continuum background. However, it is likely that this background will be measured fairly precisely at somewhat lower values of m_{ZZ} and its shape will be theoretically determined to the required accuracy once the higher order QCD corrections are computed. In addition, the quark luminosities that enter into the ZZ continuum calculation will be quite well-determined using other processes [164]. In general, the energy resolution and the efficiency of the detector play an important role in the bump-hunting technique [160,161].

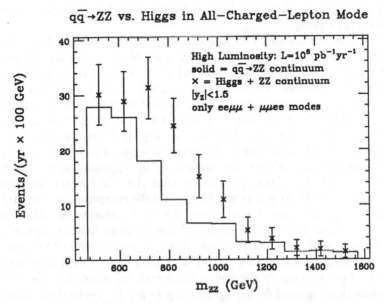

Figure 3.37 We plot the m_{ZZ} spectrum coming from the $q\bar{q} \to ZZ$ continuum background process in comparison to the sum of this background and the $\phi^0 \to ZZ$ signal process of interest. The event rates shown correspond to one Z decaying to e^+e^- and the other Z decaying to $\mu^+\mu^-$. Both Z's are required to have $|y_Z| < 1.5$. The results are for the SSC energy of $\sqrt{s} = 40$ TeV, and the ϕ^0 mass was taken to be 0.8 TeV. This graph from ref. 192 includes the effects of triggering and detector simulation. The error bars are obtained from the number of events given by the Monte Carlo in each bin.

A complementary technique that may allow access to somewhat higher ϕ^0 masses relies on analyzing the polarization of the decaying Z's in their purely leptonic decay modes [193]. As we have often stressed, the $\phi^0 \to ZZ$ decay produces longitudinally polarized Z's whereas the $q\bar{q} \to ZZ$ continuum process produces primarily transversely polarized Z's. Thus, in the region $M_{ZZ} \sim m_{\phi^0}$ one finds a large fraction f_L of longitudinally polarized Z's. A full Monte Carlo analysis is required to verify the usefulness of this approach, given the limited number of events available in an SSC year for extracting a Higgs signal.

All the above results apply to the SSC energy of $\sqrt{s} = 40$ TeV. At the LHC, with $\sqrt{s} = 17$ TeV, the Higgs cross section is substantially smaller (see fig. 3.33), and for the canonical luminosity of $10^{33}\text{cm}^{-2}\text{sec}^{-1}$ only $2m_W \lesssim m_{\phi^0} \lesssim 0.3$ TeV can be probed [151]. Thus, the LHC may be designed to run at a much higher luminosity; a number under discussion is $5 \times 10^{34}\text{cm}^{-2}\text{sec}^{-1}$. If we assume that one could only look at $\mu^+\mu^-$ Z decays at such a high luminosity, then the LHC would be able to search for ϕ^0 in this mode out to about 0.7 TeV [168].

Of course, as discussed in §1.5, the top quark probably has a mass substantially larger than 40 GeV. Since the gg fusion process has its peak cross section at $m_{\phi^0} \sim 2m_t$, it turns out that if m_t were as large as 150–200 GeV then Higgs production would always be dominated by gg fusion and would have much higher cross section than assumed in the above calculations (see fig. 3.32 [173]). This possibility was studied in ref. 164 with the conclusion that a top mass of 200 GeV would allow discovery of ϕ^0 in the four-lepton mode out to $m_{\phi^0} \sim 800$ GeV, even at standard integrated luminosity of 10^4 pb^{-1}. An additional point is that the bump-hunting approach is a relatively conservative approach, in that it assumes that the ZZ continuum background shape and normalization can not be accurately computed at the higher ZZ-pair mass regions where the ϕ^0 bump begins to disappear. There is still a significant excess of events from the ϕ^0 signal for somewhat higher m_{ϕ^0} values, even when the bump is no longer clearly apparent. If at the time the SSC or LHC is operating the lower pair mass region can be used to normalize our theoretical predictions for the higher mass regions, somewhat higher values of m_{ϕ^0} than those quoted for bump hunting might be detectable. However, for the moment, it is more prudent to be conservative and to explore additional techniques for detecting a heavy Higgs.

The Higgs mass range which can be probed by the above techniques is limited by the very small branching fraction for $\phi^0 \rightarrow ZZ \rightarrow \ell^+\ell^-\ell^+\ell^-$. One can attempt to improve this situation by examining signatures arising from other Z decays. The most promising alternative channel to study is $\phi^0 \rightarrow ZZ$ in which one Z decays to $\nu\bar{\nu}$ while the other decays to $\ell^+\ell^-$ ($l = e$ or μ) [194]. Using $\sum_\nu BR(Z \rightarrow \nu\bar{\nu}) \simeq 20\%$, we find that the raw number of signal events for $\nu\bar{\nu}\ell^+\ell^-$ is six times larger than for $\ell^+\ell^-\ell^+\ell^-$. Thus, naively, it appears that Higgs masses up to $m_{\phi^0} \sim 1$ TeV can be probed by the $\nu\bar{\nu}\ell^+\ell^-$ signature for the standard SSC yearly luminosity of 10^4 pb^{-1}. However, there are a variety of obvious backgrounds to the process of interest. These include ZZ continuum processes with one Z decaying invisibly and the WZ continuum with the W decaying to $\ell\nu$ and the ℓ being undetected. But the most serious background derives from processes giving rise to $Z+jet$ where the Z has transverse momentum similar to that expected for the Higgs decay, and much of the jet energy fails to be detected. To separate this background from the Higgs signal one first imposes a cut, on the transverse momentum of the leptonically decaying Z, which retains the enhancement at high p_T^Z ($\gtrsim m_{\phi^0}/2$) in the Higgs signal while throwing away the lower p_T^Z region where the background is largest. Secondly, it is useful to define

$$p_T^{\text{back}} = \sum_i |p_T^i|, \tag{3.99}$$

where the sum over i runs over all particles in the transverse half-plane opposite the leptonically decaying Z. The distributions for signal ($m_{\phi^0} = 800$ GeV)

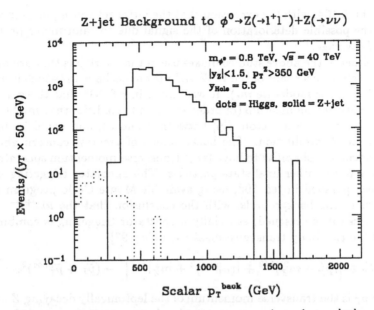

Figure 3.38 We plot at $\sqrt{s} = 40$ TeV and $L = 10^4$ pb^{-1} the p_T^{back} event number spectrum coming from the Higgs ($m_{\phi^0} = 0.8$ TeV) process $\phi^0 \to ZZ$ where one Z decays to $\ell^+\ell^-$ and the other to $\nu\bar\nu$. It is compared to that coming from $Z + jet$. Various cuts are imposed as indicated. For details of assumed electromagnetic and hadronic calorimeter resolutions and cell sizes see ref. 164. It is assumed that the detector extends to a maximum rapidity of $|y_{\text{Hole}}| = 5.5$. p_T^{back} is defined in the text.

and background in p_T^{back} are given in fig. 3.38, where a variety of calorimetry and other detector assumptions are made. In particular, it is assumed that good hadronic calorimetry is available down to $|y| = 5.5$, and that there are no cracks or dead cells. With these assumptions the background can just be conquered. A cut of $p_T^{\text{back}} < 250$ would leave 26 signal events with no background. But if there are dead cells or there is poorer coverage, the studies of ref. 164 suggest that this mode might turn out to be rather difficult to use. In addition, the direct ZZ continuum background has rather similar shape in the p_T^Z of the Z decaying to $\ell^+\ell^-$ to that obtained from the Higgs signal. As a result, the normalization of the ZZ continuum will have to be known quite accurately, even more than in the case of the $\ell^+\ell^-\ell^+\ell^-$ mode discussed previously.

These same two modes have also been explored in ref. 166. For the four charged lepton mode, conclusions are rather similar to those of ref. 164, although different cut procedures are employed. For the $\nu\bar\nu\ell^+\ell^-$ mode, somewhat different variables are used to characterize the missing energy opposite the Z that decays to $\ell^+\ell^-$. Nonetheless, the conclusions are rather similar to those of ref. 164, with the exception that the latter reference places more stress upon the possible deterioration of the signal due to calorimeter problems or lack of coverage.

Neither of the above studies makes use of the quark jets that are present in the final state as a result of the WW, ZZ fusion mechanism (which is dominant at high m_{ϕ^0} for moderate m_t). It was noted in ref. 195 that these have fairly large transverse momentum (of order m_W) and rapidities that might turn out to be accessible to detectors ($|y|$ between 3 and 5), and could be used as a trigger for ϕ^0 production. The backgrounds of greatest concern above have much smaller probability to have large transverse momentum spectator jets in addition to the other final state particles. This spectator triggering approach has been pursued in ref. 196, using available Monte Carlo program approximations to the backgrounds, with the conclusion that the $\nu\bar{\nu}\ell^+\ell^-$ signal at high mass can be rescued, especially if spectator triggering is combined with use of the two-body transverse-mass variable [197]

$$M_T^2 \equiv \left[(p_T^2 + m_Z^2)^{1/2} + \left[(p_T^{\text{miss}})^2 + m_Z^2 \right]^{1/2} \right]^2 - (\vec{p}_T + \vec{p}_T^{\text{miss}})^2, \quad (3.100)$$

where p_T is the transverse momentum of the leptonically decaying Z and p_T^{miss} is the missing transverse momentum in the event. Reference 196 concludes that a 0.8(1) TeV ϕ^0 at the SSC ($L = 10^4$ pb^{-1}) would yield 24(21) signal events for M_T between 600(700) and 1100(1500) GeV compared to 7(5) background events, after an appropriate series of cuts including tagging two jets with $p_T > 20$ GeV, $y < 5$ and requiring $p_T^{\text{miss}} > 120$ GeV. This result appears to be completely consistent with that of ref. 164 discussed above. So long as the required hadronic calorimeter coverage to large rapidity is, indeed, available, and there are no calorimeter problems, detection of the ϕ^0 up to $m_{\phi^0} \sim 0.8$ TeV and somewhat beyond using this mode should be feasible. However, the probability of detectors having sufficient rapidity coverage for spectator triggering and good p_T^{miss} and/or p_T^{back} measurement remains uncertain. The limited rapidity region covered by most proposed detectors may preclude using spectator quark triggering. In addition, spectator triggering probably cannot be usefully employed at luminosities significantly above 10^{33}cm^{-2}sec^{-1} due to radiation damage problems for the forward and backward jet tagging sections of the calorimeter. Of course, background estimates in the several studies of this mode have been based on various approximations; exact calculations of the backgrounds when extra hard spectator jets are present would be highly desirable but are very difficult. Finally, it should be noted that spectator triggering causes a significant reduction in the accepted event rate, so that at the SSC it is very unlikely to be useful above $m_{\phi^0} = 1$ TeV, while at the LHC it is almost certainly restricted to the $m_{\phi^0} \lesssim 0.8$ TeV [198].

The final techniques suggested for Higgs discovery focus on the mixed hadronic–leptonic decay modes of the WW and ZZ final states of Higgs decay. Clearly the relevant branching ratios are much larger than those appropriate to the previous channels. For instance, if we focus on the case

$$\phi^0 \rightarrow W(\rightarrow ud + cs)W(\rightarrow e\nu + \mu\nu), \tag{3.101}$$

the branching ratio for WW decay in the indicated channels is ~ 0.16. Combining all mixed hadronic–leptonic ϕ^0 decays modes yields a net effective branching ratio of more than 20%. The resulting raw event rate in the mixed mode final state is generally quite substantial. For instance, at the SSC the predicted vector-boson fusion ϕ^0 cross section is ~ 1 pb at $m_{\phi^0} = 1$ TeV; for the design yearly luminosity of 10^4 pb^{-1} we obtain of order 2000 events. Backgrounds from

$$q\bar{q} \rightarrow WW, ZZ \tag{3.102}$$

were given early consideration [143]; they do not cause major difficulty.

However, direct WW and ZZ production processes are not the only background to the mixed-mode decay of eq. (3.101). Mixed QCD-Electroweak backgrounds of the types

$$qg \rightarrow q'gW, \quad qq \rightarrow qq'W, \quad gg \rightarrow q\bar{q}'W, \text{ and } q\bar{q}' \rightarrow ggW, \tag{3.103}$$

and their analogues with Z's in the final state, present a serious challenge [199–201]. Let us focus on the case of a final state W. Simply restricting the invariant mass of the 2-jet system to a narrow bin, say

$$0.975 m_W < m_{j_1 j_2} < 1.025 m_W, \tag{3.104}$$

corresponding to 5% resolution in the jj invariant mass, is totally inadequate for obtaining a reasonable signal to background ratio [199,200]. Until very recently, techniques for discriminating between the signal and such backgrounds have relied entirely on various characteristics of the final state jet configurations. One proposal [201] relies on the longitudinal polarization of the W and Z decay products of a heavy Higgs. Longitudinally polarized vector bosons tend to decay to jets with relatively equal energies, whereas the two jets coming either from the real WW background or, especially, the Wjj background will generally have rather unequal energies. Thus, it is possible to greatly improve the signal to background ratio (without sacrificing too many signal events) by retaining only events in which both jets are reasonably hard and have fairly equal energy. The second principal proposal is that already encountered earlier [195] of triggering on the spectator jets that emerge as part of the vector-boson fusion Higgs production process. (In fact, there is nothing to prevent simultaneous use of several approaches, but event rates become more marginal.) The studies of ref. 201 find that at the parton Monte Carlo level the first type of technique allows one to achieve a very respectable signal to background ratio while retaining several hundred signal events (for an integrated luminosity of 10^4 pb^{-1} at $\sqrt{s} = 40$ TeV) all the way out to $m_{\phi^0} \sim 1$ TeV. The spectator triggering approach has been pursued for the mixed hadronic–leptonic mode in ref. 202. Although approximations are used in the calculations, for $m_{\phi^0} = 1$ TeV and the same SSC energy and luminosity it appears possible to achieve a signal to background ratio of order 1 to 1, while retaining $\sim$ 100 events, by this means as well. In both cases, the situation is less favorable at the LHC for the same integrated luminosity but with $\sqrt{s} = 17$ TeV. At the LHC, signal to background ratios of order 1 to 1 are achievable in the mixed mode but the total number of events is only of order 10 to 20 after cuts, at Higgs masses above 0.8 TeV. The Snowmass 1986 studies [155,203] began the task of Monte Carlo simulation for the Wjj and Zjj channels, and obtained preliminary conclusions that were also fairly optimistic. These studies are being continued [204]; results continue to be promising, but a conclusive demonstration of the utility of this mode must await higher statistics Monte Carlo running and further refinements of the cuts and detector simulation.

Recently it has been pointed out [205] that an additional way to separate Higgs events with a real W (or Z) decaying hadronically from background events in which the W (or Z) is simulated by a two-jet (jj) system is to take advantage of differences in the hadronic multiplicity produced in the two different types of processes. In production of the Higgs by WW fusion, there are three sources of multiplicity: two spectator systems and the W decay multiplicity itself. Each spectator system consists of a spectator quark jet and the remnant of the associated beam or target Fock state; these combine to create a minimum bias component to the multiplicity plus an additional amount set by the scale m_W of the spectator quark jet transverse energy. None of these components create a large amount of multiplicity at small rapidities. In contrast, the jj system in the backgrounds is rarely in a color singlet, and must develop multiplicity in communication with beam and target jet remnants. The invariant mass scale or allowed virtualness which determines the amount of color radiation is typically of order the Higgs mass or transverse momentum of the jj system, a much larger scale than m_W for large m_{ϕ^0}. Consequently, the average multiplicity at small rapidities which arises as a result of the hadronization is generally quite large.

Though none of the existing Monte Carlo's have the color correlations for the $2 \rightarrow 3$ subprocess backgrounds built in PYTHIA does include color correlations in the $2 \rightarrow 2$ branching approximation, and thus includes the physics relevant to dealing with multiplicities in this approximation. In refs. 205 and 162, PYTHIA is employed to develop a technique for making use of the above multiplicity differences between background and signal, in combination with the jet structure differences discussed in the preceding paragraph. A detector of rapidity extent $y = \pm 2.5$, capable of detecting individual hadrons, is used in the simulation. They first demand that the W decay products (or background equivalent) be sufficiently well-collimated that they appear as single jet when a resolution of $\Delta R = 0.5$ is employed in the jet finding algorithm ($\Delta R = \sqrt{(\Delta\eta)^2 + (\Delta\phi)^2}$). This procedure is motivated by the observation of refs. 124 and 201 that high momentum longitudinally polarized W's decay to two jets of relatively equal energy, and are thus more highly collimated than QCD jet systems of the same mass, which tend to be formed by one soft and one hard jet at large angle. The resulting distribution in the mass of the "jet" system for signal and background, fig. 3.39a, already shows a statistically significant peak in the W mass region (72 GeV $< m_{\text{jet}} <$ 92 GeV) with 580 signal events compared to 6700 background events. The associated event multiplicity, for charged hadrons with $p_T > 0.5$ GeV is plotted in fig. 3.40. A cut of $n_{ch} < 40$ produces the jet mass distribution of fig. 3.39b. The W mass region now has a signal/background event ratio of 456/490. In ref. 205 it was checked that very similar results are obtained using the ISAJET Monte Carlo. However, the theoretical underpinning of these Monte Carlo's has not yet reached the stage where the multiplicity predictions can be relied upon. Nonetheless, if this technique stands up to further scrutiny, it appears likely that it will be possible to study WW scattering into the TeV region, and have enough events left over after cuts to attempt to verify that the W decay product distributions are those characteristic of a longitudinally polarized W.

The possibility of experimentally studying the W polarization has been explored in refs. 206 and 162. As an illustration, we extract a graph from the Monte Carlo study of ref. 206. In this study, a set of cuts on hadronic multiplicity and other quantities (related to those of ref. 205, but focusing on finding *two* jets within a cone of size $\Delta R = 1$) are implemented. The Monte Carlo data are divided into bins in the two-jet system mass (M_{jj}), the two-W, i.e., Wjj, mass (M_{WW}), and the angle θ^* (in the two-jet rest frame) of one of the two jets relative to the two-jet system total momentum, as determined by an appropriate algorithm. In each M_{jj} and M_{WW} bin, the θ^* distribution of the events is fit with a polynomial form in $\cos \theta^*$. In each M_{WW} bin this fit is first performed in a M_{jj} bin that is definitely outside the W mass region. Keeping the M_{WW} mass bin fixed, the number of events in this M_{jj} region and the polynomial fit to their θ^* distribution are then used to predict the number of Wjj background events and their θ^* distribution for M_{jj} in a mass bin centered on m_W, i.e., the same normalized distribution is presumed to apply for the given M_{WW} being considered in the $M_{jj} \sim m_W$ mass bin. The remaining events are in a statistical sense only coming from real WW final states. The distribution of these remaining events is now fit by a mixture of $\sin^2 \theta^*$ (appropriate for a longitudinally polarized W decaying to two jets) and $1 + \cos^2 \theta^*$ (appropriate to a transversely polarized W), and the fraction (f_L) of WW events for which the W decaying to two jets is longitudinally polarized is determined. The results of the analysis may be seen by comparing figs. 3.41 and 3.42. In these figures, four different possible scenarios for real WW events are compared: a light ϕ^0, for which f_L should be small at high m_{WW}; a heavy Higgs, $m_{\phi^0} = 1$ TeV, where f_L should be big; and two models corresponding to a strongly interacting Higgs sector. In fig. 3.41 the theoretically computed values of f_L are given for the four different cases, while in fig. 3.42 we find the results of actually carrying out the above described procedure using the PYTHIA Monte Carlo. We see that, in one year of SSC running, the heavy Higgs scenario could be distinguished from the other three cases, but these latter cases would not

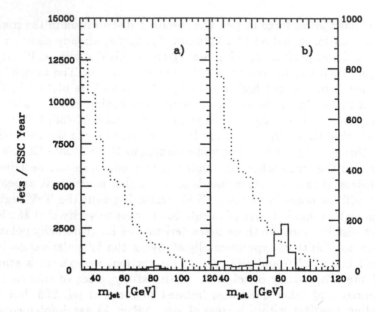

Figure 3.39 Mass of jets m_{jet} recoiling against W bosons (solid histogram: $W + W$ events from a Higgs with 1 TeV mass, dotted histogram: $W + $ QCD jets events with 1 TeV mass, both normalized to one SSC year), a) all events, b) events with $n_{ch} < 40$.

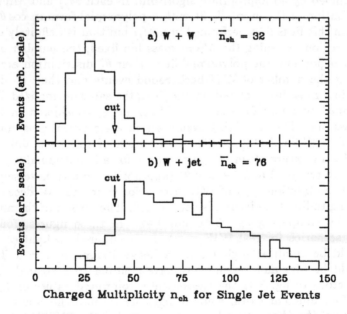

Figure 3.40 Charged multiplicity n_{ch} in events containing leptonically decaying W bosons in pp collisions (PYTHIA 4.8): a) $W + W$ events from the decay of a Higgs with 1 TeV mass, b) $W + $ QCD jets events with 1 TeV mass.

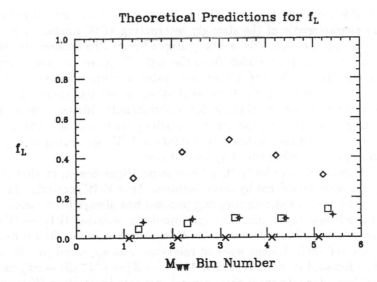

Figure 3.41 Typical theoretical predictions at $\sqrt{s} = 40$ TeV for f_L, from ref. 206. The following symbols are used: $\times$ is for a light SM Higgs (f_L is very near zero for all M_{WW} mass bins); $\diamond$ is obtained using the exact $WW \to WW$ amplitudes for $m_{\phi 0} = 1$ TeV; $\square$ and $+$ indicate strongly interacting Higgs model scenarios. The different bin numbers correspond to the following M_{WW} bins: 1, 850–950; 2, 950–1050; 3, 1050–1150; 4, 1150–1250; and 5, 1250–1350; all in GeV units.

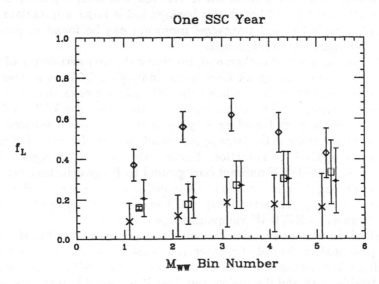

Figure 3.42 The symbols are the same as in fig. 3.41. The error bars are obtained by allowing the χ^2 of the $\cos \theta^*$ fits to vary by ± 1. The results are for one SSC year.

be separable from one another. This is because the longitudinal polarization fraction characteristic of the strongly interacting WW models is not nearly as large as that predicted for a heavy Higgs boson. To achieve separation of the strongly interacting models from the light Higgs model would require of order 10 SSC years. Of course, there are systematic uncertainties in the above predictions. For instance, the $WW \rightarrow WW$ cross sections are computed using the equivalence theorem, which yields a significantly higher cross section for $M_{WW} \lesssim m_{\phi^0}$ than a computation in unitarity gauge. But, it perhaps provides an encouraging indication that a true study of WW scattering at high M_{WW} will prove possible using the Wjj final states.

Of course, the possibility that there is no Higgs boson, or that it is very heavy, has been considered by other authors. That WW scattering in the TeV region will become a strong coupling process has always been clear. Various other models and techniques for experimentally isolating $WW \rightarrow WW$ from backgrounds have been discussed for this situation in the literature. For a review see ref. 207. Recent work of relevance has appeared in refs. 145 and 208. As discussed in ref. 164, the $W^+W^- \rightarrow Z(\rightarrow \ell^+\ell^-)Z(\rightarrow \nu\bar{\nu})$ mode (see ref. 207) has adequate event rate in most strongly interacting W models, but may have an insidious background from $Z+g$ production, where the gluon g jet largely disappears. More recently, it has been proposed [208] that one should look at like-sign W boson scattering, *e.g.* $W^+W^+ \rightarrow W^+W^+$, and search for the like-sign leptons from the decays of the two W's. An important issue is the size of the background arising from $qq \rightarrow qqW^+W^+$, in which the final W's are radiated from the quarks which have undergone strong scattering. This background, first pointed out in ref. 139, has been explicitly computed in refs. 208 and 209. Although this background is large, appropriate cuts on the lepton rapidities and transverse momenta can be found to produce an observable signal over background.

The last mentioned background, involving the bremsstrahlung of two real W's from one or more quark lines in an underlying QCD qq scattering process, is only one example of a large class of such processes that to date have not been computed fully. Other examples are $gq \rightarrow q + WW$ and various crossed versions thereof, and $gq \rightarrow gq + WW$ and crossed versions thereof. These both make use of the large gq luminosity and the latter makes use of the large $gq \rightarrow gq$ QCD cross section. Some early results [210] suggest that the former may provide a significant background to Higgs detection before spectator triggering. The latter cannot be eliminated by spectator triggering and is presently being computed [211]. Still another non-negligible background arises from $gg \rightarrow ZZ, WW$ via quark loops [212,213].

All of the above conclusions are qualitatively summarized in the overview table presented at the end of chapter 4. It should be clear that the maximum SM Higgs boson mass that can be accessed at the SSC is still a subject of considerable work and discussion, but that it is probable that m_{ϕ^0} values as high as ~ 1 TeV (and perhaps even higher) *can* be probed by the SSC. The studies that have been done for the LHC [151] are not as detailed as those performed for the SSC parameters. However, it does seem apparent that for a yearly luminosity of 10^4pb^{-1} the LHC will be considerably inferior to the SSC. For instance, in the $\ell^+\ell^-\ell^+\ell^-$ channel the maximum reachable $m_{\phi^0} > 2m_Z$ is ~ 300 GeV. It seems certain that the LHC should be designed for at least 10 times this luminosity in order to overcome some of this disadvantage. However, the ability of experiments to operate in such an environment cannot yet be regarded as established.

It should also be kept in mind that all the results of this section depend crucially on the coupling of the Higgs boson to WW and ZZ. If these couplings are suppressed, as for the heavy Higgs in many extended gauge theories to be discussed later, the above analyses must be reconsidered and modified.

Summary

Overall, we see that a high energy e^+e^- collider and the SSC each have strengths and weaknesses. An e^+e^- collider is ideal for discovering any neutral Higgs with SM couplings to VV channels, and any charged Higgs (to be discussed in chapter 4) provided the machine energy and luminosity are adequate; the exact mass reach depends upon the amount of beamstrahlung that occurs. Only the region of $m_{\phi^0} \sim m_W$ remains uncertain, due to large background from W production. Techniques for Higgs boson discovery at the SSC depend greatly on its mass. However, we hope that the preceding discussion has made it clear that most of the Standard Model Higgs mass range is accessible at the SSC. During the course of the SSC discussions we noted in several places that the LHC would have great difficulty in probing either type of Higgs sector, unless it can be designed to operate and experiments can be run at $\mathcal{L} \gtrsim 10^{34} \text{cm}^{-2}\text{sec}^{-1}$ without harming the detector's ability to study the relevant modes.

We have attempted to provide a condensed, and hence rough, overview of the comparison between a high energy e^+e^- collider and the SSC in Table 4.2, which appears at the end of chapter 4. Our hope is that both machines will eventually be built. In particular, if technology for a TeV collider will cause a large time delay, it would be highly desirable to build a ~ 350–400 GeV e^+e^- collider (with small beamstrahlung) in order to cover the region of low SM Higgs boson mass that is the most problematical for the SSC.

REFERENCES

1. A.D. Linde, *JETP Lett.* **23** (1976) 64; *Phys. Lett.* **62B** (1976) 435; S. Weinberg, *Phys. Rev. Lett.* **36** (1976) 294.
2. J. Bailey *et al.*, *Nucl. Phys.* **B150** (1979) 1.
3. R. Jackiw and S. Weinberg, *Phys. Rev.* **D5** (1972) 2396.
4. K. Fujikawa, B. Lee, and A. Sanda, *Phys. Rev.* **D6** (1972) 2923; G. Altarelli, N. Cabibbo, and L. Maiani, *Phys. Lett.* **40B** (1972) 415; W. Bardeen, R. Gastmans, and B. Lautrup, *Nucl. Phys.* **B46** (1972) 319; I. Bars and M. Yoshimura, *Phys. Rev.* **D6** (1972) 374.
5. V. Hughes and T. Kinoshita, *Comm. Nucl. Part. Phys.* **14** (1985) 341.
6. J. Leveille, *Nucl. Phys.* **B137** (1978) 63.
7. H.E. Haber, G.L. Kane, and T. Sterling, *Nucl. Phys.* **B161** (1979) 493.
8. J. Grifols and R. Pascual, *Phys. Rev.* **D21** (1980) 2672.
9. J. Ellis, M.K. Gaillard, and D.V. Nanopoulos, *Nucl. Phys.* **B106** (1976) 292.
10. I. Beltrami *et al.*, *Nucl. Phys.* **A451** (1986) 679.
11. S. Adler, R. Dashen, and S. Treiman, *Phys. Rev.* **D10** (1974) 3728.
12. R. Felst, DESY 73/65 (1973).
13. R. Barbieri and T. Ericson, *Phys. Lett.* **57B** (1975) 270.
14. D. Kohler, B. Watson, and J. Becker, *Phys. Rev. Lett.* **33** (1974) 1628.
15. L. Resnick, M. Sudaresan, and P. Watson, *Phys. Rev.* **D8** (1973) 172.
16. N.C. Mukhopadhyay, P.F.A. Goudsmit and A. Barroso, *Phys. Rev.* **D29** (1984) 565.
17. S.J. Freedman, J. Napolitano, J. Camp, and M. Kroupa, *Phys. Rev. Lett.* **52** (1984) 240.
18. M. Davier and H. Nguyen Ngoc, *Phys. Lett.* **229B** (1989) 150.
19. S. Dawson, *Phys. Lett.* **222B** (1989) 143.
20. H.-Y. Cheng and H.-L. Yu, *Phys. Rev.* **D40** (1989) 2980.
21. S. Egli *et al.*, *Phys. Lett.* **222B** (1989) 533.
22. V.A. Viktorov *et.al.*, *Sov. J. Nucl. Phys.* **33** (1981) 822; R.I. Dzhelyadin *et al.*, *Phys. Lett.* **105B** (1981) 239.
23. S. Raby and G. West, *Phys. Rev.* **D38** (1988) 3488.
24. T.N. Truong and R.S. Willey, *Phys. Rev.* **D40** (1989) 3635.
25. G.P. Yost *et al.* (Particle Data Group), *Phys. Lett.* **204B** (1988) 1.
26. N. Baker *et al.*, *Phys. Rev. Lett.* **59** (1987) 2832.
27. H.J. Lubatti, *et al.*, in *Proceedings of the XXIVth Rencontre de Moriond: Electroweak Interactions and Unified Theories*, Les Arcs, France, March 5–12, 1989, edited by J. Tran Thanh Van (Editions Frontieres, Gif-sur-Yvette, 1989) p. 349; D.M. Lazarus, *et al.*, in *Proceedings of the Third International Conference on Intersections between Particle and Nuclear Physics*, Rockport, Maine, May 14–19, 1988, edited by Gerry M. Bunce, (American Institute of Physics, New York, 1988) p. 874.
28. T. Yamazaki *et al.*, *Phys. Rev. Lett.* **52** (1984) 1089.
29. M.S. Atiya *et al.*, *Phys. Rev. Lett.* **63** (1989) 2177.

30. R. Willey and H. Yu, *Phys. Rev.* **D26** (1982) 3287.
31. R. Willey, *Phys. Lett.* **173B** (1986) 480.
32. B. Grzadkowski and P. Krawczyk, *Z. Phys.* **C18** (1984) 43.
33. F. Botella and C. Lim, *Phys. Rev. Lett.* **56** (1986) 1651.
34. B. Grinstein, L. Hall, and L. Randall, *Phys. Lett.* **211B** (1988) 363.
35. R.S. Chivukula and A.V. Manohar, *Phys. Lett.* **207B** (1988) 86, [E: **217B** (1989) 568].
36. D. Bailin *Weak Interactions* (Adam Hilger, Bristol, England, 1982).
37. F. Gilman and M. Wise, *Phys. Rev.* **D20** (1979) 2392.
38. H.-Y. Cheng, *Int. J. Mod. Phys.* **A4** (1989) 495.
39. A. Cohen and A.V. Manohar, *Phys. Lett.* **143B** (1984) 481.
40. K.R. Schubert, in *Superstrings, Unified Theories and Cosmology 1988*, the ICTP Series in Theoretical Physics, vol. 5, edited by G. Ellis *et al.* (World Scientific, Singapore, 1989) p. 523.
41. T. Pham and D. Sutherland, *Phys. Lett.* **151B** (1985) 444; *Phys. Rev.* **D34** (1986) 1634.
42. L. Wolfenstein, *Phys. Rev. Lett.* **51** (1984) 1945.
43. G. Altarelli and P. Franzini, *Z. Phys.* **C37** (1988) 271.
44. H. Burkhardt *et al.*, *Phys. Lett.* **206B** (1988) 169.
45. J.R. Cudell, F. Halzen and S. Pakvasa, *Phys. Rev.* **D40** (1989) 1562.
46. J.R. Patterson *et al.*, *Phys. Rev. Lett.* **64** (1990) 1491.
47. H. Leutwyler and M.A. Shifman, *Nucl. Phys.* **B343** (1990) 369.
48. S. Raby, G. West and C.M. Hoffman, *Phys. Rev.* **D39** (1989) 828.
49. R.S. Willey, *Phys. Rev.* **D39** (1989) 2784.
50. A.S. Carroll *et al.*, *Phys. Rev. Lett.* **44** (1980) 525.
51. K.E. Ohe *et al.*, *Phys. Rev. Lett.* **64** (1990) 2775.
52. L.K. Gibbons *et al.*, *Phys. Rev. Lett.* **61** (1988) 2661.
53. P. Bloch *et.al.*, *Phys. Lett.* **56B** (1975) 201.
54. R.J. Cence *et al.*, *Phys. Rev.* **D10** (1974) 776.
55. V. Bisi *et al.*, *Phys. Lett.* **25B** (1967) 572.
56. Y. Asano *et al.*, *Phys. Lett.* **113B** (1982) 195.
57. Y. Asano *et al.*, *Phys. Lett.* **107B** (1981) 159.
58. M.S. Atiya *et al.*, *Phys. Rev. Lett.* **64** (1990) 21.
59. L. Littenberg, private communication.
60. B. Winstein, private communication. We also thank H. Nelson for helpful conversations on this topic.
61. B. Winstein, private communication.
62. T. Shinkawa, private communication.
63. M. Zeller, private communication.
64. G.O. Barr *et al.*, *Phys. Lett.* **235B** (1990) 356.
65. R.M. Godbole, U. Turke and M. Wirbel, *Phys. Lett.* **194B** (1987) 302.
66. H.E. Haber, A.S. Schwarz, and A.E. Snyder, *Nucl. Phys.* **B294** (1987) 301.
67. A. Snyder *et al.*, *Phys. Lett.* **229B** (1989) 169.
68. M. Althoff *et al.*, *Z. Phys.* **C22** (1984) 219.

69. P. Avery *et al.*, *Phys. Rev. Lett.* **53** (1984) 1309.
70. M.S. Alam *et al.*, *Phys. Rev.* **D40** (1989) 712 [E: **D40** (1989) 3790].
71. H. Albrecht *et al.*, *Phys. Lett.* **B199** (1987) 451.
72. P. Avery *et al.*, *Phys. Lett.* **183B** (1987) 429.
73. G. Eilam and A. Soni, *Phys. Lett.* **215B** (1988) 171.
74. G. Eilam, T. Nakada, and D. Wyler, *Phys. Lett.* **231B** (1989) 184.
75. F. Wilczek, *Phys. Rev. Lett.* **39** (1977) 1304.
76. M. Vysotsky, *Phys. Lett.* **B97** (1980) 159.
77. P. Nason, *Phys. Lett.* **B175** (1986) 223.
78. M.I. Vysotskii and I.V. Polyubin, *Sov. J. Nucl. Phys.* **46** (1987) 374.
79. R. Barbieri, R. Gatto, R. Kögerler, and Z. Kunszt, *Phys. Lett.* **57B** (1975) 455.
80. H. Goldberg and Z. Ryzak, *Phys. Lett.* **218B** (1989) 348.
81. P. Franzini *et al.*, *Phys. Rev.* **D35** (1987) 2883.
82. J. Lee-Franzini, in *Proceedings of the XXIV International Conference on High Energy Physics*, Munich, 1988, edited by R. Koffhaus and J.H. Kühn (Springer-Verlag, Berlin, 1989) p. 1432.
83. R.D. Schamberger, private communication.
84. H. Albrecht *et al.*, *Z. Phys.* **C42** (1989) 349.
85. J. Polchinski, S. Sharpe, and T. Barnes, *Phys. Lett.* **148B** (1984) 493; J. Pantaleone, M. Peskin, and S.-H. Tye, *Phys. Lett.* **149B** (1984) 225.
86. S. Biswas, A. Goyal, and J. Pasupathy, *Phys. Rev.* **D32** (1985) 1844.
87. J. Pasupathy, private communication.
88. G. Faldt, P. Osland and T. T. Wu, *Phys. Rev.* **D38** (1988) 164.
89. I.G. Aznauryan, A.S. Bagdasaryan, and N.L. Ter-Isaakyan, *Sov. J. Nucl. Phys.* **36** (1982) 743.
90. I. Aznauryan, S. Grigoryan, and S. Matinyan, *JETP Lett.* **43** (1986) 646.
91. M.A. Doncheski, H. Grotch, R.W. Robinett and K. Schilcher, *Phys. Rev.* **D38** (1988) 3511.
92. H.M. Georgi, S.L. Glashow, M.E. Machacek, and D.V. Nanopoulos, *Phys. Rev. Lett.* **40** (1978) 692.
93. For the $Sp\bar{p}S$ cross section, see R. Ansari *et al.*, *Phys. Lett.* **194B** (1987) 158; C. Albajar *et al.*, *Phys. Lett.* **198B** (1987) 271. For the Tevatron, see F. Abe *et al.*, *Phys. Rev. Lett.* **62** (1989) 1005.
94. D.A. Dicus and S. Willenbrock, *Phys. Rev.* **D32** (1985) 1642; G. Altarelli, B. Mele and F. Pitolli, *Nucl. Phys.* **B287** (1987) 205.
95. G. Altarelli, in *Proceedings of the Workshop on Physics at Future Accelerators*, La Thuile, vol. 1, edited by J.H. Mulvey, CERN 87-07 (1987), p. 36.
96. A. Schwarz, *Physica Scripta* **33** (1986) 5.
97. H. Baer *et al.*, in *Physics at LEP, Vol. 1*, edited by J. Ellis and R. Peccei, CERN 86-02 (1986) p.297.
98. G. Barbiellini *et al.*, in *Physics at LEP, Vol. 2*, edited by J. Ellis and R. Peccei, CERN 86-02 (1986) p. 1.
99. J.H. Kuhn and P.M. Zerwas, *Phys. Rep.* **167** (1988) 321.

100. W. Buchmuller *et al.*, in *Physics at LEP, Vol. 1*, edited by J. Ellis and R. Peccei, CERN 86-02 (1986) p. 203.

101. I. Bigi, Y. Dokshitzer, V. Khoze, J. Kühn and P. Zerwas, *Phys. Lett.* **181B** (1986) 157.

102. L.J. Hall, S.F. King, and S.R. Sharpe, *Nucl. Phys.* **B260** (1985) 510.

103. J. Finjord, *Physica Scripta* **21** (1980) 143.

104. W. Innes, in *Proceedings of the 2nd MARK II Workshop on SLC Physics*, edited by K. Krieger, SLAC-Report-306 (1986), p. 341.

105. E. Duchovni, E. Gross, and G. Mikenberg, *Phys. Rev.* **D39** (1989) 365.

106. D. Wood, in *Proceedings of the 2nd MARK II Workshop on SLC Physics*, edited by K. Krieger, SLAC-Report-306 (1986), p. 320.

107. E. Duchovni, E. Gross, and G. Mikenberg, Weizmann preprint WIS-88/39 (1988).

108. A.S. Schwarz, in *Proceedings of the 2nd MARK II Workshop on SLC Physics*, edited by K. Krieger, SLAC-Report-306 (1986), p. 327.

109. B. Ioffe and V. Khoze, *Sov. J. Part. Nucl. Phys.* **9** (1978) 50; B.W. Lee, C. Quigg and H.B. Thacker, *Phys. Rev.* **D16** (1977) 1519.

110. R.L. Kelly and T. Shimada, *Phys. Rev.* **D23** (1981) 1940.

111. J. Fleischer and F. Jegerlehner, *Nucl. Phys.* **B216** (1983) 469; **B228** (1983) 1; J. Jegerlehner, in *Proceedings of the Trieste Workshop on Radiative Corrections in $SU(2)_L \times U(1)$, 6–8 June 1983*, edited by B. Lynn and J.F. Wheater (World Scientific, Singapore, 1984) p. 237; B.A. Kniehl, *Nucl. Phys.* **B352** (1991) 1; S. Dawson and H.E. Haber *Phys. Rev.* **D44** (1991) in press.

112. J.F. Gunion, P. Kalyniak, M. Soldate, and P. Galison, *Phys. Rev.* **D34** (1986) 101.

113. D. Atwood, A.P. Contogouris, and K. Takeuchi, *Phys. Rev.* **D38** (1988) 3437.

114. S.L. Glashow and E.E. Jenkins, *Phys. Lett.* **206B** (1988) 522.

115. J. Hilgart, M. Mermikides, S. Ritz, S.L. Wu and G. Zobernig, *Z. Phys.* **C35** (1987) 347.

116. D.R.T. Jones and S.T. Petcov, *Phys. Lett.* **84B** (1979) 440.

117. F.A. Berends and R. Kleiss, *Nucl. Phys.* **B260** (1985) 32.

118. G.J. Gounaris, D. Schildknecht and F. Renard, *Phys. Lett.* **83B** (1979) 191.

119. G. Kane, "Windows for New Particles at Super Colliders", UM-TH 83-25 (1983), invited talk at the *Conference on the Physics of the XXI Century*, Tucson, Arizona, Dec. 1983.

120. M. Chanowitz and M.K. Gaillard, *Phys. Lett.* **142B** (1984) 85; *Nucl. Phys.* **B261** (1985) 379.

121. G. Kane, W. Repko, and W. Rolnick, *Phys. Lett.* **148B** (1984) 367.

122. S. Dawson, *Nucl. Phys.* **B249** (1985) 42.

123. R. Cahn and S. Dawson, *Phys Lett.* **136B** (1984) 196 [E: **138B** (1984) 464].

124. M. Duncan, G. Kane and W. Repko, *Nucl. Phys.* **B272** (1986) 517.

125. Z. Kunszt and D. E. Soper, *Nucl. Phys.* **B296** (1988) 253.
126. C. von Weizsacker, *Z. Phys.* **88** (1934) 612; E. Williams, *Phys. Rev.* **45** (1934) 729.
127. S. Brodsky, T. Kinoshita, and H. Terazawa, *Phys. Rev.* **D4** (1971) 1532.
128. W. Rolnick, *Nucl. Phys.* **B274** (1986) 171.
129. J. Lindfors, *Z. Phys.* **C28** (1985) 427.
130. P.W. Johnson, F.I. Olness, and Wu-Ki Tung, *Phys. Rev.* **D36** (1987) 291.
131. J. P. Ralston and F. Olness, *Proceedings of the 1986 Snowmass Summer Study on the Physics of the Superconducting Super Collider*, edited by R. Donaldson and J. Marx, p. 191.
132. S. Dawson, *Proceedings of the 1984 Snowmass Summer Study on the Design and Utilization of the Superconducting Super Collider*, edited by R. Donaldson and J. G. Morfin, p. 86.
133. J. Lindfors, *Z. Phys.* **C35** (1987) 355.
134. R. Godbole and S. Rindani, *Phys. Lett.* **190B** (1987) 192; *Z. Phys.* **C36** (1987) 395.
135. S. Dawson and S. Willenbrock, *Nucl. Phys.* **B289** (1987) 449.
136. O. Eboli *et al.*, *Phys. Rev.* **D34** (1986) 771; *Phys. Lett.* **178B** (1986) 177; S. Dawson *et al.*, *Proceedings of the 1986 Snowmass Summer Study on the Physics of the Superconducting Super Collider*, edited by R. Donaldson and J. Marx, p. 235.
137. R.N. Cahn, *Nucl. Phys.* **B255** (1985) 341.
138. D. Dicus and R. Vega, *Phys. Rev. Lett.* **57** (1986) 1110.
139. J.F. Gunion, J. Kalinowski, and A. Tofighi-Niaki, *Phys. Rev. Lett.* **57** (1986) 2351.
140. J.F. Gunion, J. Kalinowski, A. Tofighi-Niaki, A. Abbasabadi, and W. Repko, *Proceedings of the 1986 Snowmass Summer Study on the Physics of the Superconducting Super Collider*, edited by R. Donaldson and J. Marx, p. 156.
141. J.F. Gunion and A. Tofighi-Niaki, *Phys. Rev.* **D36** (1987) 2671.
142. A. Abbasabadi, W.W. Repko, D. A. Dicus and R. Vega, *Phys. Rev.* **D38** (1988) 2770.
143. E. Eichten, I. Hinchliffe, K. Lane, and C. Quigg, *Rev. Mod. Phys.* **56** (1984) 579 [E: **58** (1986) 1065].
144. A. Abbasabadi and W.W. Repko, *Phys. Rev.* **D36** (1987) 289.
145. M. Chanowitz, M. Golden and H. Georgi, *Phys. Rev. Lett.* **57** (1986) 2344; *Phys. Rev.* **D36** (1987) 1490.
146. K. Hikasa, *Phys. Lett.* **164B** (1985) 385.
147. M.C. Bento and C.H. Llewellyn Smith, *Nucl. Phys.* **B289** (1987) 36.
148. J. Alexander *et al.*, in *Proceedings of the 1988 Snowmass Summer Study on High Energy Physics in the 1990's*, edited by S. Jensen (World Scientific, Singapore, 1989), p. 135.
149. C. Ahn *et al.*, SLAC-Report-329, May 1988. The experimental Monte Carlo work on the SM Higgs is presented in more detail in the following reference.

150. P. Burchat, D.L. Burke and A. Petersen, *Phys. Rev.* **D38** (1988) 2735 [E: **D39** (1989) 3515].

151. See the reports by G. Altarelli, D. Froidevaux, B. Mele, and F. Richard, *Proceedings of the Workshop on Physics at Future Accelerators*, La Thuile, vols. 1 and 2, edited by J.H. Mulvey, CERN 87-07 (1987).

152. G. Altarelli and E. Franco, *Mod. Phys. Lett.* **A1** (1986) 517.

153. E. Yehudai, *Phys. Rev.* **D42** (1990) 771.

154. E. Gabrielli, in *Proceedings of the Workshop on Physics at Future Accelerators*, La Thuile, vol. 2, edited by J.H. Mulvey, CERN 87-07 (1987), p. 1.

155. For a summary of theoretical work and references see J.F. Gunion, *Proceedings of the 1986 Snowmass Summer Study on the Physics of the Superconducting Super Collider*, edited by R. Donaldson and J. Marx, p. 142.

156. Experimental simulation studies were done by G. Alverson, *et al.*, *ibid.*, p. 93 and p. 114.

157. A. Seiden, in *Proceedings of the 1988 Snowmass Summer Study on High Energy Physics in the 1990's*, edited by S. Jensen (World Scientific, Singapore, 1989) p. 73.

158. C. Barter *et al.*, *ibid.*, p. 98.

159. J. Brau, K.T. Pitts and L.E. Price, *ibid.*, p. 103.

160. E. Wang and G.G. Hanson *et al.*, *ibid.*, p. 109.

161. I. Hinchliffe and E. Wang, *ibid.*, p. 119.

162. H.F.-W. Sadrozinski, A. Seiden and A. Weinstein, *ibid.*, p. 124.

163. J.F. Gunion and H.E. Haber, in *Proceedings of the 1987 Madison Workshop, From Colliders to Supercolliders*, edited by V. Barger and F. Halzen (World Scientific, Singapore, 1987), p. 67.

164. R. Cahn, M. Chanowitz, M. Gilchriese, M. Golden, J. Gunion, M. Herrero, I. Hinchliffe, F. Paige, and E. Wang, in *Proceedings of the 1987 Berkeley Workshop on Experiments, Detectors and Experimental Areas for the Supercollider*, edited by R. Donaldson and M. Gilchriese (World Scientific, Singapore, 1988), p. 20.

165. D.M. Atwood, J.E. Brau, J.F. Gunion, G.L. Kane, R. Madaras, D.H. Miller, L.E. Price, and A.L. Spadafora, *ibid.*, p. 728.

166. A. Savoy-Navarro, *ibid.*, p. 68.

167. *Proceedings of the University of California at Davis Workshop on Intermediate-Mass and Non-Minimal Higgs Bosons*, edited by J.F. Gunion and L. Roszkowski, (January, 1988).

168. See the reports by G. Altarelli, by W.J. Stirling, and by D. Froidevaux and B. van Eijk, in *The Feasibility of Experiments at High Luminosity at the Large Hadron Collider*, edited by J.H. Mulvey, CERN-88-02 (1988).

169. M.S. Chanowitz, in *INFN Eloisatron Project Working Group Report*, Erice, Italy, CCSEM Report EL-88/1 (1988) p. 36; J.F. Gunion and R. Vega, *ibid.* p. 42; M.S. Chanowitz, in *Higgs Particle(s): Physics Issues and Experimental Searches in High Energy Collisions*, edited by A. Ali,

(Plenum Press, New York, 1989) p. 311; J.F. Gunion, *ibid.* p. 333; R. Vega, *ibid.* p. 359.

170. Z. Kunszt, *Nucl. Phys.* **B247** (1984) 339.
171. D.A. Dicus and S. Willenbrock, *Phys. Rev.* **D39** (1989) 751.
172. R.M. Barnett, H.E. Haber, and D. Soper, *Nucl. Phys.* **B306** (1988) 697.
173. J.F. Gunion, H.E. Haber, F.E. Paige, Wu-Ki Tung, and S.S.D. Willenbrock, *Nucl. Phys.* **B294** (1987) 621.
174. J.F. Gunion and R. Vega, in *INFN Eloisatron Project Working Group Report*, Erice, Italy, CCSEM Report EL-88/1 (1988) p. 42.
175. A. Abbasabadi and W.W. Repko, in *Proceedings of the 1986 Snowmass Summer Study on the Physics of the Superconducting Super Collider*, edited by R. Donaldson and J. Marx, p. 154.
176. I. Hinchliffe and S. Novaes, *Phys. Rev.* **D38** (1988) 3475.
177. J.F. Gunion, P. Kalyniak, M. Soldate, and P. Galison, *Phys. Rev. Lett.* **54** (1985) 1226.
178. J.F. Gunion, G.L. Kane, and J. Wudka, *Nucl. Phys.* **B299** (1988) 231.
179. R.K. Ellis, I. Hinchliffe, M. Soldate and J.J. van der Bij, *Nucl. Phys.* **B297** (1988) 221.
180. M. Chaichian, I. Liede, J. Lindfors and D.P. Roy, *Phys. Lett.* **198B** (1987) 416 [E: **205B** (1988) 595].
181. B. Cox and F.J. Gilman, in *Proceedings of the 1984 Snowmass Summer Study on the Design and Utilization of the Superconducting Super Collider*, edited by R. Donaldson and J. G. Morfin, p. 87.
182. L. Price and F.J. Gilman, in *Proceedings of the 1986 Snowmass Summer Study on the Physics of the Superconducting Super Collider*, edited by R. Donaldson and J. Marx, p. 195.
183. C. Baltay, J. Huston, and B.G. Pope, in *Proceedings of the 1986 Snowmass Summer Study on the Physics of the Superconducting Super Collider*, edited by R. Donaldson and J. Marx, p. 355; D. Carlsmith, *et al.*, *ibid.*, p. 405. We note that the SLD group has measured resolution of $8\%/\sqrt{E}$ (C. Baltay, private communication) for electromagnetic calorimetry.
184. D. Dicus and S. Willenbrock, *Phys. Rev.* **D37** (1988) 1801; *Phys. Lett.* **206B** (1988) 701.
185. H. Baer and J. Owens, *Phys. Lett.* **205B** (1988) 377.
186. R. Ansari *et al.*, *Z. Phys.* **C41** (1988) 395; R. Blair (CDF Collaboration), in *Proceedings of the 7th Topical Workshop on $p\bar{p}$ Collider Physics*, Batavia, IL, June 1988, edited by R. Raja and J. Yoh (World Scientific, Singapore, 1989).
187. J.J. van der Bij and E.W.N. Glover, *Phys. Lett.* **206B** (1988) 701.
188. E.W.N. Glover, J. Ohnemus and S.S.D. Willenbrock, *Phys. Rev.* **D37** (1988) 3193.
189. J.F. Gunion and G. Kane, in *Proceedings of the 1986 Snowmass Summer Study on the Physics of the Superconducting Super Collider*, edited by R. Donaldson and J. Marx, p. 30.
190. V. Barger *et al.*, *Phys. Rev. Lett.* **57** (1986) 1672.

191. J.F. Gunion and Z. Kunszt, in *Proceedings of the 1986 Snowmass Summer Study on the Physics of the Superconducting Super Collider*, edited by R. Donaldson and J. Marx, p. 232.

192. F. Paige, in *Proceedings of the 1987 Madison Workshop, From Colliders to Supercolliders*, edited by V. Barger and F. Halzen (World Scientific, Singapore, 1987), p. 103.

193. M.J. Duncan, *Phys. Lett.* **179B** (1986) 393.

194. R.N. Cahn and M.S. Chanowitz, *Phys. Rev. Lett.* **56** (1986) 1327.

195. R.N. Cahn, S.D. Ellis, R. Kleiss, and W.J. Stirling, *Phys. Rev.* **D35** (1987) 1626.

196. V. Barger, T. Han and R.J.N. Phillips, *Phys. Rev.* **D37** (1988) 2005.

197. V. Barger, T. Han, and R.J.N. Phillips, *Phys. Rev.* **D36** (1987) 295.

198. W.J. Stirling, in *The Feasibility of Experiments at High Luminosity at the Large Hadron Collider*, edited by J.H. Mulvey, CERN-88-02 (1988).

199. J.F. Gunion, Z. Kunszt, and M. Soldate, *Phys. Lett.* **163B** (1985) 389 [E: **168B** (1986) 427].

200. W.J. Stirling, R. Kleiss, and S.D. Ellis, *Phys. Lett.* **B163** (1985) 261.

201. J.F. Gunion and M. Soldate, *Phys. Rev.* **D34** (1986) 826.

202. R. Kleiss and W.J. Stirling, *Phys. Lett.* **200B** (1988) 193.

203. S.D. Protopopescu, in *Proceedings of the 1986 Snowmass Summer Study on the Physics of the Superconducting Super Collider*, edited by R. Donaldson and J. Marx, p. 180.

204. J. Hauptman and A. Savoy-Navarro, in *Proceedings of the 1987 Berkeley Workshop on Experiments, Detectors and Experimental Areas for the Supercollider*, edited by R. Donaldson and M. Gilchriese (World Scientific, Singapore, 1988), p. 663.

205. J.F. Gunion, G. Kane, C.P. Yuan, H. Sadrozinski, A. Seiden and A.J. Weinstein, *Phys. Rev.* **D40** (1989) 2223.

206. C.P. Yuan, University of Michigan thesis; C.P. Yuan and G.L. Kane, *Phys. Rev.* **D40** (1989) 2231.

207. M. Chanowitz, *Ann. Rev. Nucl. Part. Phys.* **38** (1988) 323.

208. M. Chanowitz and M. Golden, *Phys. Rev. Lett.* **61** (1988) 1053 [E: **63** (1989) 466.

209. D. Dicus and R. Vega, *Phys. Lett.* **217B** (1989) 194.

210. U. Baur, E.W.N. Glover, and J.J. van der Bij, *Nucl. Phys.* **B318** (1989) 106.

211. U. Baur and E.W.N. Glover, *Phys. Lett.* **252B** (1990) 683.

212. J.C. Pumplin, W.W. Repko, and G.L. Kane, *Proceedings of the 1986 Snowmass Summer Study on the Physics of the Superconducting Super Collider*, edited by R. Donaldson and J. Marx, p. 211.

213. E.W.N. Glover and J.J. van der Bij, *Phys. Lett.* **219B** (1989) 488; *Nucl. Phys.* **B321** (1989) 561.

Chapter 4

Beyond the Minimal Model:
Two Higgs Doublets and the
Minimal Supersymmetric Model

We have seen that a great deal of effort has been given to considering how to search for the minimal Higgs of the Standard Model (SM) at the SSC. The phrase "minimal Higgs" means that the SU(2) × U(1) electroweak theory Higgs sector is comprised of only one complex Higgs doublet. As discussed earlier, in this case there is only one physical neutral Higgs scalar in the spectrum, and its mass is a free parameter not fixed by the theory. Although this minimal choice is completely arbitrary (as far as we know), we have seen that it provides an important benchmark in assessing our ability to detect a Higgs boson. However, given the fact that there is no experimental information concerning the Higgs sector, it is clearly prudent to explore the implications of more complicated Higgs models, both in the context of the Standard Model and in extended theories [1]. Of course, in deciding which options are worthy of discussion, we shall turn to theoretical arguments that place general constraints on the unknown Higgs sector, even in the context of the Standard Model.

There are basically two major constraints. First, it is an experimental fact [2,3] that $\rho = m_W^2/(m_Z^2 \cos^2 \theta_W)$ is very close to 1. In the Standard Model, the ρ parameter is determined by the Higgs structure of the theory. It is well known [4] that in a model with only Higgs doublets (and singlets), the tree-level value of $\rho = 1$ is automatic, without adjustment of any parameters in the model. Although the minimal Higgs satisfies this property, so does any version of the Standard Model with any number of Higgs doublets (and singlets). In fact, there are other ways to satisfy the $\rho \approx 1$ constraint. First, there are an infinite number of more complicated Higgs representations which also satisfy $\rho = 1$ at tree level [5]. The general formula is

$$\rho \equiv \frac{m_W^2}{m_Z^2 \cos^2 \theta_W} = \frac{\sum_{T,Y}[4T(T+1) - Y^2] \mid V_{T,Y} \mid^2 c_{T,Y}}{\sum_{T,Y} 2Y^2 \mid V_{T,Y} \mid^2}, \qquad (4.1)$$

where $\langle \phi(T,Y) \rangle = V_{T,Y}$ defines the vacuum expectation values of each neutral Higgs field, and T and Y specify the total $SU(2)_L$ isospin and the hypercharge of the Higgs representation to which it belongs. In addition, we have introduced the notation:

$$c_{T,Y} = \begin{cases} 1, & (T,Y) \in \text{complex representation,} \\ \frac{1}{2}, & (T, Y = 0) \in \text{real representation.} \end{cases} \tag{4.2}$$

Here, we employ a rather narrow definition of a real representation as consisting of a real multiplet of fields with integer weak isospin and $Y = 0$. The requirement that $\rho = 1$ for arbitrary $V_{T,Y}$ values is

$$(2T + 1)^2 - 3Y^2 = 1. \tag{4.3}$$

The possibilities beyond $T = 1/2$, $Y = \pm 1$ are usually discarded since the representations are rather complicated (the simplest example is a representation with weak isospin 3 and $Y = 4$). Second, one can take a model with multiple copies of "bad" Higgs representations, and arrange a "custodial" $SU(2)$ symmetry among the copies, which then naturally imposes $\rho = 1$ at tree level. Examples of this type will be considered in §6.4. Finally, one can always choose arbitrary Higgs representations and fine tune the parameters of the Higgs potential to arrange $\rho \approx 1$. We will discard this latter "unnatural" possibility from further consideration.

The definition of ρ beyond the tree level is not unique as discussed in §2.4. In a scheme advocated by Veltman [6], the tree-level definition of ρ [eq. (4.1)] is modified when radiative corrections are taken into account. For example, suppose one adds a fermion doublet (U, D) with a large mass splitting to the Standard Model. The correction to ρ at one loop is [6,7]

$$\frac{\Delta\rho}{\rho} = \frac{g^2 N_c}{64\pi^2 m_W^2} \left[m_U^2 + m_D^2 - \frac{2m_U^2 m_D^2}{m_U^2 - m_D^2} \ln\left(\frac{m_U^2}{m_D^2}\right) \right], \tag{4.4}$$

where N_c is the number of color degrees of freedom of the fermions. Note that $\Delta\rho = 0$ if $m_U = m_D$, which is again due to the existence of a "custodial" $SU(2)$ symmetry. This symmetry is violated if $m_U \neq m_D$, and affects $\Delta\rho$ at one loop due to fermion-loop corrections to the gauge boson propagators. Furthermore, $\Delta\rho \geq 0$, which is a feature common to a large class of one-loop contributions to $\Delta\rho$ [8]. This should be contrasted with the tree-level correction to ρ, due to additional (non-doublet) Higgs multiplets, which can be of either sign.

Quantitative analyses of the restrictions on multi-doublet and higher Higgs representations have appeared in several places. In particular, neutral current experiments give some limits both on models containing more than one Higgs doublet [9,10] and on the allowed vacuum expectation values of Higgs bosons in representations other than $SU(2)$ doublets [2,3]. The two-Higgs-doublet restrictions arise from corrections at the one-loop level, while those obtained for the triplet representation in ref. 2 are tree-level results.

Let us quote the triplet-Higgs restrictions first, deferring discussion of the two-doublet one-loop restrictions until §4.1. The restrictions for triplet Higgs representations arise directly from eq. (4.1). These have been given by Amaldi *et al.* [2], who assume that there is only one multiplet not in an SU(2) doublet so there are no cancellations. If the deviation from $\rho = 1$ [see eq. (2.127)] were due entirely to the existence of one Higgs triplet, then the limits of ref. 2 (at the 90% confidence level) are

$$\frac{|V_{10}|}{|V_{1/2\,1}|} \leq 0.047$$

$$\frac{|V_{1\pm2}|}{|V_{1/2\,1}|} \leq 0.081\,,$$

$$(4.5)$$

where $V_{1/2\,1}$ is the usual Higgs doublet vacuum expectation value, and V_{10} ($V_{1\pm2}$) is an SU(2) triplet with $Y = 0$ ($Y = \pm2$). Note that the addition of a $Y = \pm2$ triplet decreases ρ below 1. Hence, the above limit on $V_{1\pm2}$ would be weakened if there were other positive corrections to ρ (*e.g.*, if a heavy top quark exists). Furthermore, as we have already remarked, by combining several Higgs triplets it is possible to retain $\rho = 1$.

The second major theoretical constraint on the Higgs sector comes from the severe limits on the existence of flavor-changing neutral currents (FCNC's). In the minimal Higgs model, tree-level flavor-changing neutral currents are automatically absent, because the same operations that diagonalize the mass matrix automatically diagonalize the Higgs-fermion couplings. In general, this ceases to be true in non-minimal Higgs models. One then has two choices. First, by arranging the parameters of the model so that the Higgs masses are large (typically of order 1 TeV), tree-level FCNC's mediated by Higgs exchange can be suppressed sufficiently so as not to be in conflict with known experimental limits. The second choice is more elegant, and is based on a theorem of Glashow and Weinberg [11] concerning FCNC's in models with more than one Higgs doublet. The theorem states that tree-level FCNC's mediated by Higgs bosons will be absent if all fermions of a given electric charge couple to no more than one Higgs doublet. If we require this theorem to be satisfied, the Higgs couplings to fermions are constrained, but not unique. One example of a model satisfying this requirement is the minimal supersymmetric extension of the Standard Model. This model (discussed in the following section in more detail) possesses two Higgs doublets of opposite hypercharge; the $Y = -1$ doublet couples only to down-type quarks and leptons, and the $Y = 1$ doublet couples only to up-type quarks and leptons. Note that in supersymmetric models, this choice is not arbitrary, but is in fact required by the supersymmetry in order to give masses to both up- and down-type quarks and leptons. Furthermore, by considering the higgsino superpartners, an even number of Higgs doublets is required in order to avoid the introduction of anomalies into the theory. A second example of a two-Higgs-doublet model which avoids FCNC's is a model in which one Higgs doublet does not couple to fermions at all (due to a discrete symmetry) and the other Higgs doublet couples to fermions in the same way as in the minimal Higgs model [12]. The resulting phenomenology of Higgs-fermion interactions is quite different from that in the first example given above.

A final set of conditions that must be satisfied by any model of electroweak symmetry breaking derives from the requirement that the $V_L V_L \to V_L V_L$ and $f_+ \bar{f}_+ \to V_L V_L$ amplitudes not violate unitarity bounds. (V can be either W or Z, and the $+$ subscript on the f indicates positive helicity.) There are two aspects to this requirement. First, as an automatic consequence of the gauge structure and renormalizability of any gauge theory, no partial wave amplitude can grow with energy. This condition requires non-trivial cancellations among Feynman diagrams which contribute to a given process. For example, in $WW \to WW$ scattering, the cancellation of the growing energy terms is guaranteed in the Standard Model because of the tree-level relation $g_{\phi^0 WW} = g m_W$, where g is the gauge coupling. In models with more complicated Higgs sectors, it is no longer necessary that a single scalar boson alone cure these unitarity problems. It is only necessary that the following sum rules for the scalar boson VV and $f\bar{f}$ couplings be obeyed [13]

$$\sum_i g_{h_i^0 VV}^2 = g_{\phi^0 VV}^2 ; \qquad (4.6)$$

and

$$\sum_i g_{h_i^0 VV} g_{h_i^0 f\bar{f}} = g_{\phi^0 VV} g_{\phi^0 f\bar{f}} . \qquad (4.7)$$

It should be noted, however, that these sum rules are only appropriate in models containing doublet and singlet Higgs fields. Higgs sector extensions involving triplet or higher Higgs representations lead to more complicated sum rules. An example will be given when we discuss a model with Higgs triplets in §6.4.

The second aspect of the unitarity bound requirement deals with the size of the constant (energy independent) term of the partial wave amplitude. Tree-level unitarity would limit the size of this constant, which would then impose an upper limit on the Higgs boson mass(es). For example, as discussed in §2.6, the requirement of tree-level unitarity applied to the processes $V_L V_L \to V_L V_L$ and $f_+ \bar{f}_+ \to V_L V_L$ implies an upper bound for m_{ϕ^0} [14]. In models with more complicated Higgs sectors, a similar analysis requires that at least one neutral scalar must have mass below ~ 1 TeV [15,16] . (Additional constraints can be obtained by examining Higgs–Higgs scattering [17].) Furthermore, those Higgs bosons with mass of order 1 TeV or below must approximately saturate the coupling constant sum rules given in eqs. (4.6) and (4.7). This is an interesting result, since it allows the existence of grand unified models which possess superheavy Higgs bosons, without violating tree-level unitarity. By our observation above, such superheavy bosons must be extremely weakly coupled to $W^+ W^-$ and ZZ. Finally, it should be remarked that the requirement of tree-level unitarity (the second aspect of the unitarity bound) can only be taken as a general guide, since if it is violated then the Higgs sector is presumably strongly interacting. In this case, perturbative arguments become extremely suspect.

The main lesson to draw from the discussion so far is that there is still plenty of freedom for the Higgs sector in the Standard Model, although the choice is not totally arbitrary. It is also clear that models with multiple doublets are the preferred models for non-minimal Higgs structures, although there is still room for investigation outside this framework. Finally, the two-Higgs-doublet version of the Standard Model is particularly attractive because:

1. It is an extension of the minimal model which adds new phenomena (*e.g.* physical charged Higgs bosons).
2. It is a minimal extension in that it adds the fewest new arbitrary parameters.
3. It satisfies theoretical constraints of $\rho \approx 1$ and the absence of tree-level FCNC's (if the Higgs-fermion couplings are appropriately chosen).
4. Such a Higgs structure is required in "low-energy" supersymmetric models.

4.1   Two-Higgs-Doublet Model—General Analysis

Theoretical Structure

We begin by investigating the minimal extension of the Higgs sector—the Standard Model with two Higgs doublets [12,18–23]. We introduce two complex $Y = 1$, $SU(2)_L$ doublet scalar fields ϕ_1 and ϕ_2. The Higgs potential which spontaneously breaks $SU(2)_L \times U(1)_Y$ down to $U(1)_{EM}$ is [19]

$$
\begin{aligned}
V(\phi_1, \phi_2) = {} & \lambda_1(\phi_1^\dagger \phi_1 - v_1^2)^2 + \lambda_2(\phi_2^\dagger \phi_2 - v_2^2)^2 \\
& + \lambda_3 \left[(\phi_1^\dagger \phi_1 - v_1^2) + (\phi_2^\dagger \phi_2 - v_2^2) \right]^2 \\
& + \lambda_4 \left[(\phi_1^\dagger \phi_1)(\phi_2^\dagger \phi_2) - (\phi_1^\dagger \phi_2)(\phi_2^\dagger \phi_1) \right] \\
& + \lambda_5 \left[\mathrm{Re}\,(\phi_1^\dagger \phi_2) - v_1 v_2 \cos \xi \right]^2 \\
& + \lambda_6 \left[\mathrm{Im}\,(\phi_1^\dagger \phi_2) - v_1 v_2 \sin \xi \right]^2 ,
\end{aligned}
\tag{4.8}
$$

where the λ_i are all real parameters (by hermiticity). This potential is the most general one subject to gauge invariance and a discrete symmetry, $\phi_1 \rightarrow -\phi_1$, which is only softly violated by dimension-two terms. The latter constraint is a technical one which is related to insuring that flavor-changing neutral currents are not too large (see footnote at the bottom of p. 200). The above potential guarantees the correct pattern of electroweak symmetry breaking over a large range of parameters. That is, if all the λ_i are non-negative, then the minimum of the potential is manifestly

$$\langle \phi_1 \rangle = \begin{pmatrix} 0 \\ v_1 \end{pmatrix}, \qquad \langle \phi_2 \rangle = \begin{pmatrix} 0 \\ v_2 e^{i\xi} \end{pmatrix}, \tag{4.9}$$

which breaks the $SU(2)_L \times U(1)_Y$ down to $U(1)_{EM}$, as desired. In fact, the allowed range of the λ_i corresponding to this desired minimum is somewhat larger, corresponding to that region of parameter space that yields positive squared masses for the physical Higgs bosons, with $V(0,0) > 0$. If $\sin\xi \neq 0$ then there is CP violation in the Higgs sector. This possibility will be discussed in §6.1. Note, however, that if $\lambda_5 = \lambda_6$ then the last two terms of eq. (4.8) can be combined into a term proportional to $|\phi_1^\dagger \phi_2 - v_1 v_2 e^{i\xi}|^2$ and the phase ξ can be rotated away by a redefinition of one of the fields without affecting the other terms in the potential. This is what happens in the supersymmetry model to be discussed later. In the present discussion we henceforth set $\xi = 0$. In this case, eq. (4.8) represents the most general CP invariant potential, subject to the above remarks.

     A key parameter of the model is the ratio of the vacuum expectation values

$$\tan\beta = v_2/v_1. \tag{4.10}$$

It is straightforward to remove the Goldstone bosons and determine the physical Higgs states. In the charged sector, the charged Goldstone boson is

$$G^\pm = \phi_1^\pm \cos\beta + \phi_2^\pm \sin\beta, \tag{4.11}$$

and the physical charged Higgs state is orthogonal to $G^\pm$

$$H^\pm = -\phi_1^\pm \sin\beta + \phi_2^\pm \cos\beta, \tag{4.12}$$

with mass $m_{H^\pm}^2 = \lambda_4(v_1^2 + v_2^2)$. Due to the CP-invariance (assumed above), the imaginary parts and the real parts of the neutral scalar fields decouple. In the imaginary (CP-odd) sector, the neutral Goldstone boson is

$$G^0 = \sqrt{2}(\mathrm{Im}\,\phi_1^0 \cos\beta + \mathrm{Im}\,\phi_2^0 \sin\beta), \tag{4.13}$$

and the orthogonal neutral physical state is

$$A^0 = \sqrt{2}(-\mathrm{Im}\,\phi_1^0 \sin\beta + \mathrm{Im}\,\phi_2^0 \cos\beta), \tag{4.14}$$

with mass $m_{A^0}^2 = \lambda_6(v_1^2 + v_2^2)$. The real ($CP$-even) sector contains two physical Higgs scalars which mix through the following mass-squared matrix

$$\mathcal{M} = \begin{pmatrix} 4v_1^2(\lambda_1 + \lambda_3) + v_2^2\lambda_5 & (4\lambda_3 + \lambda_5)v_1 v_2 \\ (4\lambda_3 + \lambda_5)v_1 v_2 & 4v_2^2(\lambda_2 + \lambda_3) + v_1^2\lambda_5 \end{pmatrix}. \tag{4.15}$$

The physical mass eigenstates are

$$\begin{aligned} H^0 &= \sqrt{2}\left[(\mathrm{Re}\,\phi_1^0 - v_1)\cos\alpha + (\mathrm{Re}\,\phi_2^0 - v_2)\sin\alpha\right], \\ h^0 &= \sqrt{2}\left[-(\mathrm{Re}\,\phi_1^0 - v_1)\sin\alpha + (\mathrm{Re}\,\phi_2^0 - v_2)\cos\alpha\right]. \end{aligned} \tag{4.16}$$

The corresponding masses are

$$m^2_{H^0,h^0} = \tfrac{1}{2} \left[\mathcal{M}_{11} + \mathcal{M}_{22} \pm \sqrt{(\mathcal{M}_{11} - \mathcal{M}_{22})^2 + 4\mathcal{M}^2_{12}} \right] , \qquad (4.17)$$

and the mixing angle α is obtained from

$$\sin 2\alpha = \frac{2\mathcal{M}_{12}}{\sqrt{(\mathcal{M}_{11} - \mathcal{M}_{22})^2 + 4\mathcal{M}^2_{12}}} ,$$

$$\cos 2\alpha = \frac{\mathcal{M}_{11} - \mathcal{M}_{22}}{\sqrt{(\mathcal{M}_{11} - \mathcal{M}_{22})^2 + 4\mathcal{M}^2_{12}}} . \qquad (4.18)$$

Note that according to eq. (4.17), $m_{H^0} \geq m_{h^0}$ as suggested by the notation.

To summarize, this model possesses five physical Higgs bosons: a charged pair ($H^\pm$); two neutral CP-even scalars (H^0 and h^0, where, by convention, $m_{H^0} > m_{h^0}$); and a neutral CP-odd scalar (A^0), often called a pseudoscalar. Instead of the one free parameter of the minimal model, this model has six free parameters: four Higgs masses, the ratio of vacuum expectation values, $\tan\beta$, and a Higgs mixing angle, α. Note that $v_1^2 + v_2^2$ is fixed by the W mass $m_W^2 = g^2(v_1^2 + v_2^2)/2$.

It is, of course, critically important to examine the couplings of the physical Higgs bosons to vector bosons and fermion pairs. These couplings control the production and decay of the Higgs bosons. First, consider the couplings to vector bosons. To understand the pattern of couplings, consider the fact that the Standard Model, *in the absence of the quarks and leptons*, separately conserves C and P. Thus we can assign unique J^{PC} quantum numbers to all the bosons of the theory, if the fermions are ignored. The quantum number assignments are displayed in table 4.1 [24].

At first glance the C and P assignments for A^0 are surprising.* Formally, one could look at the bosonic sector of the Lagrangian, observe that certain terms are missing, and note that the C, P choices of table 4.1 are sufficient to explain the absence of $A^0 W^+ W^-$ and $A^0 ZZ$ couplings. Physically, one can understand the results as follows. In a one-doublet model, the imaginary part of the neutral Higgs field is the Goldstone boson which is "eaten" and becomes the longitudinal component of the Z. This field, like all Goldstone boson fields, is derivatively coupled and is, therefore, CP-odd. Since the bosonic sector conserves C and P separately, the Goldstone boson must in fact have the $C = -1$ quantum number of the Z. Its $P = +1$ quantum number is the same as that of the other scalar components, and is opposite in sign from the parity of the vector bosons due to the one unit difference in spin. In a two-doublet model, there are two neutral Higgs fields with imaginary components.

---

*The existence of the ZH^+H^- vertex implies that Z is a 1^{--} vector boson, and the $H^0 h^0 h^0$ vertex implies that H^0 is a 0^{++} scalar. It then follows from the existence of the $ZH^0 A^0$ vertex that A^0 is both C-odd and CP-odd, as indicated in table 4.1.

Table 4.1

Quantum numbers of Higgs and Gauge Bosons

	J^{PC}		J^P
When C and P are separately conserved			
γ	1^{--}	$W^\pm$	1^-
Z	1^{--}	$H^\pm$	0^+
H^0	0^{++}		
h^0	0^{++}		
A^0	0^{+-}		
When C and P are violated but CP is conserved			
γ	1^{--}	$W^\pm$	$1^-,1^+$
Z	$1^{--},1^{++}$	$H^\pm$	$0^+,0^-$
H^0	$0^{++},0^{--}$		
h^0	$0^{++},0^{--}$		
A^0	$0^{+-},0^{-+}$		

One linear combination of the imaginary components is the Goldstone boson, and the other linear combination is A^0. Both these fields must have the same C and P quantum numbers; hence, the 0^{+-} assignment for A^0 given in table 4.1.

A second argument can be given for the absence of a tree-level coupling of the A^0 to vector boson pairs. First, let us recall that the coupling of the CP-even scalar Higgs boson(s) to a pair of massive vector bosons arises from the covariant derivative $(D_\mu\phi)(D^\mu\phi)$ terms in the Lagrangian after replacing one of the ϕ's by its vacuum expectation value. However, in a CP conserving theory this mechanism does not generate a coupling for the CP-odd A^0. (This is because in the convention adopted here, where the vacuum expectation value of ϕ is taken to be real, the A^0 derives from the imaginary component of ϕ.) Nor does this tree-level mechanism generate a coupling of a CP-even Higgs to a massless vector boson pair, e.g. $\gamma\gamma$. Both types of coupling occur only at the one-loop level. More formally, since the A^0 is CP-odd, a gauge invariant interaction must take the form $\epsilon^{\mu\nu\alpha\beta}F_{\mu\nu}F_{\alpha\beta}A^0$. However, this is a dimension-five term which cannot appear in the fundamental Lagrangian, and is only generated by loop graphs. Similarly, at one loop, dimension-five couplings of vector boson pairs to a CP-even Higgs, of the form $F_{\mu\nu}F^{\mu\nu}h$, are also generated; indeed, the coupling of two photons or two gluons to a CP-even Higgs arises in exactly this way.

In common parlance, the A^0 is usually referred to as a *pseudoscalar*. This is technically incorrect, since we have seen above that in the absence of fermions, the A^0 has $P = +1$ (and $C = -1$). Incorporating the fermions into the theory, C and P are no longer separately conserved, although CP remains a good quantum number (to a very good approximation). Thus, it is more precise to refer to A^0 as being CP-odd. The reason that A^0 is sometimes referred to as being a pseudoscalar will be revealed shortly. In any case, with the assignments of table 4.1 in hand, it is easy to see which boson couplings are forbidden. Of course, the coupling of the Z to a pair of identical Higgs bosons is forbidden by Bose symmetry. For a pair of nonidentical Higgs the coupling is only present when the two Higgses have opposite CP quantum numbers; *i.e.* ZA^0H^0 and ZA^0h^0 are allowed. The ZZA^0 and $W^+W^-A^0$ couplings are absent for the reasons discussed above. In terms of the quantum number assignments in table 4.1, we would say that these couplings are forbidden by C-invariance! Of course, since C-invariance is broken when we introduce the quarks and leptons, we generate these couplings radiatively through fermion loops. There are a few other vertices forbidden at tree level for other reasons. Vertices involving neutral particles only and one or two photons clearly vanish at tree level, although they are generated at one loop. The same is true for the coupling of all neutral Higgs bosons to a pair of gluons. The radiatively generated A^0gg, H^0gg, and h^0gg vertices are important since two-gluon fusion is one of the major production mechanisms for neutral Higgses at a hadron collider. Similarly, the $\gamma\gamma$ couplings may be very important in detection of the neutral Higgs. Two other vertices, $H^+W^-\gamma$ and H^+W^-Z, also vanish at tree level. The $H^+W^-\gamma$ tree-level vertex is zero as a consequence of the conservation of the electromagnetic current. The vanishing of the H^+W^-Z vertex is more model dependent; it turns out to be a general feature of models with only Higgs doublets and singlets [25–27]. (We will return to this point again in §6.3.) Again, these vertices are radiatively generated at one loop, and lead to interesting rare decays of the charged Higgs [29,28]. All other three-point tree-level vertices involving gauge and Higgs bosons are allowed.

Probably the most important vertices for phenomenology are the couplings of H^0 and h^0 to W^+W^- and ZZ. These couplings tend to be somewhat suppressed compared to their values in the minimal-Higgs model, as is apparent from the sum rule (4.6) which reduces in the present case to

$$g^2_{H^0VV} + g^2_{h^0VV} = [g^{\text{minimal}}_{\phi^0VV}]^2 \tag{4.19}$$

and holds separately for $V = W$ or Z. In terms of the angles α and β defined earlier, we have [17,30] (see the Feynman rules given in Appendix A)

$$\frac{g_{h^0VV}}{g_{\phi^0VV}} = \sin(\beta - \alpha)$$

$$\frac{g_{H^0VV}}{g_{\phi^0VV}} = \cos(\beta - \alpha),$$

$$(4.20)$$

and the sum rule of eq. (4.19) is obviously satisfied. Without specific predictions for α and β, one might be tempted to say that the scalar Higgs coupling to vector boson pairs should be down by roughly a factor of $\sqrt{2}$ compared with the minimal Higgs model. However, we will see later that in the supersymmetric model this expectation is generally false; in particular $\cos(\beta - \alpha)$ tends to be quite small, and $\sin(\beta - \alpha)$ is near 1.

We now turn to the Higgs-fermion couplings. As discussed above, even in the two-Higgs doublet model, the Higgs-fermion coupling is model dependent. Even if one imposes the theorem of Glashow and Weinberg [11] to forbid tree-level FCNC's induced by Higgs exchange, one still has numerous choices for how to couple the quarks and leptons to the two Higgs doublets. A set of discrete symmetries can always be concocted to make a particular choice natural (in the technical sense).[†] For example, one can couple one Higgs doublet to all quarks and leptons and decouple the second Higgs doublet from the fermions [12]. A second choice is one where one Higgs doublet couples only to up-type quarks and leptons and the second Higgs doublet couples only to down-type quarks and leptons. Axion models and supersymmetric models are examples of this choice. Other choices can be made where the quark and lepton couplings are treated asymmetrically. Although the Feynman rules for the Higgs-fermion interaction differ depending on the choice made, there are a number of common features among all such models. First, the Higgs-fermion couplings can be either enhanced or suppressed compared to the minimal-Higgs model, depending on the parameters of the model. Second, introducing the fermions means that C and P are no longer separately conserved. However, in the approximation where CP is conserved, the Higgs and vector bosons can be thought of as admixtures of two eigenstates of definite C and P, as indicated in table 4.1. Consider then the coupling of the neutral Higgs to a fermion–antifermion pair. It is well known that an $f\bar{f}$ pair

---

[†]Such a discrete symmetry must necessarily be of the form $\phi_1 \to -\phi_1$, $\phi_2 \to +\phi_2$ (or vice versa) and appropriate choices for the transformations of the various right-handed quark fields. Note that the potential of eq. (4.8) violates this symmetry softly. Nevertheless, we are free to choose a Higgs fermion coupling which avoids tree-level FCNC's; the effect of the soft violation of the discrete symmetry would be to radiatively generate FCNC's which are calculable, finite, but presumably can be arranged to be small enough to avoid conflict with experiment.

has $P = (-1)^{L+1}$ and $C = (-1)^{L+S}$ for total spin S and orbital angular momentum L, and thus cannot couple to 0^{--} and 0^{+-}. Therefore, for the neutral Higgs couplings to $f\bar{f}$, the H^0 and h^0 behave as pure 0^{++} scalars, whereas A^0 behaves as a pure 0^{-+} pseudoscalar. As a result, it is common to refer to A^0 as a pseudoscalar, even though this is only a correct statement in the context of its interaction with an $f\bar{f}$ pair.

We now explicitly display the two choices for the Higgs-quark interactions which were mentioned above. We denote by Model I the case in which the quarks and leptons do not couple to the first Higgs doublet (ϕ_1), but couple to the second Higgs doublet (ϕ_2) in a manner analogous to the minimal Higgs model [12]. In Model II, we assume that ϕ_1 couples only to down-type quarks and leptons and ϕ_2 couples only to up-type quarks and neutrinos. (Note that because the neutrinos are massless, there are no couplings of neutrinos to physical neutral Higgs bosons.)

Consider a three-generation model with diagonal (positive) quark matrices M_U and M_D (for the charge 2/3 and $-1/3$ quarks respectively) and Kobayashi-Maskawa mixing matrix K. Then, in Model I, the Higgs-fermion interaction takes the following form [12,31]

$$
\begin{aligned}
\mathcal{L}_{Hf\bar{f}} = & -\frac{g}{2m_W \sin\beta}\overline{D}M_D D(H^0 \sin\alpha + h^0 \cos\alpha) - \frac{ig\cot\beta}{2m_W}\overline{D}M_D\gamma_5 DA^0 \\
& - \frac{g}{2m_W \sin\beta}\overline{U}M_U U(H^0 \sin\alpha + h^0 \cos\alpha) + \frac{ig\cot\beta}{2m_W}\overline{U}M_U\gamma_5 U A^0 \\
& + \frac{g\cot\beta}{2\sqrt{2}m_W}\left(H^+\overline{U}[M_U K(1-\gamma_5) - KM_D(1+\gamma_5)]D + \text{h.c.}\right).
\end{aligned}
$$

$$(4.21)$$

In this case, $\tan\beta \equiv v_2/v_1$, where v_2 is the vacuum expectation value of the Higgs field which couples to *both* up and down-type quarks (whereas the other Higgs field is decoupled from the quarks). In contrast, the Model II interaction is [20,31]

$$
\begin{aligned}
\mathcal{L}_{Hf\bar{f}} = & -\frac{g}{2m_W \cos\beta}\overline{D}M_D D(H^0 \cos\alpha - h^0 \sin\alpha) + \frac{ig\tan\beta}{2m_W}\overline{D}M_D\gamma_5 DA^0 \\
& - \frac{g}{2m_W \sin\beta}\overline{U}M_U U(H^0 \sin\alpha + h^0 \cos\alpha) + \frac{ig\cot\beta}{2m_W}\overline{U}M_U\gamma_5 U A^0 \\
& + \frac{g}{2\sqrt{2}m_W}\left(H^+\overline{U}[\cot\beta M_U K(1-\gamma_5) + \tan\beta KM_D(1+\gamma_5)]D + \text{h.c.}\right).
\end{aligned}
$$

$$(4.22)$$

This time, we define $\tan\beta \equiv v_2/v_1$, where v_1 (v_2) is the vacuum expectation value of the Higgs field which couples only to down-type (up-type) quarks.

In the two equations above, U and D are column matrices consisting of three generations of quark fields.[‡] (In both Models I and II, the Higgs-lepton couplings can be read off from the expressions above by replacing (U, D) with the corresponding lepton fields, replacing quark mass matrices with the corresponding *diagonal* lepton mass matrices, and setting $K = 1$.) Note that as previously advertised, the neutral Higgs interactions are flavor diagonal. In addition, the structure of the charged Higgs interactions involving the Kobayashi-Maskawa matrix is analogous to that of the ordinary charged current mediated by the W.

Since the Model II choice for the Higgs-fermion couplings is the required structure for the minimal supersymmetric model (to be discussed in great detail in the next section), we summarize the results of eq. (4.22) for the couplings of the neutral Higgs bosons relative to the canonical SM values below (using 3rd family notation)

$$H^0 t\bar{t} : \quad \frac{\sin\alpha}{\sin\beta} \qquad H^0 b\bar{b} : \quad \frac{\cos\alpha}{\cos\beta}$$

$$h^0 t\bar{t} : \quad \frac{\cos\alpha}{\sin\beta} \qquad h^0 b\bar{b} : \quad \frac{-\sin\alpha}{\cos\beta} \qquad (4.24)$$

$$A^0 t\bar{t} : \quad \cot\beta \qquad A^0 b\bar{b} : \quad \tan\beta,$$

where we must keep in mind that A^0 is coupled via a γ_5 to a $q\bar{q}$ pair. One may check that the unitarity sum rule, eq. (4.7), is satisfied by the h^0 and H^0 fermion–antifermion couplings given in eq. (4.24) and VV couplings given in eq. (4.20). We note that the $A^0 u\bar{u}$ ($A^0 d\bar{d}$) coupling is suppressed (enhanced) if $\tan\beta > 1$, and vice versa if $\tan\beta < 1$. Similar results hold for H^0 and h^0, although the couplings also involve the mixing angle α which can reduce the size of the couplings somewhat. Note, that if one of the scalars completely saturates the sum rule of eq. (4.19), and thus has Standard-Model-like WW couplings, then the fermion unitarity constraint of eq. (4.7) can be satisfied by giving this scalar Standard Model fermion couplings as well. This is the scenario that we shall find emerges in minimal supersymmetry. But, clearly this is not the only way of satisfying the sum rule. For instance, the fermion couplings of the scalar that is weakly coupled to VV could be large enough that the product of the VV and $f\bar{f}$ couplings is still significant. For the

---

[‡]The above interactions have been given in the unitary gauge where we set the Goldstone fields to zero. In the R-gauge (see Appendix A), an additional interaction term involving the Goldstone fields appears, which is identical in both models

$$\mathcal{L}_{Gf\bar{f}} = \frac{ig}{2m_W} \overline{U} M_U \gamma_5 U G^0 - \frac{ig}{2m_W} \overline{D} M_D \gamma_5 D G^0$$
$$+ \frac{g}{2\sqrt{2}m_W} \left(G^+ \overline{U} [M_U K(1-\gamma_5) - K M_D (1+\gamma_5)] D + \text{h.c.} \right) . \qquad (4.23)$$

charged Higgs boson, the Model II coupling to the $t\bar{b}$ channel is given by

$$g_{H-t\bar{b}} = \frac{g}{2\sqrt{2}m_W} \left[m_t \cot\beta(1+\gamma_5) + m_b \tan\beta(1-\gamma_5) \right]. \quad (4.25)$$

Note that the t-quark-mass piece is suppressed for $\tan\beta > 1$. The pattern of suppressed and enhanced couplings of Model I is quite distinct from that of Model II. In contrast to the Model II couplings just displayed, the Model I couplings of the pseudoscalar and charged Higgs bosons to *all* fermion types are uniformly suppressed (enhanced) if $\tan\beta > 1$ ($\tan\beta < 1$). A similar remark can be made concerning the neutral scalar Higgs-fermion couplings, although one must also take the dependence on the mixing angle α into account [see eq. (4.21)].

Finally, a very important new set of couplings appears in the two-doublet model. These are the V-Higgs–Higgs couplings. Defining the Feynman coupling for H^+W^-h as the coefficient of $-i(p+p') \cdot \epsilon_W$ (where p and p' are the four-momenta of H^+ and h, respectively) and using an analogous definition for the A^0ZH coupling, we have

$$g_{H^+W^-h^0}^2 + g_{H^+W^-H^0}^2 = \left[g_{A^0Zh^0}^2 + g_{A^0ZH^0}^2 \right] \cos^2\theta_W = g^2/4. \quad (4.26)$$

These sum rules are required in order to satisfy unitarity constraints in $H^+W^- \to H^+W^-$ and $A^0Z \to A^0Z$ [13]. In order to avoid violating unitarity in $A^0Z \to W^+W^-$, the ZA^0h^0 and ZA^0H^0 coupling strengths must have exactly the correct ratio. Similarly, unitarity for the $H^+W^- \to W^+W^-$ scattering process determines the $W^+H^-h^0$ and $W^+H^-H^0$ couplings. Thus, the unitarity constraints uniquely fix the V-Higgs–Higgs couplings to be [13]

$$g_{h^0A^0Z} = \frac{g}{2\cos\theta_W} \cos(\beta-\alpha) \qquad g_{H^+W^-h^0} = \frac{g}{2}\cos(\beta-\alpha)$$

$$g_{H^0A^0Z} = \frac{g}{2\cos\theta_W} \sin(\beta-\alpha) \qquad g_{H^+W^-H^0} = \frac{g}{2}\sin(\beta-\alpha), \quad (4.27)$$

in agreement with the result obtained directly by expanding out the covariant derivatives in the Higgs kinetic energy terms.

One-Loop Corrections to the Tree-level Higgs Potential

All of our analysis so far has been based on the tree-level Higgs potential. In some cases, however, the radiative corrections to the Higgs potential are important. For example, the Coleman-Weinberg mechanism refers to the case where all tree-level mass parameters are set to zero. Although the tree-level Higgs potential does not spontaneously break the gauge symmetry, it is found that spontaneous symmetry breaking is induced by the one-loop radiative corrections. Furthermore, Linde and Weinberg showed that it was even possible to start with a small positive tree-level Higgs mass-squared (m^2) and induce spontaneous symmetry breaking in the one-loop radiatively corrected potential. The requirement that the spontaneously broken minimum be a global minimum places an upper limit on m^2, which in turn puts a lower limit on the Higgs boson mass. These issues were discussed in §2.5 in regard to the minimal Higgs model. In models with more than one Higgs doublet, the discussion becomes more complicated. Other cases arise where radiative corrections become important. For example, it is possible to have a zero tree-level scalar mass without setting all tree-level mass parameters to zero. In this case, radiative corrections will be important in determining the correct value for the light Higgs scalar mass. We shall briefly discuss some of the new features in the context of the two-Higgs doublet model.

First, it is important to remark that it is possible to have a very light pseudoscalar in the theory. There is no Linde-Weinberg type bound for the pseudoscalar mass. This is easy to understand—if the mass of the pseudoscalar is zero at tree-level, there will be an extra U(1) global symmetry in the Higgs potential which is spontaneously broken; the pseudoscalar is then identified as the Goldstone boson corresponding to this broken U(1). If this extra U(1) global symmetry is extended to the rest of the model, then the pseudoscalar will remain massless to all orders in perturbation theory. (It may turn out that this global symmetry is anomalous. This is what happens in the axion model, in which case the pseudoscalar axion is a pseudo-Goldstone boson which acquires a small mass due to instanton effects.) Thus, we focus our attention on the Higgs *scalars* of the model.

It is convenient to rotate the fields $\mathrm{Re}\,\phi_1^0$ and $\mathrm{Re}\,\phi_2^0$ such that only one scalar field possesses a vacuum expectation value. That is, we define

$$\xi = \cos\beta\,\mathrm{Re}\,\phi_1^0 + \sin\beta\,\mathrm{Re}\,\phi_2^0$$
$$\eta = -\sin\beta\,\mathrm{Re}\,\phi_1^0 + \cos\beta\,\mathrm{Re}\,\phi_2^0,$$

(4.28)

so that

$$\langle\eta\rangle = 0, \qquad \langle\xi\rangle = \frac{v}{\sqrt{2}} \equiv \sqrt{v_1^2 + v_2^2},$$

(4.29)

where we have chosen the normalization such that $v = 2m_W/g \simeq 246$ GeV. We shall restrict our study to the direction in field space where $\eta = 0$. It is only in this direction where the radiative corrections can play a significant role. Suppose the tree-level potential in this direction is $V_{\mathrm{tree}} = a\xi^4$; *i.e.* there is no quadratic term. This is the analogue of the Coleman-Weinberg mechanism in the minimal Higgs model. The radiative corrected potential in this direction is

$$V(\xi) = a\xi^4 + \frac{\Lambda^2}{32\pi^2}\mathrm{Str}\,M_i^2(\xi)$$
$$+ \frac{1}{64\pi^2}\mathrm{Str}\left\{M_i^4(\xi)\left[\log\left(\frac{M_i^2(\xi)}{\Lambda^2}\right) - \frac{1}{2}\right]\right\} + V_{\mathrm{c.t.}},$$

(4.30)

where Λ is the ultraviolet cutoff, and $V_{\mathrm{c.t.}}$ are the counterterms consisting of a quartic polynomial in ξ (containing only even powers of ξ) chosen to satisfy some suitable boundary conditions for the full potential. The notation Str stands for

$$\text{Str} \{\cdots\} \equiv \sum_i C_i(2J_i + 1)(-1)^{2J_i} \{\cdots\}, \qquad (4.31)$$

where the sum is taken over all particles of spin J_i and mass $M_i(v/\sqrt{2})$ in the theory which couple to ξ. (In eq. (4.30), $M_i(\xi)$ indicates that $v/\sqrt{2}$ should be replaced by the field ξ.) The constant C_i counts the other degrees of freedom such as electric charge and color ($e.g.$, $C = 2$ for the W-boson and $C = 6$ for a quark, since we count both particle and antiparticle). The Higgs mass is easily computed by noting that

$$m_{CW}^2 = \frac{1}{2} \left. \frac{\partial^2 V}{\partial \xi^2} \right|_{\xi = v/\sqrt{2}}, \qquad (4.32)$$

where we use the notation m_{CW} to remind the reader of the analogy with the Coleman-Weinberg case, where the tree-level scalar mass is zero. The factor of $\frac{1}{2}$ is required according to the normalization of ξ as defined above. In order to obtain a specific result for m_{CW}, we must know how the masses $M_i(v/\sqrt{2})$ of the particles which couple to ξ depend on v. The simplest possibility which is often realized is

$$\left. M_i^2(\xi) \right|_{\xi = v/\sqrt{2}} = \mu_i^2 + \lambda_i v^2. \qquad (4.33)$$

(More complicated mass expressions can arise in supersymmetric models; for example, neutralino and chargino masses have rather complicated dependencies on v.) In a nonsupersymmetric model, the vector bosons and the fermions all have $\mu_i = 0$ since their masses are induced by the spontaneous symmetry breaking and so must be proportional to v. On the other hand, the other Higgs scalar masses can have a more general form as indicated in eq. (4.33). If we take this latter expression to be the general expression for particle masses, then the Higgs mass is easily computed to be [32]

$$m_{CW}^2 = \text{Str} \frac{\lambda_i^2 v^2}{8\pi^2} \left[1 - \frac{\mu_i^2}{\lambda_i v^2} \log \left(\frac{\mu_i^2 + \lambda_i v^2}{\mu_i^2} \right) \right]. \qquad (4.34)$$

It is instructive to examine this formula assuming the particles i can be divided into two classes j and k such that $\mu_j = 0$ and $\mu_k \gg v$. Then, we find [33]

$$m_{CW}^2 = \frac{1}{8\pi^2 v^2} \text{Str} M_j^4 + \frac{v^4}{16\pi^2} \text{Str} \frac{\lambda_k^3}{\mu_k^2}. \qquad (4.35)$$

Thus, particles with masses much larger than v decouple, and we find the familiar expression for the Higgs mass generated by the Coleman-Weinberg mechanism.*

We may also derive the analog of the Linde-Weinberg limit by adding a small positive squared-mass term $m^2 \xi^2$ to the tree-level potential. This reduces the Higgs mass to

$$m_H^2 = m_{CW}^2 - 2m^2 \, . \tag{4.36}$$

We now demand that the spontaneously broken minimum is a global minimum, *i.e.* $V(v/\sqrt{2}) < V(0)$. This condition implies that m^2 must be smaller than some maximum value, which when inserted in eq. (4.36) gives $m_H > m_{LW}$, where

$$m_{LW}^2 = \text{Str} \, \frac{\lambda_i^2 v^2}{16\pi^2} \left\{ 1 - \frac{2\mu_i^2}{\lambda_i v^2} \left[1 - \frac{\mu_i^2}{\lambda_i v^2} \log \left(\frac{\mu_i^2 + \lambda_i v^2}{\mu_i^2} \right) \right] \right\} . \tag{4.37}$$

Once again, let us assume that the particles i can be divided into two classes j and k such that $\mu_j = 0$ and $\mu_k \gg v$. Then, we find

$$m_{LW}^2 = \frac{1}{16\pi^2 v^2} \, \text{Str} \, M_j^4 + \frac{v^4}{24\pi^2} \, \text{Str} \, \frac{\lambda_k^3}{\mu_k^2} \, . \tag{4.38}$$

Note that in the absence of particles of type k (*i.e.*, all masses in the theory are proportional to v), we see that $m_{LW} = m_{CW}/\sqrt{2}$ as in the minimal Higgs model. This need not be true in more complicated situations.

In the discussion above, we have implicitly assumed that ξ is a physical Higgs boson. In general, this need not be true since ξ and η can mix. In this more general case, the above formulae for the Higgs mass refer only to one of the diagonal elements of the 2×2 scalar mass matrix. More generally, the Linde-Weinberg bound would read

$$m_{H^0}^2 \cos^2(\beta - \alpha) + m_{h^0}^2 \sin^2(\beta - \alpha) > m_{LW}^2 \, , \tag{4.39}$$

where h^0 and H^0 are the physical Higgs scalars, α is the scalar mixing angle in the $\text{Re}\,\phi_1^0$–$\text{Re}\,\phi_2^0$ basis, and m_{h^0} and m_{H^0} are the radiatively corrected physical Higgs masses. (Note that if $\alpha = \beta$, then $h^0 = \sqrt{2}\eta$ and $H^0 = \sqrt{2}\xi - v$, and the formulae derived above apply to H^0.) We remind the reader that the Linde-Weinberg bound corresponds to the assumption that the symmetry breaking vacuum is a global minimum. As emphasized by Sher [35], if one assumes a standard hot Big Bang evolution of the early universe, then a more realistic estimate requires that m_{CW} be used in place of m_{LW} in formulae

---

*In the literature, the term "Coleman-Weinberg", when applied to two-Higgs doublet models implied that all tree-level Higgs mass parameters were set equal to zero [34,35]. In our analysis above, we have been more general; we have only assumed the absence of a tree-level Higgs mass in the ξ direction. In the former case, ξ automatically corresponds to a physical Higgs direction. This is no longer true in our more general setting. Thus, we must regard m_{CW}^2 above as simply the corresponding diagonal matrix element of the Higgs mass-squared matrix in the ξ–η basis.

such as eq. (4.39). We conclude that the precise lower bound on the Higgs mass in multi-Higgs doublet models (which applies only to the neutral *scalars*), although similar in form to that of the minimal Higgs model, is more uncertain due to the unknown masses of the other Higgs bosons in the model. In the absence of any information on α and β, the best we can say is that $m_{H^0} > m_{LW}$, where H^0 is the *heavier* of the two Higgs scalars. Of course, a large t-quark mass can eliminate the bound completely just as in the minimal Higgs case. Similar effects can arise in the nonminimal model if new fermions are introduced, such as the charginos and neutralinos in supersymmetric models. The effects of radiative corrections on the Higgs sector of a supersymmetric model have been studied in refs. 32 and 36.

One can also derive upper bounds for Higgs masses using renormalization group methods similar to the ones described in §2.5. The interplay between the Higgs mass bounds and heavy fermion mass bounds is more complicated in non-minimal models, and depends on the unknown parameter $\tan\beta$. A good summary of the literature on this subject can be found in ref. 35.

One-Loop Corrections to ρ in the Two-Doublet Model

As mentioned earlier, the one-loop radiative corrections in the two-Higgs-doublet case were studied in refs. 9 and 10. These references use the renormalization scheme of Marciano and Sirlin [37], in which $m_W = m_Z \cos\theta_W$ is maintained as an exact relationship after renormalization, and radiative corrections appear in the relation between G_F and m_W [see eq. (2.126)]. We find it more intuitive to quote the corrections to ρ following the scheme of Veltman [6], in which the three free parameters of the electroweak theory are taken to be g, $\sin\theta_W$ and m_W. The effective one-loop value for m_Z^2 is then determined by computing the low-energy neutral current $\nu_\mu e$ and $\bar{\nu}_\mu e$ cross sections at one loop and writing them in the tree-level form with a corrected m_Z^2 value. In this way one obtains [6]

$$m_W^2 \Delta\rho = A_{WW}(k^2 = 0) - \cos^2\theta_W A_{ZZ}(k^2 = 0), \qquad (4.40)$$

where A_{WW} and A_{ZZ} are the propagator corrections for the W and Z and are to be evaluated at zero momentum transfer.

In the R-gauge (see Appendix A), the terms that are potentially quadratically sensitive to Higgs boson masses arise entirely from diagrams containing either two Higgs bosons or a Higgs boson and a Goldstone boson, and associated tadpole type graphs. For the two-Higgs-boson graphs we obtain[†]

[†]These results may be easily obtained from the formulae presented in the Appendices of refs. 9 and 38.

$$A_{WW}^{HH}(0) - \cos^2\theta_W A_{ZZ}^{HH}(0) = \frac{g^2}{64\pi^2}\Big[F_{\Delta\rho}(m_{H^+}^2, m_{A^0}^2)$$
$$+ F_{\Delta\rho}(m_{H^+}^2, m_{H^0}^2)\sin^2(\beta - \alpha) + F_{\Delta\rho}(m_{H^+}^2, m_{h^0}^2)\cos^2(\beta - \alpha)$$
$$- F_{\Delta\rho}(m_{A^0}^2, m_{H^0}^2)\sin^2(\beta - \alpha) - F_{\Delta\rho}(m_{A^0}^2, m_{h^0}^2)\cos^2(\beta - \alpha)\Big],$$
(4.41)

where the squares of the couplings of the various Higgs pairs to the W and Z from Appendix A are in evidence. (The first two lines of the equation are from A_{WW} and the third line is from A_{ZZ}.) In eq. (4.41) the function $F_{\Delta\rho}$ is defined as

$$F_{\Delta\rho}(m_1^2, m_2^2) \equiv \tfrac{1}{2}(m_1^2 + m_2^2) - \frac{m_1^2 m_2^2}{m_1^2 - m_2^2}\ln\frac{m_1^2}{m_2^2}.$$
(4.42)

For the Higgs-boson–Goldstone-boson graphs, we again use the couplings of Appendix A to obtain contributions to A_{WW} and A_{ZZ}. However, we must be careful to remove from these contributions pieces that correspond to the Standard Model Higgs boson contributions, in order to assess the extent to which the two-doublet structure gives rise to additional radiative corrections. It is most convenient and most natural to identify the h^0 with the SM Higgs boson for this purpose. Thus, as a schematic example, one obtains a contribution to A_{WW} of the form

$$G_{\Delta\rho}(m_W^2, m_{h^0}^2)\sin^2(\beta - \alpha) + G_{\Delta\rho}(m_W^2, m_{H^0}^2)\cos^2(\beta - \alpha),$$
(4.43)

where $G_{\Delta\rho}$ is an auxiliary function closely related to $F_{\Delta\rho}$. From this one subtracts the SM-like contribution $G_{\Delta\rho}(m_W^2, m_{h^0}^2)$ yielding a net contribution of

$$\cos^2(\beta - \alpha)\left[G_{\Delta\rho}(m_W^2, m_{H^0}^2) - G_{\Delta\rho}(m_W^2, m_{h^0}^2)\right].$$
(4.44)

A similar procedure is followed in the case of A_{ZZ}. Combining, one finds [9]

$$A_{WW}^{HG}(0) - \cos^2\theta_W A_{ZZ}^{HG}(0) = \frac{g^2}{64\pi^2}\cos^2(\beta - \alpha)$$
$$\times \left[F_{\Delta\rho}(m_W^2, m_{H^0}^2) - F_{\Delta\rho}(m_W^2, m_{h^0}^2) - F_{\Delta\rho}(m_Z^2, m_{H^0}^2) + F_{\Delta\rho}(m_Z^2, m_{h^0}^2)\right]$$
(4.45)

We are now in a position to discuss the nature of these one-loop corrections. From the structure of eq. (4.45) and the similarity of m_W^2 to m_Z^2, it is clear that the Higgs–Goldstone contribution is always small. In contrast, the Higgs–Higgs contribution to $\Delta\rho$ could be large, depending upon the mixing angles and masses.

Some analysis was performed in refs. 9 and 10. In addition to constraints on $\Delta\rho$ (or, in their scheme, G_F), information from high-energy neutral current scattering was included. It was found that experimental constraints from neutral current measurements of $\sin^2\theta_W$, and on m_W and m_Z combine to imply that the charged Higgs of a two-doublet model cannot have a mass that is much larger than the neutral Higgs bosons (excepting any neutral Higgs that has SM-like couplings). Typically the $H^\pm$ cannot be split by more than $\sim 7m_Z$ from the neutral Higgs bosons. Future experiments can be expected to reduce the allowed magnitude of this splitting to $\sim 1m_Z$. However, it should be noted that the minimal supersymmetric model predicts that $m_{H^0} \sim m_{A^0} \sim m_{H^+}$ once m_{H^+} is large, while h^0 is very SM-like. Noting that $F_{\Delta\rho}(m_{H^+}^2, m_{A^0}^2)$ vanishes as $m_{H^+} \to m_{A^0}$, while for the other terms there is approximate cancellation between Z and W contributions in this same limit, we see that the deviations from SM predictions in the case of the minimal supersymmetric model are expected to be very small. Note, however, that the presence of terms in A_{ZZ}, that cancel the otherwise large contributions to A_{WW}, is critical.

One-Loop Mediated Decays in the Two-Doublet Model

The decays we consider first are those involving two vector particles (*i.e.* g, γ, Z, or $W^\pm$) and a Higgs boson of the two-doublet model, h^0, H^0, A^0, or $H^\pm$. The h^0, H^0 and A^0 decays to gg can be easily obtained from the results of eq. (2.29) in chapter 2 for the ϕ^0. One need only include the appropriate mixing angle factors in the h^0, H^0, and A^0 couplings to $q\bar{q}$ [which depend on the choice of Model I or II, as specified in eqs. (4.21)and (4.22)] when computing the various quark-loop contributions, and generalize the functional form as follows

$$\tau\left[1 + (1 - \tau)f(\tau)\right] \to \tau\left[\xi^h + (1 - \tau\xi^h)f(\tau)\right], \qquad (4.46)$$

where $\xi^h = 1$ for $h = h^0, H^0$ and $\xi^h = 0$ for $h = A^0$, *i.e.* the functional form changes in going from scalar to pseudoscalar quark couplings.

The h^0, H^0 and A^0 decays to $\gamma\gamma$ and $Z\gamma$ are also easily obtained using the basic results of chapter 2, and explicit results appear in Appendix C. We include there also the contributions of superpartner loops in the $\gamma\gamma$ case. In the case of W-boson and quark-loop contributions for h^0 and H^0, one need only modify the SM results to account for the differences in Higgs couplings to vector bosons and fermions. For the A^0 we again have to generalize $F_{1/2}$ of eq. (2.16). For full details see Appendix C.

One-loop decays of the charged Higgs boson of interest are $H^+ \to W^+\gamma$ and $H^+ \to W^+Z$. (As discussed earlier, these vertices are absent at tree-level.) Formulae for the first decay appear in the appendix of ref. 29, while results for the second will soon be available [28]. The rates are found to be small, and these decays will not be discussed further.

Let us briefly outline the implications of the one-loop decay formulae, assuming the Model II choice for Higgs-fermion couplings [eq. (4.24)]. In the case of the gg final state, there are three important points (i) the most important contributions arise from any quark loops where $2m_q > m_h$; (ii) there are systematic coupling differences—for instance, if $\tan\beta > 1$, u,c,t loops are suppressed while d,s,b loops are enhanced in coupling strength; and (iii) loop contributions to A^0 are somewhat stronger than to h^0 and H^0 in the large m_q limit [from eq. (4.46)], $\tau[\ldots] \to 2/3$ for h^0 and H^0, while

$\tau[\ldots] \to 1$ for A^0. Obviously point (iii) implies that the A^0gg coupling is likely to be at least as strong as the h^0gg and H^0gg couplings for the same Higgs mass. Combining points (i) and (ii) we see that the h^0, if very light, will generally have gg coupling contributions from all loops, some suppressed and some enhanced, and is likely to end up with an h^0gg coupling similar to SM strength. For heavier h^0 and for the H^0, if they both have masses above $2m_b$, then the top-quark loop is dominant, and for $\tan\beta > 1$ both are likely to have suppressed $t\bar{t}$ couplings and, hence, gg couplings and decay widths somewhat below SM strength.

Turning to the $\gamma\gamma$ and $Z\gamma$ decay widths, we first remind the reader that the primary contribution in the SM $\phi^0 \to \gamma\gamma, Z\gamma$ decays comes from the W-loop diagrams, and that fermion loops enter with opposite sign to the W loop. Thus, since the h^0 and H^0 couplings to WW are generally smaller than SM strength, while there is no A^0 coupling to WW, it is clear that there is a significant potential for weaker $\gamma\gamma$ and $Z\gamma$ decays in the two-doublet model. The charged-Higgs-loop contributions to these decays tend to be quite small. This arises from two facts: a) the functions F_0 and I_1 [see Appendix C and eq. (2.16)] are small in magnitude—see for example eq. (2.20); and b) at large $m_{H\pm}^2$ the explicit factor of $m_W^2/m_{H\pm}^2$ (see Appendix C) causes the charged Higgs loop contribution to vanish. Thus, for the A^0 the only way of obtaining a $\gamma\gamma$ width that is comparable to SM strength is to have a strongly enhanced fermion coupling. For instance, if $\tan\beta \gg 1$ and $2m_b > m_{A^0}$ one might have significant $A^0 \to \gamma\gamma, Z\gamma$ decay rates. If the h^0 has nearly SM strength WW coupling it will have SM-like decay widths, while the H^0 must then have weak WW coupling and generally very small $\gamma\gamma$ and $Z\gamma$ decay widths—indeed, there is a strong potential for the weak WW loop contribution to cancel against the fermion loop contributions.

Loop graphs involving the two-doublet Higgs bosons can also lead to enhancement of otherwise highly forbidden decays of SM particles. A particularly interesting example is the charged Higgs loop contribution to the flavor-changing decays $Z \to b\bar{s}, t\bar{c}$. Such processes have been studied in ref. 39. Whereas in the one-doublet SM such decays have a branching ratio of at most $\sim 10^{-10}$, in the two-doublet model the branching ratios for such decays can be a few times 10^{-6}.

Limits on Very Light Non-minimal Neutral Higgs Bosons

As we have discussed in §3.2, there are many indirect and direct limits on Higgs bosons which are lighter than about 5 GeV. In that section the exact limits were discussed for the case of the SM Higgs boson. Here, we give a brief survey of the changes in that discussion that must be incorporated in the case of a neutral Higgs boson that is part of a more complex Higgs sector. In particular, we focus on the two-doublet extension with Higgs-fermion couplings according to Model II [eq. (4.24)]. (A more thorough discussion of the changes and the resulting limits that can be obtained in this case appears in ref. 40.) Of course, as we shall discuss in §4.2, the motivation for a very light Higgs boson is much stronger in the supersymmetry-based models with two or more doublets than it is in the Standard Model, and still more modifications to the discussion given below will arise. In our discussion below we use h^0 for our typical light scalar boson, and A^0 for a possible light pseudoscalar boson.

Clearly, the modifications in our earlier discussion are quite model dependent. The first important remark is that there is no general top-quark-mass-dependent lower bound on m_{h^0}. As we have argued earlier, in the absence of information about scalar masses and mixing angles, there is only a lower limit on the heavier of the two neutral CP-even scalars. Thus, it could happen that m_t will turn out to be somewhat lower than m_W (see §1.5) while at the same time $m_{h^0} < 5$ GeV. This would clearly adversely affect bounds based on rare K and rare B decays since rates for $K \rightarrow h^0 \pi$ and $B \rightarrow h^0 X$ both decrease rapidly with decreasing m_t (as discussed in §3.1).

Indeed, we have seen that the h^0 will generally couple to vector bosons, quarks and leptons with strength different from the SM ϕ^0. From the sum rule of eq. (4.6) and eq. (4.20), it is apparent that the $h^0 VV$ coupling is inevitably somewhat suppressed in general. Further, for $\tan \beta > 1$ (as preferred in model building, see §4.2), eq. (4.24) indicates that the h^0 couplings to $\mu^+ \mu^-$, $b\bar{b}$ and $s\bar{s}$ tend to be enhanced with respect to those of the ϕ^0, while the h^0 couplings to $t\bar{t}$ are suppressed. These changes can alter the rare K and B decay rates to h^0 substantially with respect to the rates for the ϕ^0. Let us first discuss the four-quark operator contribution which can be an important component of the K-decay rate in the SM. In the SM there is only one contributing diagram, which involves two W's fusing to form the ϕ^0. This same diagram is present in the two-doublet model, and is proportional to the $h^0 WW$ coupling squared, and will be somewhat suppressed relative to the ϕ^0 case. The additional four-quark diagrams, which arise in a two-doublet Higgs model by replacing one or both of the W's by a charged Higgs, involve $H^+ q'\bar{q}$ couplings that are suppressed by m_q/m_W where m_q is a light quark mass. These can probably be safely neglected. Turning next to the two-quark operator contributions, we find that for both rare K and B decays they become very complex in the two-doublet model case. Some general formulae appear in ref. 41, but there are some errors as delineated in ref. 40 and, further, phenomenological analysis performed there was motivated by the now defunct $\zeta(2.2)$ and assumed $\tan \beta \ll 1$. All the various modifications to the WW and $q\bar{q}$ coupling constants play a role, and new diagrams involving the charged Higgs boson (in place of one or more of the W bosons in the SM diagrams) must be included.

In addition, the decays of the h^0 itself are critical in obtaining restrictions on allowed regions of m_{h^0}, using the K and B decay data. The decays of the h^0 to $\mu^+ \mu^-$, and $s\bar{s}$ states could be significantly enhanced relative to $\pi\pi$ modes. This is because the $\mu^+ \mu^-$ and $s\bar{s}$ mode decay widths would be increased for enhanced couplings of the h^0 to these channels, while the $\pi\pi$ mode strength depends primarily on the $h^0 gg$ coupling which might be somewhat diminished. The latter is possible since the $h^0 gg$ coupling is induced by heavy quark loops, where heavy is defined by $m_{h^0} < 2m_q$. When c, t, and b are all heavy then, since both the $h^0 t\bar{t}$ and $h^0 c\bar{c}$ couplings are suppressed while only the $h^0 b\bar{b}$ coupling is enhanced, it is likely that the $h^0 gg$ coupling will be smaller than SM strength. Since, for $m_{h^0} < 2m_K$, the relative strength of the $\mu^+\mu^-$ mode compared to the $\pi\pi$ mode is so critical in excluding Higgs bosons in B and K decays, it is possible that the limits from these sources could be improved in regions of parameter space where the K and B decay rates to h^0 are similar to those obtained for the ϕ^0 in the SM as computed with m_t of order m_W.

Turning to the other limits on a light scalar Higgs boson, we note that the $\Upsilon \rightarrow h^0 \gamma$ decay rate depends only on the $h^0 b\bar{b}$ coupling. Any enhancement of this coupling would enhance the rate for $\Upsilon \rightarrow h^0 \gamma$ compared to that for the ϕ^0 (see fig. 3.14). However, the theoretical uncertainties reviewed earlier in the ϕ^0 discussion still pertain here, and definitive limits on the h^0 from Υ decays are not possible at present. Also, there are no unambiguous limits from nuclear physics. For instance, the limit from muonic atoms depends on the product $g_{\phi^0 NN} g_{\phi^0 \mu\mu}$. As we have discussed above, the latter coupling is likely to be enhanced for $\tan\beta > 1$, while the NN coupling may be diminished since it depends primarily on the $h^0 gg$ loop-induced coupling that is likely to be somewhat suppressed. Such a reduction in $g_{h^0 NN}$ would also adversely impact the $0^+ \rightarrow 0^+$ transition limits.

Let us turn now to the pseudoscalar of the two-doublet model. Of course, the "standard" axion searches are equivalent to a search for a very light pseudoscalar, the standard axion model being nothing more than a two-doublet model in which the (axion, a) pseudoscalar is taken to have zero mass at tree-level (a small non-zero mass can be generated by radiative corrections). Many axion searches have been performed, all with negative results, thereby excluding a very light axion-like pseudoscalar. Among these negative results we mention those from $\psi \rightarrow \gamma a$ [42] and $\Upsilon \rightarrow \gamma a$ [43]. Of course, these experimental constraints are not applicable in the case of "invisible axions", Majorons, and so forth. More details on axion and invisible axions will be given in §6.2.

Regarding the placing of limits on the A^0 of a more general two-doublet model, we note that many of the techniques useful for the scalar Higgs also apply with appropriate modifications. For instance, the absence of any $A^0 WW$ or $A^0 H^+ H^-$ coupling at tree-level makes the B and K decay computations somewhat simpler than in the h^0 case [31,40,44]. However, the limitations on the A^0 from these decays in light of current experimental data and top mass limits are only now being studied [40]. The $\Upsilon \rightarrow A^0 \gamma$ decay rate is subject to smaller QCD corrections than the same rate for the ϕ^0, but relativistic corrections could still be large. However, since the $\Upsilon \rightarrow A^0 \gamma$ decay rate depends only on the $A^0 b\bar{b}$ coupling, this decay would clearly be enhanced with respect to the $\Upsilon \rightarrow \phi^0 \gamma$ rate for $\tan\beta > 1$ and some of the region $m_{A^0} \gtrsim 600$ MeV might be excluded for $\tan\beta > 2$ by the CUSB and ARGUS data. A careful analysis is needed. Nuclear physics limits at lower m_{A^0} are sensitive to details of the $A^0 gg$ and $A^0 \mu^+ \mu^-$ couplings, just as discussed for the h^0, but for moderate $\tan \beta \lesssim 2$ will not be terribly different from those obtained for the ϕ^0.

Limits from $Z^* \to A^0 h^0$ are beginning to emerge from $e^+ e^-$ experiments. Published limits already exist from PEP [45,46] and PETRA [47]. For instance, in ref. 46 a region of small h^0 and A^0 masses is eliminated assuming maximal $Z h^0 A^0$ coupling and: (i) that one of the Higgs decays to $\mu^+ \mu^-$ and the other to $b \bar{b}$ or any other hadronic mode; (ii) that both Higgs decay to $\mu^+ \mu^-$; or (iii) $A^0 \to \mu^+ \mu^-$ and $h^0 \to A^0 A^0$ followed by $A^0 \to \mu^+ \mu^-$. Modes in which one or both Higgs decay to $e^+ e^-$ are difficult to detect because of QED backgrounds. As an example, in mode (i), assuming that the $Z A^0 h^0$ coupling is maximal, they are able to exclude $m_{A^0} \lesssim 11$ GeV. Even though higher A^0 masses are kinematically accessible if m_{h^0} is small, the production rate and backgrounds make it difficult to go to higher m_{A^0} values. Also if $h^0 \to \pi \pi$ decays are important, the region where $m_{h^0} \gtrsim 2 m_\pi$ is not excluded. The regions over which the necessary decays and cross sections occur are, as we shall see, highly model and mass dependent. Cross sections for $e^+ e^- \to h^0 A^0$ in the minimal supersymmetric model will be presented in §4.2. A closely related search for $Z^* \to h^0 A^0$ has been performed in which one looks for a final state in which the h^0 decays invisibly, while $A^0 \to f \bar{f}$ where $f = \tau$ or $f = q$. The idea of performing such a search was suggested in ref. 48. Assuming again that the $Z A^0 h^0$ coupling is maximal, the region $m_{A^0} \lesssim 23$ GeV is excluded, based on the PETRA limits cited above. Finally, preliminary results from the AMY Collaboration at TRISTAN [49] have recently been presented. Assuming maximal $Z A^0 h^0$ coupling, a massless stable h^0, and 100% two-body decays for A^0, AMY excludes $m_{A^0} \lesssim 28$ GeV.

Before concluding, we remind the reader that $B_s \to \tau^+ \tau^-$, $\mu^+ \mu^-$ decays are sensitive to neutral Higgs bosons with mass near m_{B_s} via the s-channel Higgs exchange diagram [50]. The sensitivity to charged Higgs bosons via loop diagrams [51] and to the s-channel exchange diagram is very much smaller when the neutral Higgs mass is substantially larger than the B_s mass. Indeed, if $\tan \beta > 1$ then the charged Higgs contributions can be neglected relative to SM-loop contributions.

Overall, we see that any limits on light Higgs bosons in a two-doublet extension of the SM depend upon specific choices of the parameters entering into the various Higgs couplings. Thus, no general limits can be stated.

Limits on the Charged Higgs Mass and $\tan \beta$

In this section we summarize existing limits on the charged Higgs boson of a two-doublet model. The first is that obtained at PEP and PETRA from the failure to observe the H^+ in $e^+ e^- \to \gamma^*$, $Z^* \to H^+ H^-$, excluding $2 m_\tau \lesssim m_{H^+} \lesssim 19$ GeV [52]. We anticipate that analysis of existing data from TRISTAN will increase this to $m_{H^+} \gtrsim 25$ GeV, if a signal is not seen, eventually pushing 30 GeV as the TRISTAN center-of-mass-energy upgrades are completed. The region of $m_{H^+} \lesssim 2 m_\tau$ is apparently excluded by failure to observe the decay $B \rightarrow H^+ X_c$ [53]. In this region they look for decays such as $H^+ \rightarrow \tau^+ \nu_\tau$ when $m_{H+} > m_\tau$. When $m_{H+} < m_c + m_s, m_\tau$, the main decay mode is $H^+ \rightarrow u\bar{s}$ which would be hard to observe directly, but since the decay mode of the B would be a two-body mode, it would overwhelm the normal weak decay mode $B \rightarrow e\nu X_c$ which is observed at the expected level. Thus, very light charged Higgs are excluded. Overall, we see that we need not consider masses below 25 GeV in what follows.

The second class of limits derives from various low-energy experiments. These include limits from K_L–K_S mixing, [54–56] B–$\bar{B}$ mixing [57–61], and various rare decays such as $b \rightarrow s\gamma$, $b \rightarrow s\ell^+\ell^-$, and $b \rightarrow sg$ [61–64], as well as $K \rightarrow \pi\nu\bar{\nu}$ and $B \rightarrow X_s\nu\bar{\nu}$ [61]. A comprehensive analysis of the limits has recently appeared in ref. 65

It will be convenient to focus first on the B–$\bar{B}$ mixing constraints. If the experimental measurements are to be explained entirely in the context of a two-doublet Higgs model, without additional new physics, then the charged Higgs mass, the value of $\tan\beta$ and the value of m_t must be appropriately related. We have performed our own analysis of these constraints.

Defining the ratio of probabilities, $r_d \equiv P(B_d^0 \rightarrow \bar{B}_d^0)/P(B_d^0 \rightarrow B_d^0)$, the ARGUS group [66] finds $r_d = 0.21 \pm 0.08$, while the CLEO group [67] finds $r_d = 0.19 \pm 0.06 \pm 0.06$. Both these results are consistent with the earlier less precise results of the UA1 experiment [68]. For purposes of illustration, let us assume that combining the various experimental results will eventually yield $r_d = 0.18 \pm 0.05$. As is well-known (e.g., see ref. 57), the value of r_d can be converted to a value for $\Delta m_{B_d - \bar{B}_d}$ via the relation

$$\Delta m_{B_d - \bar{B}_d} = \Gamma_{B_d} \sqrt{\frac{2r_d}{1 - r_d}}, \qquad (4.47)$$

where Γ_{B_d} is the decay width of the B_d^0. One may compute[‡] the contribution of charged W and charged Higgs loops to $\Delta m_{B_d - \bar{B}_d}$

$$\Delta m_{B_d - \bar{B}_d} = \tfrac{2}{3} G_F^2 f_B^2 m_B B_B |V_{td}^* V_{tb}|^2 [I_{WW} + I_{WH} + I_{HH}], \qquad (4.48)$$

where

$$
\begin{aligned}
I_{WW} &= I_0(m_t, m_W) m_t^2 \\
I_{WH} &= [8m_W^2 I_2(m_t, m_{H+}, m_W) + 2I_3(m_t, m_{H+}, m_W)] m_t^4 \cot^2\beta \quad (4.49) \\
I_{HH} &= I_1(m_t, m_{H+}) m_t^4 \cot^4\beta
\end{aligned}
$$

[‡]We use the notation of ref. 58 but with the full expression for the WW-box as given in refs. 59 and 61.

and

$$
I_0 = \frac{1}{16\pi^2} \left[1 + \frac{9m_W^2}{(m_W^2 - m_t^2)} - \frac{6m_W^4}{(m_W^2 - m_t^2)^2} - \frac{6m_W^2 \, m_t^4 \ln m_t^2/m_W^2}{(m_W^2 - m_t^2)^3} \right]
$$

$$
I_1 = \frac{1}{16\pi^2} \left[\frac{(m_{H+}^2 + m_t^2)}{(m_{H+}^2 - m_t^2)^2} + \frac{2m_{H+}^2 \, m_t^2 \ln m_t^2/m_{H+}^2}{(m_{H+}^2 - m_t^2)^3} \right]
$$

$$
I_2 = \frac{1}{16\pi^2} \left[\frac{m_{H+}^2 \ln m_t^2/m_{H+}^2}{(m_{H+}^2 - m_t^2)^2(m_{H+}^2 - m_W^2)} - \frac{m_W^2 \ln m_t^2/m_W^2}{(m_W^2 - m_t^2)^2(m_{H+}^2 - m_W^2)} \right.
$$
$$
\left. - \frac{1}{(m_{H+}^2 - m_t^2)(m_W^2 - m_t^2)} \right]
$$

$$
I_3 = \frac{1}{16\pi^2} \left[\frac{-m_{H+}^4 \ln m_t^2/m_{H+}^2}{(m_{H+}^2 - m_t^2)^2(m_{H+}^2 - m_W^2)} + \frac{m_W^4 \ln m_t^2/m_W^2}{(m_W^2 - m_t^2)^2(m_{H+}^2 - m_W^2)} \right.
$$
$$
\left. + \frac{m_t^2}{(m_{H+}^2 - m_t^2)(m_W^2 - m_t^2)} \right].
$$

$$(4.50)$$

We note that the above result assumes that the top quark mass terms are dominant in each of the loop types (as has been verified in the quoted references), so that all diagrams have the same coefficient. In addition, this coefficient for the dominant m_t terms is the same whether we adopt the Model I or Model II Higgs–fermion–antifermion couplings.

To analyze the implications of these equations, we have taken Γ_{B_d} from the B_d lifetime of 1.4×10^{-12} sec, chosen $m_B = 5.3$ GeV, $f_B = 0.16$ GeV, $|V_{tb}| \simeq 1$, and used the convention $|V_{td}| = s_1 s_2$, where $s_1 \simeq 0.22$. We then consider three cases

1. a middle-of-the-road case, with $s_2 = 0.06$, $B_B = 1$ (the vacuum saturation result), and $r_d = 0.18$;
2. a case allowing maximal charged Higgs contributions, $s_2 = 0.04$, $B_B = 1/3$ and $r_d = 0.23$; and
3. a case leaving little room for charged Higgs contributions, $s_2 = 0.08$, $B_B = 3/2$ and $r_d = 0.13$;

(In our calculations the other CKM entries will be fixed using $c_3 = 1$, $s_1 = 0.22$, and $\delta = 0$.) Case 2 minimizes the diagram coefficients and maximizes the experimental mixing result within the quoted error bars. Case 3 maximizes the diagram coefficients within reasonable limits and adopts the lowest possible experimental mixing result consistent with the quoted error. We note that for a large enough top quark mass, it is possible in each case to avoid any charged Higgs contribution. These top quark masses, such that the presumed experimental result is given by the WW loop diagrams only, are: $m_t = 88$ GeV for case 1; $m_t = 335$ GeV for case 2; and $m_t = 45$ GeV for case 3. In each case, values of m_t below these respective numbers allow for some charged Higgs contribution. We have plotted in fig. 4.1 the value of $\tan \beta$ as a function

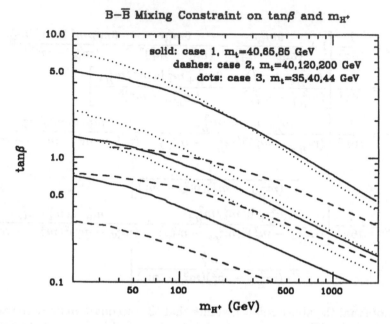

Figure 4.1 The value of $\tan\beta$ required to obtain the presumed experimental B_d–$\overline{B}_d$ mixing result for each of three cases (see text) as a function of m_{H+}, for various m_t values. Higher curves correspond to higher m_t.

of m_{H+} that yields the experimental B_d–$\overline{B}_d$ mixing assumed in each of the three cases, for several values of m_t. Much of the literature has focused on the $\tan\beta < 1$ region, in which case fig. 4.1, coupled with $m_t \gtrsim 40$ GeV, shows that fairly substantial lower bounds on m_{H+} are possible in all but case 2. But, as we shall discuss in §4.2, $\tan\beta > 1$ is preferred in model building. In this case, fig. 4.1 shows that quite light values of m_{H+} are allowed. It is especially interesting to consider the restrictions that arise if m_t turns out to be of order m_W. We see that case 3, with a low value assumed for B–$\overline{B}$ mixing, is ruled out altogether; case 1 is on the verge of being ruled out for all but a very heavy charged Higgs boson; and that only case 2, with large mixing, allows a complete range of charged Higgs boson masses. Thus, we look forward to improved error bars on the r_d value as being highly constraining for charged Higgs bosons, assuming the absence of additional new physics that could yield still more loop diagrams that compensate and cancel the loop-diagram contributions from the charged Higgs.

We should note that the contribution of these same top quark box diagrams to the K_L–K_S mass difference is given by

$$\left|\Delta m_{K_L - K_S}\right| = \frac{|V_{ts}|^2}{|V_{tb}|^2} \frac{f_K^2 m_K}{f_B^2 m_B} \left|\Delta m_{B_d - \overline{B}_d}\right|, \tag{4.51}$$

which, for $f_K = 0.12$ GeV, $m_K = 0.498$ GeV and $|V_{ts}| < 0.052$, is negligible compared to the measured $|\Delta m_{K_L-K_S}| = 3.52 \times 10^{-15}$ GeV, in all three cases discussed above.

The other rare decay processes mentioned above, including (1) K and B decays to states containing $\nu\bar{\nu}$ and (2) b decays to s plus a γ, $\ell^+\ell^-$ pair, or a g, are also potentially interesting and involve quite different combinations of the parameters. In addition, they depend strongly upon whether one uses the Model I or Model II Higgs–fermion–antifermion couplings. However, the results of refs. 62 and 61 show that the experimental constraints are not yet sufficiently strong to compete with the B-$\bar{B}$ mixing constraints discussed above, though improvements on the upper bounds for any of the decays in set (2) would make them quite powerful in the case of the Model I couplings.

Finally, we must again warn the reader that additional new physics that goes beyond extending the Higgs sector to two doublets will generally affect K_L-K_S and B-$\bar{B}$ mixing, as well as the rare b decays. For instance, in supersymmetry, there will be superpartner loops that must be included. Some discussion of these issues will appear in §4.2.

On the theoretical side, there are renormalization group-derived bounds [69] that demonstrate that $\tan\beta$ cannot be too large if the light scalar of the model is not very massive. These are conveniently reviewed in ref. 35. They are most stringent for large top quark masses. Typically, $\tan\beta \lesssim 2$ is required if $m_{h^0} \lesssim 20$ GeV and $m_t \gtrsim 80$ GeV.

The Top Quark and the Charged Higgs

The phenomenology of charged Higgs bosons at current and future colliders is intimately tied to the top quark mass. Conversely, if a charged Higgs boson of modest mass exists, it will completely alter the usual expectations for top quark discovery. In particular, the decays of the top quark will provide important information regarding the (possible) existence of a charged Higgs boson lighter than m_t. The most dramatic situation occurs if $m_t < m_W + m_b$, but $m_t > m_{H^+} + m_b$. Then $t \to H^+b$ will be the *dominant* decay mode [70]. Indeed, it is not impossible that m_t is less than the hadron collider limits quoted in §1.5, but the top has not been detected precisely because this decay is more difficult to observe than the SM decay chains [71]. Another possible scenario is $m_t > m_W + m_b$ and $m_t > m_{H^+} + m_b$. In this case, for Model II fermion couplings we find

$$\frac{BR(t \to H^+b)}{BR(t \to W^+b)} = \frac{p_{H^+}}{p_{W^+}}$$
$$\times \frac{(m_b^2 + m_t^2 - m_{H^+}^2)(m_b^2 \tan^2\beta + m_t^2 \cot^2\beta) + 4m_b^2 m_t^2 \tan\beta \cot\beta}{m_W^2(m_t^2 + m_b^2 - 2m_W^2) + (m_t^2 - m_b^2)^2}, \quad (4.52)$$

(for Model I coupling, take $\tan\beta \to -\cot\beta$ keeping $\cot\beta$ unchanged) implying roughly similar W^+ and H^+ decay rates, unless $\tan\beta$ is either very large or

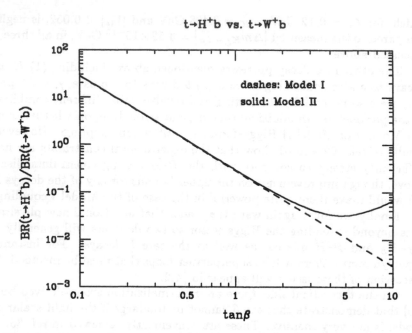

Figure 4.2 We plot $BR(t \to H^+b)/BR(t \to W^+b)$ as a function of $\tan\beta$ for $m_t = 100$ GeV, $m_b = 4.5$ GeV, and $m_{H^+} = m_W$. Results for both Model I and Model II fermion couplings are displayed.

very small. A typical example is illustrated in fig. 4.2. We observe that the $t \to H^+b$ decay falls below 10% of the $t \to W^+b$ mode only for $\tan\beta \gtrsim 2$. Of course, B_d–$\overline{B}_d$ mixing generally prefers large $\tan\beta$ when m_{H^+} is small, but certainly $m_{H^+} \simeq m_W$ is allowed in mixing case 2 so long as $\tan\beta \gtrsim 0.5$ (depending upon the precise value of m_t). Thus, there is a very substantial possibility that at a hadron collider we will discover the top quark and the charged Higgs simultaneously.

If we do not detect on-shell charged Higgs bosons in t decays, then there are three distinct possibilities: $\tan\beta$ could be very large, we could have $m_t < m_{H^+} + m_b$, or it could be that the charged Higgs does not exist. Let us focus on the second. If $m_t > m_W + m_b$ so that on-shell $t \to Wb$ decays occur, then the virtual H^+ mediated modes become very rare and we will not be able to detect the presence of the H^+ even if it has quite modest mass. An alternative possibility is $m_t < m_{H^+} + m_b$ and $m_t < m_W + m_b$, so that both modes are virtual. If B_d–$\overline{B}_d$ constraints are temporarily ignored, one can envision special circumstances which would allow one to detect the existence of the H^+. Certainly, the (virtual) charged Higgs will yield $t \to H^{+*}b$ in analogy with the usual $t \to W^{+*}b$, but the contribution to any allowed mode is generally quite small. Consider, for instance, detection of excess $t \to \tau^+\nu b$

decays relative to $t \to \mu^+ \nu b$ decays. Since the H^+ couples preferentially to heavy leptons, it will not contribute significantly to the latter channel, and the relevant ratio for Model II fermion couplings is

$$\frac{BR(t \to H^{+*}b \to \tau^+ \nu_\tau b)}{BR(t \to W^{+*}b \to \tau^+ \nu_\tau b)} \simeq \frac{m_t^2 m_\tau^2}{4 m_{H^+}^4} [\tan \beta \cot \beta]^2, \qquad (4.53)$$

where we have dropped terms of order m_τ/m_t, m_b/m_t, and $m_b \tan \beta/$ $(m_t \cot \beta)$. (For Model I couplings replace $\tan \beta$ by $- \cot \beta$, leaving $\cot \beta$ unchanged.) If, for example, $m_t = 40$ GeV, $m_{H^+} = 45$ GeV, and we use the Model II couplings, this ratio is 3×10^{-4} and clearly unobservable. But, if we have Model I couplings and $\cot \beta \gtrsim 10$, the ratio is $\gtrsim 3$ and the effect would certainly be detected. However, because of the large value of $\cot \beta$ required for such enhancement of the $\tau^+ \nu$ mode, the H^+ contribution to B_d–$\overline{B}_d$ mixing would then exceed that allowed in any of the three cases described earlier (assuming that there is no additional new physics that cancels all or part of the H^+ contribution to the mixing).

Of course, in searching for the H^+ in t decays, an important question is how the charged Higgs itself will decay. As already mentioned, one possible mode is $H^+ \to \tau^+ \nu$. When $m_{H^+} < m_t + m_b$ and $m_{H^+} < m_W + m_{h^0}$, only the $\tau^+ \nu$ and $c\bar{s}$ modes are relevant unless m_{H^+} is very small. For Model II couplings we have

$$\frac{BR(H^+ \to \tau^+ \nu)}{BR(H^+ \to c\bar{s})} = \frac{p_\tau}{3 p_c}$$
$$\times \frac{m_\tau^2 \tan^2 \beta \cos^2 \theta_c (m_{H^+}^2 - m_\tau^2)}{(m_s^2 \tan^2 \beta + m_c^2 \cot^2 \beta)(m_{H^+}^2 - m_c^2 - m_s^2) - 4 m_c^2 m_s^2 \tan \beta \cot \beta}, \qquad (4.54)$$

where θ_c is the Cabibbo angle. (The Model I result is obtained by $\tan \beta \to - \cot \beta$ keeping $\cot \beta$ unchanged.) The sensitivity of the $H^+ \to \tau^+ \nu$ branching ratio to $\tan \beta$ for Model II couplings is illustrated in fig. 4.3. One finds that the $\tau^+ \nu$ mode has a branching ratio larger than 10% so long as $\tan \beta \gtrsim 0.7$. (For Model I couplings the branching ratio is $\simeq 30\%$, independent of $\tan \beta$.) In this regard, note that small values of $\tan \beta$ in association with small m_{H^+} are probably excluded by B_d–$\overline{B}_d$ mixing.

Consider now the possibility that $m_{H^+} > m_W + m_{h^0}$. Indeed, if the top quark is sufficiently heavy that both $t \to H^+ b$ and $t \to W^+ b$ on-shell decays are possible, it is quite conceivable that m_{H^+} is, in fact, sufficiently large that on-shell $H^+ \to W^+ h^0$ decays are kinematically allowed [72]. Since $H^+ \to t\bar{b}$ decays would not be allowed in this scenario, the $W^+ h^0$ mode will be the dominant H^+ decay channel provided the $H^+ W^- h^0$ coupling (proportional to $(g/2) \cos(\beta - \alpha)$) is not *very* severely suppressed. A distinctly possible H^+ discovery scenario would then be $t\bar{t}$ production followed by a mixture of the conventional $t \to W^+ b$ final states and $t \to H^+ (\to W^+ h^0) b$ decays. Since the h^0 is most likely to decay to $b\bar{b}$, we would have to distinguish a $W^+ b\bar{b}$ final state from a $W^+ b$ final state.

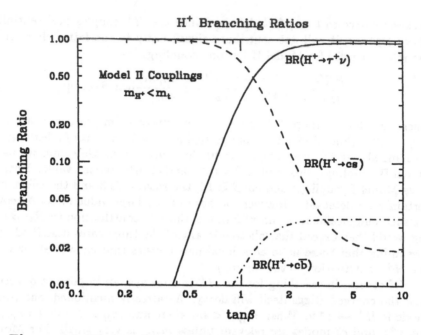

Figure 4.3 The branching ratios for charged Higgs decay as a function of $\tan\beta$ for $m_{H+} < m_t$. We assume the following values for the masses: $m_{H+} = 150$ GeV, $m_b = 4.5$ GeV, $m_c = 1.5$ GeV, $m_s = 0.15$ GeV, $m_\tau = 1.784$ GeV, and the CKM mixing angles: $V_{cs} = 0.974$, $V_{cb} = 0.044$. Results are for Model II couplings. The Model I results are independent of $\tan\beta$ and corresponds to the $\tan\beta = 1$ value shown above.

Finally, in the case where $m_{H+} > m_t + m_b$ two additional scenarios emerge. In the first, $m_{H+} < m_W + m_{h^0}$ so that $H^+ \rightarrow W^+ h^0$ is forbidden; the $H^+ \rightarrow t\bar{b}$ decay will dominate all other fermion modes. But, if the $W^+ h^0$ mode is allowed there is a competition. Because of the availability of the longitudinal W^+ polarizations, the $W^+ h^0$ final state can be important or even dominant when m_{H+} is large, unless $\cos(\beta - \alpha)$ is small. If we assume the $H^\pm t b$ coupling of Model II [eq. (4.25)], then we obtain, for the ratio $R_{Wh} \equiv BR(H^+ \rightarrow W^+ h)/BR(H^+ \rightarrow t\bar{b})$ [29],

$$R_{Wh^0} = \frac{2\cos^2(\beta - \alpha)p_W^3 m_{H+}^2}{3p_{\bar{b}}\left[(m_t^2 \cot^2\beta + m_b^2 \tan^2\beta)(m_{H+}^2 - m_t^2 - m_b^2) - 4m_t^2 m_b^2\right]} \quad (4.55)$$

where p_W and $p_{\bar{b}}$ are the center-of-mass momenta of the indicated final state particles. The corresponding formula for H^0 is obtained by replacing $\cos(\beta - \alpha)$ with $\sin(\beta - \alpha)$. As an example, let us take $m_{h^0} = 40$ GeV and assume that the h^0 saturates the allowed coupling strength [*i.e.*, $\cos(\beta - \alpha) = 1$ in eq. (4.55)]. At $\tan\beta = 1$ and $m_t = 55$ GeV the ratio R_{Wh^0} rises from ~ 0.17

at $m_{H^\pm} = 140$ GeV to ~ 1.2 at $m_{H^\pm} = 200$ GeV, passing 10 in the vicinity of $m_{H^\pm} = 460$ GeV. Clearly, for very heavy charged Higgs bosons accessible at a supercollider, the $W^+ h^0$ mode may be dominant. However, for the H^+ masses accessible up to LEP-II, the $H^+ \to W^+ h^0$ width is generally small in comparison to that for $H^+ \to t\bar{b}$.

Prospects for Non-minimal Higgs Discovery

Having summarized the general properties of the two-Higgs doublet model, and considered possible limitations placed on it by low-energy experiment, we turn to the implications for Higgs searches at $e^+ e^-$, hadron and ep colliders. For definiteness, we shall adopt Model II fermion-Higgs couplings in this subsection.

$e^+ e^-$ Colliders

We begin by noting several changes that occur in the detection of light scalar (or pseudoscalar) bosons, in the context of a two doublet model. Let us first reconsider the Higgs search at a Z factory. For the scalar bosons, we first note that the rate for $Z \to h \ell^+ \ell^-$ ($h = h^0$ or H^0) will generally be somewhat suppressed. Nonetheless, both scalar Higgs bosons would probably be detectable in this mode (presuming both are light enough) unless one Higgs completely saturates the sum rule of eq. (4.19), in which case the other scalar will not be detectable in this mode. Since a neutral pseudoscalar Higgs does not couple to VV, it cannot be produced via the above decay mode.

However, new possible decays of an on-shell Z emerge in the two doublet model, namely [30,73],

$$Z \to A^0 h^0, \qquad Z \to A^0 H^0, \qquad Z \to H^+ H^-, \qquad (4.56)$$

which lead to simultaneous production of a scalar Higgs boson and the pseudoscalar Higgs boson or of a charged Higgs pair. The decay rates normalized to the partial width of Z into one generation of neutrinos can be obtained from the Feynman rules of Appendix A

$$\begin{aligned} \frac{\Gamma(Z \to A^0 h^0)}{\Gamma(Z \to \nu\bar{\nu})} &= \tfrac{1}{2} \cos^2(\beta - \alpha) B^3 \\ \frac{\Gamma(Z \to H^+ H^-)}{\Gamma(Z \to \nu\bar{\nu})} &= \tfrac{1}{2} \cos^2 2\theta_W \, B^3, \end{aligned} \qquad (4.57)$$

where $B = 2|\vec{p}|/m_Z$ with $|\vec{p}|$ being the magnitude of the three momentum of one of the final Higgs particles. For $\Gamma(Z \to A^0 H^0)$, replace $\cos(\beta - \alpha)$ with $\sin(\beta - \alpha)$ in the first expression above. As long as the angle factors are not particularly small, the branching fraction into Higgs pairs, if kinematically allowed, can be as large as a few percent, and would constitute the dominant source of Higgs production in Z decays!

If one (or more) of the neutral Higgs bosons is lighter than some quark-onium state, then the decays of a 1^{--} quarkonium state to $h^0\gamma$, $H^0\gamma$, or $A^0\gamma$

(if allowed) can be either enhanced or suppressed over the Wilczek formula [eq. (3.60)] by the square of the relevant coupling given in eq. (4.24). Other new features can also arise. For example, ref. 74 points out that the $^3S_1(t\bar{t}) \rightarrow b\bar{b}$ rate, which is generally at least as large as the $\mu^+\mu^-$ channel rate, derives in the SM from a W exchange graph. Since the charged Higgs can couple wherever a W ,can, significant alterations in the rate are possible, especially if $\tan\beta < 1$ so that the $t\bar{b}H^-$ coupling is enhanced. New decay modes into Higgs boson pairs may also be phenomenologically important. The new modes include $^3S_1(Q\overline{Q}) \rightarrow H^+H^-$ and $^1S_0(Q\overline{Q}) \rightarrow H^0, h^0 + A^0$, where Q might be either the t or a new down quark b'. Depending upon $\tan\beta$ and (in the case of $Q = b'$) the appropriate extended CKM matrix mixing angles, these modes can even dominate all others.

Turning now to higher energy e^+e^- machines, appropriate for discovering more massive Higgs bosons, we begin by recalling that the major production mechanisms for the neutral Higgs at a e^+e^- collider are dependent upon substantial VV couplings. Thus, as for the hadron collider case, so long as the two neutral scalars share fairly equally the allowed VV couplings [see eq. (4.19)], their detection should be quite straightforward at a machine with adequate energy. If one of the neutral scalars should have weak coupling to VV channels (as in the case of the minimal supersymmetric model) its discovery is likely to be rather problematical at an e^+e^- collider. More details will be given in §4.2

Similarly, the A^0 may be particularly difficult to find at an e^+e^- machine, since it has no VV couplings. The main production mode that is available is $Z^* \rightarrow A^0h^0$ or A^0H^0. The cross sections for these processes are easily computed, and we find [30]

$$\sigma(e^+e^- \rightarrow A^0h) = \frac{g^2 f^2}{48\pi}\left(\frac{8\sin^4\theta_W - 4\sin^2\theta_W + 1}{\cos^2\theta_W}\right)\frac{\kappa^3}{\sqrt{s}\left[(s - m_Z^2)^2 + \Gamma_Z^2 m_Z^2\right]}$$
(4.58)

where $\kappa = \lambda^{1/2}(s, m_h^2, m_{A^0}^2)/(2\sqrt{s})$, with $h = H^0$ or h^0 and the kinematic factor $\lambda(a, b, c) \equiv (a + b - c)^2 - 4ab$, and

$$f = \frac{g}{2\cos\theta_W}\begin{cases} \sin(\beta - \alpha), & h = H^0 \\ \cos(\beta - \alpha), & h = h^0 . \end{cases}$$
(4.59)

While such annihilation channel pair production modes have smaller cross section than the conventional $e^+e^- \rightarrow \nu\bar{\nu}\phi^0$ or $e^+e^- \rightarrow Z^* \rightarrow Z\phi^0$ processes, detection might be possible if $m_{A^0} + m_{h^0}$ (or $m_{A^0} + m_{H^0}$) is not too large compared to the machine energy. For instance, if the ZA^0H^0 coupling saturates the strength allowed by eq. (4.26), then the cross section for $e^+e^- \rightarrow Z^* \rightarrow A^0 + H^0$ can be as large as one-tenth of a unit of R (where one unit of R corresponds to the cross section for $e^+e^- \rightarrow \gamma^* \rightarrow \mu^+\mu^-$) when not suppressed by phase space.

In fact, there is an amusing 'compensation' between the VV fusion production mode for h^0 and H^0 and the $Z^* \rightarrow A^0h^0, A^0H^0$ production processes.

Comparing eqs. (4.20) and (4.59), we see that when the H^0 is weakly produced via VV fusion, it will have near maximal production cross section via the Z^* (assuming no phase space suppression). Correspondingly, in this case the h^0 would be strongly produced via VV fusion and have suppressed production via the Z^*. Of course, exactly the opposite situation could also occur, but in either case at least one of the two production processes will be available for each of the scalar Higgs bosons. In principle, an e^+e^- collider of adequate energy and luminosity is guaranteed to find all the neutral Higgs bosons of the model. In practice, the small cross section and more rapid phase space suppression associated with the Z^* production modes will favor detection of the scalar Higgs boson with largest VV coupling.

The signature of a neutral scalar or pseudoscalar Higgs boson depends on the branching fractions of its decay modes. The heavy scalars will decay dominantly into vector boson pairs (provided the VV coupling is not unduly suppressed). The pseudoscalar (or a scalar which couples very weakly to VV) will decay into $t\bar{t}$ (or the heaviest quark pair available). These signatures have been discussed in §3.4. Of course, there are additional channels for Higgs boson decay as a result of the Higgs sector expansion. The most important of these contain one gauge boson in the final state [72,75]; some examples are $A^0 \rightarrow Zh^0$, $A^0 \rightarrow W^\pm H^\mp$, and $H^0 \rightarrow W^\pm H^\mp$. Such decay modes can be very important when m_{A^0} and m_{H^0} are large compared to m_W, due to the longitudinal modes of the final state vector boson. However, the sizes of the required couplings are very model dependent.

Due to the large parameter freedom of the most general two-Higgs-doublet model, no comprehensive Monte Carlo simulation has yet been undertaken to assess signal versus background rates for the heavy neutral scalar and pseudoscalar produced via $Z^* \rightarrow H^0 A^0$ at a very high energy e^+e^- collider. However, one such study [76] has been performed on the more constrained version of the two-doublet model that arises in the minimal supersymmetric model. Nevertheless, it seems likely that their conclusions should be applicable in more general two-doublet models. The study of ref. 76 found that a cross section of order 0.1 units of R (as predicted by eqs. (4.58) and (4.59) for f near 1 in the absence of significant phase space suppression) should prove adequate to detect the H^0 and A^0 provided there is sufficient integrated luminosity (e.g. 10 fb^{-1} at $\sqrt{s} = 1$ TeV). The principal tricks in overcoming backgrounds from $e^+e^- \rightarrow W^+W^-, q\bar{q}, ZZ$ are as follows: a) trigger on events with b quarks from the Higgs decay by requiring large multiplicity and a large number of tracks, with high impact parameter; b) eliminate events with forward/backward going clusters (the indicated backgrounds peak in the forward/backward directions); and c) require cluster masses for the two clusters to be identified with the A^0 and H^0 that are not near m_W, m_Z. Should both the A^0 and H^0 have mass near m_W, m_Z, the W and Z pair backgrounds would probably make detection of the signal difficult.

Charged Higgs boson pair production at an e^+e^- collider proceeds via virtual photon and Z exchange in the s channel. The resulting (asymptotic) cross section is given by

$$\frac{1 + 4\sin^4\theta_W}{8\sin^4 2\theta_W} \sim 0.308 \tag{4.60}$$

in units of R (for $s \gg m_Z^2, 4m_{H^\pm}^2$). The observability of the charged Higgs bosons in this production mode was studied by Komamiya in ref. 77. The result is relatively straightforward to state: a charged Higgs boson, whose decays are dominated by SM-fermion or Wh modes and with mass such that $2m_{H^\pm} \lesssim 0.8\sqrt{s}$, will be detectable at an e^+e^- machine with an integrated luminosity $\gtrsim 10^3$ inverse units of R. As in the case of a SM Higgs boson there is dependence on the beamstrahlung parameter of the machine; the smaller the beamstrahlung the higher in charged Higgs mass a given machine can reach. Of course, if $m_t > m_{H^\pm} + m_b$, then $e^+e^- \to t\bar{t}$, followed by top decay to the charged Higgs, will be an important production mechanism. This was also studied in ref. 77, which concluded that the signal for this production process could be separately detected.

However, it is clear that the case of $m_{H^+} \sim m_W$ may present special difficulties. This was examined by Komamiya in the $t \to H^+ b$ mode, with the result that the H^+ could still be detected [77]. However, detection in the pair production mode when $m_{H^+} \sim m_W$ was only recently considered in ref. 76. They considered a e^+e^- collider with $\sqrt{s} = 1$ TeV and an integrated luminosity of 10 fb^{-1}. Clearly, since the basic cross section for $e^+e^- \to W^+W^-$ is about 100 times that for $e^+e^- \to H^+H^-$ and since some of the $W^\pm$ boson decay modes are the same as the $H^\pm$ decay modes, there is a severe background problem. Fortunately, since the W^+W^- cross section peaks in the forward direction, a strong cut requiring central clusters will substantially reduce this background. Further cuts depend on whether or not $m_{H^+} > m_t + m_b$. If $m_{H^+} > m_t + m_b$, then (unless $H^+ \to W^+h^0$ is dominant) the final state contains $t\bar{b} + \bar{t}b$ and will have a very high multiplicity and a large number of high impact parameter tracks. Cuts which enhance such events can be implemented that yield a substantial signal to background ratio. If $m_{H^+} < m_t + m_b$, we have seen that there is likely to be a substantial $H^+ \to \tau^+\nu_\tau$ branching ratio. In ref. 76, detection of H^+H^- pair production in the final state mode $\tau\nu_\tau + s\bar{c}(b\bar{c})$ was examined for this case. By focusing on the $\tau \to \pi\nu$ decay mode (by requiring a single charged pion in one hemisphere of the event), the signal could be isolated from the background provided the $H^+ \to \tau^+\nu_\tau$ branching ratio was at least 30%.

There is another production mode for the Higgs bosons of a two-doublet (or other extended) Higgs sector that we should mention. This is Higgs production in association with a heavy fermion pair, in which the Higgs boson of interest is bremsstrahlunged from one or the other of the heavy fermions, e.g. a bottom or top quark or a heavy lepton. For instance, for a neutral scalar Higgs we have $e^+e^- \to F\overline{F}h^0(H^0)$. In fact, of course, there are two additional diagrams contributing to this same final state: $e^+e^- \to Z(\to F\overline{F})h^0(H^0)$ and $e^+e^- \to A^0(\to F\overline{F})h^0(H^0)$. Both of these latter processes have already been discussed; the interference among the three processes is quite small, and so the bremsstrahlung process can be treated independently to good approximation. It is of importance only if one of the other processes is kinematically forbidden as a two-body mode, or is coupling constant suppressed, or if the $F\overline{F}h^0(H^0)$ coupling constant is greatly enhanced. Analogous remarks apply to A^0 and $H^{\pm}$ production. Some exploration of the bremsstrahlung process cross section has appeared in refs. 78 and 79.

Of course, all of the above three processes ($Z^* \to A^0h^0; \gamma^*, Z^* \to H^+H^-$; and Higgs bremsstrahlung from $F\overline{F}$) are potentially of interest at energies below m_Z, such as that available at TRISTAN ($\sqrt{s} \lesssim 60$ GeV). Of these, the first two are clearly the most promising. For example, at $\sqrt{s} = 60$ GeV the $Z^* \to A^0h^0$ cross section can easily exceed 10^{-2} units of R, provided phase space suppression is not significant and $\cos(\beta - \alpha)$ is not too small [see eqs. (4.58) and (4.59)]. Without any special model assumptions, the Higgs mass limits which can be obtained are rather complicated since they depend on $\cos(\beta - \alpha)$, two Higgs masses and various branching ratio assumptions. Explicit results in the minimal supersymmetric model will appear in §4.2.

Hadron Colliders

In the discussion of hadron colliders we shall primarily use the SSC as a specific example. We begin by considering whether search techniques that worked for the SM Higgs will also be appropriate in a two-doublet model.

The scalar Higgs bosons ($h = h^0, H^0$) can be produced by gg fusion, WW/ZZ fusion, and $gg \to Q\overline{Q}h$ (where $Q = b$ or t). The cross sections of fig. 3.32 will be somewhat modified due to the modified hVV and $hQ\overline{Q}$ couplings. The vector-boson-fusion contributions are expected to be somewhat less than for the SM ϕ^0 since h^0 and H^0 will share the ϕ^0VV couplings. The gluon fusion cross sections, which derive mainly from the top quark loop, can be enhanced or reduced, depending upon whether the $t\bar{t}h$ coupling is enhanced or suppressed. Furthermore, in contrast to ϕ^0 production, $gg \to Q\overline{Q}h$ may be a substantial part of the h cross section if the $Q\overline{Q}h$ coupling is significantly enhanced [80]. Thus, in general, we expect the total h cross section to be roughly similar in magnitude to the ϕ^0 cross section of fig. 3.32.

Having produced a scalar Higgs (H^0 and h^0), it can be detected in the same manner as the minimal Higgs of the SM, so long as h^0 and H^0 share relatively equally the VV coupling strength. As explained elsewhere in §3.4, if a scalar Higgs has a mass between about $2m_W$ and 800 GeV *and its couplings to WW and ZZ are similar to SM strength*, then it should be possible to detect this Higgs at the SSC by observing its decay into a pair of vector bosons (followed by subsequent decay of the vector bosons into lepton pairs). On the other hand, for masses less than $2m_W$, we are in the regime of the "intermediate mass Higgs", in which the Higgs mainly decays into the heaviest quark pair which is kinematically allowed. In this case, we have seen that the ability to successfully observe such a Higgs at the SSC depends upon utilizing a combination of rare decay modes, such as $\gamma\gamma$ and ZZ^*, and associated production techniques (with final state $W b \bar{b}$). By combining such modes it is likely to prove possible to discover a neutral Higgs in a large region of the Higgs-mass–top-mass parameter space. Only if the Higgs mass is near m_W, m_Z, or if the top mass turns out to be fairly light (e.g., near 55 GeV) and the Higgs mass near 120 GeV, is discovery likely to prove impossible.

Consider next the pseudoscalar Higgs. For reasons that were explained above, the pseudoscalar does not couple to vector boson pairs at tree level. The phenomenological implications of this fact are potentially devastating. First, the important vector boson fusion mechanism for production of a Higgs boson is absent. Thus, the primary production mechanism will be via gg fusion, although $gg \to Q\overline{Q}A^0$, $Q = b$ (or t), can be a substantial fraction of the total cross section if $\tan\beta \gg 1$ ($\tan\beta \ll 1$), due to the enhanced $Q\overline{Q}A^0$ coupling [80]. As discussed below eq. (4.46), the $A^0 gg$ coupling strength is expected to be similar to SM strength so that the cross section will be substantial at the SSC, e.g., > 1 pb for $m_{A^0} \lesssim 500$ GeV when $m_t \sim 40$ GeV. Of course, if the top is much heavier we have learned that the gg fusion mechanism dominates the WW fusion cross section of the ϕ^0 and it would clearly yield comparable cross sections for the A^0. However, the second implication of the absence of $A^0 VV$ couplings is that the dominant decay of the pseudoscalar Higgs will be into the heaviest quark pair available, independent of the Higgs mass.* We have also learned that the $\gamma\gamma$ mode of the A^0 will have smaller branching ratio than in the case of a Standard Model Higgs boson, due to the absence of W boson loop graphs for $A^0 \to \gamma\gamma$. The fermion loop contributions (see the prescription for their inclusion in the preceding section) are generally much smaller. The absence of VVA^0 couplings also implies that the ZZ^* mode is absent at tree-level in A^0 decays. Thus, these rare decays will not be useful in searching for the A^0. However, if such an object could be found, and were shown to have a mass larger than $2m_W$, then the absence of decays into vector boson pairs would be strong evidence for the pseudoscalar nature of the object. (An exception to this rule occurs in supersymmetric models, which predict the heavy Higgs *scalar* has suppressed couplings to the vector boson channels. Nevertheless, such an observation would be definitive evidence for a non-minimal Higgs sector.)

The charged Higgs boson may also be very difficult to detect at a hadron collider. One can show [25-27] that in models with only Higgs doublets (and singlets), there is no coupling of the charged Higgs to vector boson pairs (WZ and $W\gamma$) at tree level, so that its decays are likely to be dominated by the heaviest allowed quark channel. In addition, the gluon–gluon-fusion and vector–boson-fusion mechanisms are not available for charged Higgs boson production. The dominant mechanism for $H^\pm$ production depends upon the

---

*In order to simplify the discussion, we will ignore possible A^0 decay modes involving other Higgs boson states, such as $A^0 \to Zh^0$, $A^0 \to ZH^0$, and $A^0 \to W^\pm H^\mp$.

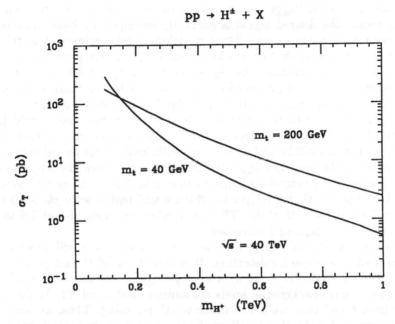

Figure 4.4 The cross section at the SSC for charged Higgs production coming from $gb \to H^-t + g\bar{b} \to H^+\bar{t}$ as a function of $m_{H\pm}$. Top quark masses of $m_t = 40$ GeV and $m_t = 200$ GeV are considered. We have taken $\tan \beta = 1$.

relative mass of the t quark and the charged Higgs boson.[†] If the top quark has a moderate mass, but $m_t > m_{H\pm} + m_b$, then the rate for $gg \to t\bar{t}$ followed by $t,\bar{t}$ decay to H^+, H^- is very large, as discussed in the previous subsection. [See eq. (4.52) and surrounding discussion.] As shown in fig. 4.2, the H^+ channel is fully competitive with the W^+ mode (as long as $\tan \beta$ is not too large). If $m_t < m_{H\pm} + m_b$, then the most important production mechanism for the $H^\pm$ arises from the subprocesses $gb \to H^-t$ and $g\bar{b} \to H^+\bar{t}$. These are computed in ref. 81 where it is also demonstrated that a naive computation of H^+ (H^-) production using $t\bar{b}$ ($b\bar{t}$) fusion would be very inaccurate in the mass region of interest [82]. The charged Higgs boson cross section is comparable in magnitude to the gg fusion cross section of the SM Higgs at the same Higgs mass. This is illustrated in fig. 4.4, from ref. 81. To shorten our notation, we shall sometimes refer to $gb \to H^\pm t$ and $H^\pm \to tb$, instead of writing out the relevant production and decay processes for H^+ and H^- separately.

The predicted cross section in the region $m_{H\pm} > m_t$ is such that the raw

---

[†] If $m_W > m_{H^+} + m_{h^0}$, then $W^+ \to H^+h^0$ provides an m_t-independent source of charged (and neutral) Higgs bosons [73]. However, this possibility is unlikely and the potential for observation is rather meagre.

number of charged Higgs events is substantial. However, for a given Higgs decay mode, the desired signal is generally swamped by huge backgrounds. To have any chance of seeing a signal, a trick must be employed. One trick that has been explored is that of a 'stiff lepton trigger', first proposed in ref. 81. In the production mechanisms, $\bar{b}g \rightarrow \bar{t}H^+$ and $bg \rightarrow tH^-$, one attempts to trigger on the t or $\bar{t}$ produced in association with the charged Higgs. One approach to doing this is to note that the final state t and $\bar{t}$ quarks are typically moving nearly parallel to the original beam. Ordinarily, they would just be lost inside the beam jets. However, if the t-quark decays semi-leptonically, the electron or muon will be kicked out with sufficiently large p_T (of order $m_t/2$), so that it can be used to trigger the desired event. Even the leptons coming from decays of the secondary b quarks that arise from t decay will contribute to this trigger, so that a trigger in which a stiff lepton with $p_T^l > 10$ GeV is required retains $\sim 45\%$ of the $H^\pm$ events, while rejecting all but 1% to 2% of most types of background processes.

Given a sufficiently large sample of $H^\pm$ bosons, one still needs a viable Higgs decay signature for detection. It is clear that QCD backgrounds to observing the H^+ via its $t\bar{b}$ decay are very large. This has been studied [83] and quantitative background levels are summarized in ref. 84. So far, no cuts have been found that make this mode worth pursuing. Thus, we concentrate on the search for the charged Higgs boson via alternative decay modes such as $H^\pm \rightarrow W^\pm\gamma$, $H^\pm \rightarrow W^\pm h^0$, $H^\pm \rightarrow W^\pm h^0\gamma$, $H^\pm \rightarrow W^\pm +$ quarkonium, $H^\pm \rightarrow \tau\nu$, and $H^\pm \rightarrow \mu\nu$. The first three modes listed were considered in ref. 29. The decay $H^\pm \rightarrow W^\pm\gamma$ occurs at one loop, and so is phenomenologically irrelevant. The tree-level processes $H^\pm \rightarrow W^\pm h^0$, and $H^\pm \rightarrow W^\pm h^0\gamma$ can be significant, depending on the parameters of the non-minimal Higgs sector. Here, we shall confine our discussion to the $W^\pm h^0$ mode, which is likely to be phenomenologically more important. Even when the tb channel of $H^\pm$ decay is open, $H^\pm$ detection in the $W^\pm h^0 (\rightarrow b\bar{b})$ final state might be possible. The problems and procedures associated with $H^\pm$ detection via such decays would be closely related to the techniques developed for the detection of the intermediate mass Higgs via $W + \phi^0 (\rightarrow b\bar{b})$ associated production. However, the prospects for discovering the $H^\pm$ by this method may be far more promising. Even if m_t is such that $H^\pm \rightarrow tb$ is allowed, the branching ratio for $H^\pm \rightarrow W^\pm h^0$ [see eq. (4.55)] combined with the $gb \rightarrow H^\pm t$ cross section yields a cross section for the $W^\pm b\bar{b}$ final state that, at moderate $m_{H^\pm}$ ($\sim$ 150–300 GeV), is typically a factor of 10 to 100 larger than the corresponding cross section for $W + \phi^0$ mentioned above. One also has the additional advantage of a Jacobian peak in the $b\bar{b}$ transverse momentum spectrum, and the ability to trigger on the spectator t quark jet. Of course, if $H^\pm \rightarrow tb$ is *not* allowed, $H^\pm \rightarrow W^\pm h^0$ becomes one of the dominate modes for $H^\pm$ decays and detection via the $W^\pm h^0$ channel might well be straightforward, either via inclusive $H^\pm$ production using the gb fusion processes, or in $gg \rightarrow t\bar{t}$ as a t quark decay product. In short, it is distinctly possible that one could use this mode for simultaneous discovery of both the $H^\pm$ and the h^0 for a substantial range of $H^{\pm}$ masses. However, much detailed Monte Carlo work is required to fully assess the practicality of this approach. See refs. 84 and 29 for more details.[‡] The b-tagging techniques studied for finding $W^{\pm}\phi^0(\to b\bar{b})$ by Brau et al. [85,86] will be very relevant here.

The $W^{\pm}$ + quarkonium mode branching ratios were considered in ref. 87. The modes $H^+ \to W^+\Upsilon$ and $H^+ \to W^+\Theta$ (where Θ is the $t\bar{t}$ $^3S_1$ bound state) were computed; both are quite sensitive to the value of m_t which enters the loop diagram calculations and controls the phase space. The conclusions of ref. 87 are easily summarized. If $H^+ \to t\bar{b}$ is not allowed, then the branching ratio for $H^+ \to W^+\Upsilon$ is quite significant [typically 1–3 $\times 10^{-2} BR(H^+ \to \tau^+\nu)$] when $m_{H\pm}$ is just below $m_t + m_b$, although it falls rapidly with increasing m_t. Together with $t\bar{t}$ production followed by $t \to H^+b$ and $\bar{t} \to W^-\bar{b}$ (or vice versa), one finds a significant rate for production of two b jets, two leptonically decaying W's and a leptonically decaying Υ. To give a specific example, suppose $m_t = 200$ GeV, $\tan\beta = 1.5$ and $m_{H\pm} = 150$ GeV. Then, there will be $\sim 5 \times 10^8$ $t\bar{t}$ pairs produced at the SSC. Combined with $BR(t \to H^+b) \sim 0.083$, $BR(H^+ \to W^+\Upsilon) \sim 3.5 \times 10^{-4}$, $BR(W^+ \to \ell^+\nu) \sim 2/9$, and $BR(\Upsilon \to \ell^+\ell^-, \ell = e, \mu) \sim 0.06$, we obtain ~ 80 $t\bar{t}$ events with a final state containing two b jets, two leptonically decaying W's, and a leptonically decaying Υ. It is difficult to imagine significant backgrounds to such a final state. In contrast, since $H^+ \to t\bar{b}$ is always allowed if $H^+ \to W^+\Theta$ is, the latter decay always has a very small branching ratio (typically $\lesssim 10^{-5}$) and is not useful for $H^{\pm}$ detection.

If the Wh^0 decay mode is absent (or highly suppressed), then it may be useful to consider the $\tau\nu$ decay mode of the $H^{\pm}$. When $H^{\pm} \to tb$ is allowed, the branching ratio is given by

$$BR(H^{\pm} \to \tau\nu) \approx \frac{m_\tau^2 \tan^2\beta}{3(m_t^2 \cot^2\beta + m_b^2 \tan^2\beta)}. \tag{4.61}$$

For example, for $\tan\beta = 1$ and $m_t > 55$ GeV, we see that $BR(H^{\pm} \to \tau\nu) < 3.5 \times 10^{-4}$. (This branching ratio can be significantly enhanced if $\tan\beta > 1$.) Of course, if the top quark mass is larger than the charged Higgs mass, then $BR(\tau\nu)$ can be substantially bigger, as shown in fig. 4.3. To evaluate whether it is feasible to detect the charged Higgs in this mode, one must carefully evaluate the charged Higgs production and the competing backgrounds to the $H^+ \to \tau\nu$ final state. A detailed discussion appears in ref. 84. Here we present a brief overview of the results.

Let us first focus on the case where $m_t > m_{H\pm} + m_b$. As an example let us consider $m_t = 200$ GeV, $m_{H\pm} = 100$ GeV and $\tan\beta = 1.5$. In this case

---

[‡]The analogous $A^0 \to Zh^0$ decay does not have as large a branching ratio as does $H^{\pm} \to W^{\pm}h^0$ when top quark channels are forbidden, and $gg \to Zb\bar{b}$ backgrounds are bigger than those relevant in the $H^{\pm}$ case. There is also no t quark trigger. Thus considerable skepticism is warranted as to the utility of the Zh^0 mode, especially when $A^0 \to t\bar{t}$ is allowed.

$gg \rightarrow t\bar{t}$ yields $\sim 5 \times 10^8$ $t\bar{t}$ pairs per standard SSC year, $BR(t \rightarrow H^+ b) \sim$ 0.21 and $BR(H^+ \rightarrow \tau^+ \nu) \sim 0.6$. In addition, we have $BR(t \rightarrow W^+ b) \sim 0.79$ and $\sum_l BR(W^+ \rightarrow \ell^+ \nu) \sim 0.3$.* The most straightforward way to detect the presence of the $H^\pm$ in t decays is to look for an enhancement in the number of isolated singly charged hadronic (H) tracks in comparison to isolated charged leptonic tracks (L). To first approximation L and H tracks arise only from the e, μ, τ decays of the W and from the τ decays of the $H^\pm$. The τ has a branching ratio of 0.35 to L states and 0.5 to H states (including ρ's which decay to a single charged pion and a single π^0), whereas e and μ are, of course, pure L. Thus W production leads on average to $(1/3)(2L + 0.35L + 0.5H) = 0.78L + 0.17H$ while τ production leads to $0.35L + 0.5H$. As a result, for pure W production in t decay we would have $H/L \sim 0.17/0.78 \sim 0.22$ whereas if charged Higgs are present at the above level we obtain

$$\frac{H}{L} \sim \frac{0.79 \times 0.33 \times 0.17 + 0.21 \times 0.6 \times 0.5}{0.79 \times 0.33 \times 0.78 + 0.21 \times 0.6 \times 0.35} = 0.44. \qquad (4.62)$$

Given the very large number of produced $t\bar{t}$ pairs, this difference is certainly detectable provided it is possible to construct sufficiently discriminating, yet efficient, triggers for isolated charged leptons and hadrons. This was examined as part of the study in ref. 84 using a transverse momentum requirement on the produced t and $\bar{t}$ quarks, with positive results. Since top quarks with $m_t \sim 150$–200 GeV can be studied at the Tevatron Collider, depending on luminosity upgrades, it may be possible to use these techniques there in the early 1990's.

Charged Higgs detection in the $\tau \nu$ decay mode is far more difficult if $m_t < m_{H^\pm} + m_b$. The most direct production mechanism is that discussed earlier: $gb \rightarrow H^- t/g\bar{b} \rightarrow H^+ \bar{t}$. Using the cross sections for these reactions, one can make a quick comparison between signal and likely backgrounds, from which it becomes immediately clear that the signal-to-noise is generally much smaller than 1. The problem here is that both real and virtual W bosons can also decay into $\tau \nu$ final states. Even when we employ the trick of triggering on the spectator t or $\bar{t}$ quark, as discussed earlier, there is an irreducible background process which is not rejected, namely $gb \rightarrow W^- t$ and its charge conjugate. The event topology for this reaction is completely identical to that of the signal, and only the lower mass of the W can be used to separate this background from the signal (using rapidity cuts and the Jacobian peak in the p_T spectrum of the single charged particles from τ decay [88]). To illustrate the magnitude of the problem, we give in fig. 4.5 the cross section for $g\bar{b} \rightarrow \bar{t}W^+$ and its charge conjugate, compared to $g\bar{b} \rightarrow \bar{t}H^+$ and its charge

---

*One might also wish to consider using the $H^+ \rightarrow \mu^+ \nu$ decay mode with branching ratio of order 10^{-3}. Unfortunately, this is likely to prove extremely difficult in the face of large backgrounds coming from $t \rightarrow W^+ b$, followed by $W^+ \rightarrow \mu^+ \nu$ (with $BR \sim 0.1$). Nonetheless, it would be important to set limits on the $H^+ \rightarrow \mu^+ \nu$ decay mode in the hope of confirming the expected suppression relative to the $\tau^+ \nu$ final state coming from the mass dependence of the couplings.

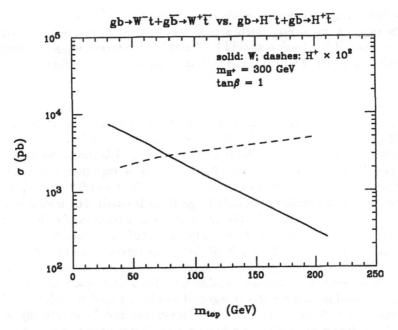

Figure 4.5 We give the cross section for $g\bar{b} \to \bar{t}W^+$ and its charge conjugate, compared to $g\bar{b} \to \bar{t}H^+$ and its charge conjugate, as a function of the t quark mass at $m_{H^\pm} = 300$ GeV. No branching ratios for the $W^\pm \to \tau\nu$ or $H^\pm \to \tau\nu$ decays have been incorporated. We have taken $\tan\beta = 1$.

conjugate, as a function of the t quark mass at $m_{H^\pm} = 300$ GeV and $\tan\beta = 1$, from refs. 90 and 84. (See also ref. 89.) The latter cross section must be adjusted for $\tan\beta \neq 1$ using

$$\sigma(gb \to tH^\pm) \propto m_t^2 \cot^2\beta + m_b^2 \tan^2\beta. \qquad (4.63)$$

No branching ratios for the $W^\pm \to \tau\nu$ or $H^\pm \to \tau\nu$ decays have been incorporated. In comparing m_t dependence it should be kept in mind that $BR(W^\pm \to \tau\nu)$ is m_t independent once the tb channel is closed, while $BR(H^\pm \to \tau\nu)$ falls like $1/m_t^2$, until m_t approaches $m_{H^\pm}$ and the tb mode decays of $H^\pm$ become phase-space suppressed. For example, the choice of $m_{H^\pm} = 300$ GeV, $\tan^2\beta = 3$ and $m_t = 70$ GeV yields (including branching ratios for the $\tau\nu$ decays of the W and H) $S/B \sim 10^{-4}$ with a signal of 150 events, where the small signal rate is due to the small $H^\pm \to \tau\nu$ branching ratio which is of order 10^{-3}. The means for discriminating between this background and the charged Higgs signal are limited and can probably never achieve better than a factor of 10 discrimination. Using such a factor it quickly becomes clear that $BR(H^\pm \to \tau\nu) \gtrsim 0.5$ is required before one could detect the charged Higgs in this manner. In general, the only way in which such a large branching ratio arises, while still having $t \rightarrow H^+ b$ forbidden, is if $m_t + m_b > m_{H\pm} > m_t - m_b$. Such a case was studied in ref. 84, with the conclusion that $H^\pm$ detection should be feasible. For larger $m_{H\pm}$ discovery of a directly produced $H^\pm$ in this channel appears to be impossible.

ep Colliders

Finally, we consider whether ep machines might allow detection of some of the Higgs bosons of the two-doublet model, despite the relatively unpromising results summarized for ϕ^0 detection in chapter 3. We choose to focus on the charged Higgs boson. As usual, if $m_t > m_{H\pm} + m_b$, then the most useful production mode is via $t\bar{t}$ production, followed by $t \rightarrow bH^+$ decay [91]. Even in this case, cross sections are only large if m_t is small. For instance, if $m_t \sim$ 30 GeV and $m_{H\pm} \sim 20$ GeV, the effective cross section is of order 10 pb and for the canonical integrated luminosity of $\gtrsim 10^2$ pb^{-1} one obtains of order 10^3 events. But if we take $m_t \gtrsim 40$ GeV, the largest possible effective cross section is of order 1 pb yielding at most 10^2 events at canonical luminosity. To date no realistic background studies have been performed, and it is unclear what production rate for the charged Higgs is required in order to allow its detection in t decays. If $t \rightarrow H^+ b$ decays are not kinematically allowed, then the primary production mechanisms are three: bremsstrahlung off heavy quarks, $\gamma Q \rightarrow Q' H^\pm$ [92]; photon–photon collisions, $\gamma\gamma \rightarrow H^+ H^-$ [93]; and gluon–photon fusion, $\gamma g \rightarrow tbH^\pm$ [91]. (This latter is clearly the analogue of the $gg \rightarrow tbH^\pm$ processes appropriate at a hadron collider.) Of these, γg yields the largest event rate, but it is almost certainly inadequate. For instance, if both the t and $H^\pm$ are light (*e.g.*, $m_t \sim m_{H\pm} \sim 25$ GeV), even a value of $\tan\beta$ as small as 0.1 (to enhance the $H^+ \bar{t} b$ coupling) leads to an effective cross section of only $\sim$0.1 pb, implying $\sim$ 10 events at canonical luminosity. In addition, the final state would be $\gamma g \rightarrow tbH^\pm \rightarrow t\bar{b}tb$ and one would have to deal with (at a minimum) backgrounds from $\gamma g \rightarrow tbW^\pm \rightarrow t\bar{b}tb$, which are likely to be much larger than the signal. (The analogous gg fusion background in the hadron collider case will be discussed in detail in the following section.) Thus, possibilities for the detection of the Higgs bosons of a two-doublet model at an ep collider are not encouraging.

Final Remarks on Non-minimal Higgs Search Techniques

To summarize, it is clear that a two-doublet model can lead to a much broader range of phenomenological implications and signatures than the single–doublet version of the SM. There is still much phenomenological analysis which remains to be done. For example, the analysis of the signatures of VH decay modes of the neutral Higgs scalars and pseudoscalar has yet to be performed. Other new decay channels in the two-doublet model where one Higgs boson decays to two other Higgs bosons should also be examined. These can have quite significant branching ratios when phase space allowed if the corresponding VV modes are absent or suppressed. They could be especially important for Higgs bosons that do not have any allowed VH channel. These will be included in our detailed study of the minimal supersymmetric model in §4.2. More general discussions of such decays have also appeared [94]. The VH and HH decay modes can also be used to distinguish between the CP-odd A^0 and the CP-even h^0 and H^0 Higgs bosons of the two-doublet model. Of course, if $h \to VV$ is observed (where h is electrically neutral), it immediately identifies h as a CP-even scalar. However, if the VV modes are suppressed or kinematically absent, additional decay modes will be needed to confirm the CP-quantum number of the h. Finally, if Higgs decays into fermion pairs dominate, then determining the CP quantum number may be difficult. Various techniques have been studied in the literature [95–99], but these will be successful only if backgrounds and rates turn out to be cooperative.

Altogether, specific predictions for the general two-Higgs-doublet model are elusive given the large number of parameters that appear. Thus, it is desirable to turn to a specific highly constrained model in order to develop a definite set of scenarios that might emerge. Supersymmetry provides such a framework since it requires that there be at least two doublets of Higgs fields. Thus, we shall turn to a detailed discussion of the minimal supersymmetric model in the next section.

4.2   The Minimal Supersymmetric Extension of the Standard Model (MSSM)

Theoretical Structure and Higgs Masses

We have seen in the previous section that it is relatively straightforward to go beyond the minimal one-doublet Higgs sector, even within the context of the SM. However, there are a number of strong theoretical reasons to suppose that the SM itself is, in fact, merely part of a much larger structure. Indeed, it is the deep-rooted problems associated with the Higgs sector which suggest that we must ultimately look beyond the SM. These are the problems of fine tuning, naturalness, and hierarchy, referred to earlier. They have been reviewed in many places (see for example refs. 100–102). Here, we merely recall that, in the SM, a calculation of the first-order correction to the Higgs boson mass squared yields a quadratically divergent expression arising from SM particle loop graphs. This implies that it is not "natural" to have a Higgs boson that is relatively light unless this divergence can be controlled by the structure of the theory. The SM provides no mechanism for this. Many different ways of regulating the divergence have been proposed, including supersymmetry and technicolor. (See §7.1 for a discussion of technicolor.) In this section, we shall focus on supersymmetry. In a supersymmetric theory the quadratic divergence is naturally cancelled by related loop graphs involving the supersymmetric partners of the SM particles contributing to the divergent loops. The result is that the tree-level mass squared of the Higgs boson receives corrections that are limited by the extent of supersymmetry breaking. In order that the naturalness and other problems be resolved, it is necessary that the scale of supersymmetry breaking not exceed $\mathcal{O}(1 \text{ TeV})$. In particular, it must be natural for the Higgs boson(s) with strong coupling to the WW channel to be lighter than this scale in order that there be no unitarity violations in WW scattering.

Supersymmetric theories are especially interesting in that, to date, they provide the only structure in which the problems of naturalness and hierarchy are resolved while retaining the Higgs bosons as truly elementary spin-0 particles. As we have stated, other theoretical approaches, such as technicolor, resolve these problems by making the Higgs boson(s) composite in nature. Supersymmetry is also the only theoretical base that provides definite models that may not encounter any phenomenological difficulties.[†] Explicit realizations of other approaches have led to contradictions with experimental facts, or have been extremely complicated formally and not thoroughly studied. For example, the simpler technicolor models contradict the observed absence of flavor-changing neutral currents (FCNC's). (However, more complicated models have been proposed to avoid such problems; see ref. 104.)

As mentioned earlier, supersymmetry avoids problems with FCNC's due to a particular requirement that is placed on the Higgs multiplet structure. Whereas in the SM only one Higgs doublet is required to give mass to the quarks and leptons, in the supersymmetric model, *two* Higgs doublets are needed to give mass to both up-type and down-type quarks (and the corresponding leptons) [105–107]. This requirement arises from a technical property of supersymmetric models. The interaction of Higgs bosons and fermions is obtained from the superpotential given by

$$W_F = \epsilon_{ij} \left[f \hat{H}_1^i \hat{L}^j \hat{R} + f_1 \hat{H}_1^i \hat{Q}^j \hat{D} + f_2 \hat{H}_2^j \hat{Q}^i \hat{U} \right], \tag{4.64}$$

where $\hat{H}_1$ and $\hat{H}_2$ are the Higgs superfields, $\hat{Q}$ and $\hat{L}$ are the SU(2) weak-doublet quark and lepton superfields, respectively, $\hat{U}$ and $\hat{D}$ are SU(2)-singlet quark superfields and $\hat{R}$ is an SU(2) weak-singlet charged lepton superfield. The SU(2) indices i, j are contracted in a gauge invariant way (with $\epsilon_{12} = -\epsilon_{21} = 1$). Supersymmetry forbids the appearance of $\hat{H}_1^*$ and $\hat{H}_2^*$ in eq. (4.64). Because of hypercharge gauge invariance, an $\hat{H}_1 \hat{Q} \hat{U}$ coupling is prohibited; hence, no up-quark mass can be generated if $\hat{H}_2$ is omitted. Another way to see the need for two Higgs doublets in a supersymmetric theory is to require that no anomalies be present, so that the theory is renormalizable [108]. It is sufficient to require the sum of all fermion charges to vanish in order to make the anomalies vanish. The fermionic partners of one Higgs doublet, $(\tilde{H}_1^0, \tilde{H}_1^-)$, must be complemented with those of the second doublet, $(\tilde{H}_2^+, \tilde{H}_2^0)$,

---

[†]One should be careful to note that supersymmetric models at present provide no insight to the flavor problem and thus could have problems from flavor-changing neutral currents. See ref. 103 for more details.

must be complemented with those of the second doublet, $(\widetilde{H}_2^+, \widetilde{H}_2^0)$, if the fermion charges are to add to zero (the fermion charges in the quark plus lepton sectors separately add to zero). Physically, the two doublets are needed because supersymmetry associates scalars with fermions of a given helicity.

Thus, the minimal supersymmetric extension of the Standard Model (MSSM) is a two-Higgs doublet model [105]. Furthermore, supersymmetry imposes powerful constraints on the Higgs boson sector of the model. Even if we assume that the supersymmetry is spontaneously or softly broken, it must be true that the dimension-four terms of the Higgs potential respect the supersymmetry. On the other hand, the most general Higgs sector of the MSSM can include all possible soft-supersymmetry-breaking terms with arbitrary coefficients (*i.e.* terms of dimension two or three which do not reintroduce quadratic divergences in the unrenormalized theory [109]). This model has been studied in detail in refs. 23 and 24.

In the minimal supersymmetric model with only two doublets of complex scalar fields, SU(2) × U(1) gauge invariance must be spontaneously broken by the soft supersymmetry-breaking terms in the Higgs potential; the appropriate soft supersymmetry-breaking potential parameters might arise, for instance, as a result of renormalization group evolution down from the Planck mass m_{Pl} to energies of order m_W [110]. However, it is possible to extend the Higgs sector in such a way that the SU(2) × U(1) gauge symmetry is spontaneously broken at tree-level, even in the supersymmetric limit. The simplest extension is to include a complex scalar field which is an SU(2) × U(1) gauge singlet [105]. We shall retain the possibility of such a singlet field briefly below, and will discuss such models in more detail in chapter 5.

Including a gauge singlet scalar field, denoted by N, in addition to the two complex doublet fields, the most general superpotential (which conserves baryon and lepton number) is

$$W = \lambda \epsilon_{ij} \widehat{H}_1^i \widehat{H}_2^j \widehat{N} - \mu \epsilon_{ij} \widehat{H}_1^i \widehat{H}_2^j - \Lambda \widehat{N} + \tfrac{1}{2} M \widehat{N}^2 - \tfrac{1}{3} k \widehat{N}^3 + W_F, \qquad (4.65)$$

where W_F is given in eq. (4.64). The scalar field potential, V, arising from this superpotential may now be computed as the sum of 'D' and 'F' terms (see ref. 23). In our notation, we use the same letter without a hat to denote the scalar field component of the corresponding superfield. In analyzing V we shall make the following assumptions. First, we assume that the scalar-quark and scalar-lepton fields do not acquire vacuum expectation values. (This is so that color and lepton number remain unbroken, and implies that we can ignore W_F when studying the Higgs boson mass matrix.) Second, we note that we can make a shift in the N field such that the parameter M in eq. (4.65) disappears. We shall simply set $M = 0$ with no loss of generality. For more details see ref. 23. Including all possible soft supersymmetry breaking terms we arrive at a scalar potential for the Higgs sector of the form

$$V = \tfrac{1}{8}g^2 \left[4|H_1^{i*}H_2^i|^2 - 2(H_1^{i*}H_1^i)(H_2^{j*}H_2^j) + (H_1^{i*}H_1^i)^2 + (H_2^{i*}H_2^i)^2 \right]$$
$$+ \tfrac{1}{8}g'^2 (H_2^{i*}H_2^i - H_1^{i*}H_1^i)^2 + |\lambda H_1^i H_2^i \epsilon_{ij} - \Lambda - kN^2|^2$$
$$+ |\lambda|^2 (H_1^{i*}H_1^i + H_2^{i*}H_2^i)N^*N + |\mu|^2(H_1^{i*}H_1^i + H_2^{i*}H_2^i) \qquad (4.66)$$
$$- (H_1^{i*}H_1^i + H_2^{i*}H_2^i)(\mu^*\lambda N + \text{h.c.}) + V_{\text{soft}},$$

with

$$V_{\text{soft}} = m_1^2(H_1^{i*}H_1^i) + m_2^2(H_2^{i*}H_2^i) - (m_{12}^2 \epsilon_{ij} H_1^i H_2^j + \text{h.c.}) + m_N^2 N^* N$$
$$+ (m_N'^2 N^2 + \text{h.c.}) - (\epsilon_{ij}\lambda A_\lambda H_1^i H_2^j N + \tfrac{1}{3}kA_k N^3 + \text{h.c.}). \qquad (4.67)$$

The parameters m_1, m_2, m_{12}, m_N, m_N', A_λ, and A_k, have dimensions of mass, Λ has dimensions of mass-squared, and λ and k are dimensionless. The Higgs field notation in the above equations is related to that used earlier in describing the general two-doublet model structure by

$$H_1 = \begin{pmatrix} H_1^1 \\ H_1^2 \end{pmatrix} = \begin{pmatrix} \phi_1^{0*} \\ -\phi_1^- \end{pmatrix}$$
$$H_2 = \begin{pmatrix} H_2^1 \\ H_2^2 \end{pmatrix} = \begin{pmatrix} \phi_2^+ \\ \phi_2^0 \end{pmatrix}. \qquad (4.68)$$

In analyzing the physical Higgs boson spectrum predicted by this potential we shall assume that the Higgs doublet fields H_1 and H_2 acquire vacuum expectation values

$$\langle H_1 \rangle = \begin{pmatrix} v_1 \\ 0 \end{pmatrix} \qquad \langle H_2 \rangle = \begin{pmatrix} 0 \\ v_2 \end{pmatrix}. \qquad (4.69)$$

As noted below eq. (4.8), supersymmetry requires that $\lambda_5 = \lambda_6$ (in the notation of that equation), and we may choose the phases for the Higgs doublet fields so that v_1 and v_2 are real and non-negative, and there is no CP violation in the Higgs sector.[‡] For $v_1, v_2 > 0$, we define $\tan\beta = v_2/v_1$ as in eq. (4.10), with $0 \leq \beta \leq \pi/2$. Since the Higgs sector phenomenology is sensitive to the value of $\tan\beta$ it is of interest to consider to what extent $\tan\beta$ is predicted in supersymmetric models. In the early days of supergravity model building, a large t-quark Yukawa coupling was used to trigger SU(2) × U(1) breaking in the low-energy theory [110]. This invariably led to the result that $\tan\beta > 1$. In the many models constructed subsequently, which contained a rather light top quark ($m_t \sim 40$ GeV), $\tan\beta$ tended to be of order 1. We shall typically consider a value of $\tan\beta = 1.5$. However, it has been noted [111] that if the top mass is very large, certain types of models, when evolved from an appropriate large mass scale to energy scales of order m_W, yield consistent solutions with fairly large $\tan\beta$ values.

[‡]In general, the phase rotation required to accomplish this may allow CP-violating phases to reappear in the interaction of the neutral Higgs bosons with other supersymmetric particles, unless reality conditions are imposed on parameters in the supersymmetric sector.

Turning to the Higgs mass spectrum, we note that, in general, one must resort to numerical methods to compute some of the resulting physical Higgs masses and eigenstates. In doing so one must look for an extremum of the potential V. The condition that this extremum be a true minimum of the potential is simply that the squared masses of all the physical Higgs bosons be positive.

For the remainder of this chapter, we will focus on the simplest supersymmetric model in which there is no N field. This is the minimal supersymmetric extension of the Standard Model (MSSM), which contains the minimal possible Higgs structure that can be considered in the context of supersymmetry. This model has the advantage of being very straightforward to analyze. Furthermore, it has been argued (see ref. 110 and the references therein) that if an N field were present it may destabilize the mass hierarchy between the electroweak and Planck scales. This occurs when the soft-supersymmetry-breaking terms that connect the low mass sector being considered to the high mass sector of the underlying supergravity theory include terms $V_{\text{soft}} \propto NB^2$, where B is a scalar member of the heavy sector. Such terms are analogous to the A terms introduced in eq. (4.67). It is possible to arbitrarily take the offending A terms to be zero, but this should probably be regarded as unnatural. However, in a supersymmetric context the heavy B scalars are generally accompanied by heavy fermion partners, F. If the splitting between m_B and m_F is not too great, the dangerous 'tadpole' diagrams of the B scalars tend to cancel with those of the F fermions, and destabilization will not occur. This is what happens in the context of those superstring theories that have been analyzed to date, where it is natural for N fields to arise. Thus, in chapter 5 we shall analyze the impact of adding an N field, paying particular attention to the way in which the results of the minimal supersymmetric model with no N field are modified. Typically the superstring motivated models do not correspond to the completely general case with $\mu \neq 0$ and with the N field present. Those analyzed to date imply restrictions on the allowed superpotential and soft-supersymmetry-breaking terms, that prevent the analysis from becoming unmanageable.

In the minimal model with no N field, obtained by taking $\lambda = m_N = m'_N = \Lambda = k = 0$ in eqs. (4.66) and (4.67), the simplest possibility might seem to be to take μ equal to zero also. This would be acceptable provided m_{12}^2 is kept non-zero. Otherwise, the pseudoscalar Higgs boson A^0, which has mass proportional to m_{12}^2, would be an axion. However, in many models

$$m_{12}^2 = B\mu, \tag{4.70}$$

where B is a new mass parameter, and taking μ to zero keeping $m_{12}^2 \neq 0$ would not be possible. The need for a non-zero value of m_{12} creates a new naturalness problem, as it is not easy to provide an explanation as to why this new mass parameter should not be very large, say of order the grand unification mass scale. We shall see in §5.1 that introduction of a singlet N field can allow one to avoid introduction of such ad hoc mass scales.

The resulting potential for the minimal model takes the form*

$$V = (m_1^2 + |\mu|^2)H_1^{i*}H_1^i + (m_2^2 + |\mu|^2)H_2^{i*}H_2^i - m_{12}^2(\epsilon_{ij}H_1^iH_2^j + \text{h.c.})$$
$$+ \tfrac{1}{8}(g^2 + g'^2)\left[H_1^{i*}H_1^i - H_2^{j*}H_2^j\right]^2 + \tfrac{1}{2}g^2|H_1^{i*}H_2^i|^2. \tag{4.71}$$

It is convenient to reexpress the doublet fields in terms of the physical Higgs boson degrees of freedom and the Goldstone boson fields. The relations are

$$H_2^1 = H^+ \cos\beta + G^+ \sin\beta$$
$$H_1^2 = H^- \sin\beta - G^- \cos\beta$$
$$H_1^1 = v_1 + \frac{1}{\sqrt{2}}(H^0 \cos\alpha - h^0 \sin\alpha + iA^0 \sin\beta - iG^0 \cos\beta) \tag{4.72}$$
$$H_2^2 = v_2 + \frac{1}{\sqrt{2}}(H^0 \sin\alpha + h^0 \cos\alpha + iA^0 \cos\beta + iG^0 \sin\beta),$$

where α is the mixing angle, referred to earlier, that arises in the process of diagonalizing the 2×2 neutral scalar Higgs mass matrix. Also recall that v_1 and v_2 can be taken real and non-negative. The Goldstone boson degrees of freedom are, as usual, absorbed in giving the W and Z their masses. One may now work out the minimization constraints that guarantee non-zero values for v_1 and v_2 and obtain various relations among the parameters appearing in eq. (4.71), v_1 and v_2, and the physical Higgs boson masses. The results of this procedure can be formulated as a special case of the more general two doublet potential structure obtained earlier. This results in constraints among the λ_i of the general two-doublet potential [23], and various relations between the λ_i and the parameters appearing in eq. (4.71)

$$\lambda_2 = \lambda_1$$
$$\lambda_3 = \tfrac{1}{8}(g^2 + g'^2) - \lambda_1$$
$$\lambda_4 = 2\lambda_1 - \tfrac{1}{2}g'^2$$
$$\lambda_5 = \lambda_6 = 2\lambda_1 - \tfrac{1}{2}(g^2 + g'^2) \tag{4.73}$$
$$m_1^2 = -|\mu|^2 + 2\lambda_1 v_2^2 - \tfrac{1}{2}m_Z^2$$
$$m_2^2 = -|\mu|^2 + 2\lambda_1 v_1^2 - \tfrac{1}{2}m_Z^2$$
$$m_{12}^2 = -\tfrac{1}{2}v_1 v_2(g^2 + g'^2 - 4\lambda_1).$$

The masses of the physical Higgs bosons and the mixing angle α are easily obtained. For example, eq. (4.73) combined with the mass formula below eq. (4.14) yields

$$m_A^2 = m_{12}^2(\tan\beta + \cot\beta) . \tag{4.74}$$

Note that we may choose $0 \leq \beta \leq \pi/2$ since we have chosen to take $v_1, v_2 > 0$. It is shown in Appendix A of ref. 24 that this implies that $-\pi/2 \leq \alpha \leq 0$. In

---

*The identity $|H_1^{i*}H_2^i|^2 + |\epsilon_{ij}H_1^iH_2^j|^2 = (H_1^{i*}H_1^i)(H_2^{j*}H_2^j)$ has been used in deriving the potential below.

this same Appendix various choices for possible independent parameters for the Higgs sector of this minimal model are outlined. It is often convenient to adopt $\tan\beta$ and $m_{H^\pm}$ as our independent parameters. (As we shall see below, $m_{H^\pm}$ must be chosen such that $m_{H^\pm} \geq m_W$.) Once these two quantities are specified (and no N field is present) all the Higgs tree-level masses can be computed according to

$$
\begin{aligned}
m_{A^0}^2 &= m_{H^\pm}^2 - m_W^2 \\
m_{H^0,h^0}^2 &= \tfrac{1}{2}\left[m_{A^0}^2 + m_Z^2 \pm \sqrt{(m_{A^0}^2 + m_Z^2)^2 - 4m_Z^2 m_{A^0}^2 \cos^2 2\beta} \right],
\end{aligned}
\tag{4.75}
$$

while the mixing angle α can be computed using

$$
\cos 2\alpha = -\cos 2\beta \left(\frac{m_{A^0}^2 - m_Z^2}{m_{H^0}^2 - m_{h^0}^2} \right)
\tag{4.76}
$$

$$
\sin 2\alpha = -\sin 2\beta \left(\frac{m_{H^0}^2 + m_{h^0}^2}{m_{H^0}^2 - m_{h^0}^2} \right).
\tag{4.77}
$$

There are several particularly crucial predictions following from eq. (4.75). We find $m_{H^\pm} \geq m_W$, $m_{H^0} \geq m_Z$, $m_{A^0} \geq m_{h^0}$, and

$$
\begin{aligned}
m_{h^0} &\leq m|\cos 2\beta| \leq m_Z, \\
m &\equiv \min\{m_Z, m_{A^0}\}.
\end{aligned}
\tag{4.78}
$$

It is also useful to note that as $m_{A^0} \to 0$ at fixed $\tan\beta$, $m_{h^0} = m_{A^0}\cos 2\beta$. Thus, the theory must possess at least one light Higgs boson.[†] Unless $\cos 2\beta$ is near its maximum of 1, and $m_{A^0}, m_{H^+}, m_{H^0}$ are large, h^0 will be observable at SLC, LEP or LEP-II. The spectrum of Higgs boson masses (at tree-level) in the minimal supersymmetric model is illustrated in fig. 4.6. From this graph, we note an interesting limiting case of this model when $m_{A^0} \to \infty$ (at fixed $\tan\beta$). In this limit, A^0, H^0 and $H^\pm$ decouple from the theory and we are left with a Higgs sector (consisting of a single physical CP-even scalar, h^0) which is identical to the Higgs boson of the minimal Standard Model. Moreover, as shown in Appendix A, the interactions of h^0 with the Standard Model gauge bosons and fermions are equivalent to those of the minimal Higgs boson of the Standard (nonsupersymmetric) Model. From fig. 4.6 we also observe that

---

[†]More stringent Higgs mass bounds can be obtained in the context of specific supergravity theories (with a minimal low-energy supersymmetric content). See, for example, the review by P. Roy [112] and references therein.

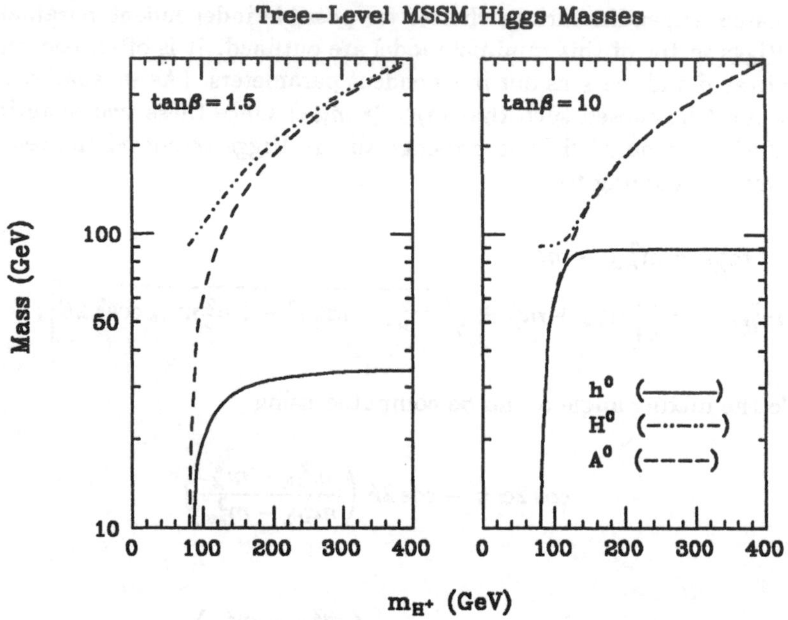

Tree–Level MSSM Higgs Masses

Figure 4.6 We display the masses of H^0, h^0 and A^0 as functions of m_{H+} at fixed $\tan \beta = 1.5$ and 10 for the minimal supersymmetric model. At large m_{H+}, we have $m_{h^0} \simeq m_Z |\cos 2\beta|$ and $m_{H^0} \simeq m_{A^0} \simeq m_{H+}$.

at large $\tan \beta$ either H^0 or h^0 is roughly degenerate with the Z, there being a rapid crossover at a fairly low value of m_{H+}.

It is important to emphasize that the Higgs boson masses above are *tree-level* masses, which will be modified when radiative corrections are taken into account. In general, the relative size of the corrections will be small, except in the case of a very light scalar Higgs boson (h^0). For example, if $\tan \beta = 1$, then eq. (4.75) implies that $m_{h^0} = 0$ at tree-level (independent of m_{A^0}). It follows that the radiatively corrected scalar mass in this case is $m_{h^0} = m_{CW}$, where m_{CW} is computed as described in §4.1. (Note that in the case considered here, eq. (A.12) implies $\cos(\beta - \alpha) = 0$, so that we may identify $h^0 = \sqrt{2}\xi - v$ in the notation of §4.1.) For general supersymmetric parameters, there is no closed-form analytic expression for m_{CW}, due to the rather complicated dependence of the chargino and neutralino masses on v, as noted below eq. (4.33). Numerical results for m_{CW} in particular supersymmetric models have been given in refs. 32 and 36. Note that in the limit where the supersymmetry-breaking parameters become much larger than v (the scale of electroweak physics), we may use eq. (4.35) to compute the radiatively corrected mass for h^0. In this limit, $m_{A^0} \to \infty$, so all other Higgs bosons decouple and we see that m_{h^0} approaches the standard Coleman-Weinberg mass of (roughly) 10 GeV.[‡]

Finally, we note that we can arrange $m_{h^0} = 0$ at tree-level by setting $m_{A^0} = 0$ independent of $\tan \beta$. In this limit, eq. (A.12) implies that $\alpha = -\beta$, so that eq. (4.39) reads

$$m^2_{H^0} \cos^2 2\beta + m^2_{h^0} \sin^2 \beta > m^2_{LW}, \qquad (4.79)$$

where m_{h^0} and m_{H^0} are the radiatively corrected Higgs masses. If $m^2_{H^0} \cos^2 \beta > m^2_{LW}$ (which must be satisfied over a large range of $\tan \beta$ since $m_{H^0} \gtrsim m_Z$), then no bound exists for m_{h^0}; presumably the parameters of the supersymmetric model can be chosen such that m_{h^0} is arbitrarily small. Only for $\tan \beta$ near 1 do we get a nontrivial bound on m_{h^0} due to radiative corrections.

It is also of interest to investigate how stable the mass relations of eq. (4.75) are under one-loop renormalization. A numerical study was carried out in ref. 113, and more recently an analytical evaluation of the sum rule corrections has been obtained in ref. 38. Both these studies focus on the gauge bosons, the Higgs bosons and their superpartners, the neutralinos and charginos, $\tilde{\chi}^0_i$ and $\tilde{\chi}^{\pm}_j$. A priori, one might expect that corrections to Higgs boson masses might be large when the superpartner masses become large; in this limit, the natural limitations coming from the supersymmetric structure of the theory become far removed from the tree-level Higgs boson mass scale. However, both references find very small corrections to the mass relations, once the standard infinities found in renormalization are absorbed in two physical parameters, e.g. the values of $\tan \beta$ and $m_{H^\pm}$ that we find it convenient to use here. This has been understood in refs. 38 and 114 as a type of screening effect. Corrections to the Higgs boson masses involve quadratically and logarithmically divergent loop graphs containing all these particles. After cancellation of explicit infinities in the dimensional regularization approach, all Higgs boson masses squared have shifts from the one-loop diagrams proportional to the Higgs boson masses squared themselves, m^2_W, m^2_Z, and the chargino-neutralino masses squared, denoted generically by $m^2_{\tilde{\chi}}$. Thus, as $m^2_{\tilde{\chi}}$ becomes large, these finite shifts become very large, as expected. However, it is found that all such terms proportional to $m^2_{\tilde{\chi}}$ cancel in the mass relations of eq. (4.75); i.e. the large $m^2_{\tilde{\chi}}$ shifts can all be absorbed into $m^2_{H^\pm}$, for example. The remaining terms grow with $m^2_{\tilde{\chi}}$ only logarithmically! However, the screening theorem can be evaded if one includes the quark/squark sector of the supersymmetric theory, and allows the Higgs–squark–squark Yukawa couplings to be large. This was demonstrated in ref. 115, where corrections from this sector to the $m^2_{H^\pm} = m^2_{A^0} + m^2_W$ sum rule were computed. It was found that these corrections could be large (without contradicting observed limits on corrections to the electro-weak ρ) in a special region of parameter space where

---

[‡]Of course, in the limit where all supersymmetric-breaking parameters become large, the theory becomes unnatural and one must fine tune the bare Higgs mass in order to obtain electroweak symmetry breaking at the appropriate energy scale. Thus this particular limit is instructive for academic purposes only.

there is a large splitting between the masses of the up and down squark eigen-states *and* the Higgs–squark–squark Yukawa couplings are big. Thus, in the minimal supersymmetric model, deviations from the tree-level mass relations could provide important information regarding the squark/slepton sector of the theory, even if the W and Z masses have the ratio predicted by the SM.

As a final note, we examine the possibility of taking the supersymmetric limit, $m_1 = m_2 = B = 0$ in this minimal model case. We find that the results of eq. (4.73) are inconsistent unless $v_1 = v_2 = 0$. The reason for the problem is simply that the potential V of eq. (4.71) does not spontaneously break SU(2) × U(1) in this limit. Only by including an N field is spontaneous SU(2) × U(1) breaking possible in the supersymmetric limit. The supersymmetric limit with the N field present is of some theoretical interest. The simplest example is one in which $\mu = M = k = 0$ in eq. (4.65). This example was originally proposed in ref. 105. In this model, one finds that $v_1 = v_2$, and that the ground state [which spontaneously breaks SU(2) × U(1) to U(1)$_{EM}$] is supersymmetric. The supersymmetry of the ground state implies that the states of the theory must lie in supersymmetric multiplets, and this is indeed the case. In particular, two of the Higgs states ($H^\pm$ and one of the CP-even scalars) are associated with the massive vector boson ($W^\pm$ and Z^0) supermultiplets; the remaining Higgs scalars and pseudoscalars are combined in scalar multiplets, with the appropriate spin-1/2 superpartners [105,116]. This explains why the limit $m_{A^0} = 0$ and $\tan\beta = 1$ [which implies that $\cos(\beta - \alpha) = 0$] applied to the minimal supersymmetric model (without the N field) often yields a supersymmetric-looking result.

Higgs Boson Couplings

We now turn to the couplings of the various Higgs bosons of the supersymmetric model, focusing on the minimal model outlined above. A complete summary of all Higgs boson couplings in the minimal supersymmetry model is given in Appendix A. For the moment we shall discuss only couplings to SM particles, in particular to quarks, W's and Z's. It is these couplings which are crucial in determining the production of the Higgs bosons, and in the absence of a light superpartner sector would also completely determine their decays.

We have already seen that the A^0 has no VV couplings, while those of h^0 and H^0 are related by eq. (4.20) to the $\phi^0 VV$ coupling. Thus, in the supersymmetric model these important couplings can be computed using the results for β and α that have been given in detail above. It is obviously of interest to consider the behavior of these angle suppression factors as a function of the parameters. A dramatic suppression emerges. Using eqs. (4.76) and (4.77), we may derive

$$\cos^2(\beta - \alpha) = \frac{m_{h^0}^2(m_Z^2 - m_{h^0}^2)}{(m_{H^0}^2 - m_{h^0}^2)(m_{H^0}^2 + m_{h^0}^2 - m_Z^2)}. \tag{4.80}$$

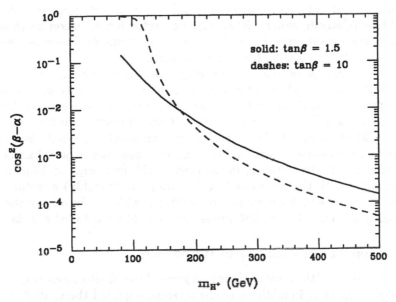

Figure 4.7 $\cos^2(\beta - \alpha)$ as a function of the charged Higgs mass, m_{H^+}, at $\tan \beta = 1.5$ and 10.

It is easily verified that the maximum possible value for $\cos^2(\beta - \alpha)$ at fixed $\tan \beta$ occurs in the limit $m_{H^+} \to m_W$ and is $\cos^2 2\beta$. Note that the smaller the maximum value of m_{h^0} [eq. (4.78)], the smaller the maximum possible value of $\cos^2(\beta - \alpha)$. As m_{H^+} increases, $\cos^2(\beta - \alpha)$ decreases further. For example, for $m_{H^0} \gtrsim 2m_W$, we see that m_{H^+} is large enough so that $\cos^2(\beta - \alpha)$ decreases rapidly as $1/m_{H^0}^4$. This behavior is illustrated in fig. 4.7 for the cases of $\tan \beta = 1.5$ and $\tan \beta = 10$.

For $\tan \beta = 1.5$, $\cos^2(\beta - \alpha)$ is $\lesssim 0.15$ at $m_{H^+} = m_W$, and is $\lesssim 0.01$ by the time $m_{H^0} > 2m_Z$. To have $\cos^2(\beta - \alpha) \gtrsim 0.5$ requires $\tan \beta \gtrsim 4$ and m_{H^+} near m_W. As an example of a large $\tan \beta$ case, we plot $\cos^2(\beta - \alpha)$ for $\tan \beta = 10$ in fig. 4.7. Comparing to the $\tan \beta = 10$ curve of fig. 4.6, we see that the scalar Higgs (h^0 or H^0) that is roughly degenerate with the Z always has nearly maximal coupling to the VV channels. In this case, over the small region (in m_{H^+}) where the h^0 is much lighter than the Z it will have very weak couplings to WW and ZZ channels. This contrasts to the $\tan \beta = 1.5$ case in which the h^0, no matter how light, always has large couplings to the VV channels. In any case, *i.e.* regardless of $\tan \beta$, once m_{H^0} is significantly larger than m_Z we see that the H^0 will have very weak coupling to VV channels in comparison to SM expectations. This result will obviously have significant consequences for H^0 production and decay, as will be discussed later. Note that the H^0 width to two vector bosons will, for most parameters, be small compared to the width of a heavy SM Higgs boson. Since the H^0 events are not spread out over a large mass region, one might think that detection would be relatively easier. However, the vector boson fusion mechanism is simultaneously suppressed and fermion decay channels dominate, with the result that detection is, in fact, more difficult.

The couplings of the various Higgs bosons to quarks were given in eqs. (4.24) and (4.25) as a function of α and β. A survey of the parameter space of the minimal model case we have outlined above reveals that when $\tan\beta > 1$ the H^0 and h^0 couplings to $t\bar{t}$ are somewhat suppressed relative to the SM ϕ^0, while the $b\bar{b}$ couplings are somewhat enhanced. For $\tan\beta < 1$ the exact opposite occurs. In general, the suppression or enhancement is relatively mild. For instance, in the case of H^0, once $m_{H^0} \gtrsim 2m_W$ the couplings agree with the corresponding SM values to (roughly) a factor of $\cot\beta$ or $\tan\beta$, for up and down quarks, respectively, while in the case of the h^0 the couplings are very close to SM values. See ref. 24 for detailed graphs.

One-Loop Corrections to ρ in the MSSM

The discussion of these corrections in a general two-doublet model has already been given in §4.1. In addition to the references quoted there, studies of such corrections specifically appropriate to supersymmetric models have appeared in refs. 115—120. We define the MSSM Higgs sector contribution to the ρ parameter to be $\Delta\rho \equiv \Delta\rho_{MSSM} - \Delta\rho_{SM}$, where $\Delta\rho_{MSSM}$ is the contribution from all Higgs bosons of the MSSM, and $\Delta\rho_{SM}$ is the contribution of a Standard Model Higgs boson with $m_{\phi^0} = m_{h^0}$. Due to the approximate degeneracy of m_{H^+}, m_{A^0}, and m_{H^0} at large m_{H^+} (see fig. 4.6), there are large cancellations in eqs. (4.41) and (4.45). To see just how complete these cancellations are, we present in fig. 4.8 the numerical result for $\Delta\rho$ as a function of m_{H^+} at three different values of $\tan\beta$. The value of $\Delta\rho$ is extremely small over the entire region of MSSM Higgs parameter space, and therefore leads to no useful constraints on the model. The fact that $\Delta\rho \to 0$ as m_{H^+} becomes large is expected on general grounds. In the limit of large m_{H^+}, the effects of H^+, H^0 and A^0 must smoothly decouple, leaving only h^0 (with Standard Model couplings) in the effective low energy theory. Since $\Delta\rho$ is a measure of the effect of the supersymmetric Higgs sector relative to that of the Standard Model Higgs boson, it follows that the MSSM contribution to $\Delta\rho$ must vanish as m_{H^+} becomes large. It is clear from this general argument that $\Delta\rho$ from the Higgs sector of a more general (non-minimal) supersymmetric model would also be expected to be rather small.

One-Loop Mediated Decays of the MSSM Higgs

In addition to the modifications previously outlined in the corresponding discussion of §4.1, we have now to include loops involving various superpartners of SM particles. In the case of the gg coupling to neutral Higgs, we must include the squarks. This was considered in ref. 24 where it was found that the

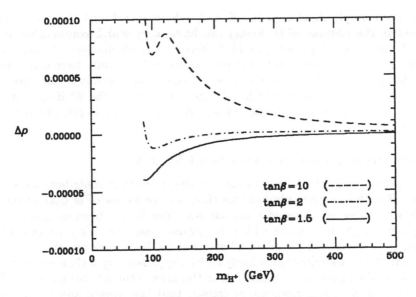

Figure 4.8 The MSSM Higgs sector contribution to $\Delta\rho$.

modifications to the quark-loop contributions were generally quite small except for $m_h \sim 2m_{\tilde{q}}$. When $m_{\tilde{q}}$ is much larger than m_h squark loops become unimportant since the squark couplings to the Higgs do not contain a term behaving as $m_{\tilde{q}}^2/m_W^2$ [see eq. (2.14)] as required to get a non-vanishing result for the squark loop in the large mass limit. Even at moderate $m_{\tilde{q}}$, the contributions are modest simply because of the small size of the controlling function F_0 as compared to $F_{1/2}$, [see eqs. (2.16) and (2.20)].

The $\gamma\gamma$ decay modes in the MSSM have been considered in refs. 121 and 122 and the $Z\gamma$ mode in ref. 123. In the $\gamma\gamma$ case, the results of ref. 122 for the $\gamma\gamma$ case are further compactified and explicitly presented in Appendix C. As in the gg case discussed above, the squark loops are not very important, and as discussed earlier in the two-doublet SM presentation, charged Higgs loops also tend to be small for similar reasons. The big new contribution is from the chargino loops, where the charginos are the superpartners of the $W^\pm$ and $H^\pm$. However, these chargino-loop contributions do not have a non-vanishing contribution in the limit of large chargino mass. In the $\gamma\gamma$ case this is apparent from the formulae of Appendix C, which show that they vanish as a single inverse power of the chargino mass at amplitude level (note that this is not as fast as the quadratic vanishing for the scalar-particle loops). The important point is that the $h\tilde{\chi}^\pm\tilde{\chi}^\mp$ couplings do not grow as $m_{\tilde{\chi}^\pm}/m_W$, unlike the normal fermion couplings which behave as m_f/m_W. But, considering for instance the $\gamma\gamma$ width, the $F_{1/2}$ function is large, and so the charginos can make substantial contributions for small to moderate chargino masses. This has been amply illustrated in ref. 122, where it is found that even the A^0 (despite the absence of W loops) can have a $\gamma\gamma$ width comparable to that of the ϕ^0 at similar mass, provided there is a light chargino. Certainly, the supersymmetric model neutral Higgs bosons will generally have very different $\gamma\gamma$ widths from a SM ϕ^0 of the same mass, except possibly in the case of the h^0 which will have a $\gamma\gamma$ width quite near to that of the ϕ^0 if charginos are heavy and $\tan\beta \lesssim 1.5$. Similar results have been found in ref. 123 in the case of the $Z\gamma$ mode.

Rare Decay Limits on a Very Light h^0 or A^0

Not surprisingly, the MSSM yields new diagrams that contribute to $K \to h\pi$ and $B \to hX_s$. The only analysis that we are aware of is that of ref. 124. They consider two types of contributions. The first is from diagrams involving squark-gluino loops in which the gluino–quark–squark vertices lead to a change in flavor and the A^0 or h^0 attaches to the squark or external b. They find that this contribution is likely to be suppressed by a chance cancellation between the above two attachments in the case of the h^0, but quite significant for a light A^0. This assumes, of course, that the squark and gluino masses are not too large. They find that for squark and gluino masses $\lesssim 100$ GeV one can obtain $BR(b \to sA^0) \sim 0.01$ from this source computed on its own. Of course, there are other diagrams they did not consider so firm conclusions are not possible at this time. Certainly, we feel that their conclusion that the CLEO data places limits on a A^0 for such a small branching ratio is premature. Indeed, we saw in §3.1 that to rule out the SM ϕ^0 from either inclusive $B \to \phi^0 X_s$ or exclusive $B \to \phi^0 K$ decays generally required the full SM branching ratio, $BR(b \to s\phi^0) \sim 0.36$, as well as additional theoretical assumptions.

The second class of contribution is that from a chargino-squark loop. This would be important in the case of the h^0 when the gluino–squark loop cancellation occurs as discussed in the preceding paragraph. For light charginos and other supersymmetry mass scales, these diagrams computed on their own can again give $BR(b \to sh^0) \sim 0.01$. However, since in the MSSM the h^0 is very likely to have SM-like couplings when it is very light, it seems clear that the more standard two-quark operator diagrams would dominate the branching ratio, and lead to a BR of order 0.36, so long as diagrams involving charged Higgs bosons are not important. (This latter question has not been clearly addressed in the literature.) The SM discussion would then apply, and such a very light h^0 can probably be ruled out using B decays in the mass ranges delineated in §3.1. In addition, for a very light h^0 with SM-like couplings, the K-decay and nuclear limits will also apply.

Of course, if the gluino, squark and chargino mass scales are all substantially above m_W, loop contributions to rare B and K decays involving these superparticle states will be suppressed and the discussion given under the two-doublet extension of the SM in §4.1 will become appropriate also for the MSSM.

Limits on the Charged Higgs and $\tan\beta$ in the MSSM

As in the preceding subsection, the discussion of the two-doublet extension of the SM in §4.1 must now be extended to account for the presence of new loop diagrams involving superpartners that contribute to K_L–K_S mixing, B–$\overline{B}$ mixing, and the rare decays $b \rightarrow s + \gamma$, $b \rightarrow s + \ell^+\ell^-$, and $b \rightarrow s + g$. The B–$\overline{B}$ mixing contributions can be comparable to those arising in the SM so long as the gluino and squark masses are $\lesssim 100$ GeV. For a review see ref. 57. They are completely negligible in K_L–K_S mixing, which however is irrelevant compared to B–$\overline{B}$ mixing in constraining the charged Higgs and $\tan\beta$. Superpartner loops also enhance the decays $b \rightarrow s + \gamma$, $b \rightarrow s + \ell^+\ell^-$, and $b \rightarrow s + g$ [126] if the supersymmetry mass scales are $\lesssim 100$ GeV. In this case the interpretation of a positive measurement of any of the above branching ratios in terms of restrictions on the charged Higgs mass would be more difficult. But, once again, if the supersymmetry-breaking scale is large, we may revert to our two-doublet SM extension discussion in §4.1.

Higgs Boson Search Techniques in Minimal Supersymmetry: SUSY Decays Kinematically Forbidden

In this section we will explore the impact of the minimal supersymmetry model (with just two Higgs doublets and no singlet) upon our ability to experimentally probe the Higgs sector. The discussion is based on refs. 84, 29, 90, 127, and 128. In this subsection we shall assume that all superpartners are too heavy to appear in the decays of the Higgs bosons. In the following subsection we shall take seriously the possibility of a light supersymmetric sector and discuss its impact on Higgs detection.

Of course, one of the most important ingredients in assessing the detectability of the various neutral bosons of the supersymmetric theory is a thorough understanding of their decays. Here we summarize the dominant decay modes in the case that all superpartners are heavy. The relevant decay width formulae appear in Appendix B. In the present case only SM particle modes, and kinematically allowed modes involving the scalar particles themselves need be considered. In the latter category only four modes are allowed by the mass relations of the minimal supersymmetry model. These are

$$H^0 \rightarrow h^0 h^0, \tag{4.81}$$
$$H^0 \rightarrow A^0 A^0, \tag{4.82}$$
$$A^0 \rightarrow Z h^0, \tag{4.83}$$

and

$$H^+ \rightarrow W^+ h^0. \tag{4.84}$$

Of these, $H^0 \rightarrow A^0 A^0$ is only allowed over a very tiny range of possible m_{H^+}

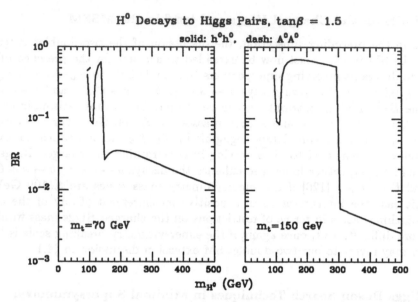

Figure 4.9 Branching ratio for the H^0 decays (4.81) and (4.82) as a function of m_{H^0}, for $\tan\beta = 1.5$ and two choices of the top-quark mass, $m_t = 70$ GeV and $m_t = 150$ GeV. The $h^0 h^0$ mode is given by the solid curve, while the $A^0 A^0$ mode is indicated by a single very short dash and is zero elsewhere. Modes containing supersymmetric particles are assumed to be absent.

values (keeping in mind the correlated h^0, H^0, and A^0 masses of fig. 4.6). Branching ratios for these decays for several choices of the top quark mass are illustrated in figs. 4.9–4.11. In the case of H^+ decays we show also a number of the fermion–antifermion modes. We note that when decays to fermion–antifermion channels containing the top are not allowed, the decay channels containing a lower mass Higgs boson are very important. Indeed, the only reason the channels (4.83) and (4.84) containing a final gauge boson are not dominant even when top-quark channels are open is the suppression by the $\cos^2(\beta - \alpha)$ factor illustrated in fig. 4.7. (See the couplings in Appendix A.)

h^0 Detection at $e^+ e^-$ and Hadron Colliders

We begin by focusing on the light scalar h^0 with $m_{h^0} \leq m_Z$. As discussed above, it is very likely to have couplings that are essentially the same as those of a SM Higgs boson. If h^0 is light enough, it will be discovered at SLC or LEP-I, whereas if m_{h^0} is near m_Z, discovery at an $e^+ e^-$ collider will have to await future higher energy colliders. At the SSC techniques for its discovery that do not rely upon simultaneous discovery of other non-Standard-Model particles fall into the intermediate mass category discussed earlier, and are essentially

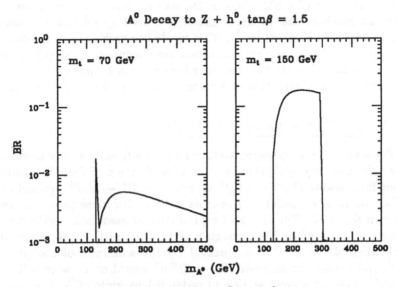

Figure 4.10 Branching ratio for the $A^0 \to Zh^0$ decay as a function of m_{A^0}, for $\tan\beta = 1.5$ and top-quark mass choices of $m_t = 70$ GeV and $m_t = 150$ GeV. Modes containing supersymmetric particles are assumed to be absent.

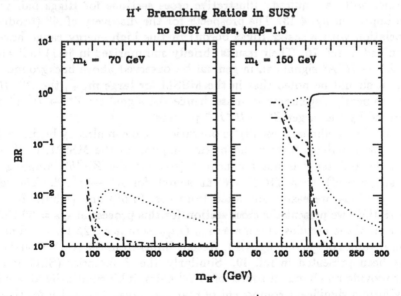

Figure 4.11 Branching ratios for various H^+ decay channels as a function of m_{H^+}, in the minimal supersymmetric model. We take $\tan\beta = 1.5$ and consider $m_t = 70$ GeV and 150 GeV. The curves are: solid = $t\bar{b}$; dashes = $c\bar{s}$; dotdash = $\tau^+\nu$; and dots = W^+h^0. Modes containing supersymmetric particles are assumed to be absent.

identical to those of a SM Higgs of the same mass. Since its mass may be below the reach of the $\gamma\gamma$ decay mode, we need to rely on the associated production mode $W + h^0(\to b\bar{b})$. This mode has received some study in this mass range, and preliminary indications are encouraging [85,86]. Thus, it is entirely possible that the first evidence for or against supersymmetric models could come from observation or failure to observe the predicted light Higgs scalar, h^0.

H^0, A^0 and $H^\pm$ Detection at e^+e^- Colliders

Because of the suppression factor of eq. (4.80), only a very light H^0 with mass quite near m_Z could share a reasonable fraction of the VV couplings required for sizeable $e^+e^- \to \nu\bar{\nu}H^0$ or $e^+e^- \to Z^* \to ZH^0$ production cross sections at an e^+e^- collider. The cross sections for these processes are illustrated in fig. 4.12. The A^0 can be produced in association with one of the CP-even scalars (h^0 or hh) through a virtual (or real) Z. However, if A^0 is heavy, then $Z^* \to h^0A^0$ is suppressed by the smallness of $\cos^2(\beta - \alpha)$, as mentioned earlier. In contrast, $Z^* \to H^0A^0$ would occur with full coupling strength, yielding a cross section of order 0.1 in units of $R \equiv \sigma/\sigma_{pt}$, where σ_{pt} is the cross section for $e^+e^- \to \gamma^* \to \mu^+\mu^-$, assuming no phase space suppression. As we shall discuss shortly, this level of cross section will probably prove adequate, but once $m_{A^0} + m_{H^0}$ becomes large very high machine energies will be required. Illustrative cross sections for Higgs pair production appear in fig. 4.13. The prospects for the discovery of A^0 (produced in association with a scalar Higgs boson) in this high energy range have been examined in ref. 76. Their results (briefly summarized in §4.1) indicate that the $Z^* \to H^0A^0$ signal can in general be observed above background. However, it should be noted that in the MSSM for large $m_{H^\pm}$, the A^0, H^0 and $H^\pm$ are nearly degenerate in mass; hence the signal for $Z^* \to H^0A^0$ will be obscured by the larger $Z^* \to H^+H^-$ process!

At e^+e^- colliders presently in operation, one can already begin to deduce limits on the Higgs masses and mixing angles. In the MSSM, if A^0 and h^0 are light and $\tan\beta$ is not too near 1 (so that the ZA^0h^0 coupling is not too suppressed), then TRISTAN can search for $Z^* \to h^0A^0$. The relevant cross sections are easily computed from eqs. (4.58) and (4.59). By way of illustration, we present the cross section for this process at $\sqrt{s} = 60$ GeV as a function of the pseudoscalar mass m_{A^0} ($m_{h^0} \sim m_{A^0}\cos 2\beta$ for small m_{A^0}) for several values of $\tan\beta$ in fig. 4.14. Recent limits from the AMY Collaboration have been presented in ref. 49. Similarly, the Z factories (SLC and LEP) may provide an abundant source of A^0h^0 pairs; if kinematically allowed, they may form a significant component of real Z decays. The width for this decay mode was given relative to that for a single neutrino channel in eq. (4.57). The resulting ratio is given in fig. 4.15. Clearly, the search for $Z^0 \to A^0h^0$ at SLC and LEP can lead to interesting constraints on the parameters of the minimal supersymmetric model [129].

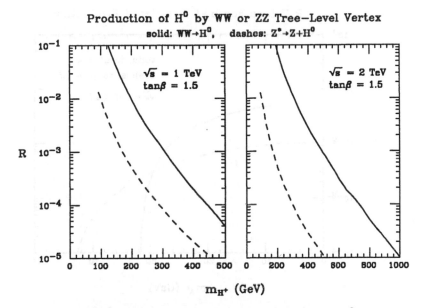

Figure 4.12 We give the cross sections for $W^+W^- \to H^0$ fusion and $Z^* \to Z + H^0$ associated production, at $\sqrt{s} = 1$ and 2 TeV for $\tan\beta = 1.5$, in units of $R = \sigma/\sigma_{pt}$, from ref. 128. To find m_{H^0} for a given m_{H^+}, use fig. 4.6.

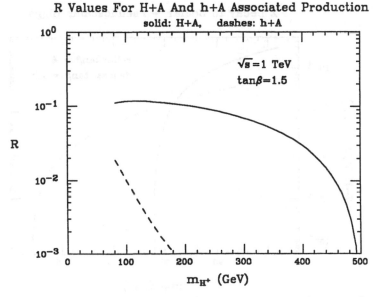

Figure 4.13 The cross section in units of $R \equiv \sigma/\sigma_{pt}$ for $e^+e^- \to H^0, h^0 + A^0$ as a function of the charged Higgs mass. We take $\tan\beta = 1.5$ and $\sqrt{s} = 1$ TeV. From ref. 128.

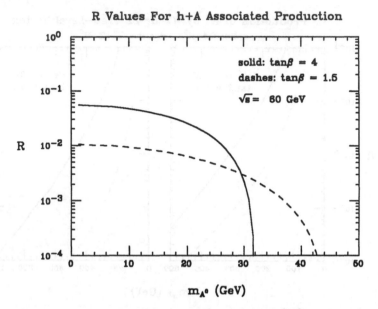

Figure 4.14 The cross section for $e^+e^- \to Z^* \to h^0 A^0$ as a function of m_{A^0} at $\sqrt{s} = 60$ GeV in the MSSM. Two values of $\tan\beta$ are considered: 1.5 and 4.

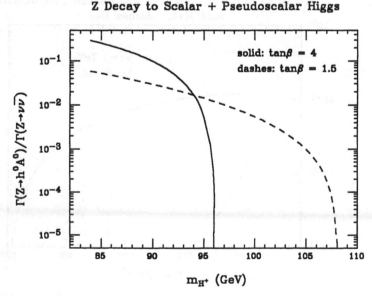

Figure 4.15 $\Gamma(Z \to A^0 h^0)/\Gamma(Z \to \nu\bar{\nu})$ as a function of m_{H^+} (see fig. 4.6 for the correlated A^0 and h^0 masses). We give results for $\tan\beta = 1.5$ and 4.

Regarding the charged Higgs, we first note that $Z \to H^+ H^-$ is not possible in minimal supersymmetry since $m_{H^\pm} \geq m_W$. Thus, the primary production mechanism is via Drell-Yan processes, with a cross section as given in eq. (4.60). As described in our discussion of the general two-Higgs-doublet model, the studies of refs. 130, 77, and 76 show that backgrounds to its detection are generally quite controllable whenever it has a reasonable production rate via the Drell-Yan processes. However, the phase space for pair production is rather limited and high masses require machines of significant energy.

The survey of ref. 128 indicates that other possible production modes for neutral Higgs, such as inclusive two-photon production of H^0 or A^0, γW and WW fusion production of Higgs pairs, and associated production via $e^+ e^- \to W^\mp + H^\pm + H^0, A^0$, all yield lower rates for H^0 and A^0 production than does $Z^* \to A^0 + H^0$, until machine energies $\gtrsim 4$ TeV. In the case of the two-photon production processes, this statement assumes a standard two-photon spectrum; it has been noted [131] that a high energy $e^+ e^-$ collider could be run in such a way as to enhance the spectrum by one to two orders of magnitude by making use of the beamstrahlung phenomenon. Two-photon production processes then become competitive with the primary production modes already discussed.

Finally, we again mention the process in which a neutral or charged Higgs boson is produced via bremsstrahlung off of a heavy fermion, e.g. a bottom or top quark. This is explored in the MSSM in refs. 78 and 79. Usable cross sections from this process require significant enhancements in the appropriate fermion coupling.

H^0 and A^0 Detection at Hadron Colliders

Turning next to hadron colliders, we shall focus on results for the SSC. It is still true that H^0 and A^0 are generally much more difficult to detect than a SM Higgs boson of similar mass. The A^0 has no VV couplings while those of H^0 are severely suppressed unless m_{H^0} is very near m_Z. Thus, VV fusion and $V + h$ ($h = H^0$ or A^0) associated production processes are not likely to be relevant. Production cross sections using gg fusion can, however, still be very significant, especially if the top quark is fairly massive, as discussed with respect to SM Higgs boson cross sections (see ref. 81). Another possible production mode for $h = H^0$ or A^0 is via $gg \to b\bar{b}h$. For moderate values of $\tan\beta$, this process generally has a smaller cross section than that from direct gg fusion, especially for large m_t. However, if $\tan\beta > 4$ it has been demonstrated [80] that the former process could be dominant and significantly enhance the expected cross sections. This occurs because of the enhanced coupling of $b\bar{b}$ to H^0 and A^0 at large $\tan\beta$. Regarding decays, VV modes are, of course, suppressed by $\cos^2(\beta - \alpha)$. Decays such as $H^0 \to ZA^0$ and $A^0 \to ZH^0$ are kinematically forbidden [see eq. (4.75)]. A decay mode requiring further investigation is $A^0 \to Zh^0$. Even though it is suppressed by the severe $\cos^2(\beta - \alpha)$ factor of eq. (4.80), the longitudinal Z polarization states can lead to a possibly interesting ($\gtrsim 10\%$ for m_{A^0} far enough above m_Z) branching ratio for this mode when the $t\bar{t}$ decay channel of the A^0 is closed. However, backgrounds are bigger and triggering techniques fewer than in the analogous situation for the $H^\pm \rightarrow W^\pm h^0$ decay that will be discussed shortly in some detail. As for the SM Higgs, if the $t\bar{t}$ decay channel is kinematically forbidden, the $\gamma\gamma$ decay modes of the neutral Higgs should be considered. The branching ratios for A^0 and H^0 to decay in the $\gamma\gamma$ mode can be comparable to [121,122] that of a SM Higgs if one artificially assumes that decays to supersymmetric partners are forbidden. However, in order to obtain useful branching ratios it is necessary to have loops containing light charginos coupling A^0 and H^0 to the two photons. This is because the WW loops are either absent or suppressed, and these were the main contribution in the case of the Standard Model Higgs; the charginos provide a substitute if light enough. Unfortunately, the calculations of ref. 122 show that the chargino and associated neutralino final states would then dominate the A^0 and H^0 decays, so that the $\gamma\gamma$ mode is never in fact useful. Continuing on, if the H^0 should have mass very near to m_Z then its VV coupling will be sufficiently large that the $W + H^0(\rightarrow b\bar{b})$ associated production technique might prove usable. Like the real VV decays, the ZZ^* mode is absent for the A^0 and is suppressed for H^0 in the mass range where it might have become useful. Thus, for essentially all masses, if the H^0 and A^0 decay only to SM channels their decays will be dominated by $b\bar{b}$ or $t\bar{t}$ and there is no known technique for observing them at the SSC. However, before concluding that the picture is totally bleak, it is important to consider non-SM channels. Because we are discussing a supersymmetric theory, there are new supersymmetric particles in the spectrum, which can couple to the Higgs bosons. In fact, as already mentioned above, the neutralinos and charginos that are automatically present in a supersymmetric model could be particularly important in their decays. We shall return to this point shortly, after considering the accessibility of the charged Higgs boson of the model in the absence of such new modes. Because $H^\pm$ detection would represent an unambiguous signal for a non-minimal Higgs sector, it has been more closely examined than detection of A^0 and H^0.

$H^\pm$ Detection at Hadron Colliders

In §4.1, we demonstrated that the charged Higgs boson could be discovered at a hadron collider only in certain restricted circumstances. First, if $m_t > m_{H^\pm} + m_b$, then the charged Higgs will be produced in t-decays, and can be detected by various means. Second, if $m_t < m_{H^\pm} + m_b$, then $H^\pm$ can be detected only if $m_{H^\pm} \lesssim 300$ GeV (in order to insure sufficient charged Higgs production) and if various rare decay modes have sufficient branching ratio. In order to give an overall impression of the most important of the SM decay channels for the H^+, we presented in fig. 4.11 the branching ratio for H^+ decay to $t\bar{b}$, $c\bar{s}$, $\tau^+\nu$, and W^+h^0, for two extreme top-quark mass choices—$m_t = 70$ and 150 GeV. It was assumed that decays to channels containing supersymmetric particles are not allowed. (Possible supersymmetric final states will be considered in the next section.) This should be compared with the conclusions of §4.1, where we showed that charged Higgs discovery may be possible if $BR(H^\pm \to W^\pm h^0) \gtrsim$ 1–10% or if $BR(H^\pm \to \tau^\pm \nu) \gtrsim$ 0.5. In the minimal supersymmetric model, $BR(H^\pm \to W^\pm h^0)$ is found to be much smaller than the maximal results quoted in §4.1. This is because the associated coupling squared (as well as that for $H^\pm \to W^\pm h^0 \gamma$) turns out to be proportional to the $\cos^2(\beta - \alpha)$ suppression factor of eq. (4.80) [see eq. (4.55)] and is thus very small unless m_{H^0} is very near m_Z (in which case $H^\pm$ has mass very near m_W and the decays are phase space suppressed). Despite this suppression, $H^\pm \to W^\pm h^0(\to b\bar{b})$ will dominate $H^\pm$ decays when $H^\pm \to tb$ decays are not allowed, and can still be significant even when the tb modes are present. For instance, if m_t is such that $H^\pm \to tb$ is allowed, the minimal SUSY branching ratios of fig. 4.11 combined with the $gb \to H^\pm t$ cross section yield a cross section for the $W^\pm b\bar{b}$ final state that, at moderate $m_{H^\pm}$ ($\sim 150-300$ GeV), is large enough for $H^\pm$ detection. On the other hand, $BR(H^\pm \to \tau^\pm \nu)$ is too small unless both the tb and Wh^0 decay modes are absent. We conclude that over most of the parameter space of the minimal supersymmetric model, the discovery of $H^\pm$ at a hadronic collider is impossible, if it is not produced in top-quark decay.

Higgs Boson Search Techniques in Minimal Supersymmetry: SUSY Decays Allowed

Even if the $H^\pm$, H^0 and A^0 cannot be detected through SM decay modes, there is still hope. In a supersymmetric model, Higgs bosons can also decay into squark and slepton pairs or into chargino/neutralino pairs (the latter are the charged and neutral mass eigenstates deriving from the gauginos and higgsinos and tend to be relatively light). Appropriate formulae for these modes are given in Appendix B. If the relevant decays are kinematically allowed, then the corresponding branching ratios can be large. Indeed, over a substantial region of the supersymmetric model parameter space, one finds total branching ratios into supersymmetric final states which are larger than 10%, and can easily approach 100%. As an example, we show in fig. 4.16 (taken from ref. 132) the charged Higgs branching ratio into chargino and neutralino final states as a function of the supersymmetric parameters. For reasonable choices of certain SUSY parameters M and μ, described in detail in refs. 23, 24, and 132, we see in fig. 4.16 that a charged Higgs with mass of order 500 GeV can decay more than 80% of the time into chargino and neutralino modes, even when the tb decay mode is allowed. Similar results are found for the H^0 and A^0.

The relatively large branching ratios into charginos and neutralinos can be explained by the fact that the relevant mass parameters which set the scale of the Higgs couplings to these particles are m_W, m_Z, and the parameters of the neutralino and chargino mass matrices, which are of the same order. In addition, unlike the coupling of H^0 to WW and ZZ, the Higgs couplings to the neutralinos and charginos are not suppressed, in general. As a result,

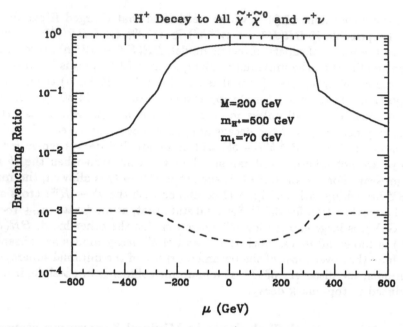

Figure 4.16 The branching ratio for $H^\pm$ to decay to any channel containing a neutralino-chargino pair, compared to the $\tau\nu$ branching ratio. We take $m_t = 70$ GeV, $\tan\beta = 1.5$ and $m_{H^\pm} = 500$ GeV, and a reasonable value of $M = 200$ GeV for the gaugino mass parameter of the model and plot the branching ratio as a function of the higgsino mass parameter, μ. Squark and slepton modes are assumed to be forbidden. The curves are: solid, sum over all neutralino+chargino channels; and dashes, $\tau\nu$.

it is not surprising that these final states can be dominant. The decay into squarks and sleptons can, in principle, also be an important fraction of the total Higgs widths. In evaluating various possible scenarios, we note that it seems more probable that some light charginos or neutralinos exist which would be accessible to Higgs decay. On the other hand, the general mass scale which controls the squark and slepton masses (and is *a priori* unrelated to the neutralino and chargino parameters) may be large enough so that Higgs decay into squark and sleptons would be forbidden. Clearly, no one can definitively predict, at present, which supersymmetric final states (if any) will dominate.

Supersymmetric decays of the Higgs present the possibility of completely novel signatures for Higgs searches. In order to establish what types of signatures will be important it is necessary to consider the decay modes of the neutralinos and charginos that are the primary Higgs decay products. These decay modes were explored in ref. 133. The lightest neutralino (usually denoted by $\tilde{\chi}_1^0$) is presumed to be the lightest supersymmetric particle (LSP) and will create missing energy. The heavier charginos and neutralinos can decay via two-body modes to lighter charginos and neutralinos plus a W and Z; the lighter inos will then, in turn, decay, often to three-body modes such as $q'\bar{q}\tilde{\chi}_1^0$ or $\ell^+\ell^-\tilde{\chi}_1^0$. The possibilities are numerous and the relative importance of different decay modes for the Higgs and the neutralinos and charginos is strongly dependent upon the Higgs spectrum and the M and μ parameters, mentioned earlier, that determine the ino mass spectrum.

A few examples may help to illustrate the possible departures from standard signatures for Higgs discovery. Consider first a Z factory. We have already seen that the h^0 is likely to be sufficiently low in mass that the decays $Z \rightarrow h^0\ell^+\ell^-$ and $Z \rightarrow h^0\gamma$, can occur. Further, over a certain region of parameter space, it is possible for the $\tilde{\chi}_1^0$ to be light enough that $h^0 \rightarrow \tilde{\chi}_1^0\tilde{\chi}_1^0$ can be an important decay mode of the h^0, leading to invisible Higgs decays and missing energy in association with the γ or $\ell^+\ell^-$ trigger [134]. This should serve as a warning to experimentalists who might be tempted to discard $Z \rightarrow \mu^+\mu^- X$ events with missing energy, under the assumption that the Higgs decay products will be visible. Also a new decay mode of the Z emerges when one has neutralinos and charginos: $Z \rightarrow h^0\tilde{\chi}\tilde{\chi}$ [98], where the $\tilde{\chi}$'s involved should have large higgsino components and be light enough for the branching ratio to be substantial. For the heavier H^0, H^+, and A^0, events with substantial missing transverse energy will now play an important role in the search for Higgs bosons. Previously, missing transverse energy was relevant in Higgs searches only in the search for leptonically decaying W bosons in the final state. When neutralinos and charginos are present in the final state of a Higgs boson decay, a search for events having high-p_T leptons and missing energy will once again be appropriate, but it is also possible to have final states with large missing energy but no leptons. A sample study [135] based on looking for the H^0 at the SSC in the decay mode $H^0 \rightarrow \tilde{\chi}_i^0\tilde{\chi}_j^0$ (where $\tilde{\chi}_i^0 \rightarrow \ell^+\ell^-\tilde{\chi}_1^0$ and $\tilde{\chi}_j^0$ was either the LSP $\tilde{\chi}_1^0$ or decayed to $\tilde{\chi}_1^0\gamma$) was performed at Snowmass 1986. For a specific choice of parameters for which the neutralinos are very light, encouraging conclusions were reached.

But, for large Higgs boson masses, more complex signatures may be most relevant. As mentioned above, W's and Z's can emerge from two-body decays of the heavier neutralinos and charginos into which the Higgs decay. To more quantitatively assess the potential for tagging Higgs bosons via the W's and Z's, we shall combine the results of ref. 133 for chargino and neutralino decays to W's and Z's, with the results for Higgs decay to charginos and neutralinos. There are a variety of possible W and Z signatures. The final state may have one, two, three, or, in some extremes of parameter space, four vector bosons. Following ref. 132 we shall confine ourselves to presenting the probabilities for: a) one or more vector bosons; and b) two or more vector bosons. In case b) the two vector bosons may come from the same neutralino or chargino (we do not exclude additional vector bosons from the other $\tilde{\chi}^0$ or $\tilde{\chi}^\pm$), or one (and only one) can come from each of the two chargino/neutralino

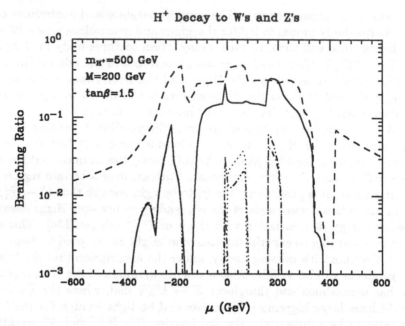

Figure 4.17 Branching ratios for H^+ decay to $Z +$ anything (solid), W + anything (dashes), $ZZ +$ anything (dashdot), $WZ +$ anything (dash-dotdot), and $WW +$ anything (dots). Results are plotted as function of μ taking $m_{H\pm} = 500$ GeV, $M = 200$ GeV and $\tan\beta = 1.5$. These are the same values used in the previous figure. The branching ratios are inclusive in that additional vector bosons are not excluded.

states of the primary Higgs decay. Numerical results are given in fig. 4.17,* for the case of H^+ decays and for $m_{H\pm} = 500$ GeV, $M = 200$ GeV, and $\tan\beta = 1.5$ (the parameter choices for the earlier figure). For this choice of parameters the H^+ (and also the H^0 and A^0, see ref. 132) has quite substantial probabilities [$\mathcal{O}(10\%)$] to decay to at least one W or Z (the inclusive W branching ratio being generally larger), over a large range of μ. The probability of finding two or more vector bosons is only significant in quite specific regions of μ, but where it can occur the inclusive branching ratio can be as large as 5%. Of course, one must keep in mind that triggering on a W or Z in its leptonic mode will result in a substantial additional branching fraction penalty. It is clear that the fraction of decays yielding vector bosons is substantially less for the heavy Higgs bosons of the minimal supersymmetric

---

*W's and Z's can also arise from direct Higgs decays to VV or VH final states, and have been included in fig. 4.17. However, the VV modes are completely negligible and the VH modes never contribute more than 3×10^{-3} to the branching ratios shown in fig. 4.17.

model as compared to the Standard Model Higgs boson (for which *direct* decays into vector bosons are not suppressed). It is unclear at present whether the vector bosons which arise indirectly in the Higgs boson decays studied here can be phenomenologically useful. A systematic survey of the minimal supersymmetric model parameter space, along with appropriate Monte Carlo studies, is highly desirable. Nevertheless, the W and Z bosons in the final states of Higgs decay may increase the potential for discovery of the Higgs bosons of the minimal supersymmetric model. For example, in the case of a 500 GeV charged Higgs [136] there is a region of parameter space where the branching ratio for $H^+ \rightarrow \tilde{\chi}_2^+ \tilde{\chi}_1^0$ is of order 20%, and that for $\tilde{\chi}_2^+ \rightarrow Z\tilde{\chi}_1^+$ is around 16%. The $\tilde{\chi}_1^+$ then decays to $q\bar{q}'\tilde{\chi}_1^0$, while we imagine triggering on the Z in the e^+e^- or $\mu^+\mu^-$ mode. The number of events in a typical SSC year (with an integrated luminosity of 10^{40} cm^{-2}) is of order 20 to 30. There is substantial missing energy in every event because of the $\tilde{\chi}_1^0$'s. The primary background is probably from Z + gluon production, in which there is substantial missing transverse energy generated by mismeasurement, accompanying heavy quarks (which lose energy into neutrinos) and energy lost in the beam holes. Despite the very large cross section for this latter process, we have identified a variable, $(p_T^{\mathrm{miss}} - p_T^Z)/(p_T^{\mathrm{miss}} + p_T^Z)$, which can be used (at least at the partonic level) to eliminate the background while retaining most of the Higgs signal. Further progress requires a serious Monte Carlo effort to closely examine this and other promising signatures.

It should be emphasized that the search for Higgs bosons in the supersymmetric context cannot proceed in a vacuum; Higgs bosons are part of the overall picture which would emerge should supersymmetric particles be discovered. Experimental searches will have to look for correlated signals in a large variety of channels.

Supersymmetry also predicts a host of new production modes for Higgs bosons. For example, it has been shown [137,138] that heavy gluinos prefer to decay to heavy neutralinos and charginos that in turn have a high probability [133] to decay to Higgs bosons, when phase space is available. Thus the high rate for the strong interaction production of gluino pairs can potentially result in the production of a large number of Higgs bosons; in addition, there are generally associated W's and Z's which can be used as a trigger. While such a mechanism is probably relevant only to Higgs bosons that are relatively light, it is clearly of great interest. In general, we see that if low-energy supersymmetry is correct, Higgs physics could become a branch of SUSY phenomenology.

4.3   Tabular Comparison of e^+e^- and Hadron Colliders

We present below our summary table, comparing the relative abilities of e^+e^- and hadron colliders (focusing on the SSC) to detect the SM Higgs boson and the Higgs bosons of the minimal supersymmetric model.

Table 4.2

Potential For Higgs Discovery

Qualitative comparison between an $\mathcal{O}(1\,\text{TeV})$ e^+e^- collider and the SSC. The comparison is made for both the Higgs boson of the Standard Model and for the Higgs bosons of a minimal two-doublet supersymmetry model. Integrated luminosity of $10^4\,\text{pb}^{-1}$ is assumed for the SSC and of $30\,\text{fb}^{-1}$ for the e^+e^- collider. Our notation is the following. For the e^+e^- collider we assign 2 points if sufficient energy ($\sim 2m_h$) is likely to be technically feasible in the next 10 years and the h has SM couplings. Fewer or fractional points are assigned as the machine probability decreases. Non-SM Higgs that are weakly produced at an e^+e^- collider because of small or absent HVV couplings are assigned fractional points for the $Z^* \to A^0 h$ mechanism. For the SSC 2 points implies that discovery in the indicated mode is certain. Fractional points imply that backgrounds or luminosity may be a problem or, in the case of $\tilde{\chi}\tilde{\chi}$ modes, that the $\tilde{\chi}\tilde{\chi}$ decay channel is not certain to be present. The table applies only to Higgs bosons with mass $\gtrsim 0.5m_Z$. Any Higgs that can have mass smaller than this will be found at LEP-I and SLC.

	e^+e^- Collider	SSC
Standard Model Higgs Boson		
$m_{\phi^0} \lesssim 2m_Z$	2	$1[W + \phi^0(\to b\bar{b})$ or $\phi^0 \to \gamma\gamma, ZZ^*]$
$2m_Z \lesssim m_{\phi^0} \lesssim 0.6\,\text{TeV}$	1	$2(\phi^0 \to \ell^+\ell^-\ell^+\ell^-)$
$0.6\,\text{TeV} \lesssim m_{\phi^0} \lesssim 0.8\,\text{TeV}$	1/2	$3/2(\phi^0 \to \ell^+\ell^-\ell^+\ell^- + $ big $\mathcal{L}$ or $m_t)$
$0.8\,\text{TeV} \lesssim m_{\phi^0} \lesssim 1\,\text{TeV}$	0	$1(\phi^0 \to \nu\nu\ell\ell$ or $j_1 j_2 \ell\nu)$
Total Points	3 1/2	5 1/2
Two-Doublet SUSY Higgs Bosons		
$h^0(m_{h^0} < m_Z)$	2	$1[W + h^0(\to b\bar{b})]$
$H^0(2m_t > m_{H^0} > m_Z)$	1/2	$1[W + H^0(\to b\bar{b})$ or $H^0 \to \tilde{\chi}\tilde{\chi}]$
$H^0(m_{H^0} > 2m_t)$	1/3	$1/2(H^0 \to \tilde{\chi}\tilde{\chi})$
$A^0(80\,\text{GeV} \lesssim m_{A^0} \lesssim 2m_t)$	1/3	$3/4(A^0 \to \tilde{\chi}\tilde{\chi}, Zh^0)$
$A^0(m_{A^0} \gtrsim 2m_t)$	1/3	$1/2(A^0 \to \tilde{\chi}\tilde{\chi}, Zh^0)$
$H^\pm(m_{H^\pm} \lesssim m_t)$	2	$3/2(t \to H^\pm b, H^\pm \to \tau\nu, W^\pm\Upsilon, W^\pm h^0)$
$H^\pm(m_{H^\pm} > m_t)$	3/2	$3/4(H^\pm \to \tilde{\chi}\tilde{\chi}, W^\pm h^0)$
Total Points	7	6

REFERENCES

1. Early surveys of relevant phenomenology and theory appear in *Proceedings of the 1984 Snowmass Summer Study on the Design and Utilization of the Superconducting Super Collider*, edited by R. Donaldson and J. G. Morfin: see P. Langacker *et al.*, p 771, and J. Ellis *et al.*, p. 782.

2. U. Amaldi, A. Bohm, L.S. Durkin, P. Langacker, A. K. Mann, W.J. Marciano, *Phys. Rev.* **D36** (1987) 1385.

3. G. Costa, J. Ellis, G.L. Fogli, D.V. Nanopoulos and F. Zwirner, *Nucl. Phys.* **B297** (1988) 244.

4. B.W. Lee, *Proceedings of the XVI International Conference on High Energy Physics*, Batavia, IL (1972), ed. J.D. Jackson, A. Roberts and R. Donaldson, Vol. 4; D.A. Ross and M. Veltman *Nucl. Phys.* **B95** (1975) 135.

5. H.-S. Tsao, in *Proceedings of the 1980 Guangzhou Conference on Theoretical Particle Physics*, edited by H. Ning and T. Hung-yuan, (Science Press, Beijing, 1980) p. 1240.

6. M. Veltman, *Nucl. Phys.* **B123** (1977) 89.

7. M. Chanowitz, M. Furman and I. Hinchliffe, *Phys. Lett.* **78B** (1985) 285; *Nucl. Phys.* **B153** (1979) 402.

8. M.B. Einhorn, D.R.T. Jones and M. Veltman, *Nucl. Phys.* **B191** (1981) 146.

9. S. Bertolini, *Nucl. Phys.* **B272** (1986) 77.

10. W. Hollik, *Z. Phys.* **C32** (1986) 291; *Z. Phys.* **C37** (1988) 569.

11. S. Glashow and S. Weinberg, *Phys. Rev.* **D15** (1977) 1958.

12. H.E. Haber, G.L. Kane, and T. Sterling, *Nucl. Phys.* **B161** (1979) 493.

13. J.F. Gunion, H.E. Haber, and J. Wudka, *Phys. Rev.* **D43** (1991) 904.

14. B.W. Lee, C. Quigg and G.B. Thacker, *Phys. Rev. Lett.* **38** (1977) 883; *Phys. Rev.* **D16** (1977) 1519.

15. P. Langacker and H.A. Weldon, *Phys. Rev. Lett.* **52** (1984) 1377; H.A. Weldon, *Phys. Rev.* **D30** (1984) 1547; *Phys. Lett.* **146B** (1984) 59.

16. R. Casalbuoni, D. Dominici, F. Feruglio, and R. Gatto, *Nucl. Phys.* **B299** (1988) 117.

17. H. Huffel and G. Pocsik, *Z. Phys.* **C8** (1981) 13.

18. N.G. Deshpande and E. Ma, *Phys. Rev.* **D18** (1978) 2574.

19. H. Georgi, *Hadronic J.* **1** (1978) 155.

20. J.F. Donoghue and L.-F. Li, *Phys. Rev.* **D19** (1979) 945.

21. L.F. Abbott, P. Sikivie and M.B. Wise, *Phys. Rev.* **D21** (1980) 1393.

22. B. McWilliams and L.-F. Li, *Nucl. Phys.* **B179** (1981) 62.

23. J.F. Gunion and H.E. Haber, *Nucl. Phys.* **B272** (1986) 1.

24. J.F. Gunion and H.E. Haber, *Nucl. Phys.* **B278** (1986) 449.

25. J.A. Grifols and A. Mendez, *Phys. Rev.* **D22** (1980) 1725.

26. A.A. Iogansen, N.G. Ural'tsev, and V.A. Khoze, *Sov. J. Nucl. Phys.* **36** (1983) 717.

27. G. Keller and D. Wyler, *Nucl. Phys.* **B274** (1986) 410.

28. A. Mendez and A. Pomarol, *Nucl. Phys.* **B349** (1991) 361; M. Capdequi Peyranere, H.E. Haber and P. Irulegui, *Phys. Rev.* **D44** (1991) in press.
29. J.F. Gunion, G.L. Kane, and J. Wudka, *Nucl. Phys.* **B299** (1988) 231.
30. G. Pocsik and G. Zsigmond, *Z. Phys.* **C10** (1981) 367.
31. L. Hall and M. Wise, *Nucl. Phys.* **B187** (1981) 397.
32. E. Franco and A. Morelli, *Nuovo Cim.* **96A** (1986) 257.
33. H.E. Haber and Y. Nir, *Nucl. Phys.* **B335** (1990) 363.
34. E. Gildener and S. Weinberg, *Phys. Rev.* **D13** (1976) 333; K. Inoue, A. Kakuto and Y. Nakano, *Prog. of Theor. Phys.* **63** (1980) 235.
35. M. Sher, *Phys. Rep.* **179** (1989) 273.
36. K. Tabata, I. Umemura and K. Yamamoto, *Phys. Lett.* **B129** (1983) 80.
37. A. Sirlin, *Phys. Rev.* **D22** (1980) 971; W.J. Marciano and A. Sirlin, *Phys. Rev.* **D22** (1980) 2695 [E: **D31** (1985) 213].
38. J.F. Gunion and A. Turski, *Phys. Rev.* **D39** (1989) 2701.
39. C. Busch, *Nucl. Phys.* **B319** (1989) 15.
40. S. Dawson, *Nucl. Phys.* **B339** (1990) 19; Markus E. Lautenbacher, *Nucl. Phys.* **B347** (1990) 120.
41. R.M. Barnett, G. Senjanovic and D. Wyler, *Phys. Rev.* **D30** (1984) 1529.
42. C. Edwards *et al.*, *Phys. Rev. Lett.* **48** (1982) 903.
43. M. Sivertz *et al.*, *Phys. Rev.* **D26** (1982) 717 (E: **D26** (1982) 2534).
44. J. Frere, M. Gavela, and J. Vermaseren, *Phys. Lett.* **103B** (1981) 129.
45. W.W. Ash *et al.* (MAC Collaboration), *Phys. Rev. Lett.* **54** (1985) 2477; G.J. Feldman *et al.* (MARK II Collaboration), *Phys. Rev. Lett.* **54** (1985) 2289; C. Akerlof *et al.* (HRS Collaboration), *Phys. Lett.* **156B** (1985) 271.
46. S. Komamiya *et al.* (MARK II Collaboration), *Phys. Rev.* **D40** (1989) 721.
47. W. Bartel *et al.* (JADE Collaboration), *Phys. Lett.* **155B** (1985) 288; H.-J. Behrend *et al.* (CELLO Collaboration), *Phys. Lett.* **161B** (1985) 182.
48. S. Glashow and A. Manohar, *Phys. Rev. Lett.* **54** (1985) 526.
49. E.H. Low *et al.* (AMY Collaboration), *Phys. Lett.* **228B** (1989) 548.
50. G. Eilam and A. Soni, *Phys. Lett.* **215B** (1988) 171.
51. J.L. Hewett, S. Nandi, and T. Rizzo, *Phys. Rev.* **D39** (1989) 250.
52. W. Bartel *et al.* (JADE Collaboration), *Z. Phys.* **C31** (1986) 259; H.-J. Behrends *et al.* (CELLO Collaboration), *Phys. Lett.* **193B** (1987) 376; W. Braunschweig *et al.* (TASSO Collaboration), in *XXIV International Conference on High Energy Physics*, Munich, West Germany, August 1988, edited by R. Koffhaus and J.H. Kühn (Springer-Verlag, Berlin, 1989) p. 1432.
53. A. Chen *et al.* (CLEO Collaboration), *Phys. Lett.* **122B** (1983) 317.
54. L.F. Abbott, P, Sikivie and M. Wise, *Phys. Rev.* **D21** (1980) 1391.
55. G.G. Athanasiu and F.J. Gilman, *Phys. Lett.* **153B** (1985) 274.
56. P.A.S. de Sousa Gerbert, *Nucl. Phys.* **B272** (1986) 581.
57. G. Altarelli and P. Franzini, *Z. Phys.* **C37** (1988) 271.

58. G.G. Athanasiu, P.J. Franzini and F.J. Gilman, *Phys. Rev.* **D32** (1985) 3010.

59. S.L. Glashow and E.E. Jenkins, *Phys. Lett.* **196B** (1987) 233.

60. F. Hoogeveen and C.N. Leung, *Phys. Rev.* **D37** (1988) 3340.

61. C.Q. Geng and J.N. Ng, *Phys. Rev.* **D38** (1988) 2857.

62. W.-S. Hou and R.S. Willey, *Phys. Lett.* **202B** (1988) 591. See also T. Rizzo, *Phys. Rev.* **D38** (1988) 820; care with the definition of $\tan\beta$ is necessary in this latter reference.

63. X.-G. He, T.D. Nguyen and R.R. Volkas, *Phys. Rev.* **D38** (1988) 814.

64. M. Ciuchini, *Mod. Phys. Lett.* **A4** (1989) 1945.

65. A. Buras, P. Krawczyk, M. Lautenbacher and C. Salazar, *Nucl. Phys.* **B337** (1990) 284.

66. H. Albrecht *et al.* (Argus Collaboration), *Phys. Lett.* **192B** (1987) 245.

67. M. Arturo *et al.* (CLEO Collaboration), *Phys. Rev. Lett.* **62** (1989) 2233.

68. C. Albajar *et al.* (UA1 Collaboration), *Phys. Lett.* **186B** (1987) 247.

69. H. Georgi, A. Manohar and G. Moore, *Phys. Lett.* **149B** (1984) 407; M. Cvetic, C.R. Preitschopf and M. Sher, *Phys. Lett.* **164B** (1985) 90.

70. E. Golowich and T.C. Yang, *Phys. Lett.* **80B** (1979) 245; L.N. Chang and J.E. Kim, *Phys. Lett.* **81B** (1979) 233; D.R.T. Jones, G.L. Kane and J.P. Leveille, *Phys. Rev.* **D24** (1981) 2990.

71. S.L. Glashow and E.E. Jenkens, *Phys. Lett.* **196B** (1987) 233; V. Barger and R.J.N. Phillips, *Phys. Lett.* **201B** (1988) 553; *Phys. Rev.* **D40** (1989) 2875; *Phys. Rev.* **D41** (1990) 884.

72. G. Pocsik and G. Zsigmond, *Phys. Lett.* **112B** (1982) 157.

73. A. Grau, J.A. Grifols and N. Lupón, *Phys. Rev.* **D25** (1982) 105.

74. R.W. Robinett, *Phys. Rev.* **D33** (1986) 23.

75. G. Zsigmond, *Acta Phys. Hung.* **56** (1984) 73.

76. J. Alexander, D.L. Burke, C.K. Jung, S. Komamiya, and P.R. Burchat, in *Proceedings of the Summer Study on High Energy Physics in the 1990s*, Snowmass 1988, edited by S. Jensen (World Scientific, Singapore, 1989) p. 135.

77. S. Komamiya, *Phys. Rev.* **D38** (1988) 2158.

78. J. Kalinowski and S. Pokorski, *Phys. Lett.* **219B** (1989) 116.

79. J.F. Gunion and J. Lewis, unpublished.

80. D. A. Dicus and S. Willenbrock, *Phys. Rev.* **D39** (1989) 751.

81. J.F. Gunion, H.E. Haber, F.E. Paige, Wu-Ki Tung, and S.S.D. Willenbrock, *Nucl. Phys.* **B294** (1987) 621.

82. R.M. Barnett, H.E. Haber, and D. Soper, *Nucl. Phys.* **B306** (1988) 697.

83. J.F. Gunion and D. Millers, unpublished.

84. J.F. Gunion, H.E. Haber, S. Komamiya, H. Yamamoto, and A. Barbaro-Galtieri, in *Proceedings of the 1987 Berkeley Workshop on Experiments, Detectors and Experimental Areas for the Supercollider*, edited by R. Donaldson and M. Gilchriese (World Scientific, Singapore, 1988), p. 110.

85. D.M. Atwood, J.E. Brau, J.F. Gunion, G.L. Kane, R. Madaras, D.H. Miller, L.E. Price, and A.L. Spadafora, in *ibid.*, p. 728.

86. J. Brau, K.T. Pitts and L.E. Price, in *Proceedings of the 1988 Snowmass Summer Study on High Energy Physics in the 1990's*, edited by S. Jensen (World Scientific, Singapore, 1989) p. 103.

87. J.A. Grifols, J.F. Gunion, and A. Mendez, *Phys. Lett.* **197B** (1987) 266.

88. J.F. Gunion and H.E. Haber, in *Proceedings of the 1984 Snowmass Summer Study on the Design and Utilization of the Superconducting Super Collider*, edited by R. Donaldson and J. G. Morfin, p. 150.

89. F. Halzen and C.S. Kim, in *Proceedings of the 1987 Madison Workshop, From Colliders to Supercolliders*, edited by V. Barger and F. Halzen (World Scientific, Singapore, 1987), have also computed the $gb \to W^{\pm}t$ cross section, including some QCD corrections.

90. J.F. Gunion and H.E. Haber, in *ibid.*, p. 67.

91. B. Grzadkowski and W.-S. Hou, *Phys. Lett.* **210B** (1988) 233.

92. T. Han and H.C. Liu, *Z. Phys.* **C28** (1985) 295.

93. I.S. Choi, B.H. Cho, B.R. Kim and R. Rodenberg, *Phys. Lett.* **200B** (1988) 200.

94. L.F. Li, Y. Liu and L. Wolfenstein, *Phys. Lett.* **158B** (1985) 45; Y. Liu, *Z. Phys.* **C30** (1986) 631.

95. G.J. Gounaris and A. Nicolaidis, *Phys. Lett.* **102B** (1981) 144; *Phys. Lett.* **109B** (1982) 221.

96. H.E. Haber and G.L. Kane, *Nucl. Phys.* **B250** (1985) 716.

97. J. Pantaleone, *Phys. Lett.* **172B** (1986) 261.

98. H.E. Haber, I. Kani, G.L. Kane, and M. Quiros, *Nucl. Phys.* **B283** (1987) 111.

99. C.A. Nelson, *Phys. Rev.* **D30** (1984) 1937 (E: **D32** (1985) 1848); J.R. Dell'Aquila and C.A. Nelson, *Phys. Rev.* **D33** (1986) 80, 93; *Nucl. Phys.* **B320** (1989) 86.

100. G. 't Hooft, in *Recent Developments in Gauge Theories, Proceedings of the NATO Advanced Summer Institute*, Cargese, 1979, edited by G. 't Hooft *et al.* (Plenum, New York, 1980) p. 135; L. Suskind, *Phys. Rep.* **104** (1984) 181.

101. H.E. Haber and G.L. Kane, *Phys. Rep.* **117** (1985) 75.

102. *The Standard Model Higgs Boson*, edited by M.B. Einhorn, (North-Holland, Amsterdam, 1991).

103. L.J. Hall, V.A. Kostelecky and S. Raby, *Nucl. Phys.* **B267** (1986) 415.

104. S. Dimopoulos and H. Georgi, *Phys. Lett.* **127B** (1983) 101; B. Holdom, *Phys. Lett.* **143B** (1984) 227; A.E. Nelson, *Phys. Rev.* **D38** (1988) 2875; K.-I. Aoki, M. Bando, H. Mino, T. Nonoyama, H. So and K. Yamawaki, *Prog. Theor. Phys.* **82** (1989) 388.

105. P. Fayet, *Nucl. Phys.* **B90** (1975) 104.

106. E. Witten, *Nucl. Phys.* **B231** (1984) 419; S. Dimopoulos and H. Georgi, *Nucl. Phys.* **B193** (1981) 150; N. Sakai, *Z. Phys.* **C11** (1981) 153.

107. K. Inoue, A. Komatsu and S. Takeshita, *Prog. Theor. Phys.* **67** (1982) 927 [E: **70** (1983) 330]; **71** (1984) 413.

108. D.J. Gross and R. Jackiw, *Phys. Rev.* **D6** (1972) 477; C. Bouchiat, J. Iliopoulos and Ph. Meyer, *Phys. Lett.* **38B** (1972) 519; H. Georgi and S. Glashow, *Phys. Rev.* **D6** (1972) 429; L. Alvarez-Gaume and E. Witten, *Nucl. Phys.* **B234** (1983) 269.

109. L. Girardello and M.T. Grisaru, *Nucl. Phys.* **B194** (1984) 419.

110. H.P. Nilles, *Phys. Rep.* **110** (1984) 1.

111. S. Pokorski, in *New Theories in Physics, Proceedings of the XI Warsaw Symposium on Elementary Particle Physics*, Kazimierz, Poland, 1988, edited by Z. Ajduk, S. Pokorski and A. Trautman (World Scientific, Singapore, 1989) p. 36.

112. P. Roy, *Comm. Nucl. Part. Phys.* **17** (1987) 293.

113. S.P. Li and M. Sher, *Phys. Lett.* **140B** (1984) 339.

114. J.F. Gunion and A. Turski, *Phys. Rev.* **D39** (1989) 2325.

115. J.F. Gunion and A. Turski, *Phys. Rev.* **D40** (1989) 2333.

116. P. Fayet, *Nucl. Phys.* **B237** (1984) 367.

117. L. Alvarez-Gaume, J. Polchinski, and M.B. Wise, *Nucl. Phys.* **B221** (1983) 495.

118. R. Barbieri and L. Maiani, *Nucl. Phys.* **B224** (1983) 32.

119. C.S. Lim, T. Inami, and N. Sakai, *Phys. Rev.* **D29** (1984) 1488.

120. R. Barbieri, M. Frigeni, F. Giuliani, and H.E. Haber, *Nucl. Phys.* **B341** (1990) 309.

121. P. Kalyniak, R. Bates and J. N. Ng, *Phys. Rev.* **D33** (1986) 755; R. Bates, J.N. Ng, and P. Kalyniak, *Phys. Rev.* **D34** (1986) 172.

122. J.F. Gunion, G. Gamberini, and S.F. Novaes, *Phys. Rev.* **D38** (1988) 3481.

123. T.J. Weiler and T.-C. Yuan, *Nucl. Phys.* **B318** (1989) 337.

124. S. Bertolini, F. Borzumati and A. Masiero, *Nucl. Phys.* **B312** (1989) 281.

125. J.F. Donoghue, H.P. Nilles and D. Wyler, *Phys. Lett.* **128B** (1983) 55; M.J. Duncan, *Nucl. Phys.* **B221** (1983) 285; M. Dugan, B. Grinstein and L. Hall, *Nucl. Phys.* **B255** (1985) 413; A. Bouquet, J. Kaplan and C.A. Savoy, *Phys. Lett.* **148B** (1984) 69.

126. See, for example, S. Bertolini, F. Borzumati and A. Masiero, *Phys. Lett.* **192B** (1987) 457; *Phys. Rev. Lett.* **59** (1987) 180; *Nucl. Phys.* **B294** (1987) 321.

127. J.F. Gunion, in *Superstrings, Unified Theories and Cosmology 1987*, the ICTP Series in Theoretical Physics, vol. 4, edited by G. Furlan, J.C. Pati, D.W. Sciama, E. Sezgin, and Q. Shafi, (World Scientific, Singapore, 1988) p. 483.

128. J.F. Gunion, L. Roszkowski, A. Turski, H.E. Haber, G. Gamberini, B. Kayser, S.F. Novaes, F. Olness, and J. Wudka, *Phys. Rev.* **D38** (1988) 3444.

129. G.F. Giudice, *Phys. Lett.* **208B** (1988) 315.

130. C. Ahn *et al.*, SLAC-Report-329, (May 1988).

131. R. Blankenbecler and S. Drell, *Phys. Rev. Lett.* **61** (1988) 2324 [E: **62** (1989) 116].

132. J.F. Gunion and H.E. Haber, *Nucl. Phys.* **B307** (1988) 445.

133. J.F. Gunion and H.E. Haber, *Phys. Rev.* **D37** (1988) 2515.

134. K. Griest and H. E. Haber, *Phys. Rev.* **D37** (1988) 719.

135. R.M. Barnett, J.A. Grifols, A. Mendez, J.F. Gunion, and J. Kalinowski, in *Proceedings of the 1986 Snowmass Summer Study on the Physics of the Superconducting Super Collider*, edited by R. Donaldson and J. Marx, p. 188.

136. R.M. Barnett, J.F. Gunion and H.E. Haber, in *Proceedings of the 1988 University of California at Davis Workshop on Intermediate Mass and Non-Minimal Higgs Bosons*, edited by J.F. Gunion and L. Roszkowski.

137. R.M. Barnett, J.F. Gunion, and H.E. Haber, *Phys. Rev. Lett.* **60** (1988) 401 and *Phys. Rev.* **D37** 1892 (1988).

138. G. Gamberini, *Z. Phys.* **C30** (1986) 605; H. Baer, V. Barger, D. Karatas, and X. Tata, *Phys. Rev.* **D36** (1987) 96.

Chapter 5

Beyond the Minimal Supersymmetric Model: Higgs Doublets and Singlets

There are many possible extensions of the minimal supersymmetric model (MSSM) that contain more than just two SU(2) doublets. The simplest extension is one already mentioned in §4.2, namely the introduction of an SU(2) singlet field N. There is actually a non-trivial motivation for this extension that we alluded to earlier, namely the ad hoc nature of the parameter μ of the minimal supersymmetric model. In the following sections we consider this and several other possible elaborations of the minimal two-doublet extension of the SM, focusing exclusively on those that have the strongest aesthetic and theoretical motivations.

5.1 The Mixing Hierarchy Problem*

As we have already mentioned, even though the minimal supersymmetric model is attractive due to the small number of parameters required to fully specify the theory, there is no fundamental understanding of the origin of the μ parameter. Many supergravity and superstring models do not yield a superpotential term of the form $\mu \widehat{H}_1 \widehat{H}_2$ which is present in the minimal supersymmetric model. In addition, the soft-supersymmetry-breaking parameter m_{12}^2 is often related to μ [see eq. (4.70)], and so in such models, there is also no natural explanation for having $m_{12}^2 = \mathcal{O}(m_W^2) \ll m_{\rm Pl}^2$. Giudice has argued [2] that, in some models, μ of order the electroweak scale is automatically generated by the supergravity-breaking mechanism. Here, we would like to focus on a more elegant scenario for Higgs doublet mixing which often appears quite naturally in the context of superstring theories in which one or more Higgs singlet field is present. Let us consider the simplest case of one additional

---

* The discussion in this and the following section is based on ref. 1.

SU(2) singlet Higgs field, first studied many years ago by Fayet [3]. A simple way of generating the required Higgs doublet mixing term is through a contribution to the superpotential of the form

$$\delta W = \lambda \widehat{H}_1 \widehat{H}_2 \widehat{N}. \tag{5.1}$$

Such a trilinear superpotential term emerges automatically in string theories. Indeed, those explored to date have *only* trilinear terms in the superpotential. In this case, the parameter μ of the MSSM is generated as

$$\mu = \lambda \langle N \rangle \equiv \lambda n, \tag{5.2}$$

where the vacuum expectation value n of the N field is expected to be $\mathcal{O}(v_1, v_2)$ in most theories. Since λ is also in the perturbative domain, we have a natural explanation for keeping $\mu \ll m_{\text{Pl}}$, given that $v_1, v_2 \ll m_{\text{Pl}}$. Of course, once a term of the form (5.1) is present in the superpotential, evolution from m_{Pl} to m_W will, in general, automatically generate a contribution to V_{soft} of the form

$$V_{\text{soft}} \ni -\lambda A_\lambda H_1 H_2 N. \tag{5.3}$$

With $\langle N \rangle = n$ we may then isolate a H_1–H_2 mixing term of the form given in V_{soft} for the minimal supersymmetric model, using the identification $m_{12}^2 = B\mu = \lambda A_\lambda n$.

It has been noted in the literature (see ref. 4 and references therein) that there is some danger that the introduction of an SU(2) singlet field could in itself destabilize the Higgs boson mass hierarchy. The dangerous graphs that appear with the introduction of the N field are tadpole-like graphs in which the N couples via the interaction term of (5.3) to the H_1, H_2 propagators, and the N line then is attached to a heavy fermion or heavy scalar loop (*e.g.*, of color triplet Higgs). However, destabilization requires a large mass splitting between a heavy scalar and its fermionic superpartner. In a large class of superstring models, including those that we shall mention briefly later, the splitting is not so big as to present a problem.

5.2   The Minimal Supersymmetric Model with Higgs Doublets and Singlets

Let us discuss in more detail the structure of the supersymmetric model containing exactly one Higgs-singlet field in addition to the required two doublets of the MSSM. We shall call the resulting model the minimal non-minimal supersymmetric model (MNMSSM). This model is also 'minimal' in the sense that we do not allow for any additional low energy gauge group structure beyond the SU(2) × U(1) groups of the SM. The full superpotential form that we shall consider for the pure Higgs field sector is

$$W = \lambda \hat{H}_1 \hat{H}_2 \hat{N} - \frac{k}{3} \hat{N}^3. \tag{5.4}$$

This is a particularly simple version of the most general potential given earlier that is possible in the presence of an $\hat{N}$ superfield. Our discussion will be based on ref. 1. Closely related work has appeared in refs. 5 and 6. These references have in common the feature that no bilinear terms are included in the superpotential, eq. (5.4); as discussed above they lead to naturalness problems and, furthermore, do not appear in a large class of superstring inspired models. In ref. 7 bilinear terms are allowed. The models without bilinear terms also do not yield linear terms in $\hat{N}$. In addition, the corresponding soft-supersymmetry-breaking terms will therefore not occur. One should note that if these types of terms are absent at the grand unification scale, as suggested by the models referred to above, then the renormalization group equations for the associated parameters imply that they are absent for all mass scales [1]. The full scalar potential of the model then takes the form

$$V = |\lambda H_1 H_2 - kN^2|^2 + \lambda^2 \left(|H_1|^2 + |H_2|^2 \right) |N|^2$$
$$+ \tfrac{1}{8}(g^2 + g'^2) \left(|H_2|^2 - |H_1|^2 \right)^2 + V_{\text{soft}}, \tag{5.5}$$

where

$$V_{\text{soft}} = m_1^2 |H_1|^2 + m_2^2 |H_2|^2 + m_N^2 |N|^2$$
$$- [\lambda A_\lambda N H_1 H_2 + \text{h.c.}] - \left[\frac{k A_k}{3} N^3 + \text{h.c.} \right]. \tag{5.6}$$

In the above potential there are seven parameters:

$$\lambda, k; \quad A_\lambda, A_k; \quad m_1^2, m_2^2, m_N^2. \tag{5.7}$$

As in the MSSM, gauge symmetry breaking leads to non-zero values for the vacuum expectation values of the neutral Higgs fields: v_1, v_2, and n. We use the minimization conditions to eliminate m_1^2, m_2^2, and m_N^2 in favor of the three VEV's. However, just as in the case of the MSSM, not all three VEV's are independent; we must require again that $m_W^2 = g^2(v_1^2 + v_2^2)/2$, leaving six independent parameters.

After absorption of those degrees of freedom required to give the W and Z bosons mass, there remain seven physical Higgs particles: the charged Higgs bosons $H^\pm$; three neutral scalar Higgs bosons, h^0, H_1^0, and H_2^0; and two pseudoscalar Higgs boson, A_1^0 and A_2^0. (As before, increasing index corresponds to increasing mass.) There is no mixing between the scalars and pseudoscalars if CP is conserved, as we assume. The third scalar and second pseudoscalar states appear due to the introduction of the complex N field. The scalar Higgs boson mass eigenstates are determined by diagonalizing a 3×3 mass matrix, $\mathcal{M}_S^2$, while the pseudoscalar Higgs boson mass eigenstates are obtained by diagonalizing a 2×2 mass matrix, $\mathcal{M}_P^2$. An important point is that it is necessary that k be non-zero in order that $m^2_{A^0_1} \neq 0$. The six independent parameters referred to above may be taken to be $\tan\beta = v_2/v_1$, $r = n/\sqrt{v^2_1 + v^2_2}$, λ, k, $m_{H\pm}$ (which fixes the combination $A_\Sigma \equiv A_\lambda + kn$), and A_k. For all other parameters held fixed at physically allowed values, a range of A_k values will typically be possible. This allowed range is determined by requiring that all Higgs bosons have positive mass-squared at the local symmetry breaking minimum, and by the requirement that this local minimum be the true global minimum. As A_k varies through this allowed range, the h^0, H^0_1, H^0_2 and A^0_1, A^0_2 masses will vary over some range, as will all Higgs boson couplings. For fixed $\tan\beta$, r, λ and k, not all $m_{H\pm}$ values will lead to a non-zero range of allowed A_k values; as a result there will always be a minimum and maximum allowed $H^\pm$ mass for fixed values of the former four parameters.

Clearly there is considerably more freedom in the MNMSSM than in the MSSM, due to the larger number of parameters. As a result there are no simple mass sum rules, though we may give generalizations of the relations that appeared in the MSSM. We have, for instance,

$$
\begin{aligned}
m^2_{H\pm} &= \mathrm{Tr}\mathcal{M}^2_P + m^2_W - \lambda^2 v^2_2 - \lambda A_\Sigma \frac{v_2 v_1}{n} - 3k(\lambda v_1 v_2 + A_k n) \\
\mathrm{Tr}\mathcal{M}^2_S &= m^2_Z + \mathrm{Tr}\mathcal{M}^2_P + 4k(kn^2 - A_k n - \lambda v_1 v_2),
\end{aligned}
\tag{5.8}
$$

where $\mathrm{Tr}\mathcal{M}^2_S$ is the sum of squares of the masses of the three scalars, and $\mathrm{Tr}\mathcal{M}^2_P$ is the sum of squares of the masses of the two pseudoscalars. These sum rules may be compared to the MSSM results of eq. (4.75).

It is also useful to note that

$$
m^2_{H\pm} = m^2_W - \lambda^2(v^2_1 + v^2_2) + \lambda A_\Sigma \frac{2n}{\sin 2\beta}.
\tag{5.9}
$$

This latter formula makes it clear that, unlike the MSSM, it is possible in the MNMSSM for the charged Higgs to have mass smaller than m_W. In practice, however, the $\tan\beta$, r, λ and k parameters for which $m_{H\pm}$ is substantially smaller than m_W are not favored by the renormalization group analysis.

Clearly, a full survey of Higgs boson masses over all of parameter space is not feasible. Instead, we choose to illustrate the Higgs boson mass spectrum for choices of $\tan\beta$, r, λ and k that are likely to emerge from renormalization group evolution from the grand unification scale, M_X, down to m_W. Of course, there is sensitivity to the boundary conditions assumed at M_X. In addition, the full theory including squarks, quarks, gauginos, etc., must be considered. We cannot detail the complete analysis in this short presentation. We simply summarize results. One extreme is to imagine that the couplings λ, k and h (h specifies the strength of the coupling of the $\widehat{H}_2$ to the top-quark fields) all start out large, and that the boundary conditions at M_X on other parameters are adjusted so as to yield an acceptable low energy scenario for sneutrino, chargino and other experimentally constrained masses. In this case, the renormalization group equations tend to drive the couplings toward well-determined fixed point values. For the pure Higgs sector couplings, the fixed point values are

$$\lambda = 0.87 \quad k = 0.63 \,. \tag{5.10}$$

When considering these fixed point values, it will be useful to consider three choices of the parameter r describing the relative size of the vacuum expectation value of the N field compared to those of the neutral doublet fields. We shall consider $r = 0.1$, $r = 1$, and $r = 10$, as representative. Note that there is no experimental constraint on r in the MNMSSM. Another extreme possibility that leads to well-defined low energy couplings, but an extreme that is very attractive theoretically, is to assume that at M_X all soft-supersymmetry-breaking terms are zero other than a universal gaugino mass which acts as the seed for all supersymmetry breaking. In addition, at M_X one chooses values for λ, k, and the top quark Yukawa coupling to the $\widehat{H}_2$ field, h. As one evolves down to the low energy scale, the other soft-supersymmetry-breaking terms are automatically generated by the renormalization group equations; all mass scales are determined by the initial gaugino mass scale, M_U, which is therefore fixed by requiring that at low energy the mass of the W boson have the experimentally observed value. One finds that the starting parameters, λ_U, k_U, and h_U at M_X, must lie within a relatively well-defined range in order that the renormalization group equations lead to an experimentally acceptable scenario at low energy. In particular, most starting points do not yield correct symmetry breaking; only for a very restricted range of λ_U, k_U, and h_U values is the global minimum one for which (i) all the Higgs bosons have positive mass-squared and (ii) color and electric charge breaking are both absent (that is scalar fields carrying electric charge or color do not have non-zero vacuum expectation values). Further restrictions on the allowed parameter range derive from requiring that squark, gluino, sneutrino, slepton, and chargino masses are not so light as to have already been excluded by experiment. We will illustrate the Higgs boson mass spectrum for a sample resulting set of values for $\tan\beta$, λ, k and n at low energy. These values we call the renormalization group solution (RGS). They are given by

$$\tan\beta = 2.04 \quad r = 0.639 \quad \lambda = 0.128 \quad k = 0.097 \,. \tag{5.11}$$

Of course, the renormalization group equations actually yield definite values for all parameters, and hence all Higgs masses, for a given solution. However, it will be useful to see how much variation in the Higgs masses is possible if one fixes only the above four parameters.

The results for the four different cases outlined above are presented in fig. 5.1. A number of remarks are in order. First we note, in agreement with our earlier discussions, that at small r the relatively large fixed-point λ value leads to an allowed range for $m_{H\pm}$ that lies below m_W. For the moderate and large choices of r this is no longer true; at high r the charged Higgs must be quite heavy. The renormalization group solution sample case yields a range for $m_{H\pm}$ that is quite near to m_W. This is not atypical; once both r and λ are

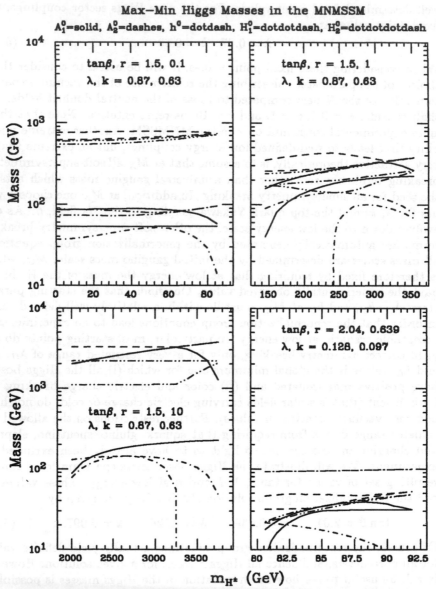

Figure 5.1 The masses of the neutral Higgs bosons of the MNMSSM as functions of the charged Higgs mass in the four cases outlined in the text. Only $m_{H^\pm}$ values in the plotted range lead to positive squared masses for all the Higgs (in particular, for $m_{h^0}^2$). At each $m_{H^\pm}$ an upper and a lower bound is indicated for each neutral Higgs mass, obtained by varying A_k over the entire range for which: (1) $V < 0$; and (2) $m_{h^0}^2 > 0$. When the limits of the allowed A_k range are determined by condition (2) the lower bound on m_{h^0} is 0.

small, as is typical of the renormalization group solutions, the possible range for $m_{H^\pm}$ is rather restricted [as might be anticipated from eq. (5.9)].

The neutral Higgs masses in the four cases illustrated also exhibit a number of interesting features. First, analogous to the MSSM, the lightest scalar Higgs boson is never much heavier than m_Z. For the RGS parameter choice, h^0 must be substantially lighter than the Z. The other neutral Higgs masses show more variety. If r is either large, $r = 10$, or small, $r = 0.1$, then there is a large splitting among the neutral Higgs masses, and $m_{A_2^0}$ and $m_{H_2^0}$ will be large and approximately degenerate. At large r, m_{H^+} is also nearly equal to $m_{A_2^0}$ and $m_{H_2^0}$. For moderate values of r, i.e., the $r = 1$ fixed point case and the RGS case, there is not very much splitting between the Higgs boson masses, and all have relatively moderate masses, substantially below 1 TeV.

Similar results have been obtained in refs. 5–7. In particular, ref. 5 focuses on the light h^0 and finds that $m_{h^0} \lesssim 155$ GeV, a result also found even if the bilinear superpotential terms are included as in ref. 7. (An earlier upper limit on the m_{h^0} mass of 134 GeV was obtained in the context of the potential of eq. (4.66) in ref. 8, neglecting the supersymmetric particle sector's influence on the renormalization group evolution.) The analytic form of the bound is

$$m_{h^0}^2 \leq m_Z^2 \left[\cos^2 2\beta + \frac{2\lambda^2 \cos^2 \theta_W}{g^2} \sin^2 2\beta \right], \qquad (5.12)$$

showing the importance of renormalization group limitations on λ in order to obtain the upper bounds quoted above. Reference 7 also finds approximate mass degeneracies and patterns similar to those of fig. 5.1. Among the analytic results obtained there are: $m_{h^0}^2 \leq m_{H^\pm}^2 + m_Z^2 \sin^2 \theta_W$; $m_{H^\pm}^2 \lesssim m_{H_2^0}^2 + m_W^2$; and $m_{H_1^0}^2 \leq m_{A_2^0}^2 + m_Z^2$. These results are most useful for moderate r (e.g., $r = 1$); when r is small ($r \sim 0.1$) or large ($r \sim 10$) then the mass-squared on the right-hand side of these relations can be much larger than the mass-squared appearing on the left-hand side.

Clearly, a full exploration of the Higgs boson spectrum below 1 TeV will be necessary in order to reveal the nature of the gauge symmetry-breaking mechanism and to provide guidance as to the correct theoretical structure and grand unification group. However, detection of all the Higgs bosons of the MNMSSM, even at a high energy e^+e^- collider, is far from easy. We will not discuss this particular model in detail in this respect, but there are many similarities to the MSSM. See ref. 1 for a complete treatment. Certainly, we must find the h^0 neutral scalar Higgs boson with mass $\lesssim m_Z$ (or a little heavier), or the supersymmetric structure common to the MSSM and the MNMSSM will be in severe doubt. This first Higgs boson has an excellent chance of being discovered at SLC, LEP, or LEP-II. If all the other scalar Higgs bosons are much heavier, then the h^0 couplings should be very similar to those of a Standard Model Higgs boson of similar mass, simply by virtue of the unitarity constraints discussed in chapter 4. (The specific models discussed here always yield such a result.) Thus it could be observed with non-trivial rate in $Z \rightarrow \ell^+\ell^- h^0$ decays, and so forth. Of course rates for this decay and quarkonia decays are likely to be modified if there is more than one light scalar (as in the example to follow shortly). The effects on quarkonia decays in this general type of two-doublet plus one-singlet model have been explored in ref. 9. Next on the most critical list are the charged Higgs bosons, $H^\pm$, which are likely to have a mass in the vicinity of m_W (although, as we have seen, this is not a theorem). The best hope for their discovery would be at an e^+e^- collider in the $Z^*, \gamma^* \rightarrow H^+ + H^-$ production mode. If $m_{H\pm}$ is significantly above m_W, then discovery would require substantial e^+e^- collider energy. The other neutral scalar and pseudoscalar Higgs bosons may well have masses below the 1 TeV scale, but the detection of all of them might prove a great challenge.

It is perhaps useful to give one detailed sample. Let us take the renormalization group solution of eq. (5.11), including the precise values of A_t and A_λ that emerge from the renormalization group evolution for this particular case. This special case is discussed in detail in ref. 1. The Higgs boson and sparticle masses obtained for this solution are given in table 5.1. Note how light all particles and sparticles are. The couplings of the Higgs bosons of the model to Standard Model particles are given in this special case in table 5.2. The $R_{...}$ are the ratios relative to SM values in the case of VV or $q\bar{q}$ couplings. The $R_{A_i^0 H_j^0}$ values refer to the $ZA_i^0 H_j^0$ couplings of interest for $A_i^0 H_j^0$ production in e^+e^- collisions, and are defined relative to $g/2$, which is the maximum possible value as exhibited by the sum rule of eq. (4.26). Note that the VV coupling is shared quite equally among the three-scalar Higgs, that all have somewhat suppressed couplings to $u\bar{u}$ and somewhat enhanced couplings to $d\bar{d}$, and that $ZA_1^0 H_i^0$ couplings are very weak. The A_1^0 would thus be very difficult to produce at a e^+e^- collider. In contrast, the light masses for all the scalars, combined with their having at least 1/4 the VV fusion and $Z^* \rightarrow ZH_i^0$ squared couplings, imply significant production cross sections in e^+e^- collisions. At a hadron collider, gg fusion would be the primary production mode for h^0, H_1^0, H_2^0, A_2^0, but A_1^0's weak coupling to $t\bar{t}$ implies that it is very weakly produced, just as in e^+e^- collisions. The weak A_1^0 couplings are due to the fact that it has a large Im N component. We now turn to a discussion of the decays of the Higgs bosons for this special case.

The predicted branching ratios for various channels in the fully determined renormalization group special case are given in table 5.3. We see that among the neutral Higgs bosons, $b\bar{b}$ is the dominant mode except in the case of H_2^0 where HH modes are dominant. For the H_2^0, among the HH modes the $A_2^0 A_2^0$ mode is the largest with a BR of 0.38, followed by $h^0 h^0$ with $BR \sim$ 0.19. The H_2^0 also has some $\tilde{\chi}\tilde{\chi}$ decays, with total $BR \sim 0.05$. None of the neutral Higgs bosons is heavy enough to have any VH or VV decays. In the case of the H^+, its mass is too small to allow the $t\bar{b}$ mode, and, as a result, its decays are dominated by $\tilde{\chi}_1^+ \tilde{\chi}_1^0$.

Let us briefly consider prospects for detecting the Higgs bosons in the special renormalization group case, given the branching ratios and production cross sections outlined above. Consider first e^+e^- collisions. Detection

Table 5.1

Higgs boson and sparticle masses in the special case.

Particle	Mass (GeV)	Particle	Mass (GeV)
h^0	15	$\tilde{\nu}$	23
H_1^0	38	$\tilde{e}_L$	67
H_2^0	95	$\tilde{e}_R$	46
A_1^0	31	$\tilde{\chi}_1^+$	49
A_2^0	39	$\tilde{\chi}_2^+$	123
$H^\pm$	86	$\tilde{\chi}_1^0$	23
t	83	$\tilde{\chi}_2^0$	33
$\tilde{g}$	204	$\tilde{\chi}_3^0$	36
$\tilde{q}$	177	$\tilde{\chi}_4^0$	78
$\tilde{l}_1$	130	$\tilde{\chi}_5^0$	120
$\tilde{l}_2$	229		

Table 5.2

Higgs boson reduced couplings in the special case.

Channel	h^0	H_1^0	H_2^0	A_1^0	A_2^0
$R_{VVH_i^0}$	0.56	0.6	0.57		
$R_{u\bar{u}H_i^0,u\bar{u}A_j^0}$	0.35	0.42	0.97	0.12	0.48
$R_{d\bar{d}H_i^0,d\bar{d}A_j^0}$	1.45	1.36	-1.1	0.49	1.98
$R_{A_1^0H_i^0}$	0.1	0.09	0.2		
$R_{A_2^0H_i^0}$	0.42	0.36	0.8		

Table 5.3
Higgs boson branching ratios in the special case.

Channel	h^0	H_1^0	H_2^0	A_1^0	A_2^0	H^+
$q\bar{q}$	0.92	0.81	0.13	0.96	0.96	0.01
$\ell\bar{\ell}$	0.08	0.03	0.005	0.04	0.04	0.006
HH	0	0.16	0.82	0	0	0
$\tilde{\chi}\tilde{\chi}$	0	0	0.05	0	0	0.99

of the h^0, H_1^0 Higgs bosons at an e^+e^- collider would be straightforward; they have large cross sections as discussed in the previous section, and backgrounds to the $b\bar{b}$ mode are not severe. H_2^0 would be even more unique because of the dominance of HH modes leading to $b\bar{b}b\bar{b}$ final states. A_1^0 is essentially hopeless because of its small production rate, and A_2^0 would be difficult because its production rate would be roughly at the level of 5×10^{-3} units of R_{pt} (combining the $Z^* \to A_2^0 h^0$ and $Z^* \to A_2^0 H_1^0$ cross sections, with phase space suppression included) and one would have to reconstruct both the A_2^0 and the h^0 or H_1^0 in $b\bar{b}$ channels. Detection of the charged Higgs boson would also not be straightforward. Even though the $\gamma^* \to H^+H^-$ cross section at an e^+e^- collider with $\sqrt{s} \gtrsim 300$ GeV would be of order 1/4 unit of R_{pt}, the $\tilde{\chi}_1^+\tilde{\chi}_1^0$ decays for both the produced charged Higgs would make direct mass reconstruction impossible (the $\tilde{\chi}_1^0$ would be the LSP). Turning to a hadron collider, all the neutral scalar Higgs definitely fall into the category of "Intermediate Mass" approaches. Their mass is such that only the production-detection mode of $W^* \to W(\to \ell\nu)h(\to b\bar{b})$ could be employed. While current studies of this mode have yielded optimistic results for scalar Higgs with full Standard Model VV couplings, it is likely that the reduction in cross section by a factor of ~ 3 for the h^0, H_1^0, H_2^0 will make these Higgs more difficult to detect. The $A_{1,2}^0$, which have no VV couplings, will be essentially impossible to observe directly at a hadron collider; however, the A_2^0 might be found in the decays of the H_2^0. For $m_t < m_{H^+} + m_b$, the largest charged Higgs production mode would be $g\bar{b} \to H^+\bar{t}$, followed by a trigger on the spectator $\bar{t}$ quark [10]. The $\tilde{\chi}_1^+\tilde{\chi}_1^0$ decay might provide a sufficiently clean missing energy trigger to allow detection; a detailed study is needed.

In short, Higgs boson decays in this theory are complex and many channels must be examined before one can be certain of discovering or excluding a Higgs boson of any given type in any given mass range. In addition, even though the generally expected Higgs boson masses are fairly modest, the most useful production cross sections are often not as large as in the corresponding Standard Model Higgs boson case. Thus we typically find that one or more of the Higgs bosons will be difficult to detect at both an e^+e^- collider and a hadron collider.

As a final point, we wish to reemphasize that this type of model, when examined in the context of grand unification, is highly constrained when the only source of supersymmetry breaking at the grand unification scale M_X is a gaugino mass. The renormalization group procedure allows one to solve for the masses of all physical particles. The constraint imposed by requiring that the vacuum structure be that observed in nature (*i.e.*, no charge or color breaking, but standard electroweak symmetry breaking) is very powerful indeed. Only a narrow range of choices for the basic trilinear couplings of the superpotential yield allowed solutions. The predicted range of masses of all the physical particles is small. We find a top-quark mass between 75 GeV and 93 GeV,* a gluino mass between 140 GeV and 260 GeV, and light charginos, sleptons, and sneutrinos. In addition, all Higgs bosons have mass below 100 GeV, with the lightest scalar certainly accessible at LEP and SLC. Indeed, if there is any truth to this general approach, the next few years should provide an abundance of new and exciting physics, in particular allowing our first experimental probes of the Higgs boson sector. However, in exploring the Higgs boson sector, we have seen that it will be necessary to consider a variety of different decay channels, with, in particular, channels containing lighter Higgs boson pairs, a vector boson plus a lighter Higgs, or neutralino–chargino pairs. The resulting final states should all be explored more thoroughly with regard to important backgrounds and experimental cuts. Overall, this study makes explicit the fact that we must not rely on studies conducted purely in the context of the Standard Model in assessing the ability of new colliders and detectors to study the Higgs sector.

Finally, it is of interest to note that the MNMSSM model is realized [11] in a four-dimensional superstring model with grand unification gauge group of 'flipped' SU(5) × U(1). As reviewed elsewhere [12] this unification group avoids many of the problems associated with high-dimensional massive Higgs representations that appear in the simple SU(5) case. As typical of phenomenologically viable string models, there are actually three generations of particles, including the scalar fields. However, it is possible to work in a basis in which only one generation of the scalar fields participates in the usual gauge symmetry breaking. This set behaves like those we have been discussing. It is usually referred to as the third generation set, since the $\widehat{H}_2$ of this generation must have a large Yukawa coupling to the top quark. The other generations of scalar fields are sometimes termed 'unhiggses'. They do not acquire vacuum expectation values and do not participate in either gauge

---

*In ref. 5, a more general upper bound of $m_t \leq 200$ GeV is obtained in the context of a renormalization group analysis of this model, with less specific grand unification boundary conditions.

boson or fermion mass generation. The properties of such scalar fields are highly model dependent; see ref. 13 for further discussion.

5.3   Superstring-Inspired Models Based on E_6

One of the superstring models with possible phenomenological viability is that provided by the Calabi-Yau compactification of the heterotic string [14], with unification group based on E_6. A representative and attractive example is that based on the simplest E_6-based low energy group, $SU(2)_L \times U(1)_Y \times U(1)_{Y_1}$, resulting from compactification of ten-dimensional $E_8 \times E_8'$ super-string theory to four dimensions on a manifold with a $SU(3)$ holonomy. Since this model provides a nice but relatively simple example of the types of new physics that could enter in a typical superstring model, we describe it in detail. The resulting theory has many phenomenologically interesting features: it includes a new neutral gauge boson, Z_2; full $N = 1$ low energy supersymmetry; a number of exotic new fermions; and an extensive Higgs boson sector. The features of the latter were first outlined in refs. 15–17, followed by more detailed treatments in refs. 18–24. The summary presented here will be based largely on the survey of the Higgs boson sector presented in ref. 23, which focused on the production and detection of the Higgs bosons of the theory. Related results can be found in ref. 22. However, the phenomenology of the Higgs bosons is closely tied to the other sectors of the theory and, in particular, it will be necessary to remark briefly on Z_1–Z_2 mixing, Z_2 decays, and the mass spectrum of the neutralinos and charginos of the theory.

The compactification scheme of the model predicts that the matter fields occur in three (we adopt the minimal number required) $N = 1$ supersymmetric chiral multiplets, each transforming according to the quantum numbers of the fundamental **27** of E_6. Among the fields associated with each such multiplet are five colorless neutral superfields. Of these, one is usually assigned non-zero lepton number. The spin-zero components of the remaining four neutral superfields can potentially acquire non-zero vacuum expectation values as part of the breakdown of the $SU(2)_L \times U(1)_Y \times U(1)_{Y_1}$ symmetry. Of these, two belong to doublets of the residual $SU(2)$, and the other two are singlets under $SU(2)$. Within a given **27** multiplet labelled by a $(a = 1, 2, 3)$, we denote the two doublet fields by $H_1^{(a)}$ and $H_2^{(a)}$, and the two singlets by $N_1^{(a)}$ and $N_2^{(a)}$. The breaking of the $SU(2)_L \times U(1)_Y \times U(1)_{Y_1}$ symmetry down to $U(1)_{EM}$ occurs when some of these "Higgs" fields acquire vacuum expectation values. It is possible to work in a basis in which the Higgs fields of only one of the three chiral multiplets acquires a non-zero VEV. By convention we define these to be the "third-family" Higgs fields, i.e., only $N_1^{(3)}$, $N_2^{(3)}$, and the neutral components of $H_1^{(3)}$ and $H_2^{(3)}$ can acquire VEV's.

The relation of the third-family defined in this way to the normal fermions contained in the three **27** E_6 multiplets is more model dependent. In the context of the theory, the symmetry breaking occurs as a result of evolution of one or more of the mass-squared terms, appearing in the potential for the above Higgs fields, from a positive value at the grand-unification scale to a negative value at the electroweak scale. The renormalization group equations that control this evolution imply that a large Yukawa coupling to fermions is required to generate such a negative mass squared in the Higgs potential. Thus the fermion members of the third-family multiplet defined above should also be those having the largest Yukawa couplings. In other words, the third-family Higgs fields are, by definition, the first (and only) ones to acquire vacuum expectation values and participate in the electroweak symmetry breaking, and the fermions belonging to the same matter multiplet should have the largest Yukawa couplings to these Higgs. An additional argument for this point of view is based on the requirement that flavor-changing neutral currents (FCNC's) are adequately suppressed. To achieve this suppression of FCNC's one must assume that only one of the three generations of Higgs bosons couples significantly to the Standard Model quarks and leptons [25].

As usual, several of the degrees of freedom of these third-family Higgs that acquire VEV's are eaten by the W, Z_1, and Z_2 in acquiring mass, leaving a certain number of physical Higgs boson states. The phenomenology of these (third-family) Higgs bosons has close similarities to that of the Higgs sector in the minimal supersymmetric model described earlier. They will have trilinear tree-level couplings to vector boson pairs and their Yukawa couplings to fermions of all generations will be tied to the fermion masses (since only those Higgs which acquire vacuum expectation values can produce masses for the fermions). Indeed, many features of their interactions and decays will be analogous to those of the Standard Model (SM) Higgs. It is these Higgs bosons upon which we shall focus. Altogether, their phenomenology is controlled by only five parameters: these can be taken as the mass of the Z_2, the ratio of the vacuum expectation values of the neutral components of the doublet fields $H_2^{(3)}$ and $H_1^{(3)}$, the masses of two of the physical Higgs bosons, and the scale of gaugino masses (as set by the gluino mass). As a result, relatively well-defined predictions for third-family Higgs production and decay emerge.

In contrast, the Higgs fields of the first and second families by definition do not have vacuum expectation values and the associated particle degrees of freedom have been given the name "unhiggses" in ref. 20. They possess only quartic couplings to vector boson pairs, and their Yukawa couplings to fermions of their own and other generations cannot be very large. (Since, by definition, the unhiggses do not acquire VEV's the latter couplings should be substantially smaller than the Yukawa couplings of the third-generation Higgs bosons to the heavier third-generation fermions.) If the Yukawa couplings are weak, then they will be mainly pair produced via Drell-Yan type processes. The unhiggses might be relatively light [20], but in the case of the charged unhiggses they cannot be lighter than the experimental bounds on charged Higgs boson masses [26]. The phenomenology of the unhiggses has been further examined in ref. 13.

We shall divide our discussion into a number of subsections. In the first, we give additional details concerning the theoretical structure of the (third-family) Higgs sector and give results for their masses. In the second we make a few brief remarks concerning the neutralinos and charginos of the third family, which are the ones that will be important in Higgs decays. In particular, we discuss their mass spectrum, which is closely tied to that of the Higgs bosons once the gluino mass is specified. In the third we discuss couplings of the Higgs bosons to vector boson pairs and to SM fermion pairs. In the fourth subsection we discuss the implications of these couplings for the production phenomenology of the Higgs. In the fifth subsection we briefly comment on the remaining Higgs couplings that, along with those to SM fermions and vector bosons, control the Higgs boson decays. Finally, in the sixth subsection the implications of the decays with regard to Higgs detection are reviewed. From a more general point of view, the discussion of this section can be understood as describing a particular model with two doublets, a singlet, and some constraints on parameters.

Theoretical Preliminaries, Constraints, and the Higgs Mass Spectrum

As we have said it is possible to define the Higgs bosons in such a way that only the Higgs bosons associated with the third generation **27** multiplet of E_6 participate in the electroweak symmetry breaking. Generally there is mixing with the Higgs fields of the other generations in the superpotential through trilinear couplings that have a matrix form, $\lambda_{abc} \widehat{H}_1^{(a)} \widehat{H}_2^{(b)} \widehat{N}_1^{(c)}$, where $a, b, c = 1, 2, 3$ are generation indices. The fact that we have defined the third generation by those Higgs that acquire non-zero VEV's implies [20] that it is technically natural to assume that $\lambda_{i33} = \lambda_{3j3} = \lambda_{33k} = 0$ for $i, j, k = 1, 2$. With this convention the masses of the true Higgs bosons can be determined by considering only the portion of the scalar field potential that contains the third-generation Higgs fields. It is to this that we now turn, dropping all reference to the third family index; in particular the λ appearing below is λ_{333}.

Focusing on third-family superfields only, we note that the quantum numbers of $\widehat{N}_1$ and $\widehat{N}_2$ are the same under the residual low energy symmetry group, $\mathrm{SU}(2)_L \times \mathrm{U}(1)_Y \times \mathrm{U}(1)_{Y_1}$, but are different under the larger E_6. As a result, in our notation (described more fully in ref. 19), only $\widehat{N}_1$ can couple to the $\widehat{H}_1$ and $\widehat{H}_2$, via a term of the form $W = \lambda \widehat{H}_1 \widehat{H}_2 \widehat{N}_1$ in the superpotential. Other possible terms, such as $\widehat{H}_1 \widehat{H}_2 \widehat{N}_2$ and $m_{N_{12}}^2 (\widehat{N}_1^\dagger \widehat{N}_2 + \mathrm{h.c.})$, are omitted since they would not be allowed by the underlying E_6 gauge symmetry. The resulting scalar Higgs potential,

$$
\begin{aligned}
V = {} & m_1^2 H_1^\dagger H_1 + m_2^2 H_2^\dagger H_2 + m_{N_1}^2 N_1^\dagger N_1 + m_{N_2}^2 N_2^\dagger N_2 \\
& - \lambda A_\lambda (H_1 H_2 N_1 + \text{h.c.}) \\
& + \lambda^2 \left(H_1^\dagger H_1 H_2^\dagger H_2 + H_1^\dagger H_1 N_1^\dagger N_1 + H_2^\dagger H_2 N_1^\dagger N_1 \right) \\
& + \tfrac{1}{8}(g^2 + g'^2) \left(H_1^\dagger H_1 - H_2^\dagger H_2 \right)^2 \\
& + \tfrac{1}{72} g_1^2 \left(H_1^\dagger H_1 + 4 H_2^\dagger H_2 - 5 N_1^\dagger N_1 - 5 N_2^\dagger N_2 \right)^2 \\
& + \left(\tfrac{1}{2} g^2 - \lambda^2 \right) |H_1^\dagger H_2|^2 ,
\end{aligned}
\tag{5.13}
$$

where g_1 is the $U(1)_{Y_1}$ coupling constant, is completely fixed in terms of only two parameters (in addition to the vacuum expectation values of the Higgs fields described below): λ, defined above; and λA_λ, which specifies the strength of the soft-supersymmetry-breaking scalar potential term, $\lambda A_\lambda H_1 H_2 N_1$, appearing above. Other soft-supersymmetry-breaking terms in the potential can be eliminated in terms of these two parameters and the Higgs field vacuum expectation values by employing the minimization conditions.

The desired minimum of the potential [eq. (5.13)] is such that λA_λ is real and positive. By further phase redefinitions, the vacuum expectation values of the neutral fields can also be chosen to be real and positive. Thus, the Higgs potential is CP–invariant, and the physical scalar Higgs fields derive from the real parts of the neutral components of H_1 and H_2 and the real parts of N_1 and N_2; the pseudoscalar field derives from the imaginary parts. For simplicity we shall assume that only the first three fields acquire vacuum expectation values— v_1, v_2 and n, respectively. (Results are not sensitive to this assumption. It has also been argued [27] that a non-zero VEV for N_2 would lead to a variety of phenomenological and theoretical difficulties.) In this case the first three of the above neutral Higgs fields mix to form three physical mass eigenstates, while the real part of N_2 remains an unmixed mass eigenstate, N^0. This last neutral scalar may only be pair produced, and will not play any role in the phenomenology to be discussed. In addition to the three remaining Higgs neutral scalars, there is a single pseudoscalar Higgs (since two of the imaginary parts of H_1, H_2, N_1 are absorbed in giving mass to Z_1, Z_2), denoted by A^0, and a charged Higgs, $H^\pm$. This notation follows the closely related minimal two-doublet supersymmetry model explored previously in chapter 4. (As before, the terms scalar and pseudoscalar refer to the way in which a given neutral Higgs boson couples to a fermion–antifermion pair.)

Let us also fix our notation for the vector boson sector of the theory. In addition to the charged $W^\pm$ and the photon, the theory possesses two massive neutral gauge bosons, denoted by Z_1 and Z_2, with $m_{Z_1} < m_{Z_2}$. The mixing angle between them is denoted by δ. We fix the gauge coupling constants by $g = e/\sin\theta_W$, $g_1 = g' = g\tan\theta_W$ and $\sin^2\theta_W = 0.23$. (The condition $g_1 \approx g'$ is expected from renormalization group arguments [15,16], where g_1 is the coupling associated with the $U(1)_{Y_1}$ subgroup.) In addition, we take $m_W^2 = g^2(v_1^2 + v_2^2)/2$, and define $m_Z \equiv m_W/\cos\theta_W$.

With this background it is now possible to specify a convenient independent set of physical parameters for the theory. In addition to m_W, g, g' and g_1, these are m_{Z_2}, $\tan\beta = v_2/v_1$, m_{A^0}, and $m_{H^\pm}$. Of these, the value of m_W and choices for m_{Z_2} and $\tan\beta$ completely determine v_1, v_2, n, and the Z_1–Z_2 mixing angle δ, mentioned above. It is useful to note that the Z_2 mass is proportional to n at large n

$$m_{Z_2}^2 = \frac{25}{18}g_1^2 n^2 + \frac{g_1^2}{9g^2}m_W^2(\cos^2\beta + 16\sin^2\beta) + \mathcal{O}\left(\frac{1}{n^2}\right). \qquad (5.14)$$

The experimental constraints on m_{Z_2} and δ are outlined in refs. 28–30, and possible future bounds considered in ref. 31. (The $\theta_{\rm mix}$ of this latter reference is related to our mixing angle by $\theta_{\rm mix} = -\delta$.) However, if $\tan\beta$ is known then the constraints on m_{Z_2} and δ are linked, which can significantly reduce the allowed region of parameter space. As discussed earlier, supersymmetric models based on $N = 1$ supergravity invariably require $\tan\beta \gtrsim 1$. This is also true of all the superstring motivated models discussed in the literature (see refs. 16 and 17, for example). Basically, due to renormalization effects, one finds that at the weak scale $m_2^2 \ll m_1^2$ (where m_2^2 and m_1^2 are the soft mass-squared parameters for the H_2 and H_1 fields), which leads to $v_2 > v_1$. One can probably devise schemes to circumvent this conclusion. However, as a heavier top quark becomes more likely, the tendency to have $\tan\beta > 1$ becomes more difficult to avoid.

A restriction such as $\tan\beta \geq 1$ leads to potentially powerful constraints on m_{Z_2} and δ [18]. This is because the ordering $v_2 \gtrsim v_1$ requires relatively large values of δ, especially for small m_{Z_2}. For instance, the choice

$$\tan\beta \equiv v_2/v_1 = 1.1 \qquad (5.15)$$

yields δ values that do not fall within present neutral current data bounds unless $m_{Z_2} \gtrsim 350$ GeV [30]. Probable future bounds, as computed in ref. 31, would force $m_{Z_2} \gtrsim 0.5$ TeV. Significant improvements in any of several experimental measurements involving the Z_1 could easily require large Z_2 masses in order to maintain $v_2/v_1 \gtrsim 1$. Alternatively, observation of a light Z_2 in combination with improved Z_1 information could require values of v_2/v_1 that appear difficult to accommodate within the context of the theory.

We shall survey results for the choice (5.15). The remaining free Lagrangian parameters are λ and λA_λ, mentioned earlier. These are completely fixed (once v_1, v_2 and n have been determined by specifying m_{Z_2} and $\tan\beta$) by choosing values for $m_{H^\pm}$ and m_{A^0}:

$$\lambda A_\lambda = \frac{m_{A^0}^2 v_1 v_2 n}{v_1^2 n^2 + v_2^2 n^2 + v_1^2 v_2^2}$$

$$\lambda^2 = \frac{m_W^2 - m_{H^\pm}^2}{v_1^2 + v_2^2} + \frac{n^2 m_{A^0}^2}{v_1^2 n^2 + v_2^2 n^2 + v_1^2 v_2^2}. \qquad (5.16)$$

We note that the Higgs mass spectrum and couplings are quite insensitive to the value of v_2/v_1, when considered as functions of m_{Z_2}, m_{A^0} and $m_{H^\pm}$. Extreme values of v_2/v_1 (either very large or very small) would be required to significantly alter the results regarding the Higgs sector as presented below. Note also that the sign of λ is not determined in terms of the above two Higgs masses, although the sign of λA_λ is. However, for convenience, we take $\lambda > 0$ for the ensuing discussion in this section.

We begin by presenting in fig. 5.2 the upper and lower bounds for $m_{H^\pm}$, the masses of the heavier two neutral scalars, and the upper bound for m_{h^0} (the lower bound is zero) as a function of m_{A^0}. Upper and lower bounds are obtained by varying the free parameters with the constraint that all physical Higgs squared masses are positive. Note that, with the exception of m_{h^0}, the value of m_{A^0} almost completely determines the other Higgs masses for a given choice of m_{Z_2} and $\tan\beta$. Indeed, $H^\pm$ and one of the heavier neutral Higgs are always nearly degenerate with A^0, while the other heavy scalar Higgs boson is always nearly degenerate with the Z_2. The neutral scalar degenerate with A^0 and $H^\pm$ is denoted by H^0_{deg} and that degenerate with Z_2 by $H^0_{Z_2}$. "Level crossing" obviously takes place in the vicinity of $m_{A^0} \approx m_{Z_2}$. The tightly constrained nature of the mass spectrum arises through the mass matrix diagonalization, from which we find that $m^2_{h^0}$ is only positive for a very narrow range of $m_{H^\pm}$—or, equivalently, λ—at a fixed value of m_{A^0}. The neutral scalar Higgs mass spectrum given in fig. 5.2 has been computed numerically by diagonalizing the Higgs mass matrix. One can also obtain approximate analytic formulae for the masses which are accurate at the $\sim$ 10% level for all values of m_{A^0}, so long as $m_{Z_2} \gtrsim 0.4$ TeV. These can be found in ref. 23. For large m_{A^0} these formulae are valid to $\sim 1\%$.

As can be seen from fig. 5.2, in the absence of restrictions on λ and λA_λ, the maximum value of m_{h^0} can be arbitrarily large; one need only take an appropriately large value for m_{A^0}. This is illustrated in fig. 5.3, where we see that (at fixed m_{Z_2}, i.e., at fixed n) λ must grow with m_{A^0}. Indeed, at large m_{A^0} only a small wedge of λ values is allowed, as seen in the figure. It can be demonstrated [23] that A_λ (which is related to the scale of supersymmetry breaking) also grows proportionally to m_{A^0}, as seen in fig. 5.3. The increase of λ and A_λ with m_{A^0} implies that we can obtain additional constraints on all Higgs masses by applying various theoretically motivated restrictions on λ and A_λ. First, we should demand that the $SU(2)_L \times U(1)_Y \times U(1)_{Y_1}$ vacuum be a true minimum of the low energy theory. (In particular, electric charge and color must be conserved.) Early on, suggestions were made [17] that this requires that A_λ be no larger than the gravitino mass, $A_\lambda \lesssim \mathcal{O}(m_{3/2})$. However, the results of ref. 32 suggest that these arguments are not valid. The specific analysis in the case of the MNMSSM considered in the previous section seems to bear this out. Thus we will not include any constraints on A_λ. A second possible requirement is that λ (evaluated at the weak boson mass scale) should be in the perturbative domain. In the renormalization group analysis of the MNMSSM we saw that this is likely to be true even if

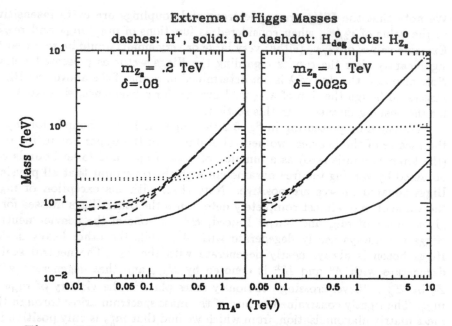

Figure 5.2 Plots of upper and lower bounds for Higgs boson masses, as a function of m_{A^0}, for $m_{Z_2} = 0.2$ TeV and $m_{Z_2} = 1$ TeV. The dashed curve is buried under the dashdot curve for $m_{A^0} \gtrsim m_{Z_2}$. The solid curve corresponds to the *upper* bound for m_{h^0}; the lower bound for m_{h^0} is zero. The allowed mass regions are mapped out as λ (or, equivalently, $m_{H^\pm}$) is varied at any fixed m_{A^0}, and are determined by requiring that the lightest neutral scalar Higgs, h^0, have positive mass squared.

the value of λ at the grand unification scale were rather large. In the present case, various authors have obtained [15,19,21]

$$\lambda \lesssim 0.65 - 1.0, \tag{5.17}$$

depending upon the precise definition of perturbative unification. By examining the maximum and minimum values of λ as a function of m_{A^0}, as well as the corresponding extrema of A_λ, at each value of m_{A^0}, the impact of this constraint can be determined.

To first approximation fig. 5.3 shows that a given value of λ corresponds to a value for the maximum h^0 mass, which is independent of m_{Z_2}, and to values of A_λ, $m_{H^0_{deg}}$, $m_{H^\pm}$, and m_{A^0} that scale with m_{Z_2}. In particular, to restrict m_{h^0} to lie below about 108 GeV it is sufficient to require $\lambda \lesssim 0.6$, independent of m_{Z_2}. In fact, explicit models typically result in values of $\lambda \lesssim 0.1$ [15,17], which restricts m_{h^0} even further. In contrast, $m_{A^0} \gtrsim m_{Z_2}$ is

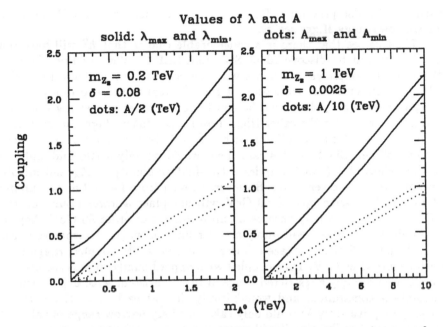

Figure 5.3 Plots of λ and A_λ. Solid curves give the maximum and minimum values of λ for a given m_{A^0}, and correspond to $m_{h^0} = 0$. The dotted curves give the maximum and minimum values of A_λ (corresponding to $\lambda_{\min}$ and $\lambda_{\max}$, respectively). The allowed regions shown correspond to parameter values such that all Higgs have positive mass squared.

generally allowed by eq. (5.17), and for $m_{Z_2} \gtrsim 1$ TeV all Higgs bosons other than h^0 could lie above 1 TeV in the absence of any upper bound on $|A_\lambda|$.

   In the absence of bounds on λ or A_λ, it is still necessary to verify that the potential V has a lower value when the Higgs fields acquire VEV's than the $V = 0$ value appropriate to the unbroken symmetry phase. This was done in ref. 23 with the result that at $m_{Z_2} = 0.2$ TeV $\langle V \rangle$ becomes positive for m_{A^0} just below the 2 TeV limits of the plot in fig. 5.2, while for $m_{Z_2} = 0.6$ TeV and above $\langle V \rangle$ remains negative to very high m_{A^0} values.

   It is interesting to compare the above results with those that were discussed for the minimal supersymmetry model. Despite the fact the E_6 model that we are discussing contains an $SU(2)_L$ singlet field, very extreme values for m_{A^0} (and λ and A_λ) are required before m_{h^0} exceeds m_Z. This result is analogous to that found in the minimal non-minimal extension of supersymmetry with a singlet field. The reason here is simple: the N_1 field is not a singlet with respect to the larger E_6 group. This results in a restrictive set of soft supersymmetry breaking terms, equivalent to those that would survive in the MNMSSM in the limit of $k = 0$. This minimal soft-supersymmetry-breaking structure prevents rapid growth of m_{h^0} with λ. Note that the non-E_6-singlet nature of N_1 also prevents it from destroying the naturalness of light masses for the low energy Higgs sector.

Thus we see that there is a considerable chance that h^0 will have mass $\lesssim 50$ GeV and be discovered at SLC and LEP-I. (This assumes Standard Model (SM) couplings, which we will demonstrate are characteristic of the h^0.) Experimental determination of the h^0 mass would enormously reduce the parameter space that must be considered in order to test a given theory through searches for the other Higgs bosons at a future high energy collider, such as an e^+e^- linear collider with $\sqrt{s} \sim 1$ TeV or the Superconducting Super Collider (SSC). In this case there would be only a two-fold ambiguity in the values of λ (which implies a two-fold ambiguity in A_λ and all other parameters) at a given m_{A^0}. However, as seen from fig. 5.2 a measurement of m_{h^0} in the range $m_{h^0} \lesssim 50$ GeV will not place a lower bound on m_{A^0}. LEP-II could potentially observe h^0 up to masses of order 80 GeV. Any m_{h^0} mass above ~ 50 GeV will imply a lower limit on m_{A^0}. Alternatively, SLC, LEP-I and LEP-II may place a lower bound on a SM-coupled Higgs boson, such as $m_{h^0} \gtrsim 80$ GeV. This, coupled with upper bounds on m_{A^0}, as discussed above, would imply a restricted range of m_{A^0} consistent with experiment and theoretical constraints, and would imply that at each m_{A^0} the parameter $m_{H^\pm}$ (or equivalently λ) could only take on a very narrow range of values. To illustrate the possibilities, it will sometimes be convenient to discuss results in the case that one has a bound of $m_{h^0} \geq 40$ GeV. Often, as will be mentioned when appropriate, very similar results obtain if the m_{h^0} mass were known to be 40 GeV.

Neutralinos and Charginos in the E_6 Model

In addition to the Higgs sector there are many other particles associated with the E_6 based supersymmetric theories. These include the gluino, squarks, sleptons, gauginos and Higgsinos. The gauginos and Higgsinos are the spin 1/2 partners of: the W; the γ, Z_1 and Z_2; and the Higgs fields $H_1^{(a)}$, $H_2^{(a)}$, $N_1^{(a)}$, and $N_2^{(a)}$ ($a = 1, 2, 3$ is the multiplet generation index as before). They mix to form the particles termed neutralinos and charginos. Generally speaking, Higgs boson tree-level decays will include supersymmetric final states, such as squark pairs, slepton pairs, and neutralino/chargino pairs. However, as detailed in ref. 23, neutralino/chargino final states are likely to be the most important in Higgs decays. First, the squarks and sleptons are expected to be significantly more massive than the neutralinos and charginos that will dominate Higgs decays. Second, the couplings of the squarks and sleptons are such that, even when phase space allowed, a Higgs boson prefers to decay to a neutralino/chargino pair.

The mass spectrum and mixing of the neutralinos and charginos is critical in determining their importance as final states in Higgs decays. The gaugino–third-generation-higgsino mass matrix is very closely tied to the Higgs boson sector and the parameters that occur there. An important question is whether the first and second generation higgsinos can be important in Higgs decays. In the basis we employ, such decays arise through the λ_{ij3}-type couplings, and through possible inter-generational mixing in the gaugino-higgsino-Higgs couplings. The sizes of these two types of generation-mixing terms are uncertain. However, as discussed in detail in ref. 20 it is not unreasonable to assume that they are significantly smaller than those determining the intra-generational decays we retain. However, this is only a first approximation. In this approximation, the mass matrix for the neutralinos and charginos of interest is completely determined in terms of the parameters already discussed for the Higgs bosons, and one primary additional parameter. This can be taken to be the gluino mass of the theory. This is because the various gaugino masses, which are equal at the grand unification scale, can be related at the electroweak scale by

$$\frac{3}{5}\frac{M_1}{g_1^2} = \frac{3}{5}\frac{M'}{g'^2} = \frac{M}{g^2} = \frac{M_{\tilde{g}}}{g_s^2}, \tag{5.18}$$

where M_1, M', M, and $M_{\tilde{g}}$ are the soft-supersymmetry-breaking masses associated with the $U(1)_{Y_1}$, $U(1)_Y$, $SU(2)_L$, and $SU(3)$ subgroups. Thus, to determine the neutralino and chargino sector, we need only specify the sign of λ, and values for $M_{\tilde{g}}$ and g_s^2. We shall take $\alpha_s \equiv g_s^2/(4\pi) = 0.136$ and $\lambda > 0$ when presenting specific results. Since renormalization group investigations tend to suggest a large gluino mass [16,17] we will take $M_{\tilde{g}} = 0.5$ TeV as representative. Also, we will henceforth adopt the value of $m_{Z_2} = 0.6$ TeV (a value intermediate between those for which Higgs masses were plotted in fig. 5.2—recall that Higgs masses scale completely with m_{Z_2}.) We note that the negative sign for λ leads to slightly smaller masses and somewhat greater prominence of neutralinos and charginos in Higgs decays. (For details see refs. 33 and 34.) In this brief survey we will consider the case where we require $m_{h^0} \geq 40$ GeV which substantially restricts the range of variation of chargino and neutralino masses as a function of m_{A^0}. We label states as $\tilde{\chi}_i^+$ ($i = 1,2$) and $\tilde{\chi}_i^0$ ($i = 1,6$) where i increases with mass. The charginos and neutralinos tend to have quite well determined eigenstate content in terms of the underlying gauginos and higgsinos. In an obvious notation we find that the $\tilde{\chi}_{\tilde{B}'}^0$ has mass $\sim M'$, $\tilde{\chi}_{\tilde{W}_3}^0$ and $\tilde{\chi}_{\tilde{W}^+}^+$ have mass $\sim M$, $\tilde{\chi}_{\tilde{N}_1+\tilde{B}_1}^0$ and $\tilde{\chi}_{\tilde{N}_1-\tilde{B}_1}^0$ have mass $\sim m_{Z_2}$, and $\tilde{\chi}_{\tilde{H}_1+\tilde{H}_2}^0$, $\tilde{\chi}_{\tilde{H}_1-\tilde{H}_2}^0$ are approximately degenerate at a mass that asymptotes to $\sim m_{A^0}/2$ (for $\tan\beta \sim 1$).

Altogether, we have a neutralino/chargino mass spectrum that is closely tied to the Z_2 mass and the Higgs mass spectrum. Since these neutralinos and charginos of the third generation are crucially important channels for Higgs boson decays, this close inter-connection will lead to interesting patterns in the Higgs branching ratios.

Couplings of Higgs bosons to $q\bar{q}$ and VV' Channels

Couplings of Higgs bosons to quark–antiquark channels

Here we summarize the couplings of the Higgs bosons of this model to the various channels involving Standard Model fermions. In addition, we include here the couplings of the various Higgs to the extra exotic charge $-1/3$ quark, D, that occurs in the **27** representation in which each SM generation resides. As stated earlier we assume that only the Higgs of the third generation acquire vacuum expectation values, and that it is these VEV's that lead to the quark and lepton masses. The results are given in third-family notation. The couplings of the (third-generation) Higgs to quarks and leptons of other generations are obtained, in the usual way by using the appropriate values for the quark masses. (For simplicity, we will neglect the off-diagonal elements of the Kobayashi-Maskawa mixing matrix which are present in the charged Higgs–quark couplings.)

Taking $\tan\beta \sim 1$ as an example, we may summarize the results of ref. 23.

1. First, the $H^+ b\bar{t}$ Feynman coupling is exactly as obtained in the minimal supersymmetry model in eq. (4.25), for any value of $\tan\beta$.
2. For the $H^0_{\mathrm{deg}} q\bar{q}$, $h^0 q\bar{q}$ and $A^0 q\bar{q}$ couplings we have SM-like strength of order $g m_q/(2 m_W)$.
3. For the $H^0_{Z_2} D\overline{D}$ coupling we have a value controlled by m_{Z_2}:

$$g_{D\overline{D}H^0_{Z_2}} \sim (5/6) g_1 m_D/m_{Z_2}. \tag{5.19}$$

We restrict q above to be either b or t, while D denotes the exotic quark described previously. Couplings not listed are approximately zero. In particular, h^0, H^0_{deg}, and A^0 have ~ 0 coupling to $D\overline{D}$, while $H^0_{Z_2}$ has ~ 0 coupling to any $q\bar{q}$. In principle, there are three generations of D quarks, each with the indicated coupling strength to $H^0_{Z_2}$. We will assume that the mass of the third-generation D quark, which we take to be $m_D = 250$ GeV, is significantly larger than those of the first and second generation. If this is not true the branching ratios for the $D\overline{D}$ decay mode of the $H^0_{Z_2}$, presented later, would have to be increased.

Couplings of Higgs bosons to vector boson pairs

As we have discussed in earlier sections, one of the important roles of the Higgs boson of the SM is to cure the bad high energy behavior of the scattering amplitudes for longitudinally polarized W's and Z's. As a consequence, the Higgs coupling to WW and ZZ is very substantial and the SM Higgs decays will be dominated by vector boson pair channels whenever they are phase space allowed. In the E_6 model we are discussing here, there are three neutral scalar Higgs bosons and it is important to know which (or what combination) of them is responsible for preserving unitarity in the WW, $Z_1 Z_1$, $Z_1 Z_2$, and $Z_2 Z_2$ scattering amplitudes. Similar considerations apply to the $Z_2 W^\pm$ scattering channels, whose bad high energy behavior is cured by the

$H^\pm$. A priori, one only knows that

$$\sum_i g^2_{H_i VV'} = C_{VV'},\tag{5.20}$$

where $C_{VV'}$ has a definite value for every VV' scattering channel. We give below the results for these sum rules in the limit of large m_{Z_2}, *i.e.*, large n,

$$
\begin{aligned}
C_{W^+W^-} &= g^2 m^2_W\\[4pt]
C_{Z_1 Z_1} &= \frac{g^2 m^2_Z}{\cos^2\theta_W}\\[4pt]
C_{Z_1 Z_2} &= \frac{g^2_1 m^2_Z}{9}(\cos^2\beta + 16\sin^2\beta)\\[4pt]
C_{Z_2 W^\pm} &= \frac{25}{36}g^2_1 m^2_W \sin^2 2\beta\\[4pt]
C_{Z_2 Z_2} &= \frac{25}{9}g^2_1 m^2_{Z_2}.
\end{aligned}\tag{5.21}
$$

The W^+W^- sum rule is exact, independent of the value of m_{Z_2}; in the other channels it is only the mixing of the neutral gauge bosons that would make exact formulas complicated.

The explicit couplings of the various Higgs can be found in ref. 23. Here we shall only summarize crucial results. The HVV' couplings will be key ingredients in determining the decay patterns of these Higgs bosons. First, we give several key couplings in the limit of large n and small neutral gauge boson mixing, δ

$$
\begin{aligned}
g_{h^0 W^+ W^-} &\simeq gm_W\\[4pt]
g_{H^0_{\text{deg}} W^+ W^-} &\simeq 0;
\end{aligned}\tag{5.22}
$$

$$
\begin{aligned}
g_{h^0 Z_1 Z_1} &\simeq \frac{gm_Z}{\cos\theta_W}\\[4pt]
g_{H^0_{\text{deg}} Z_1 Z_1} &\simeq 0;
\end{aligned}\tag{5.23}
$$

$$
\begin{aligned}
g_{h^0 Z_1 Z_2} &\simeq \frac{g_1 m_Z}{3}(\cos^2\beta - 4\sin^2\beta)\\[4pt]
g_{H^0_{\text{deg}} Z_1 Z_2} &\simeq -\tfrac{5}{6}g_1 m_Z \sin 2\beta;
\end{aligned}\tag{5.24}
$$

$$
g_{H^0_{Z_2} Z_2 Z_2} \simeq \tfrac{5}{3}g_1 m_{Z_2};\tag{5.25}
$$

$$
\begin{aligned}
g_{H^\pm W^\mp Z_1} &\simeq 0\\[4pt]
g_{H^\pm W^\mp Z_2} &\simeq \tfrac{5}{6}g_1 m_W \sin 2\beta.
\end{aligned}\tag{5.26}
$$

The couplings of the charged Higgs to $Z_2 W^\pm$ and of the neutral Higgs to $Z_2 Z_1$ are of particular phenomenological importance, in that they lead to decays of the type $Z_2 \to HV$. It will be helpful to define certain normalized couplings

$$f_{h^0 Z_1} \equiv \frac{g_{Z_2 h^0 Z_1}}{gm_Z} \qquad f_{H^\pm W^\mp} \equiv \frac{g_{Z_2 H^\pm W^\mp}}{gm_W} \qquad f_{H^0_{\deg} Z_1} \equiv \frac{g_{Z_2 H^0_{\deg} Z_1}}{gm_Z}. \quad (5.27)$$

The dependence of these ratios on m_{Z_2} and m_{A^0} was considered in ref. 18. For example, for the case of a very light Z_2, $m_{Z_2} = 0.2$ TeV, $f_{h^0 Z_1}$ ranged from ~ 0.2 to ~ 0.4 as m_{A^0} was varied up to 1 TeV. For higher m_{Z_2}, the range of values found for $f_{h^0 Z_1}$ quickly narrows, and the asymptotic result of eq. (5.24) can be employed, yielding (at $\tan \beta \sim 1$) $f_{h^0 Z_1} \sim 0.5 \tan \theta_W \sim 0.27$, except for very small m_{A^0} values. (To see what happens for small m_{A^0}, the reader is referred to fig. 4 of ref. 18.) Regarding the latter two, $f_{H^\pm W^\mp}$ is always close to the large m_{A^0} asymptotic result, $\sim 5/6 \tan \theta_W \sim 0.45$, while $f_{H^0_{\deg} Z_1}$ lies in the range ~ 0.45 to ~ 0.6 at low m_{Z_2} and approaches the same asymptotic limit (at $\tan \beta \sim 1$). We shall see that the squares of the f_{HV}'s control the decay of the Z_2 to a HV channel. Hence, if phase space effects are ignored, one finds at large m_{Z_2} (and $\tan \beta \sim 1$) a ratio of $9 : 25 : 50$ for $h^0 : H^0_{\deg} : H^\pm$ in Z_2 decays.

Of course, the mass spectrum of the Higgs bosons is such that it is also possible to have (at high m_{A^0}) $m_{H^0_{\deg}} > m_{Z_2} + m_{Z_1}$ and $m_{H^\pm} > m_{Z_2} + m_W$. In this case the significant size of the $Z_1 Z_2 H^0_{\deg}$ and $W^\pm Z_2 H^\mp$ couplings will imply that the $Z_1 Z_2$ and $W^+ Z_2$ channels will be important in $H^0_{\deg}$ and H^+ decays, respectively. (Note, that, to the extent that Z_1–Z_2 mixing is small, the H^+ does not couple to $Z_1 W^+$. The absence of this latter coupling is a common feature of models with only doublets, *e.g.*, the minimal supersymmetric model described in ref. 35.)

Another important class of couplings for both production and decay are those of the neutral Higgs bosons to $W^+ W^-$ and $Z_1 Z_1$, for which we have given asymptotic results above. Asymptotically, we see that the WW and ZZ channels couple primarily to h^0 (which, given any reasonable bounds on λ and/or λA_λ, will be too light to decay in these modes), but can have small couplings to $H^0_{Z_2}$. However, these asymptotic results are not accurate for small m_{A^0} and/or small m_{Z_2}. Thus, we plot in fig. 5.4 the extrema of these couplings relative to the SM coupling strength of gm_W, as we vary the parameters within their allowed domains. (The $Z_1 Z_1$ couplings have essentially the same ratio to the SM strength of $gm_Z / \cos \theta_W$.) Note that outside the asymptotic domain there is considerable variation with m_{Z_2} and m_{A^0}; for instance, at small m_{A^0} the $H^0_{\deg}$ (which is relatively light for such m_{A^0} values) can be moderately coupled to $W^+ W^-$ when m_{Z_2} is small, even though this coupling vanishes asymptotically. Also worthy of note are the extrema of the $H^0_{Z_2} W^+ W^-$ coupling. This coupling actually changes sign as $m_{H^\pm}$ is varied at fixed m_{A^0}. In the allowed $m_{H^\pm}$ range it takes its maximum value when $m_{H^\pm}$ is at its minimum allowed value (where $m_{h^0} = 0$), falls to 0 as $m_{H^\pm}$ (and also m_{h^0}) increases, and goes to its minimum value of the opposite sign as $m_{H^\pm}$ increases further towards its upper allowed value (and m_{h^0}

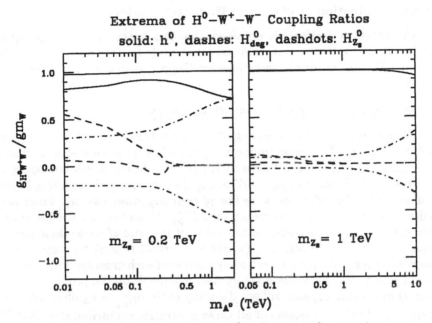

Figure 5.4 Extrema of couplings of h^0, H^0_{deg} and $H^0_{Z_2}$ to W^+W^- as a function of m_{A^0} for $m_{Z_2} = 0.2$ TeV and $m_{Z_2} = 1$ TeV. We have normalized relative to the SM coupling of gm_W. Results for the Z_1Z_1 couplings of these Higgs bosons are identical if normalized to $gm_Z/(2\cos\theta_W)$.

falls back to 0). When this coupling is maximum in absolute value, it can be a non-negligible fraction of SM strength, and $H^0_{Z_2} \to W^+W^-$ (and Z_1Z_1) will turn out to be the dominant decays. Further, even though its maximum values occur for $m_{h^0} \sim 0$, a restriction such as $m_{h^0} \geq 40$ GeV does not visibly affect the extreme values unless m_{A^0} is fairly small. (This last statement also applies to H^0_{deg}.)

Regarding the Z_2Z_2 channel, we see asymptotically that the only Higgs boson which couples significantly is the $H^0_{Z_2}$. This remains true non-asymptotically, and this channel is never important in Higgs boson decays since $m_{H^0_{Z_2}} \approx m_{Z_2}$ and $H^0_{Z_2} \to Z_2Z_2$ is never possible.

Finally, we mention that, on the theoretical side, it will be interesting to determine if any of the VV' scattering channels become strongly interacting if the Z_2 or A^0 masses are taken too large. Unlike the minimal supersymmetric model with no SU(2) singlet Higgs fields, we see from eqs. (5.24)–(5.26) that vector boson couplings to heavy Higgses do not all vanish, and eventually tree-level unitarity will be violated.

Higgs Production at e^+e^- and Hadron Colliders

In the previous section we obtained the two types of couplings, quark–antiquark and vector–vector, that are responsible for the production of Higgs bosons. We now outline expectations for Higgs production mechanisms using these results.

Higgs production via quark–antiquark couplings

The quark–antiquark coupling of a neutral Higgs boson to a heavy quark determines one of its most important production processes at a hadron collider, namely gluon–gluon fusion to Higgs via a heavy quark triangle graph. Indeed, in the limit of large quark mass, the effective ggH coupling becomes independent of quark mass, since the $q\bar{q}$ coupling grows like m_q. From the results of the previous section we see that $H^0_{\rm deg}$, h^0 and A^0 are produced at full SM rate. All are dependent on the value of m_t, and cross sections are maximal when m_H is of order $2m_t$ (where the virtual t-quark triangle loop is at maximal strength). For a recent computation of such cross sections see ref. 10. Analogously, $H^0_{Z_2}$ production via gg fusion depends on the probably large exotic D quark masses, and will not die away until $m_{H^0_{Z_2}}$ is significantly beyond $2m_D$. In addition, D quarks of all three generations contribute, so that if the D quarks are all heavy, one gets a factor of 3 at amplitude level. However, the scale of the $H^0_{Z_2}D\bar{D}$ coupling is set by g_1/m_{Z_2} (compared to g/m_W in the SM case), and the resulting gg fusion cross section will not be very large unless the Z_2 is quite light. To quote a specific example, suppose that there are three D quarks of mass 200 GeV, and that we take $m_{H^0_{Z_2}} = 0.4$ TeV. (The D mass choice corresponds to maximizing the gg fusion D-quark triangle graph for this choice of $m_{H^0_{Z_2}}$.) Then the results of ref. 10 may be rescaled to the coupling constant given in eq. (5.19), and we find a cross section of 0.25 pb at the SSC.

The charged Higgs is produced at a hadron collider via the subprocess $g\bar{b} \rightarrow \bar{t}H^+$. This reaction turns out to have a remarkably large cross section [10]. Even for a charged Higgs mass of order 1 TeV, $\sigma \sim 1$ pb, corresponding to 10^4 events in a standard SSC year of $L = 10^4$ pb^{-1}. And, of course, if the t quark is heavy $t\bar{t}$ production, followed by the decay $t \rightarrow H^+b$, will yield a substantial rate for charged Higgs production at a hadron collider.

Higgs production using vector–vector–Higgs couplings

Phenomenologically, Higgs couplings to vector boson pairs are crucial to three types of Higgs production processes. These are production by vector boson fusion, bremsstrahlung from a virtual vector boson, and Z_2 decays. In the SM, WW and ZZ fusion dominate Higgs production once the Higgs mass is $\gtrsim 0.3$ TeV. This is true at high energies for both hadron colliders and e^+e^- colliders. A second process, which also makes use of $VV'H$ vertices, is virtual V^* production followed by $V^* \rightarrow V'H$; this is of particular importance at an e^+e^- collider with center of mass energy only moderately larger than the Higgs mass of interest. Let us first focus on the fusion production reaction.

In the E_6 theory such processes could be of dramatically reduced importance. First, the h^0 which has the dominant couplings to WW and Z_1Z_1 for large m_{Z_2} is also sufficiently light that its production is dominated by gg fusion at a hadron collider, with WW and ZZ fusion providing $\lesssim 20\%$ of the total (depending on the top quark mass). Secondly, the W^+W^- and Z_1Z_1 couplings to H^0_{deg} are only likely to be significant when both m_{Z_2} and m_{A^0} are small (recall that small m_{A^0} also implies small $m_{H^0_{\mathrm{deg}}}$) where, again, gg fusion will be dominant at a hadron collider. Thirdly, since the $WWH^0_{Z_2}$ coupling squared is always $\lesssim 0.1$ of SM strength, the WW fusion cross section for $H^0_{Z_2}$ production is also of no practical importance. Combined with the small size of the gg fusion cross section discussed earlier, we see that the $H^0_{Z_2}$ is very hard to produce, at both e^+e^- and hadron colliders. The importance of Z_1Z_2 fusion for H^0_{deg} production and of Z_2Z_2 fusion for $H^0_{Z_2}$ production is critically dependent upon m_{Z_2}. Computations of Z_1Z_2 and Z_2Z_2 fusion production processes were performed for the Superconducting Super Collider (SSC) in a different model in ref. 36. Rescaling the couplings to those appropriate in the present case shows that only for $m_{H^0_{Z_2}} \sim m_{Z_2}$ below ~ 0.4 TeV will one find $H^0_{Z_2}$ cross sections at the SSC above 0.1 pb, and that $m_{H^0_{Z_2}} \sim m_{Z_2}$ of order ~ 0.2 TeV is required for cross sections of order 1 pb. Similarly, the Z_1Z_2 fusion process for H^0_{deg} is only capable of yielding 1 pb level cross sections for $m_{Z_2} \lesssim 0.3$ TeV. Clearly, if $m_{Z_2} \gtrsim 0.5$ TeV, these processes are not useful. At an e^+e^- collider, it is well known that the Z_2 is too weakly coupled to electrons for Z_1Z_2 or Z_2Z_2 fusion to be significant. In fact, even the Z_1Z_1 fusion process for the SM Higgs boson is very small [37].

Finally, we recall that the coupling of H^+ to W^+Z_1 is proportional to the very small Z_1Z_2 mixing angle δ, while the A^0 has no VV couplings at all. As a result, vector boson fusion processes do not contribute to H^+ or A^0 production.

Production of Higgs bosons via bremsstrahlung from a virtual vector boson suffers much the same fate as do the fusion processes. For a heavy Z_2, only the h^0 will have a significant cross section coming from this source.

Finally, we remind the reader that when the production mechanisms requiring a significant hVV coupling are suppressed, at an e^+e^- collider one can often turn to processes of the type $Z_1, Z_1^* \rightarrow A^0h$ ($h = h^0, H^0_{\mathrm{deg}}$) for which the required couplings are often near maximal when the hVV coupling is suppressed. We have not explored this option here.

Higgs production in Z_2 decays

As we have seen above, h^0 and H^0_{deg} both couple to Z_1Z_2. This opens up the possibility of finding these two Higgs as decay products of the Z_2 [18,22,23,38-40]. Also, the large $H^+Z_2W^-$ coupling leads to H^+ production in Z_2 decays [18]. Thus we consider

$$Z_2 \rightarrow Z_1 H^0_{2,deg} \tag{5.28}$$

and

$$Z_2 \rightarrow W^{\mp} H^{\pm}. \tag{5.29}$$

We shall describe the results of ref. 18, which employ the full mass matrix diagonalization machinery. Among the possible decays the process $Z_2 \rightarrow h^0 Z_1$ is of greatest interest since it is almost always kinematically allowed, even for m_{Z_2} as low as 0.2 TeV.

In assessing the possibilities for Higgs detection in the modes (5.28) and (5.29) we require the branching ratio for a generic $Z_2 \rightarrow HV$ decay. It can be easily expressed in terms of the $f_{HV} \equiv g_{Z_2HV}/(gm_V)$, defined in eq. (5.27), as

$$B^{Z_2}_{HV} = \frac{9 f^2_{HV}}{40 \tan^2 \theta_W} K(m^2_{Z_2}, m^2_H, m^2_V) B^{Z_2}_{e^+e^-}, \tag{5.30}$$

where the kinematical factor is given by $K(a, b, c) = L^{3/2} + 12(c/a)L^{1/2}$, with $L(a, b, c) = [(a + b - c)^2 - 4ab]/a^2$, and $B^{Z_2}_{e^+e^-}$ is the branching ratio for $Z_2 \rightarrow e^+e^-$ decay. The value of $B^{Z_2}_{e^+e^-}$ depends upon how many fermion-antifermion channels are open. Ignoring supersymmetric partner modes, it can vary from 0.036, if only standard model fermion decays are allowed, to 0.009, if three full families of E_6 fermion modes are allowed [41]. As stated earlier, asymptotically in m_{Z_2} we find (at $\tan \beta \sim 1$) that

$$f_{h^0 Z_1} \simeq \tfrac{3}{5} f_{H^0_{deg} Z_1} \simeq \tfrac{3}{5} f_{H^{\pm} W^{\mp}} \simeq \frac{\tan \theta_W}{2}, \tag{5.31}$$

so that h^0, H^0_{deg} and $H^{\pm}$ are produced in Z_2 decays in the ratio 9 : 25 : 50 (the extra factor of 2 in the $H^{\pm}$ modes coming from the two different charge states). Non-asymptotically there is some variation of the f_{HV}'s but it is useful in the following discussions to refer to a "typical" case:

$$f_{HV} = 0.5, \quad K = 1, \quad B^{Z_2}_{e^+e^-} = 0.009 \Rightarrow B^{Z_2}_{HV} = 1.8 \times 10^{-3}. \tag{5.32}$$

We also note that the Z_1 or W, produced in association with a Higgs, might have to be detected in a leptonic decay mode in order to keep backgrounds under control. Keeping e and μ modes, the branching ratios for $Z_1 \rightarrow \ell^+\ell^-$ and $W \rightarrow \ell$ are $B^{Z_1}_{\ell^+\ell^-} \sim 0.06$ and $B^W_{\ell\nu} \sim 0.2$, respectively. Altogether, we require $\gtrsim 10^5$ Z_2's to have a detectable number of $Z_2 \rightarrow HV$ decays in the leptonic V decay channels, while only $\gtrsim 10^4$ Z_2's would be required if the hadronic V decay channels can be employed.

In e^+e^- collisions the peak rate for Z_2 production (in the absence of beam energy smearing) is given in units of $\sigma_{point} = (4\pi\alpha^2)/(3s)$ by

$$R^{Z_2}_{peak} = \frac{9 B^{Z_2}_{e^+e^-}}{\alpha^2}. \tag{5.33}$$

For the parameters of eq. (5.32) we obtain $R^{Z_2}_{peak} \approx 1327$ and $R^{Z_2}_{peak} B^{Z_2}_{HV} \approx 2$. Assuming that a typical e^+e^- collider will achieve an integrated luminosity $\gtrsim$

$10^3/\sigma_{\text{point}}$, we find a large number of events for discovery of several of the E_6 Higgs bosons in Z_2 decays, especially if some of the extra E_6 $f\bar{f}$ decay modes of the Z_2 are kinematically forbidden. Even if we must include $B^{Z_1}_{\ell^+\ell^-}$ or $B^W_{\ell\nu}$, we are left with hundreds of events. Of course, it could happen that $m_{A^0} \gtrsim m_{Z_2}$ and $m_{Z_2} \gtrsim 0.5$ TeV, in which case the only kinematically allowed decay of Z_2 to a Higgs would be $Z_2 \to h^0 Z_1$. As discussed previously the $f_{h^0 Z_1}$ value is such that this decay would generally be observable.

Turning to the situation at the Superconducting Super Collider (SSC), the cross sections for Z_2 production have been summarized in ref. 41. For small $Z_1 Z_2$ mixing they vary from $\sim 2.2 \times 10^3$ pb at $m_{Z_2} = 0.2$ TeV to ~ 22 pb at $m_{Z_2} = 1$ TeV and ~ 1.7 pb at $m_{Z_2} = 2$ TeV. The typical parameters (5.32) correspond to ~ 35 events, before including $B^{Z_1}_{\ell^+\ell^-}$ or $B^W_{\ell\nu}$, for every 2 pb of Z_2 cross section, at an integrated yearly luminosity of 10^4 pb^{-1}. At a hadron collider backgrounds will be severe unless we detect the Z_1 or W in a leptonic decay mode. Thus we include $B^{Z_1}_{\ell^+\ell^-}$ or $B^W_{\ell\nu}$ and find that event rates at the SSC for an E_6 Higgs produced via Z_2 decays are only likely to be reasonable if $m_{Z_2} \lesssim 1$ TeV, although searches at higher m_{Z_2} values might be possible if $B^{Z_2}_{e^+e^-}$ is larger than 0.009.

The major backgrounds to Higgs detection in Z_2 decays will depend upon the precise secondary decay modes of the Higgs boson produced. Typically, backgrounds are most difficult if the Higgs decays primarily to standard model fermions, as will be the case for a light Higgs. A first consideration of the backgrounds in this case appears in ref. 18, with the conclusion that they should not be a problem in e^+e^- collisions, but that they may be fairly severe at a hadron collider such as the SSC.

Comparison of e^+e^- and Hadron Colliders

Using the above results, we give a first level comparison of the abilities of e^+e^- and hadron colliders to probe the E_6 Higgs sector.

e^+e^- Collisions

1. An e^+e^- collider remains a good place for finding a light h^0, whether produced via vector boson fusion, virtual vector bremsstrahlung, or in the decays of the new Z_2.

2. e^+e^- colliders are also ideal for finding a light H^0_{deg} or $H^\pm$ using the decays of the new Z_2, assuming, of course, that the Z_2 can be produced and that the decays are kinematically allowed.

3. If m_{Z_2} and m_{A^0} (which is $\sim m_{H^0_{\text{deg}}}$) are small then H^0_{deg} can also have sizeable W^+W^- and $Z_1 Z_1$ couplings. (The h^0 and H^0_{deg} share the SM coupling strength squared.) The H^0_{deg} production rate at an e^+e^- collider might be as large as 25% of that for an h^0 of similar mass, although it can also be quite small, depending upon the precise λ value that is appropriate.

4. If $s_{e^+e^-} > 4m_{H^\pm}^2$ then the charged Higgs are easily pair produced at an e^+e^- collider and backgrounds should not prevent studying them even if they decay primarily to SM heavy quark channels, assuming sufficient luminosity.

5. The $H_{Z_2}^0$ is generally totally inaccessible at an e^+e^- collider, due to the smallness of its couplings to W^+W^- and Z_1Z_1. In addition, its production via $Z_1^* \to H_{Z_2}^0 A^0$ is very small due to the suppression of the relevant coupling, to be discussed. However, if the Z_2 turns out to be very light, the $H_{Z_2}^0$ can be produced via $Z_2^* \to Z_2 H_{Z_2}^0$.

6. Similarly the A^0 will normally be difficult to produce in e^+e^- collisions due to the absence of any VV couplings. However, analogous to the situation for the minimal supersymmetric model, $Z_1^* \to H_{\rm deg}^0 A^0$ yields a non-negligible cross section of the order of 0.1 unit of R. (We shall see that the required coupling is not suppressed.) In addition, it may be possible to produce A^0 in association with either h^0 or $H_{\rm deg}^0$ in Z_2 decay if the corresponding decay is kinematically allowed.

7. If the $H^\pm$ and $H_{\rm deg}^0$ cannot be found in Z_2 decays (either because $m_{H^\pm} \sim m_{H_{\rm deg}^0} \gtrsim m_{Z_2}$ or because the Z_2 is too heavy to be produced), then one must turn to $\gamma^*, Z_1^* \to H^+ + H^-$ and $Z_1^* \to H_{\rm deg}^0 A^0$ for their production—the $H^\pm$ has no couplings to $Z_1 W^\pm$ and the $H_{\rm deg}^0 VV$ couplings are likely to be strongly suppressed relative to SM type strength (see fig. 5.4). The resulting production rates may be adequate depending upon the machine energy and luminosity relative to the final state masses.

8. Generally speaking, any Higgs with significant production cross section will be observable even if it decays primarily to SM fermions, since backgrounds to such channels at an e^+e^- collider are relatively mild, though further background studies are needed.

Hadron Collisions

1. A hadron collider will not have a significant gg fusion cross section to $H_{Z_2}^0$ [via the D loop(s)], unless $m_{H_{Z_2}^0}$ and, hence, m_{Z_2} is $\lesssim 0.4$ TeV. Since its W^+W^- and Z_1Z_1 couplings are quite small, these fusion processes will not be important. Finally, the Z_2Z_2 fusion process can only be significant when $m_{H_{Z_2}^0} \sim m_{Z_2} \lesssim 0.4$ TeV, where the gg fusion process would dominate in any case. Thus, if m_{Z_2} is heavy, the $H_{Z_2}^0$ will not be easily produced, let alone detected at a hadron machine.

2. The A^0, h^0, and $H_{\rm deg}^0$ cross sections from gg fusion should be of typical SM strength, and will be significant, depending upon the size of m_t relative to the Higgs mass in question.

3. VV fusion processes will not play a significant role in the production of $H_{\rm deg}^0$, since it decouples from W^+W^- and Z_1Z_1 when it is sufficiently massive that VV fusion processes could have become important relative to gg fusion.

4. A^0 does not couple to VV at tree-level so that only the gg fusion process is relevant for its production.
5. The $H^\pm$ will be produced via processes of the type $g\bar{b} \to \bar{t}H^+$. The resulting cross section is comparable in strength to the gg fusion cross sections for A^0 and H^0_{deg} at the same mass.
6. Unfortunately, even though the production cross sections are significant, Higgs bosons that decay primarily to heavy SM fermions will be difficult to detect at a hadron collider due to large QCD backgrounds. Only when the masses of $H^\pm$, A^0 and H^0_{deg} are small enough that the gg fusion cross sections are very large is there hope; in this instance rare decay modes of the Higgs appear to be usable [42,43]. However, as we show below, supersymmetric modes and other exotic modes characteristic of a complicated Higgs sector tend to be quite important in the decays of the heavy Higgs bosons, in which case a hadron machine could allow more complete access to the Higgs sector than an e^+e^- collider.

Higgs couplings to HV, HH, and $\tilde{\chi}\tilde{\chi}$ Channels

The Higgs–quark–antiquark and Higgs–vector–vector couplings, while dominant in considerations involving the production of the Higgs bosons, are not the only couplings of importance when it comes to the Higgs boson decays. In this section we consider the remaining couplings that must be included in fully assessing the branching ratios of the E_6 model Higgs bosons to various different final states. There are three crucial sets of couplings which arise at tree level and yield two-body final states:
   1. the couplings of Higgs bosons to other Higgs bosons plus a vector boson;
   2. the trilinear self-couplings of the Higgs bosons, which allow decay of one Higgs to two others;
   3. the couplings of the Higgs bosons to supersymmetric particle pairs.
In considering these couplings we shall continue to make the approximation that we can neglect inter-generational couplings. As well as affecting the Higgs to $\tilde{\chi}\tilde{\chi}$ decays, as discussed earlier, such inter-generational couplings could also result in such decays as Higgs to unhiggs plus vector boson, and Higgs to Higgs plus unhiggs, *etc.* By neglecting (in first approximation) inter-generational mixing, we are assuming that such decays have relatively small branching ratios.

As outlined earlier, which pairs of supersymmetric particles are allowed in the decay of a particular Higgs boson is model dependent. In the minimal supersymmetry model studies of ref. 44 both squark/slepton pairs and neutralino/chargino pairs were studied, for several different mass scale possibilities. It was found that when both types of pair states were allowed, the neutralino/chargino pair states were dominant. In addition, it is likely that E_6 models prefer a rather large gluino mass and that the squarks and sleptons will be even more massive than the gluino. In chapter 3 we demonstrated that even if $M_{\tilde{g}} \sim 0.5$ TeV there would still be many relatively light chargino and neutralino states, since their masses tended to be set by the scale of M and M', which are of order $M_{\tilde{g}}/4$ and $M_{\tilde{g}}/8$, or by m_{Z_2}, which could be significantly smaller than $M_{\tilde{g}}$. Thus, the only pairs of supersymmetric particles that we shall incorporate are those containing neutralinos and/or charginos, with masses computed for representative values of $M_{\tilde{g}}$ and m_{Z_2}.

In the following three subsections we will give brief overviews of the three different types of couplings enumerated above. More detail can be found in ref. 23. Here we simply state which of these couplings are important in Higgs decays.

HHV Couplings

The $Z_1 A^0 h^0$ and $Z_1 A^0 H^0_{Z_2}$ couplings are suppressed once m_{A^0} is larger than about m_{Z_1}. The entire Z_1–Higgs–Higgs coupling is saturated in this region by the $Z_1 A^0 H^0_{\text{deg}}$ coupling. Since $m_{A^0} \simeq m_{H^0_{\text{deg}}}$ when m_{A^0} is substantial this coupling is never important in any decay processes. But we have stressed the fact that it leads to the process $Z_1^* \to A^0 H^0_{\text{deg}}$ at an $e^+ e^-$ collider, which may provide the only non-negligible source for A^0 and H^0_{deg} production at such a machine if $m_{A^0} \gtrsim m_{Z_1}$.

Only a very few HHV couplings are of real importance in decays: the $H^0_{Z_2}$ decays to VH final states will turn out to be an important part of the $H^0_{Z_2}$ total width when the $H^0_{Z_2}$ is substantially heavier than H^+ and A^0 (due to the presence of longitudinal V polarization states in these modes); decays of the type $A^0 \to Z_2 h^0$ are allowed when m_{A^0} is large, and will be very important due to the longitudinal Z_2 modes and the full strength coupling; and, at small m_{A^0}, the $A^0 \to Z_1 h^0$ decay can be significant since the corresponding coupling is not negligible in that region.

HHH Couplings

Only couplings for $H^0_{Z_2}$ and H^0_{deg} are of potential importance, and, in the latter case, only those to the possibly phase space allowed channels of $h^0 h^0$, and (at large $m_{H^0_{\text{deg}}}$) $H^0_{Z_2} H^0_{Z_2}$. Due to the low mass of the h^0, the kinematic inaccessibility of other HH modes, and the smallness of other classes of channels, it turns out that the $H^0_{Z_2} \to h^0 h^0$ decays can be dominant at large m_{A^0}. At small m_{A^0} all the HH channels become phase space allowed and are significant. Regarding the H^0_{deg}, it turns out that its HHH couplings are not significant except at very small m_{A^0} where the $H^0_{\text{deg}} \to h^0 h^0$ mode can be significant.

$H\tilde{\chi}\tilde{\chi}$ Couplings

Only a few general points regarding the Higgs couplings to neutralino/chargino channels are worth noting here. First, over much of parameter space there are actually relatively few important couplings, but those that are significant will be of order λ, g or $g' = g_1$. In fact, for any Higgs decay there is almost always one or more allowed neutralino/chargino decay channel with substantial coupling. Recalling that W^+W^- and Z_1Z_1 channels are not important for our heavy Higgs, and that only the $t\bar{t}$ (or, for the $H^0_{Z_2}$, the $D\bar{D}$) channel will have coupling of order g, we find that the neutralino/chargino modes will be an extremely important component of Higgs boson decays, and will almost always dominate SM fermion modes.

Higgs boson decays and detection

In order to examine the possibilities for detecting the E_6 model Higgs bosons, either in the inclusive production modes or through Z_2 decays, a full assessment of all the decay modes of each of the Higgs bosons is required. In particular, we have argued that, in the case of the heavier Higgs bosons, supersymmetric decay modes will be important, due to the substantial couplings and the relatively low mass scale of the neutralino/chargino sector. (As discussed earlier we will not include any squark or slepton decay modes in the computations that follow. Their masses are generally larger than the mass of the gluino which, in the scenario we investigate has a mass of 0.5 TeV. Thus squark and slepton decay channels are likely to have a threshold substantially above that for the neutralinos and charginos. As discussed earlier, even when allowed they will not be as important as the neutralino/chargino channels at high Higgs masses.) Also, modes in which a heavy Higgs decays to a pair of lighter Higgs bosons or to a vector boson in association with a lighter Higgs must be considered. In contrast, the decays of the light h^0 will be dominated by SM channels for moderate $M_{\tilde{g}}$ values. Of course, if a small $M_{\tilde{g}}$ scale is appropriate (*e.g.*, $M_{\tilde{g}} \lesssim 200$ GeV), resulting in some very light neutralinos and charginos, h^0 decays to neutralinos and charginos could become important. However, one must be careful that the lightest chargino is not so light as to violate known experimental bounds [45]. In this section we will give an overview of the branching ratios for Higgs boson decays to both Standard Model channels and the above mentioned more exotic channels. We shall consider the $m_{Z_2} = 0.6$ TeV case, and compute masses and mixings in the neutralino/chargino sector assuming $M_{\tilde{g}} = 0.5$ TeV, the case considered earlier in discussing the neutralino/chargino masses. In a certain sense, this latter assumption may be a pessimistic one in that it leads to smaller branching ratios for these supersymmetric channels than would obtain for lighter $M_{\tilde{g}}$ values. We complete our parameter specification by taking $m_t = 70$ GeV and $m_D = 250$ GeV (for the third-generation D—the first- and second-generation D's are assumed to be considerably lighter). This m_D value has been deliberately taken to be fairly small so that we can display the role of the $D\bar{D}$ mode in $H^0_{Z_2}$ decays, should it be allowed.

We begin by noting that, with regard to decays, the results for the vector boson pair channels can be summarized as follows: whenever a VV' channel (with $V,V' = Z_1$ or $W^\pm$) is phase space allowed for a given Higgs, the coupling of this Higgs to the VV' channel is very small. In addition, couplings to modes containing Z_2 are proportional to g_1 and thus such modes naturally contribute less strongly than the low mass vector boson modes, even when phase space allowed. Since it is primarily the longitudinal modes of VV' decay channels that can lead to a large Higgs width, it is not surprising that *all the Higgs bosons of this theory remain quite narrow*. Only the A^0, H^0_{deg} and $H^\pm$ can become massive enough that VV' modes containing the Z_2, as well as modes containing a vector boson and a lighter Higgs, become important and total widths begin to become significant. Thus, for the $m_{Z_2} = 0.6$ TeV, $M_{\tilde{g}} = 0.5$ TeV case that we are focusing on, h^0 always remains very narrow, $H^0_{Z_2}$ has $\Gamma \lesssim 15$ GeV at $m_{A^0} = 3$ TeV (recall that this is our basic parameter) while A^0, $H^\pm$ and H^0_{deg} (all approximately degenerate in mass) have $\Gamma \lesssim 100$ GeV at $m_{A^0} = 3$ TeV. The delayed growth in the Higgs widths is, of course, closely related to the fact that the Higgs sector of this theory does not become strongly interacting, and bad high energy behavior of various scattering amplitudes involving scalar modes does not emerge until very high energy scales.

With this background, it is not surprising that non-SM modes could play a major role in the decays of these Higgs over this region of m_{A^0} parameter space. In particular, the neutralino/chargino channels provide a significant fraction of the total decays of the heavier Higgs. For instance, for a light gluino, $M_{\tilde{g}} = 0.2$ TeV it is not uncommon for the $\tilde{\chi}\tilde{\chi}$ modes to have a net branching ratio of order 50%, especially at large $m_{A^0} \sim m_{H^0_{\text{deg}}} \sim m_{H^\pm}$. In fact, if $M_{\tilde{g}}$ is as small as 200 GeV, many of the neutralinos and charginos, in particular $\tilde{\chi}^0_{\tilde{B}'}$, $\tilde{\chi}^0_{\tilde{W}_3}$ and $\tilde{\chi}^\pm_{\tilde{W}^+}$, can become light enough that these decay modes can have substantial importance in A^0, H^0_{deg}, and $H^\pm$ decays at low Higgs masses. They even become a substantial component of h^0 decays when the h^0 takes on its maximum mass allowed at high m_{A^0} values.

To give a general survey of decays is beyond the scope of this review, but to give one example we illustrate in fig. 5.5 the channels of importance in the decays of the various Higgs bosons. For this graph we shall adopt the moderate $M_{\tilde{g}} = 0.5$ TeV, $m_{Z_2} = 0.6$ TeV values considered in several previous graphs. We shall also take $\tan\beta = 1.1$ and $\lambda > 0$. We present graphs for $H^0_{Z_2}$, A^0, H^0_{deg} and H^+, in turn, which display branching ratios for all general classes of modes that are important in the decays of the given Higgs boson. For this sampling, we plot branching ratios computed at a unique λ value at each m_{A^0}. This unique value is $\lambda^{[40]}_{\text{min}}$, the minimum $\lambda > 0$ value that yields $m_{h^0} \geq 40$ GeV. (Of course, at $\lambda = \lambda^{[40]}_{\text{min}}$, m_{h^0} is exactly 40 GeV.) It is easily seen that the combined branching ratios for the various modes yield close to 100% of a given Higgs' decays. We note that for the choice of $M_{\tilde{g}} = 0.5$ TeV that we have made, the h^0 decays are entirely similar to those of a Standard Model Higgs boson of the same mass. For additional graphs of branching ratios at different m_{Z_2} and $M_{\tilde{g}}$ values, and for the $\lambda < 0$ choice at the same values of m_{Z_2} and $M_{\tilde{g}}$ considered here, see refs. 23 and 33. The graphs presented here turn out to be quite representative.

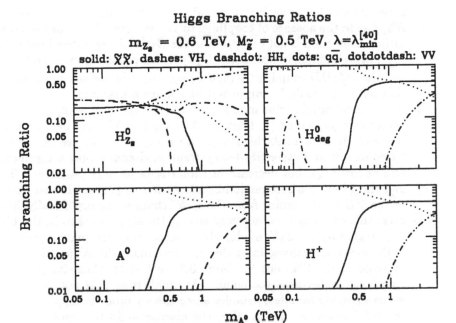

Higgs Branching Ratios

$$\text{m}_{Z_2} = 0.6 \text{ TeV, } M_{\widetilde{g}} = 0.5 \text{ TeV, } \lambda = \lambda_{\min}^{[40]}$$

solid: $\widetilde{\chi}\widetilde{\chi}$, dashes: VH, dashdot: HH, dots: $q\bar{q}$, dotdotdash: VV

Figure 5.5 We present branching ratios for $H_{Z_2}^0$, A^0, H_{deg}^0, and $H^\pm$ decay to the following channels: $\widetilde{\chi}\widetilde{\chi}$; HH; VH; VV; and $q\bar{q}$. We sum over all allowed channels of a given type. In $H_{Z_2}^0$ decays the $q\bar{q}$ channel is dominated by $D\overline{D}$, while for the other Higgs the $q\bar{q}$ channel is dominated by channels containing the top quark, when kinematically allowed. In the sample graphs shown here, we have chosen a unique λ value at each m_{A^0}: $\lambda_{\min}^{[40]}$, defined as the minimum value of λ that yields $m_{h^0} = 40$ GeV. For all plots $M_{\widetilde{g}} = 0.5$ TeV, $m_{Z_2} = 0.6$ TeV, and $\tan\beta = 1.1$.

We may now survey the prospects for Higgs decays and detection.

1. h^0: We remark that the h^0 decays entirely to light SM fermion channels until $m_{A^0} \gtrsim 3$ TeV (beyond the range allowed by the imprecise bounds mentioned earlier—see eq. (5.17) and the discussion which follows), at which point it can become massive enough (for λ appropriately chosen) to decay, as well, to $t\bar{t}$, WW and $Z_1 Z_1$ modes. Of course, as mentioned if the neutralinos and charginos are light (*i.e.*, $M_{\widetilde{g}}$ is small), then $\widetilde{\chi}\widetilde{\chi}$ decays will become important at the higher m_{h^0} values.

2. $H_{Z_2}^0$: The decays of this Higgs exhibit considerable complexity. But, despite the many possible modes with good signatures, the $H_{Z_2}^0$ may still be very difficult to find, due to its small production cross sections. At a hadron collider the gg fusion cross section will be substantially smaller than that for a SM Higgs boson, even if the D quark has large

mass. At an e^+e^- machine, it will also be very difficult to produce the $H^0_{Z_2}$ due to the absence of couplings to W^+W^- and Z_1Z_1, as discussed earlier. Detection is not likely unless the Z_2 (and, correspondingly, the $H^0_{Z_2}$) mass is small, in which case Z_2Z_2 fusion could become significant, and the $H^0_{Z_2}D\overline{D}$ coupling ($\sim (5/6)g_1m_D/m_{Z_2}$) is not as suppressed so that gg fusion could be important at a hadron machine.

3. A^0: Neutralino/chargino modes tend to be important except for small values of m_{A^0}, where (at least for $M_{\tilde{g}} = 0.5$ TeV) they become kinematically forbidden. At high m_{A^0}, a variety of modes are important. Of course, $t\bar{t}$ and other SM heavy fermion channels play a significant role at low m_{A^0}, but decrease in importance as m_{A^0} passes ~ 1 TeV and the $\tilde{\chi}\tilde{\chi}$ modes become dominant. At low m_{A^0} there is also a peak in the VH mode coming from the Z_1h^0 channel; asymptotically, however, this coupling becomes very small. Its place is taken in the high m_{A^0} region by the Z_2h^0 mode, which becomes increasingly significant. If the neutralino/chargino modes are dominant, the A^0 should be detectable at the SSC out to about 0.5 TeV after which its gg fusion cross section becomes too small. At small m_{A^0} it is not likely that we can employ rare decay modes since the $\gamma\gamma$ mode is suppressed and the V^*V modes are absent. (See the discussion for the minimal supersymmetric model case.) At an e^+e^- collider, much as in the minimal supersymmetric model case, one will look for $Z_1^* \to A^0 + H^0_{\text{deg}}$.

4. H^0_{deg}: There are many similarities to A^0. Neutralino/chargino modes are important. The Z_1Z_2 mode becomes significant at high $m_{H^0_{\text{deg}}}$; it can be as large as ~ 0.3 of the total at $m_{H^0_{\text{deg}}} \sim 3$ TeV. This mode has been studied in some detail in ref. 46. The $t\bar{t}$ mode tends to be dominant (if allowed) for small m_{A^0} decreasing slowly as the $\tilde{\chi}\tilde{\chi}$ modes turn on at high m_{A^0}. VH modes are never important since the H's to which H^0_{deg} couples are more or less degenerate with the H^0_{deg} itself. Conclusions are similar to those for the A^0: the gg fusion mode cross section is possibly sufficient for detection of the H^0_{deg} at the SSC out to $m_{H^0_{\text{deg}}} \sim .5$ TeV, provided the $\tilde{\chi}\tilde{\chi}$ decay modes are significant. When SM fermion decay modes are dominant at small m_{A^0}, detection will be difficult at a hadron collider.

5. H^+: For low masses it is clear that H^+ decays primarily to $t\bar{b}$ making detection very difficult except at an e^+e^- collider. As $m_{H^\pm}$ passes ~ 0.3 TeV the $\tilde{\chi}^+\tilde{\chi}^0$ modes can become up to 50% of the decays and may allow detection of the H^+ at a hadron collider as well as at an e^+e^- collider; further study is required to be more definite. Finally, we note the increasing importance of VV modes (dominated by W^+Z_2) in H^+ decays as $m_{H^\pm} \sim m_{A^0}$ becomes large. These would have a clean signature at a hadron collider if at least one of the V's decayed leptonically, but the associated event rate would be very low, due to the $\lesssim 0.05$ pb cross section at the SSC [10] for $m_{H^\pm} \gtrsim 2$ TeV, where the

VV modes become significant. Overall, detection at an e^+e^- collider is probably only limited by the phase space and luminosity for H^+H^- Drell-Yan type pair production, while at a hadron collider it is limited by the region of $m_{H^\pm}$ over which the $gb \to H^-t$ and $g\bar{b} \to H^+\bar{t}$ cross section is substantial. (At the SSC, the latter cross sections are of reasonable size, $i.e.$, $\gtrsim 1$ pb, out to about $m_{H^\pm} \sim 0.8$ TeV.)

That detection of the heavy Higgs bosons will be a challenge is obvious. It will be necessary to full explore the myriad of possible channels. For all the heavy Higgs we must note that many of their decay products—$e.g.$, neutralinos, charginos, Z_1, Z_2, and other Higgs—will in turn decay. The resulting final states are clearly very complex and backgrounds must be carefully explored. This task will be difficult and will require detailed Monte Carlo work.

At either an e^+e^- or hadron collider, the importance of supersymmetric decay modes for all Higgs bosons other than the h^0 implies that the search for Higgs bosons at future colliders may well become a branch of supersymmetric phenomenology. Other evidence for supersymmetry may already be available, although, if squarks and gluinos should turn out to be very heavy, it is certainly possible that supersymmetric decay products of the Higgs may be the first direct evidence for supersymmetry! At a minimum, information on the Higgs sector will provide extremely useful consistency checks and will be crucial to pinning down the details of the underlying supersymmetric theory. A careful evaluation of Higgs detection possibilities in the plethora of exotic modes awaits further work, but we are cautiously optimistic for those situations where adequate production rate is available. Unless an e^+e^- collider has sufficient energy and luminosity to produce a large number of Z_2's and the Higgs bosons other than the h^0 are light enough to be produced in its decays, it is likely that a hadron collider will provide more complete access to the full Higgs spectrum of an E_6-based model.

REFERENCES

1. J. Ellis, J.F. Gunion, H.E. Haber, L. Roszkowski, and F. Zwirner, *Phys. Rev.* **D39** (1989) 844.
2. G.F. Giudice and A. Masiero, *Phys. Lett.* **206B** (1988) 480.
3. P. Fayet, *Nucl. Phys.* **B90** (1975) 104.
4. H.P. Nilles, *Physics Reports* **110** (1984) 1.
5. L. Durand and J.L. Lopez, *Phys. Lett.* **217B** (1989) 463.
6. J. Ellis, J.S. Hagelin, S. Kelley and D.V. Nanopoulos, *Nucl. Phys.* **B311** (1988) 1.
7. M. Drees, *Int. J. Mod. Phys.* **A4** (1989) 3635.
8. K.S. Babu and E. Ma, *Phys. Rev.* **D33** (1986) 3337.
9. R. Rodenberg, *Nuovo Cim.* **97A** (1987) 277.
10. J.F. Gunion, H.E. Haber, F.E. Paige, Wu-Ki Tung, and S.S.D. Willenbrock, *Nucl. Phys.* **B294** (1987) 621.

11. I. Antoniadis, J. Ellis, J.S. Hagelin, and D.V. Nanopoulos, *Phys. Lett.* **205B** (1988) 459; *Phys. Lett.* **208B** (1988) 209.
12. I. Antoniadis, J. Ellis, J.S. Hagelin, and D.V. Nanopoulos, *Phys. Lett.* **194B** (1987) 231.
13. K. Griest and M. Sher, CfPA-TH-89-007 (1989).
14. E. Witten, *Phys. Lett.* **155B** (1985) 151; *Nucl. Phys.* **B258** (1985) 75; P. Candelas, G.T. Horowitz, A. Strominger and E. Witten, *Nucl. Phys.* **B258** (1985) 46.
15. P. Binetruy, S. Dawson, I. Hinchliffe, and M. Sher, *Nucl. Phys.* **B273** (1986) 501.
16. J. Ellis, K. Enqvist, D.V. Nanopoulos and F. Zwirner, *Nucl. Phys.* **B276** (1986) 14.
17. L.E. Ibanez and J. Mas, *Nucl. Phys.* **B286** (1987) 107.
18. J.F. Gunion, H.E. Haber, and L. Roszkowski, *Phys. Lett.* **189B** (1987) 409.
19. H.E. Haber and M. Sher, *Phys. Rev.* **D35** (1987) 2206.
20. J. Ellis, D.V. Nanopoulos, S.T. Petcov, and F. Zwirner, *Nucl. Phys.* **B283** (1987) 93.
21. M. Drees, *Phys. Rev.* **D35** (1987) 2910.
22. H. Baer, D. Dicus, M. Drees, X. Tata, *Phys. Rev.* **D36** (1987) 1363.
23. J.F. Gunion, L. Roszkowski, and H.E. Haber, *Phys. Rev.* **D38** (1988) 105.
24. J. Hewett and T. Rizzo, *Phys. Rep.* **183** (1989) 193.
25. J.-P. Derendinger, L.E. Ibanez and H.P. Nilles, *Nucl. Phys.* **B267** (1986) 365.
26. W. Bartel *et al.* (JADE Collaboration), *Z. Phys.* **C31** (1986) 259; H.-J. Behrend *et al.* (CELLO Collaboration), *Phys. Lett.* **193B** (1987) 376; W. Braunschweig *et al.* (TASSO Collaboration), in *XXIV International Conference on High Energy Physics*, Munich, West Germany, August 1988, edited by R. Kotthaus and J.H. Kühn (Springer-Verlag, Berlin, 1989) p. 1432.
27. B. Campbell, J. Ellis, M.K. Gaillard, D.V. Nanopoulos, K.A. Olive, *Phys.Lett.* **180B** (1986) 77.
28. U. Amaldi, A. Bohm, L.S. Durkin, P. Langacker, A. K. Mann, W.J. Marciano, *Phys. Rev.* **D36** (1987) 1385; G. Costa, J. Ellis, G.L. Fogli, D.V. Nanopoulos and F. Zwirner, *Nucl. Phys.* **B297** (1988) 244.
29. L.W. Durkin and P. Langacker, *Phys. Lett.* **166B** (1986) 436; V. Barger, N.G. Deshpande, and K. Whisnant, *Phys. Rev. Lett.* **56** (1986) 30.
30. F. del Aguila, M. Quiros and F. Zwirner, *Nucl. Phys.* **B284** (1987) 530; **B287** (1987) 419; P. Franzini, J. Ellis and F. Zwirner, *Phys. Lett.* **202B** (1988) 417.
31. P.J. Franzini and F.J. Gilman, *Phys. Rev.* **D35** (1987) 855.
32. J.F. Gunion, H.E. Haber, and M. Sher, *Nucl. Phys.* **B306** (1988) 1.
33. L. Roszkowski, University of California at Davis, Thesis (1987).
34. L. Roszkowski, in *Proceedings of the 1987 Cargese Summer School.*

35. J.F. Gunion and H.E. Haber, *Nucl. Phys.* **B272** (1986) 1.
36. J.F. Gunion, B. Kayser, R.N. Mohapatra, N.G. Deshpande, J. Grifols, A. Mendez, F. Olness, and P.B. Pal, in *Proceedings of the 1984 Snowmass Summer Study on the Design and Utilization of the Superconducting Super Collider*, edited by R. Donaldson and J. G. Morfin, p. 197.
37. See, for example, J.F. Gunion and A. Tofighi-Niaki, *Phys. Rev.* **D36** (1987) 2671.
38. S. Nandi, *Phys. Lett.* **181B** (1986) 375.
39. C. Dib and F.J. Gilman, *Phys. Rev.* **D36** (1987) 1337.
40. V. Barger and K. Whisnant, *Phys. Rev.* **D36** (1987) 3429.
41. H.E. Haber, in *Supercollider Physics* edited by D.E. Soper (World Scientific, Singapore, 1986), p 194.
42. J.F. Gunion, P. Kalyniak, M. Soldate and P. Galison, *Phys. Rev.* **D34** (1986) 101.
43. J.F. Gunion, G.L. Kane, and J. Wudka, *Nucl. Phys.* **B299** (1988) 231.
44. J.F. Gunion and H.E. Haber, *Nucl. Phys.* **B278** (1986) 449.
45. See, for example, the limits found at PETRA—W. Bartel *et al.*, *Z. Phys.* **C29**, 505 (1985); H.J. Behrend *et al.*, *Z. Phys.* **C35**, 181 (1987). These limits have recently been improved at TRISTAN [see, *e.g.*, Y. Sakai *et al.* (AMY Collaboration), KEK Preprint 89-134 (1989)] and at LEP [see, *e.g.*, B. Adeva *et al.* (L3 Collaboration), *Phys. Lett.* **233B** (1989) 530].
46. G. Pocsik and T. Torma, *Int. J. Mod. Phys.* **A4** (1989) 3629.

Chapter 6

Other Approaches with Non-Minimal Higgs Sectors

6.1 CP Violation and Higgs Bosons

CP violation could occur in several different ways that originate in the Higgs sector. In this section we provide a partial survey of the literature in this area, with some description of the underlying physics and of the kinds of predictions that help distinguish these mechanisms from others. Our treatment is not meant to fully review this subject, but to provide a starting place for interested readers. For a recent review of CP-violating phenomena and a summary of models in which CP violation arises from the Higgs sector, see ref. 1.

That CP might not be a good symmetry in the presence of spontaneous symmetry breaking was first pointed out by T.D. Lee [2]. For such an effect to occur, two or more complex SU(2) doublets of Higgs fields are required. In addition, the Higgs exchanges must lead to flavor-changing neutral interactions [3,4]. However, the mere presence of more than one-doublet does not guarantee that CP violation in the Higgs sector is possible. For instance, a CP-violating phase in the case of two-doublet models was introduced explicitly in eq. (4.8), where we remarked that $\lambda_5 \neq \lambda_6$ is required. But in the case of supersymmetry we found that $\lambda_5 = \lambda_6$ automatically, and CP-violating phases could be rotated out of the Higgs sector entirely. But, if such phases cannot be rotated away, this approach leads to CP violation from neutral Higgs boson interactions, from charged Higgs boson interactions, and perhaps still from the presence of a non-vanishing phase in the CKM matrix; the effect of the neutral Higgs bosons can be the dominant one. Such models have been studied several times [5–7], most recently by Liu and Wolfenstein [8]. In all cases, the various types of CP violation are related at a fundamental level, but because of our ignorance of the Higgs sector, in practice the parameters of each type of interaction are independent. Reference 8 studies CP violating effects in the K^o system, including the values of ϵ and ϵ'; the latter is non-zero in these models, but often smaller than in the standard approach. They also consider the neutron electric dipole moment and the muon polarization in $K_L \to \mu^+\mu^-$.

Somewhat after ref. 2, Weinberg [3] pointed out that flavor-changing neutral currents could be avoided by introducing at least three complex SU(2) doublets of Higgs fields. Then there are three charged fields $\phi_1^+, \phi_2^+, \phi_3^+$. The matrix relating them to the mass eigenstates is a 3×3 unitary matrix, and by essentially the same argument [9] as for the CKM matrix it depends on three real angles and on a phase that cannot be rotated away. The phase then leads to CP-violating couplings of fermions to charged Higgs bosons, analogous to the charged current couplings of the Standard Model. One general prediction of this approach is that the CP-violating couplings will occur for the leptons as well as for the quarks if the same Higgs scalars couple to both sectors; another is that the strengths of CP violating effects will be mass dependent.

In the early 1980's it was argued [10,11] that this approach gave too large a value of ϵ'/ϵ. That was thought to occur because the box diagram with charged Higgs bosons is small, so the CP violation occurs mainly via $\Delta S = 1$ diagrams, which relate ϵ and ϵ' so that $\epsilon'/\epsilon = 1/20$. Later, however, it was noted [12,13] that long distance effects can induce effects that decrease ϵ'/ϵ, and currently it is believed that having CP violation originate with charged Higgs boson couplings to fermions is consistent with all constraints. In the current view [13], chiral symmetry effects would require CP violation due to the Higgs sector to be quite small, so the expected size depends on chiral symmetry-breaking effects.

Another general feature of putting the CP violation in the Higgs sector is that larger effects are expected in heavier quark systems (compared to the usual Standard Model approach), because Higgs boson couplings increase with quark mass. This makes the b-quark system a good testing ground. Recently Donoghue and Golowich [14] have normalized the charged Higgs contribution to give CP violation in kaon decays, and found that processes such as $b \to s +$ gluon are expected to be quite large (10–15%) and CP-violating, although it is not easy to think of a way to observe the CP-violating effects even if they are there. The process $b \to s\gamma$ should be appropriately enhanced as well [15].

Because the Higgs sector can lead to CP violation that is either spontaneous (via complex vacuum expectation values), or from complex couplings, induced from complex Higgs fields, and because the parameters of the Higgs sector are too poorly understood and not related to other data, it is not possible to evaluate the role of the Higgs sector in this area at the present time. There is room for considerable analysis here. CP-violating effects that originate in the Higgs sector will lead to a number of predictions that are different from CP-violating effects that originate from the CKM matrix phase and $W^\pm$ interactions.

6.2   Axions and Majorons

There are two classes of extended Higgs models which we will not discuss in any detail in this review: axion models and majoron models. Briefly, axion models were invented to solve the strong CP problem and Majoron models were invented in the study of models with spontaneously broken lepton number. The original axion model [16] was just an ordinary two-Higgs doublet model in which the pseudoscalar was massless at tree level. This can be easily arranged by choosing the Higgs potential to be invariant under independent phase rotations of the two Higgs fields [e.g., by setting $\lambda_5 = \lambda_6 = 0$ in eq. (4.8)]. Such a model will then exhibit an extra axial U(1) symmetry. This symmetry is desirable since it allows the QCD θ parameter to be rotated to zero, which "explains" why its observed value is so small ($\theta < 10^{-9}$, based on the current upper limit on the neutron electric dipole moment [17]) hereby solving the strong CP problem. In addition, we note that the massless pseudoscalar in the model is just the Goldstone boson of the axial U(1) global symmetry which is spontaneously broken when the Higgs fields acquire vacuum expectation values. In fact, on closer examination, one discovers that this axial U(1) symmetry is anomalous. The end result is that the pseudoscalar does acquire a small mass (from instanton effects) of order

$$m_A \simeq \frac{f_\pi m_\pi}{v}, \tag{6.1}$$

where $f_\pi \simeq 93$ MeV and $v = 246$ GeV is the scale of electroweak symmetry breaking. That is, A is a pseudo–Goldstone boson, which has been dubbed the "axion". Unfortunately, an axion in this mass range (which has very predictable properties) has been ruled out experimentally. Thus, theorists have looked for models which preserved the axion as the solution to the strong CP problem, without running into conflict with experiment. The most popular modification to the model was to add a complex singlet Higgs field with vacuum expectation value $V \gg v$. Models can be constructed in which $m_A \simeq f_\pi m_\pi / V$, and the couplings of the axion to all matter is suppressed by a factor of m_π / V. By taking V large enough (say, of order 10^{10} GeV), the resulting axion is extremely light and very weakly coupled to ordinary matter [17,18]. Such an axion is called the "invisible axion", and is very difficult to rule out experimentally. The couplings of the invisible axions to ordinary matter are rather model dependent; see, e.g., refs. 19 and 20. For further details, the reader should consult two recent excellent reviews by Kim [17] and Cheng [21].

Majoron models are somewhat different. The Majoron is the Goldstone boson of a spontaneous broken lepton number. This broken symmetry is not anomalous, so the Majoron is exactly massless. There are a number of different realizations of this model which have rather different properties. In one version due to Chikashige, Mohapatra and Peccei [22], a complex scalar field which explicitly carries lepton number is added to the one-Higgs-doublet model.

This scalar then acquires a vacuum expectation value $V \gg v$. In this model, the Majoron is very weakly coupled to matter (with couplings suppressed by a factor of v/V); like the invisible axion, this Majoron is also invisible. (In such a model, one must introduce a right-handed neutrino field in order to generate neutrino masses.) A second model introduced by Gelmini and Roncadelli [23] introduces a $Y = 2$ triplet Higgs field instead of a singlet field. The field can couple to the left-handed neutrino, so that the neutrino acquires a Majorona mass when the triplet Higgs field develops a vacuum expectation value (v_3). There are a number of phenomenological constraints on the size of v_3 and the strength of Majoron couplings to matter [24]. The strictest constraints arise from astrophysical considerations; namely, the requirement that the energy loss in red giants due to Majoron emission not be too large so as to adversely affect the star's evolution. The constraint implies that $v_3 \lesssim 10$ keV [25], a result far stricter than the bounds on v_3 quoted in eq. (4.5) which arise from the requirement that $\rho \simeq 1$.

The triplet Majoron model also has implications that can be immediately tested at the Z factories. This model possesses both a massless Majoron (χ) and a very light scalar partner (h) whose mass is of order v_3 or less.* . Thus, there is no phase-space suppression to the decay $Z \rightarrow h\chi$ and we find

$$BR(Z \rightarrow h\chi) \simeq 2BR(Z \rightarrow \nu\bar{\nu}) \tag{6.2}$$

Both χ and h would escape collider detectors, so that this model would predict that five "effective" neutrinos (N_ν) should be observed in Z decay. This model is now ruled out by the initial results from SLC and LEP: $N_\nu = 3.2 \pm 0.2$ (averaged over the five experiments reporting data [27]).

Meanwhile, other variants of the Majoron model have been developed. For example, a model introduced by Bertolini and Santamaria [28] involves two Higgs doublets and a charged Higgs singlet. In this model, the second doublet and singlet carry two units of lepton number. Lepton number is spontaneously broken when the neutral field of the second doublet acquires a vacuum expectation value, resulting in the Majoron. This model presents a clever way for producing neutrino masses, without introducing a right-handed neutrino or a Higgs triplet field (so that $\rho \simeq 1$ is automatically retained). Such models also possess a very light Higgs scalar and predict $BR(Z \rightarrow h\chi) \simeq 0.5BR(Z \rightarrow \nu\bar{\nu})$, which cannot be ruled out by present data. The last class of models which we shall mention here are supersymmetric models in which R-parity is spontaneously broken by a sneutrino vacuum expectation value [29,30]. This breaks lepton number, so these models must possess a Majoron, which turns out to be the supersymmetric partner of the neutrino. Implications of these models have been investigated in ref. 31.

One of the more amusing possibilities is that of invisible Higgs decay. For example, in the triplet and doublet models just mentioned, it is very

---

*The existence of such a light Higgs scalar is common to all nonsinglet Majoron models [26]

possible that the dominant decay of one of the neutral Higgs bosons of the models would be into two Majorons [32,33,28]. Since Majorons are weakly interacting, such a decay would appear "invisible". Experimentalists should be alerted to such a possibility, particularly in their Higgs searches at a Z^0 factory. When searching for $Z^0 \to h\mu^+\mu^-$, one can detect the Higgs boson by observing a peak in the recoil $\mu^+\mu^-$ invariant mass spectrum. It may be tempting to trigger on events with hadronic activity, since the Higgs boson (if it were heavier that 10 GeV) would be expected to decay dominantly into $b\bar{b}$ pairs. However, if most Higgs decays are invisible, then such a cut would be very unwise.

Other effects of note in Majoron models are the possibility of fast non-radiative neutrino decay [34] and new sources for $K \to \pi + nothing$ [35].

6.3   The $H^{\pm}W^{\mp}Z$ Vertex: A Signature of Exotic Higgs Representations

We have seen in §4.1 that models containing only Higgs doublets and singlets do not have $H^{\pm}W^{\mp}\gamma$ and $H^{\pm}W^{\mp}Z$ tree-level vertices [36,37]. However, quantum number arguments alone are not sufficient to eliminate these vertices. In fact, it is easy to show that all models require the absence of a tree-level $H^{\pm}W^{+}\gamma$ vertex based on the conservation of the electromagnetic current. The latter requirement forbids any off-diagonal photon couplings to physical particles in the fundamental Lagrangian of the theory. On the other hand, tree-level $H^{\pm}W^{\mp}Z$ vertices generally do arise in more complicated models with triplet and/or higher Higgs representations; such vertices vanish only in very specially constructed models. In this section, we would like to investigate this matter more closely and deduce requirements which guarantee the existence or vanishing of these vertices.

The general analysis is straightforward, and we shall present a brief exposition here [37] for the case of an $SU(2)_L \times U(1)$ electroweak gauge group. (See ref. 37 for an extension of the analysis to $SU(2)_L \times SU(2)_R \times U(1)$ models.) The HVV vertices (H = physical Higgs bosons and V = vector bosons) arise from the term

$$\sum_k (D^\mu \phi_k)^* (D_\mu \phi_k) + \tfrac{1}{2}\sum_i (D^\mu \eta_i)^T (D_\mu \eta_i) \tag{6.3}$$

in the Lagrangian, where

$$D_\mu \equiv \partial_\mu + igW_\mu^a T^a + \tfrac{i}{2}g'B_\mu Y \tag{6.4}$$

is the covariant derivative, T^a are the SU(2) generators appropriate to the field on which they act (normalized to $\text{Tr}T^aT^b = \tfrac{1}{2}\delta^{ab}$), and Y is the corresponding hypercharge generator. W^a and B are the SU(2) and U(1) gauge bosons, respectively; the γ and Z are obtained as usual by a rotation by θ_W [see eq. (A.2)]. Finally, we sum over all Higgs multiplets which may include both complex representations ϕ_k and real representations η_i. Here, we employ a rather narrow definition of a real representation as consisting of a real multiplet of fields with integer weak isospin and $Y = 0$. The generators in a real representation can be chosen to be purely imaginary and antisymmetric. We then shift the Higgs fields with non-zero vacuum expectation value: $\phi_k \rightarrow \phi_k + v_k$ and $\eta_i \rightarrow \eta_i + u_i$, and we read off the desired vertices. In the simplification of the resulting expression, we make use of the fact that $U(1)_{EM}$ is preserved by the spontaneous symmetry breaking. This implies that $Qv_k \equiv (T^3 + 1/2Y)v_k = 0$, and $Qu_i = 0$. For complex representations, we work in a basis where T^3 and Y are diagonal (the eigenvalue of Y is denoted by Y_k below).[†] After a bit of manipulation, we find the following interaction of $H^\pm W^\mp$ with the neutral vector bosons (V^0)

$$
\mathcal{L}_{H^\pm W^\mp V^0} = \frac{g^2}{\sqrt{2}} W_\mu^+ \left\{ \sum_k \left[\phi_k^\dagger T^+ v_k \left(W_\mu^3 - \frac{Y_k}{\cos\theta_W} Z_\mu \right) \right. \right.
$$

$$
\left. -(T^- v_k)^\dagger \phi_k \left(W_\mu^3 + \frac{Y_k}{\cos\theta_W} Z_\mu \right) \right]
$$

$$
\left. + \sum_i \eta_i^T T^+ u_i W_\mu^3 \right\} + \text{h.c.}, \tag{6.5}
$$

where Y_k is the hypercharge of ϕ_k, $T^\pm \equiv T^1 \pm iT^2$, and $W_\mu^3 \equiv A_\mu \sin\theta_W + Z_\mu \cos\theta_W$. At this point, we note that the charged Goldstone boson (which is eaten by the W thereby giving it mass) is equal to the following linear combination of scalar fields

$$
G^- = \frac{g}{\sqrt{2}m_W} \left\{ \sum_k \left[\phi_k^\dagger T^+ v_k - (T^- v_k)^\dagger \phi_k \right] + \sum_i \eta_i^T T^+ u_i \right\}. \tag{6.6}
$$

This is easily derived by noting that it is precisely $W_\mu^+ \partial^\mu G^-$ which appears in the expansion of eq. (6.3) (after shifting the fields) as described above. We have normalized the state G^- so that it has unit norm. Thus, we can immediately rewrite eq. (6.5) as follows:

$$
\mathcal{L}_{H^\pm W^\mp V^0} = em_W \left(W_\mu^+ A^\mu G^- + \text{h.c.} \right)
$$

$$
+ g m_Z \left[W_\mu^+ Z^\mu \left\{ G^- \cos^2\theta_W - \frac{g}{\sqrt{2}m_W} \sum_k Y_k \right. \right.
$$

$$
\left. \left. \times \left[\phi_k^\dagger T^+ v_k + (T^- v_k)^\dagger \phi_k \right] \right\} + \text{h.c.} \right]. \tag{6.7}
$$

[†] For real representations, it may appear somewhat inconvenient to be using generators which are purely imaginary and antisymmetric, since T^3 is not diagonal for this choice of basis. However, this will not cause us any problems in practice; since we have restricted our definition of real representations to multiplets with $Y = 0$, it follows from $Qu_i = 0$ that $T^3 u_i = 0$. As a result, the contributions of real representations in our formulae below are rather compact.

As expected, the photon does not couple to W^+ and a physical charged Higgs scalar. On the other hand, the existence of a tree-level H^+W^-Z coupling is model dependent. For example, in a model with only complex $Y = 1$ Higgs doublets, we note that $T^- v_k = 0$. Furthermore, we omit the terms involving η_i in eq. (6.6) in models with no real Higgs multiplets. Then, using eqs. (6.6) and (6.7), we see that W^+Z couples precisely to the charged Goldstone field G^-. Thus, there are no tree-level H^-W^+Z couplings in multi-Higgs doublet models (which contain no other non-singlet Higgs representations), as claimed in §4.1. More generally, the above equations imply that a model *can* contain a non-zero tree-level H^-W^+Z coupling unless one of the following two conditions is satisfied:

1. the model contains no representations with $v_k \neq 0$ and non-zero hypercharge;

2. the model contains only (complex) representations with $Y_k \neq 0$, such that all the Y_k are equal to a common value, and either $T^+ v_k = 0$ for all k or $T^- v_k = 0$ for all k.

Without loss of generality, we may choose $Y_k \geq 0$ for all k, so that $T^+ v_k \neq 0$. Then $T^- v_k = 0$ only if the neutral member of ϕ_k has the minimum value of T^3, which implies that the weak isospin of ϕ_k is $Y_k/2$. Thus, condition 2 above can be replaced by

2′. the model contains only complex representations, with the property that all multiplets have a common value of hypercharge Y and weak isospin equal to $Y/2$.

From eq. (6.7), we can derive an explicit expression for the $H^\pm W^\mp Z$ coupling, by defining $H^\pm$ to be a charged-state orthogonal to $G^\pm$. (In models with more than one charged physical Higgs boson pair, the notation $H^\pm$ below may refer to some linear combination of physical states. Which charged states couple to WZ is a model-dependent question whose answer depends on the charged Higgs boson mass matrix.) By using $T^3 v_k = -(Y/2)v_k$, it is possible to express the final result in terms of the eigenvalues of $T^2 \equiv T^a T^a$ and Y. A simple calculation then gives:

$$\mathcal{L}_{H^\pm W^\mp Z} = -g m_Z \xi [W_\mu^+ Z^\mu H^- + \text{h.c.}], \tag{6.8}$$

with

$$\xi^2 = \frac{g^2}{4m_W^2} \left\{ \sum_{T,Y} Y^2 \left[4T(T+1) - Y^2 \right] |V_{T,Y}|^2 \right\} - \frac{1}{\rho^2}, \tag{6.9}$$

where $\rho \equiv m_W^2/(m_Z^2 \cos^2 \theta_W)$ is given in eq. (4.1), and m_W^2 can be expressed as

$$m_W^2 = \tfrac{1}{4}g^2 \sum_{T,Y} \left[4T(T+1) - Y^2 \right] |V_{T,Y}|^2 c_{T,Y}. \tag{6.10}$$

Note that we have used the same notation as in eq. (4.1); *i.e.*, $V_{T,Y} \equiv v_k$ (u_i) and $c_{T,Y} = 1$ (1/2) for complex (real) representations; the above sums extend over all Higgs representations of weak isospin T and hypercharge Y.

Note that $\xi = 0$ if either of the two conditions mentioned earlier is satisfied, by virtue of eqs. (4.1) and (6.10).

The simplest model with a non-zero tree-level $H^- W^+ Z$ vertex is one with a single complex Higgs triplet with $Y \neq 0$. Of course, such a model would give $\rho \neq 1$, which is not phenomenologically acceptable. To generalize the discussion, consider a class of models with an arbitrary number of $Y = 1$ Higgs doublets and multiple copies of one additional Higgs field with weak isospin T and hypercharge Y. Furthermore, we define

$$v^2 \equiv \sum_{\text{doublets}} v_k^2, \qquad V^2 \equiv \sum_{k \in (T,Y)} v_k^2. \tag{6.11}$$

We can use the expressions for m_W^2 and ρ to eliminate v^2 and V^2 from eq. (6.9); the result is

$$\xi^2 = \left(1 - \frac{1}{\rho}\right) \frac{1}{T(T+1) - \frac{3}{4}Y^2}$$
$$\times \left\{ [T(T+1) - \tfrac{1}{4}Y^2] \left(Y^2 + \frac{1}{\rho}\right) - \tfrac{1}{2}Y^2 \left(1 + \frac{1}{\rho}\right) \right\}, \tag{6.12}$$

where,

$$1 - \frac{1}{\rho} = \frac{g^2 V^2}{m_W^2} \left[T(T+1) - \tfrac{3}{4}Y^2\right]. \tag{6.13}$$

The connection between $\rho \neq 1$ and $\xi \neq 0$ in this simple class of models should be clear. We see that V^2 determines the deviation of ρ from 1. If this deviation is small, then ξ^2 is also small, since it is proportional to $\rho - 1$. Note however that the inverse is not true; namely, $\rho = 1$ does *not* imply that $\xi = 0$. Moreover, using the experimental fact that $\rho \simeq 1$ implies that any model with additional Higgs multiplets beyond (singlets and) doublets will have $\xi \neq 0$. In the class of models considered here, it is sufficient to choose $T(T+1) = \frac{3}{4}Y^2$ in order to guarantee $\rho = 1$; in this case, eqs. (6.12) and (6.13) imply that

$$\xi^2 = \frac{g^2 V^2}{2m_W^2} Y^2 (Y^2 - 1). \tag{6.14}$$

In this case, there is no phenomenological restriction on V^2; although as noted in the discussion following eq. (4.3), the Higgs representations required are quite baroque. For more complicated Higgs models with three or more different representations, there is no relation between ρ and ξ. It is then rather easy to construct a fairly simple model which has $\rho = 1$ and a non-zero $H^{\pm} W^{\mp} Z$ tree-level vertex (*i.e.*, $\xi \neq 0$). For example, consider a model with one $Y = 1$ Higgs doublet, one $Y = 0$ real triplet, and one $Y = 2$ complex triplet. If the vacuum expectation values of the two neutral components of the triplet Higgs fields are equal, then from eq. (4.1) it follows that $\rho = 1$. Nevertheless, it is easy to see from eq. (6.9) that $\xi \neq 0$ in this model. We will explore such a model in more detail in the next section.

6.4   A Model with Higgs Triplets

As outlined at the beginning of chapter 4, models with only Higgs doublets (and, possibly, singlets) provide the most straightforward extensions of the SM that satisfy constraints deriving from $\rho \approx 1$ and the absence of flavor-changing neutral currents. However, there are many more complicated possibilities. We shall see one example of Higgs triplets in our discussion of the conventional left–right symmetric models, in the next section. In that case, we shall find that it is necessary to assign a very small vacuum expectation value to the neutral member of the left-handed triplet in order to avoid unacceptable corrections to the W–Z mass ratio. However, it is certainly not necessary to go to left–right symmetric extensions of the SM in order to consider Higgs triplet fields. Even within the context of the SM a Higgs sector with Higgs triplet as well as doublet fields can be considered. Large corrections to the electroweak ρ parameter can be avoided by two means: (i) the neutral triplet fields can be given vacuum expectation values that are much smaller than those for the neutral doublet fields [see eq. (4.5) for current experimental limits]; or (ii) the triplet fields and the vacuum expectation values of their neutral members can be arranged so that a custodial SU(2) symmetry is maintained. By custodial SU(2) at the tree level we mean simply that the hypercharges Y and vacuum expectation values V of all the Higgs multiplets are chosen so that the net result from eq. (4.1) is $\rho = 1$. More generally, one might hope that a model could be constructed that maintains a custodial SU(2) when loop corrections are included.

Models of the type (i) include the models of ref. 26, considered in the previous section. A number of models of type (ii), with a custodial SU(2) symmetry, have also been proposed in the literature. In particular, we focus on the model constructed by Georgi and collaborators [38–40]. This model was considered in greater depth by Chanowitz and Golden [41], who showed that a Higgs potential for the model could be constructed in such a way that it preserves the tree-level custodial SU(2) symmetry. This has the important implication that the custodial SU(2) is maintained after higher-order loop corrections from Higgs self-interactions. Thus, the model provides an attractive example of an extension of the SM Higgs sector which contains Higgs triplets but no other new physics. We shall examine it with regard to the signatures and production mechanisms for the various Higgs bosons, focusing in particular on the singly- and doubly-charged Higgs bosons. A more detailed exposition of the phenomenology of this model appears in ref. 42.

In the model of ref. 38, the Higgs fields take the form

$$\phi = \begin{pmatrix} \phi^+ \\ \phi^0 \end{pmatrix} \quad \chi = \begin{pmatrix} \chi^0 & \xi^+ & \chi^{++} \\ \chi^- & \xi^0 & \chi^+ \\ \chi^{--} & \xi^- & \chi^{0*} \end{pmatrix}, \tag{6.15}$$

i.e., one $Y = 1$ complex doublet, one real ($Y = 0$) triplet, and one $Y = 2$ complex triplet. Following ref. 38, we shall choose phase conventions for the fields such that $\phi^- = -(\phi^+)^*$, $\chi^{--} = (\chi^{++})^*$, $\chi^- = -(\chi^+)^*$, $\xi^- = -(\xi^+)^*$, and $\xi^0 = (\xi^0)^*$. [Note that we only use this convention in this subsection, since it differs from the convention used elsewhere in this review, where $H^- = (H^+)^*$.] Tree-level custodial SU(2) is arranged by giving the χ^0 and ξ^0 the same vacuum expectation value, $\langle\chi^0\rangle = \langle\xi^0\rangle = b$; we also take $\langle\phi^0\rangle = a/\sqrt{2}$. It is convenient to define

$$v^2 \equiv a^2 + 8b^2, \quad c_H \equiv \frac{a}{\sqrt{a^2 + 8b^2}}, \quad s_H \equiv \sqrt{\frac{8b^2}{a^2 + 8b^2}}, \tag{6.16}$$

where c_H and s_H are the cosine and sine of a doublet–triplet mixing angle. We will also employ the notation

$$\phi^0 \equiv \sqrt{\tfrac{1}{2}}(\phi_R^0 + i\phi_I^0), \qquad \chi^0 \equiv \sqrt{\tfrac{1}{2}}(\chi_R^0 + i\chi_I^0),$$
$$\psi^\pm \equiv \sqrt{\tfrac{1}{2}}(\chi^\pm + \xi^\pm), \qquad \zeta^\pm \equiv \sqrt{\tfrac{1}{2}}(\chi^\pm - \xi^\pm), \tag{6.17}$$

for the complex neutral and charged fields, respectively. The $W^\pm$ and Z are given mass by absorbing the Goldstone bosons

$$G_3^\pm = c_H\phi^\pm + s_H\psi^\pm, \qquad G_3^0 = c_H\phi^{0i} - s_H\chi^{0i}. \tag{6.18}$$

The gauge boson masses so obtained are

$$m_W^2 = m_Z^2\cos^2\theta_W = \tfrac{1}{4}g^2v^2. \tag{6.19}$$

The remaining physical states can be classified according to their transformation properties under the custodial SU(2). One finds a five-plet $H_5^{++,+,0,-,--}$, a three-plet $H_3^{+,0,-}$ and two singlets, H_1^0 and $H_1^{0\prime}$. The compositions of the H states are

$$H_5^{++} = \chi^{++}$$
$$H_5^+ = \zeta^+ \qquad\qquad H_3^+ = c_H\psi^+ - s_H\phi^+ \qquad H_1^0 = \phi_R^0$$
$$H_5^0 = \sqrt{\tfrac{1}{3}}\left(\sqrt{2}\xi^0 - \chi_R^0\right) \qquad H_3^0 = c_H\chi_I^0 + s_H\phi_I^0 \qquad H_1^{0\prime} = \sqrt{\tfrac{1}{3}}\left(\sqrt{2}\chi_R^0 + \xi^0\right).$$

$$\tag{6.20}$$

According to our phase conventions, $H_5^{--} = (H_5^{++})^*$, $H_5^- = -(H_5^+)^*$, and $H_3^- = -(H_3^+)^*$. However, not all these states need be mass eigenstates. Only the doubly-charged $H_5^{++,--}$ and, for appropriately chosen phases, the H_3^0 cannot mix. In general, the remaining neutral Higgs can mix with one another, as can the singly-charged Higgs, depending upon the precise structure of the Higgs potential. The masses and compositions of the mass eigenstates are determined by the quartic interactions among the Higgs fields ϕ and χ. However, as we have already mentioned, it is desirable to choose the Higgs potential in such a way that it preserves the custodial SU(2) symmetry, as done in ref. 41. In this case, the 5-plet and 3-plet states cannot mix with one another or with the singlets; the only possible mixing is between H_1^0 and $H_1^{0\prime}$. This latter mixing depends upon the parameters of the Higgs potential, and can range from zero to maximal. We shall adopt the language of zero mixing. Thus, we shall give results for couplings using the fields defined in eq. (6.20).

From the Higgs boson couplings to fermions and vector bosons we can determine the basic phenomenological features of the Higgs sector of the model. The fermion–antifermion couplings have not been thoroughly studied in this model. Let us discuss the simplest case in which only the doublet field has a Yukawa interaction with the fermions. The only other possible coupling is one closely analogous to that appearing in and providing one of the motivations for left–right symmetric models; namely, a coupling of the triplet Higgs fields (with $Y = 2$) to the lepton–lepton channel. Here, we shall analyze the phenomenological situation assuming that it is absent. Then, all Higgs boson couplings to fermions must arise from the overlap of the mass eigenstate Higgs fields with the doublet field. One finds that the $H_5^{++,--}$, $H_5^{+,-}$, H_5^0, and $H_1^{0\prime}$ states have no such overlap, and that only the $H_3^{+,-}$, H_3^0 and H_1^0 will have fermion–antifermion couplings. (This is obviously very different from the conventional left–right symmetric model in which the triplet fields to first approximation remain unmixed with the doublet fields; in this limit, at tree level they can couple only to lepton–lepton channels, but cannot couple to fermion–antifermion channels.) The Feynman rules for the various couplings are given below (to be multiplied by an overall factor of i)

$$g_{H_1^0 q\bar{q}} = -\frac{gm_q}{2m_W c_H} \quad (q = t, b),$$

$$g_{H_3^0 t\bar{t}} = \frac{igm_t s_H}{2m_W c_H}\gamma_5,$$

$$g_{H_3^0 b\bar{b}} = -\frac{igm_b s_H}{2m_W c_H}\gamma_5,$$

$$g_{H_3^- t\bar{b}} = \frac{gs_H}{2\sqrt{2}m_W c_H}\left[m_t(1+\gamma_5) - m_b(1-\gamma_5)\right],$$

$$(6.21)$$

where third-generation notation is employed for the quarks. Analogous expressions hold for the couplings to leptons. As pointed out in ref. 41 it is possible that $b \gtrsim a$, so that most of the mass of the W and Z comes from the triplet vacuum expectation values. In this case, the doublet vacuum expectation value $a/\sqrt{2}$ is much smaller than in the SM, and the Yukawa couplings of the doublet to the fermions must be much larger than in the SM in order to obtain the experimentally determined quark masses. In this case, the Higgs bosons that do couple to fermions have much larger fermion–antifermion pair couplings and decay widths than in the SM.

Most interesting, however, are the couplings to vector bosons. The Feynman rules for these are specified for the states of eq. (6.20) as follows (we drop an overall factor of $ig_{\mu\nu}$)

$$H_5^{++}W^-W^- : \sqrt{2}gm_W s_H$$
$$H_5^+W^-Z : -gm_W s_H/c_W$$
$$H_5^+W^-\gamma : 0$$
$$H_5^0W^-W^+ : (1/\sqrt{3})gm_W s_H$$
$$H_5^0ZZ : -(2/\sqrt{3})gm_W s_H c_W^{-2}$$

$$H_1^0W^-W^+ : gm_W c_H$$
$$H_1^0ZZ : gm_W c_H c_W^{-2}$$
$$H_1^{0\prime}W^-W^+ : (2\sqrt{2}/\sqrt{3})gm_W s_H$$
$$H_1^{0\prime}ZZ : (2\sqrt{2}/\sqrt{3})gm_W s_H c_W^{-2}$$

$$(6.22)$$

where s_W and c_W are the sine and cosine of the standard electroweak angle, respectively. There are no non-zero couplings of the H_3 Higgs multiplet members to vector bosons. Obviously, the SM is regained in the limit where $s_H \to 0$, in which case the H_1^0 plays the role of the SM Higgs and has SM couplings. However, in this model with custodial SU(2) symmetry, there is no intrinsic need for s_H to be small. Also note the presence of a $H_5^+W^-Z$ coupling, in contrast to the absence of such a coupling of the charged Higgs in any model containing only Higgs doublets (and singlets). The $H_5^+W^-Z$ coupling is easily computed using the general formulae derived in the previous subsection [see eqs. (6.8)–(6.10)]; the result is given in eq. (6.22) above. In summary, we have a remarkable dichotomy between the H_5 and the H_3 multiplets: ignoring for the moment the HV and HH type channels, the former couple and decay only to vector boson pairs, while the latter couple and decay only to fermion–antifermion pairs.

However, to be complete, we must consider the couplings of the Higgs bosons to a vector boson and another Higgs, or to two other Higgs bosons. The latter couplings are model dependent and will not be considered here. The Feynman rules for the former couplings are specified below in the convention where we remove an overall factor of $ig(p-p')^\mu/2$ for the W couplings and a factor of $ig(p-p')^\mu/(2c_W)$ for the Z couplings [with p (p') being the incoming momentum of the Higgs boson listed first (second)]

$$H_1^0H_3^-W^+ : -s_H$$
$$H_1^0H_5^-W^+ : 0$$
$$H_1^{0\prime}H_3^-W^+ : \frac{2\sqrt{2}}{\sqrt{3}}c_H$$
$$H_1^{0\prime}H_5^-W^+ : 0$$
$$H_5^0H_3^-W^+ : \frac{1}{\sqrt{3}}c_H$$
$$H_5^0H_5^-W^+ : -\sqrt{3}$$
$$H_3^0H_5^-W^+ : -ic_H$$
$$H_3^0H_3^-W^+ : -i$$
$$H_5^+H_5^{--}W^+ : -\sqrt{2}$$
$$H_3^+H_5^{--}W^+ : -\sqrt{2}c_H$$

$$H_3^0H_1^0Z : is_H$$
$$H_3^0H_1^{0\prime}Z : -\frac{2\sqrt{2}}{\sqrt{3}}ic_H$$
$$H_3^0H_5^0Z : \frac{2}{\sqrt{3}}ic_H$$
$$H_5^-H_3^+Z : c_H$$
$$H_3^-H_3^+Z : 2s_W^2 - 1$$
$$H_5^-H_5^+Z : 2s_W^2 - 1$$
$$H_5^{--}H_5^{++}Z : 2 - 4s_W^2$$

$$(6.23)$$

Couplings of a pair of charged Higgs to the photon are, of course, diagonal and determined entirely by the magnitude of the charge.

Before turning to phenomenology, it is quite amusing to examine the manner in which high energy unitarity is preserved for longitudinal vector boson scattering processes in this model. As discussed in the introduction of chapter 4, for Higgs sector extensions involving only doublets and singlets good high energy behavior for longitudinal vector boson scattering is guaranteed by the sum rule of eq. (4.6). However, the manner in which unitarity is preserved in Higgs sector extensions containing triplets and higher representations is much more complicated. We give two examples in the context of the model being discussed in this section. Consider $ZW^- \rightarrow ZW^-$. In the SM there is one t-channel graph involving the exchange of the ϕ^0, with effective strength proportional to $g^2 m_Z^2$. In our triplet model the couplings of eq. (6.22) make it clear that we have three t-channel graphs for the neutral Higgs bosons, and an s-channel and a u-channel graph for the singly-charged H_5^-. Asymptotically, the latter s- and u-channel graphs combine together to give the same result as a t-channel graph except for an overall sign difference. Thus, the four contributions have effective strength proportional to

$$
\begin{aligned}
H_1^0 &: g^2 c_H^2 m_Z^2 \\
H_1^{0\prime} &: (8/3)g^2 s_H^2 m_Z^2 \\
H_5^0 &: -(2/3)g^2 s_H^2 m_Z^2 \\
H_5^- &: -g^2 s_H^2 m_Z^2,
\end{aligned}
\tag{6.24}
$$

where the minus sign in the H_5^- case is introduced to account for the sign difference alluded to above. Clearly the sum of all four terms gives back the original $g^2 m_Z^2$ of the SM Higgs t-channel exchange graph. However, a non-zero vertex for $W^- H_5^+ Z$ was crucial. Such a vertex cannot appear in a multi-doublet model, and this is why the unitarity sum rule takes the simple form given in eq. (4.6) in such models. It is also amusing to consider the case of $W^+ W^+ \rightarrow W^+ W^+$ scattering. In the SM there are two ϕ^0-exchange graphs: one is a t-channel and the other a u-channel graph. They can be thought of as combining together and having effective strength $g^2 m_W^2$. In our triplet model we have three t-channel and three u-channel neutral Higgs graphs, and an s-channel graph, the latter involving the H_5^{++}. An s-channel graph is equivalent to the sum of a t- and u-channel graph except for an overall sign. Thus the effective strengths of the various contributions are

$$
\begin{aligned}
H_1^0 &: g^2 c_H^2 m_W^2 \\
H_1^{0\prime} &: (8/3)g^2 s_H^2 m_W^2 \\
H_5^0 &: (1/3)g^2 s_H^2 m_W^2 \\
H_5^{++} &: -2g^2 s_H^2 m_W^2,
\end{aligned}
\tag{6.25}
$$

and again these sum to give the SM result.

Let us now leave aside the theoretical structure of the model and briefly survey the phenomenology of the Higgs bosons in light of their couplings. This survey presumes rough familiarity with the earlier survey for the two-doublet extension of the SM, and with the minimal supersymmetric model discussion. It is meant only to provide a crude guide and is by no means complete.

If any of the neutral and singly-charged Higgs bosons of the model have very low masses, then many of the rare decay processes considered for a very light SM ϕ^0 or very light Higgs bosons from a two-doublet model would again be relevant. However, appropriate adjustments for differences in couplings must be made. For instance, Υ decays would be an excellent place to search for H_3^0 and H_1^0 (which probably have enhanced $b\bar{b}$ couplings, as discussed earlier), but would not be useful in searching for a very light H_5^0 or $H_1^{0\prime}$ (which have no tree-level $f\bar{f}$ couplings). Similarly, $B_d - \overline{B}_d$ mixing would be very sensitive to the H_3^+, but not to the H_5^+. And, of course, low-energy experiments are not sensitive to the H_5^{++}, since it too has no fermion couplings.

Moving to slightly higher Higgs boson masses, we note that eq. (6.23) makes it clear that on-shell Z decays would be a copious source of Higgs pairs of all kinds. Also, from eq. (6.22), we see that $Z \to \ell^+ \ell^- H_i^0$ is possibly useful for $H_i^0 = H_5^0$, H_1^0 and $H_1^{0\prime}$, but not for $H_i^0 = H_3^0$. On the other hand, searches for a neutral Higgs in $\Theta \to \gamma H_i^0$ would only be possible for $H_i^0 = H_3^0$ and H_1^0.

Finally, if the Higgs bosons of the model have larger masses, high energy colliders become appropriate. Consider first the doubly-charged Higgs. For moderate s_H, we find that at hadron colliders the H_5^{++} could be made at near SM strength via $W^+ W^+$ fusion, whereas the Δ^{++} of a conventional left–right symmetric model has suppressed $W^+ W^+$ couplings. (See the next section.) At an $e^+ e^-$ collider, $W^+ W^+$ fusion is not possible and one would turn to $H_5^{++} H_5^{--}$ pair production via virtual Z^* or γ^* exchange. The H_5^{++} decays would, in general, yield a mixture of $W^+ W^+$, $H_3^+ W^+$, and $H_3^+ H_3^+$ states. (Since all the H_5 states are degenerate when custodial SU(2) is assumed for the Higgs potential, decays of one charge of H_5 to another charge state of the same multiplet are forbidden. Similar remarks apply to the H_3 states. However, the relative mass of the H_5 and H_3 states is parameter dependent so that either, but not both, decays of H_5 states to states containing an H_3 or the reverse might be allowed.) Or, if the real channels are not allowed, such channels as $W^+ l^+ \nu$ when one real W is possible, or when the H_5^{++} mass is below m_W, $l^+ \nu l^+ \nu$ become important. Of course, if there were significant direct lepton–lepton couplings for the doubly charged Higgs (a possibility as mentioned earlier), then the $l^+ l^+$ modes would almost certainly be dominant when the real $W^+ W^+$ decay is kinematically forbidden. The influence of the full Higgs sector of this model (in particular, of $W^+ W^+ \to H_5^{++}$ followed by $H_5^{++} \to W^+ W^+$ decay to real W's) upon the $W^+ W^+ \to W^+ W^+$ scattering process has been examined in ref. 43. For moderate values of s_H and H_5^{++} masses below 1 TeV, they find that the presence of the H_5^{++} should be detectable above the $qq \to qq W^+ W^+$ background at the SSC. Finally, we note that, as in the conventional left–right symmetric model, pair production of

$H_5^{++} H_5^{--}$ via Drell-Yan type processes would also occur, followed by both the doubly-charged Higgs decaying to real or virtual like-sign W pairs or other channels. Whether there are substantial backgrounds at a hadron collider to pair production and decay remains an open question.

There could also be substantially new phenomenology for the singly-charged Higgs sector. We see that in this model the singly-charged H_5^+ could be made with substantial rate by $e^+ e^- \to Z^* \to W^- H_5^+$ at an electron-positron collider, and via $ZW^+ \to H_5^+$ fusion at a hadron collider. And (ignoring lepton–lepton couplings) it could decay, at tree level, only into $W^+ Z$, $H_3^0 W^+$, $H_3^+ Z$ and $H_3^+ H_i^0$, where $H_i^0 = H_3^0$, H_1^0 or $H_1^{0\prime}$. Particularly interesting is the $W^+ Z$ channel. As already emphasized, this channel of production and decay is not possible for the singly-charged Higgs bosons found in previously discussed models. In contrast to the H_5^+, the H_3^+ has no tree-level $W^+ Z$ coupling; instead, as we have noted, its fermion couplings could be greatly enhanced relative to those of the two–Higgs-doublet model charged Higgs boson if $b \gtrsim a$. Thus, at an $e^+ e^-$ collider, in addition to the usual Drell-Yan pair-creation process, bremsstrahlung from a heavy top quark could also become an important source of H_3^+ production. Meanwhile, H_3^+ production at a hadron collider via $g\bar{b} \to H_3^+ \bar{t}$ could occur at a much larger rate than in ordinary two–Higgs-doublet models. Of course, if the H_3^+ is light enough to appear in $t \to H_3^+ b$ decays, then its enhanced couplings would imply that such decays would totally dominate over the SM $t \to W^+ b$ decays. In this case, i.e., when the tb channel is closed, the decays of the H_3^+ might not be totally dominated by fermion channels; depending upon the exact size of b compared to a, channels with Higgs pairs or a Higgs boson plus a vector boson could become important, but the ZW^+ channel would not be present.

Turning to the neutral Higgs states, the H_3^0, a pseudoscalar with no VV couplings but (enhanced) fermion couplings, would be produced at an $e^+ e^-$ collider in association with neutral scalar Higgs bosons via the typical mechanism we have discussed in earlier sections: $e^+ e^- \to Z^* \to H_3^0 H_i^0$ with $H_i^0 = H_1^0$, $H_1^{0\prime}$, or H_5^0. At a hadron collider the gg fusion mechanism would probably be enhanced over SM expectations due to enhanced $t\bar{t}$ coupling. The decays of the H_3^0 would be to fermions, Higgs pairs, or HV states. The H_5^0, with no fermion couplings, would be produced using its $H_5^0 VV$ couplings at both $e^+ e^-$ and hadron colliders, and its decays would be to VV, HV or HH type channels, or virtual versions thereof, if no real two-body channel is open. The H_1^0 and $H_1^{0\prime}$ would generally be fairly SM-like in their behavior since they typically have both VV and $f\bar{f}$ couplings. However, as for all the Higgs considered in extensions of the SM Higgs sector, one must be alert to the possibility of important HH and HV decay channels in addition to the SM modes.

Thus, this extension of the Higgs sector, like all that we have previously discussed, leads to a considerable complication in Higgs phenomenology. Clearly, the cleanest signature for a Higgs sector with triplet fields would be the discovery of a doubly-charged Higgs boson. As we have outlined, relatively clean signatures and copious production mechanisms are available for the H_5^{++} that appears in the model discussed here.

6.5   Left–Right Symmetric Models

Models in which the breaking of parity is a result of a spontaneous symmetry breaking have long had appeal. Several different versions of these 'left–right symmetric' extensions of the Standard Model can be constructed. First, one can have left–right symmetric extensions that merely expand the gauge boson and Higgs boson sectors. These will be reviewed in the first subsection below. Second, one can combine the left–right symmetric structure with a supersymmetric or superstring structure. One such model will be briefly discussed in the second subsection. In all left–right symmetric models, a greatly expanded Higgs boson sector is an unavoidable requirement. We will focus on the phenomenological implications and signatures of the additional Higgs bosons.

Conventional Left–Right Symmetric Models

In this section we shall consider conventional left–right symmetric theories, outside the context of either supersymmetry or superstrings. These left–right (LR) symmetric theories [44–46] are an attractive example of models that predict physics which is beyond the standard model and which may be accessible at future colliders. The LR theories are appealing to some physicists for a number of reasons:

  i) they restore parity to the status of an exact short distance symmetry of the weak interactions [44], and thus provide a more aesthetic description of electroweak phenomena as well as a renormalizable framework for describing the origin of parity violation;

  ii) they incorporate full quark-lepton symmetry for the weak interactions and give the U(1) generator of electroweak symmetry a new meaning in terms of the $B-L$ quantum number [47];

  iii) for an appropriately chosen Higgs sector, they suggest a natural explanation of the smallness of neutrino masses [46] by relating this smallness to the observed suppression of $V+A$ currents in low energy weak processes.

The LR theories contain two W bosons, W_1 and W_2. The W_1 is the already-discovered W which couples mostly to left-handed currents. The W_2, which must be heavier than the W_1, couples mainly to right-handed currents. These theories also predict two neutral gauge bosons, Z_1 and Z_2, the lighter of which is the familiar 93 GeV particle. In the fermion sector, they contain the usual quarks and charged leptons, three light neutrino mass eigenstates, ν_i ($i=1,2,3$), and three heavy neutrino mass eigenstates, N_i ($i=1,2,3$). The former couple predominantly to W_1 while the latter couple mainly to W_2.

Of greatest interest to us is the exotic Higgs sector which is required in order to accomplish the LR symmetry breaking. There are two distinct alternatives for the Higgs sector. All models contain a bi-doublet field ϕ; the masses of the standard W_1 and Z_1 derive primarily from the vacuum expectation values, κ and κ', of the two neutral members of this doublet. Since experimental constraints from K_L–K_S mixing force the W_2 to be very heavy [48] ($\gtrsim 1.6$ TeV),[‡] an additional Higgs representation, with large vacuum expectation value (v_R, with $v_R \gg \kappa, \kappa'$) for its neutral member, is required that couples primarily to the W_2. To preserve LR symmetry, there must be a corresponding Higgs representation coupling to the W_1, but the VEV (v_L) of its neutral member must be small [$v_L \ll max(\kappa, \kappa')$] in order to preserve the SM relation between the W_1 and Z_1 masses. If the additional Higgs fields are members of doublets, then the above criteria can be met, but the theory then fails to incorporate a natural explanation of the smallness of neutrino masses. In contrast, if the extra neutral Higgs are members of triplets, then the VEV v_R can induce a large Majorana mass term for the N, in addition to the Dirac mass terms induced by the VEV's of the neutral bi-doublet fields that mix the N and ν. Since the ratio of Majorana to Dirac mass terms is of order $v_R/max(\kappa, \kappa')$, we are naturally led to the standard "see-saw" mechanism yielding a very small Majorana mass for the left-handed neutrinos. Because of the attractiveness of this latter alternative, we will focus on models containing extra triplet Higgs fields, Δ_R and Δ_L. (For a discussion of the case with extra doublet fields, that considers some of the same issues discussed below in the triplet case, see ref. 50.) The resulting Higgs sector has many exotic features. It must include neutral, singly-charged, and doubly-charged Higgs particles. We wish to consider the theoretical structure and constraints upon this Higgs sector, and assess possibilities for detection of one or more of the resulting physical Higgs bosons. Our discussion will be based largely on the analysis of the Higgs sector appearing in refs. 51, 52, and 53.

Structure of the Higgs Sector

The left–right symmetric models are based on the gauge group $SU(2)_L \times SU(2)_R \times U(1)_{B-L}$, with the quarks and leptons assigned to multiplets with quantum numbers as follows [here $Q \equiv \binom{u}{d}$, $L \equiv \binom{\nu}{e^-}$, and the quantum numbers are indicated in the order $(I_L, I_R, B-L)$]:

$$Q_L : \left(\tfrac{1}{2}, 0, \tfrac{1}{3}\right) \qquad Q_R : \left(0, \tfrac{1}{2}, \tfrac{1}{3}\right)$$
$$L_L : \left(\tfrac{1}{2}, 0, -1\right) \qquad L_R : \left(0, \tfrac{1}{2}, -1\right). \qquad (6.26)$$

Note that we have chosen to place the quarks and leptons in LR symmetric multiplets. There are two free gauge couplings in this model: $g_L \equiv g$ and g_R (generally similar in magnitude to g_L) for the $SU(2)$ groups and g' for the

---

[‡] This conclusion is not absolutely inescapable, as discussed in ref. 49, but is certainly strongly preferred.

$U(1)_{B-L}$ group. The electric charge formula for the model is

$$Q = I_{3L} + I_{3R} + \frac{B-L}{2}. \tag{6.27}$$

The minimal Higgs sector that leads to the symmetry breaking pattern

$$SU(2)_L \times SU(2)_R \times U(1) \rightarrow SU(2)_L \times U(1)_Y \rightarrow U(1)_{EM} \tag{6.28}$$

consists of the multiplets

$$\phi \equiv (\tfrac{1}{2}, \tfrac{1}{2}, 0) \tag{6.29}$$

and

$$\Delta_L \equiv (1, 0, 2) \qquad \Delta_R \equiv (0, 1, 2). \tag{6.30}$$

In component form they can be expressed as

$$\phi \equiv \begin{pmatrix} \phi_1^0 & \phi_1^+ \\ \phi_2^- & \phi_2^0 \end{pmatrix},$$

$$\Delta_L \equiv \begin{pmatrix} \Delta_L^+/\sqrt{2} & \Delta_L^{++} \\ \Delta_L^0 & -\Delta_L^+/\sqrt{2} \end{pmatrix}, \tag{6.31}$$

$$\Delta_R \equiv \begin{pmatrix} \Delta_R^+/\sqrt{2} & \Delta_R^{++} \\ \Delta_R^0 & -\Delta_R^+/\sqrt{2} \end{pmatrix}.$$

We have written not only the bi-doublet, but also the triplet fields in a 2×2 representation (*i.e.* $\Delta \equiv \vec{\tau} \cdot \vec{\Delta}$) that is common notation. The I_{3L} increases by 1 unit as we move upward in a given column of the matrix (different columns can have different I_{3L} central values), while I_{3R} increases by one unit as we move to the right in a given row. Only a bi-doublet such as ϕ can preserve left–right symmetry and be used to give masses to the quarks when the latter are placed in a LR symmetric multiplet (the conventional choice). This is accomplished through a term in the Lagrangian of the form

$$\mathcal{L}_Y = \overline{Q}_L^i \left(F_{ij}\phi + G_{ij}\widetilde{\phi} \right) Q_R^j + \text{h.c.} , \tag{6.32}$$

where $\widetilde{\phi} = \tau_2 \phi^* \tau_2$ and F and G are the Yukawa coupling matrices with i, j running over family indices.

The Higgs potential involving these multiplets can be constructed in such a way that the minimum of the potential corresponds to the following vacuum expectation values for ϕ, Δ_L and Δ_R:

$$\langle\phi\rangle = \begin{pmatrix} \kappa & 0 \\ 0 & \kappa' \end{pmatrix},$$

$$\langle\Delta_L\rangle = \begin{pmatrix} 0 & 0 \\ v_L & 0 \end{pmatrix}, \tag{6.33}$$

$$\langle\Delta_R\rangle = \begin{pmatrix} 0 & 0 \\ v_R & 0 \end{pmatrix}.$$

In terms of these vacuum expectation values and the gauge couplings, the masses of the gauge bosons are given by

$$m_{W_1}^2 \simeq \tfrac{1}{2}g^2(\kappa^2 + \kappa'^2 + 2v_L^2),$$

$$m_{W_2}^2 \simeq \tfrac{1}{2}g^2(\kappa^2 + \kappa'^2 + 2v_R^2),$$

$$m_{Z_1}^2 \simeq \frac{g^2}{2\cos^2\theta_W}(\kappa^2 + \kappa'^2 + 4v_L^2), \tag{6.34}$$

$$m_{Z_2}^2 \simeq 2(g^2 + g'^2)v_R^2.$$

Here, θ_W is the conventional Weinberg angle, and is related to the gauge couplings in a manner different than in the SM

$$\tan\theta_W = g'/(g^2 + g'^2)^{1/2} \tag{6.35}.$$

The bosons W_1 and W_2 are mixtures of W_L and W_R, which couple purely to LH and RH currents, respectively. The mixing angle is given by $\tan 2\zeta = 2\kappa\kappa'/v_R^2$.

Constraints and Phenomenology

Naively, it would appear that low energy physics constraints on the relative magnitudes of κ and κ' are weak, since even for the minimal allowed value of v_R the maximal mixing angle ζ obtained for $\kappa = \kappa'$ is consistent with limits on W_1–W_2 mixing [53]. However, a closer look reveals that constraints on the relative size of κ and κ' may emerge for specific Higgs potentials. To see this, let us examine the coupling of the neutral Higgs fields to the quarks and leptons. Starting from eq. (6.32) we can solve for the F and G matrices in terms of the quark masses. The resulting Lagrangian describing the interaction of the down-type quarks with the neutral bi-doublet Higgs fields is

$$\mathcal{L}_D = \frac{1}{\kappa^2 - \kappa'^2}\bar{d}_L^i\left[V_{ij}^{L\dagger}M_{jj}^u V_{jk}^R\left(\kappa\phi_2^0 - \kappa'\phi_1^{0*}\right) + M_{ik}^d\left(\kappa\phi_1^{0*} - \kappa'\phi_2^0\right)\right]d_R^k + \text{h.c.}, \tag{6.36}$$

where $V^{L,R}$ are the left and right CKM matrices, and $M^{u,d}$ are diagonal matrices containing the up- and down-quark masses, respectively. The eigenvector d represents the down-type *mass* eigenstates. As we shall discuss in more detail below, it was shown in ref. 52 that it is not possible in the context of three families to avoid FCNC interactions coming from the term in eq. (6.36) involving the V matrices. Thus, physical eigenstates that have large ϕ_1^0 or ϕ_2^0 components are in danger of violating FCNC constraints unless they have mass in the multi-TeV range. In addition, in ref. 53 it was found that, for the simplest form of the Higgs potential, all neutral Higgs mass eigenstates have significant overlap with either ϕ_1^0 or ϕ_2^0, or both. As a result, if both κ and κ' are non-zero, all neutral Higgs bosons would have to be very massive to avoid FCNC, and none of them could play the usual role of the SM Higgs boson in curing unitarity and related problems in the $W_1 W_1$ sector. In contrast, if, say, κ' is zero, a typical solution to the mass matrix diagonalization is such that ϕ_1^0 is part of different mass eigenstates than ϕ_2^0. Hence, only the latter mass eigenstates need be very heavy, and one or more of those containing the ϕ_1^0 can play the role of the SM Higgs. Thus, for such a model one is forced to take $\kappa' = 0$. While this conclusion need not hold for the most general left–right symmetric Higgs potential, we will simplify our discussion by taking $\kappa' = 0$ in the following.

In order to have near maximal parity violation at low energies, as observed, we must have $\kappa \ll v_R$. Furthermore, the structure of the Higgs potential is such that, if finely tuned relations among potential parameters are to be avoided, the VEV's must be related by $v_L = \gamma \kappa^2 / v_R$, where γ is a typical coupling constant. Thus $v_L \ll \kappa$ and we shall neglect it.* This also guarantees that corrections to the ρ parameter coming from the LR Higgs sector, which are of order $\Delta\rho = (1 + 2x^2)/(1 + 4x^2)$ with $x = v_L/\kappa$, are negligible.

Thus, W_1 receives all its mass from κ, while the W_2 mass is sensitive to v_R, and similarly for Z_1 and Z_2. As we have already noted, the strongest constraint on the vector boson masses arises from an analysis of the K_L–K_S mass difference, which is very sensitive to the W_2 mass. The result [48,54] is that we must have $m_{W_2} \gtrsim 1.6$ TeV, assuming the right-handed CKM matrix is the same as the left-handed CKM matrix. (As noted earlier, this constraint is weakened if we do not demand that the left and right CKM matrices be the same [49]. We do not discuss these less natural possibilities here.) From eq. (6.34) we see that for the simple Higgs sector of eq. (6.31) this means that the Z_2 is even heavier. Whether or not in a more complicated model the Z_2 mass can be disconnected from the W_2 mass will be discussed below.

With this background it is now possible to discuss the expectations for the Higgs bosons in the simplest model. Altogether, we have in eq. (6.31) 20 real degrees of freedom in the Higgs fields. Of these, 6 unphysical Goldstone fields are absorbed in giving masses to the $W_1^\pm$, $W_2^\pm$, Z_1 and Z_2. We are left with 4 doubly-charged, 4 singly-charged, and 6 neutral physical Higgs bosons. They correspond to the quanta of the fields

---

*Discussion of a $v_L \neq 0$ scenario appears in ref. 53.

$$H^+ = \phi_1^+ + \frac{\kappa}{\sqrt{2}v_R}\Delta_R^+,$$

$$h^0 = \sqrt{2}\left[\cos\theta^0 \text{Re}\phi_1^0 + \sin\theta^0 \text{Re}\Delta_R^0\right],$$

$$H^0 = \sqrt{2}\left[-\sin\theta^0 \text{Re}\phi_1^0 + \cos\theta^0 \text{Re}\Delta_R^0\right],$$

(6.37)

$$\Delta_R^{++},\Delta_L^{++},\Delta_L^+,\sqrt{2}\text{Re}\Delta_L^0,\sqrt{2}\text{Im}\Delta_L^0,\sqrt{2}\text{Re}\phi_2^0,\sqrt{2}\text{Im}\phi_2^0,$$

together with the negatively charged versions of the non-neutral Higgs. The mixing angle θ^0 is given by

$$\tan 2\theta^0 = \frac{\alpha_H \kappa v_R}{\rho_H v_R^2 - \lambda_H \kappa^2},$$

(6.38)

where α_H, ρ_H, and λ_H are linear combinations of the scalar boson self-couplings of the Higgs potential. Note, in particular, that the experimental observation of doubly-charged scalars would strongly support the ideas of a left–right symmetric approach.

The masses of the Higgs bosons depend upon both the symmetry breaking scales κ and v_R and on the scalar self-couplings. The latter are rather arbitrary and in general do not allow us to predict the masses of the physical Higgs bosons. However, if all quartic couplings were of order unity, it would be most natural for all but one of the Higgs to have masses of order v_R, while there will be one Higgs, h^0, with low mass (of order κ) and phenomenology similar to the SM ϕ^0. [The only difference would be that h^0 can decay to a pair of heavy neutrinos N through the $\sin\theta^0$ mixing term in eq. (6.37).]

The fact that both the Z_2 and many of the Higgs bosons are forced to be heavy for the simple Higgs sector of eq. (6.31) has led to attempts to decouple these masses from the W_2 mass scale. Certainly, there are very weak phenomenological constraints on the Z_2 mass [55,56]. Indeed, only $m_{Z_2} \gtrsim 275$ GeV is required by current data. In this modified version of LR models, one breaks SU(2)$_L \times$ SU(2)$_R \times$ U(1)$_{B-L}$ in two stages

$$\text{SU(2)}_L \times \text{SU(2)}_R \times \text{U(1)} \rightarrow \text{SU(2)}_L \times \text{U(1)}_{I_{3R}} \times \text{U(1)}_{B-L}$$
$$\rightarrow \text{SU(2)}_L \times \text{U(1)}_Y,$$

(6.39)

where the first breakdown occurs at the W_2 mass scale, while the second breakdown occurs at the Z_2 mass scale. To accomplish this one must introduce additional Higgs multiplets

$$\Sigma_L : \quad (1,0,0) \qquad \Sigma_R : \quad (0,1,0)$$

(6.40)

that are triplets under SU(2)$_L$ and SU(2)$_R$, respectively. If one takes

$$\langle \Sigma_L \rangle = 0 \qquad \langle \Sigma_R \rangle = \begin{pmatrix} \sigma_R & 0 \\ 0 & -\sigma_R \end{pmatrix},$$

(6.41)

then $m_{W_2}^2 = \frac{1}{2}g^2(\kappa^2 + \kappa'^2 + 2v_R^2 + 2\sigma_R^2)$, whereas the mass formulas for the other vector bosons remain unchanged. Thus W_2 can be made heavy by having a large value for σ_R (implying, probably, large masses for the Higgs bosons associated with $\Sigma_{R,L}$), while keeping the Z_2 and the Higgs bosons in eq. (6.37) relatively light, regardless of the scalar self-coupling magnitudes so long as they are not anomalously large.

However, we have already seen that we must be careful that there are no FCNC's coming directly from the Higgs sector. In particular, for $\kappa' = 0$, it was realized that if the CKM matrices obey $V^R = V^L$ in eq. (6.36) then both the real and imaginary components of ϕ_2^0 lead to FCNC effects in the K_L-K_S system that violate experimental constraints unless $m_{\text{Re}\phi_2^0}$ and $m_{\text{Im}\phi_2^0}$ are $\gtrsim 5 - 10$ TeV [57,58]. In a more recent investigation [52] the $V^R = V^L$ assumption was relaxed and a search for a V^R that did not lead to such a strong mass bound on the ϕ_2 Higgs bosons was made. However, it was found that within the context of three generations it is impossible to find any V^R that is both unitary (as required) and such as to adequately suppress FCNC matrix elements in eq. (6.36) so as to allow the $\text{Re}\phi_2^0$ and $\text{Im}\phi_2^0$ masses to be below 1 TeV. It is important to note that the introduction of two bi-doublet fields does not escape this problem, even if one bi-doublet couples only to up quarks and the other only to down quarks (as in the standard FCNC avoiding procedure discussed in non-left–right-symmetric models in chapter 4). If there is a fourth generation, then evasion is possible if the up-down quark associations of the right sector are crossed relative to the left sector: *i.e.*, $[u,t']$ are paired with $[d,s]$ and $[c,t]$ are paired with $[b,b']$ (t' and b' being the fourth generation's up- and down-type quarks). Clearly, this is a rather *ad hoc* possibility. Thus, since the masses of the Higgs bosons containing the ϕ_2^0 are set by the scale v_R multiplied by Higgs potential couplings, we see that (barring unnaturally large couplings) v_R must be as large or larger than σ_R, and there was no point in considering the above discussed complication of the Higgs sector in order to separate the Higgs masses from the W_2 mass.

In any case, the masses of the other physical Higgs bosons of eq. (6.37), and of course of Z_2 (as well as W_2 in the simpler model), are rather closely related to those of $\text{Re}\phi_2^0$ and $\text{Im}\phi_2^0$. Following the notations of refs. 51 and 59, we find in particular that

$$m_{H^\pm} \simeq m_{\text{Re}\phi_2^0, \text{Im}\phi_2^0} \pm \mathcal{O}(\kappa), \tag{6.42}$$

independent of the form of the Higgs self-couplings. This is easily understood from the fact that $H^+ \sim \phi_1^+$ and ϕ_2^0 are not only in the same bi-doublet multiplet, but they are in a singlet relative to the right-handed generators (*i.e.*, they have the same RH quantum numbers). Thus it is only the left-handed generators which can distinguish between the ϕ_1^+ and the ϕ_2^0 fields, and the mass splitting is constrained to be of order κ.

While the other Higgs of eq. (6.37) have masses that are somewhat less closely related to those of the two ϕ_2^0 Higgs, these masses are still proportional to v_R and could be large. In particular, using the constraints on $m_{\phi_2^0}$ due to the absence of FCNCs mentioned above and applying the condition of vacuum stability, one can derive a Linde-Weinberg type bound for m_{H^0}; *e.g.*, for $\theta^0 = 0$, the bound is $m_{H^0} \gtrsim$ 1–2 TeV [60]. The Higgs bosons most easily kept below 1 TeV in mass are those belonging to the left-handed triplet field. For them to have mass significantly below the TeV scale, the Higgs self-couplings that determine the coefficients of v_R in their masses must be in the range $\lesssim 0.2$ (in the conventions of refs. 51 and 53). This is not necessarily unreasonable if grand unification is perturbative. But, more generally it must be admitted that a certain amount of fine tuning of the Higgs self-couplings becomes necessary to keep the left-handed triplet members below the 1 TeV scale. However, if produced their signatures are likely to be unmistakable and relatively background free. This is because the triplet Higgs have $L = 2$, and thus do not couple to quarks. Their direct couplings are of two types: (a) to lepton pairs; and (b) to vector boson pairs. The couplings to vector boson pairs are suppressed relative to SM type expectations by the ratio of v_L/κ, which must be small and may be zero. The couplings to lepton pairs are not directly related to the lepton masses (which are provided by the bi-doublet Higgs VEV), but they are constrained by a variety of experimental limits from Bhabha scattering, $(g - 2)_\mu$, and $\mu^+ \rightarrow e^+e^-e^+$, etc. (see ref. 53 for a survey and further references). The most striking signature for these triplet Higgs would arise from pair production via Drell-Yan processes of $\Delta^{++}\Delta^{--}$. Since the doubly-charged triplet has lepton number $L = 2$, each would decay to a pair of like-sign leptons (or possibly like-sign W pairs, which could then decay to like-sign leptons). For Δ^{++} masses below a few TeV, the Drell-Yan cross sections are significant at the SSC, and at a e^+e^- collider of sufficient energy [61]. There are no backgrounds in the case where both the Δ^{++} and Δ^{--} decay directly to leptons, but the case where both decay to W pairs would require further study. These possibilities and signatures are explored in detail in ref. 53.

One should ask if a large mass for most of the Higgs bosons of the LR model creates problems for unitarity, similar to those that are encountered if the SM Higgs boson mass is taken too large. The problems with unitarity encountered in the WW scattering channels when $s \gg m_H^2 \gg m_W^2$ were explored in ref. 59. In the present case there are several different Higgs and W mass scales. One finds that the couplings of the different W's to the various Higgs bosons are such that a problem is never encountered. For instance, in W_1W_1 (or Z_1Z_1) scattering the h^0 (which is allowed to be light) is the only strongly coupled Higgs and constraints are analogous to those found in the SM situation. The coupling of the other neutral Higgs of the LR model to W_1W_1 and Z_1Z_1 are suppressed by the smallness of $\sin\theta^0$. In Z_1Z_2, Z_2Z_2, W_1W_2 and W_2W_2 scattering, the large W_2 masses implied by our FCNC limits imply that any constraints on the heavier neutral Higgs masses are moved to quite high masses and there is no contradiction with the FCNC limits on these masses.

Of course, in the absence of a supersymmetric extension of the LR model, there will be the conventional problems of hierarchy and naturalness associated with quadratic divergences in the one-loop corrections to the Higgs boson masses, including that of the light h^0. No detailed exploration of this issue has appeared in the literature.

The only way that one can avoid the severe FCNC constraints on the Higgs masses is to not use the bi-doublet structure to generate quark masses. However, we have already remarked that this is the only type of Higgs field that can be used to give quark masses *if* we insist that the quarks be placed in an LR symmetric multiplet. Placing the quarks in non-left–right-symmetric multiplets drastically changes the structure of both the gauge boson and the Higgs boson sectors. We shall consider an example of such a model in the context of superstring motivated theories. Such a model can be made free of FCNC's without requiring either large W_2 or large Higgs masses. However, we conclude that, in the conventional approach to left–right symmetric gauge groups, some of the Higgs that are uniquely characteristic of the left–right extension and symmetry breaking are likely to be too heavy to be observable at the next generation of machines. Generally, only the left-handed triplet members (other than the Higgs boson which plays the role of the SM Higgs) are allowed to be light, and for this to happen some of the Higgs potential parameters must be moderately small. However, the spectacular like-sign dilepton decays of the doubly-charged members of this triplet are relatively background free and clearly worth searching for.

A Superstring-Motivated Left–Right Symmetric Model

As we have already discussed, the $E_8 \times E_8$ heterotic string has low energy manifestations in which the matter fields, including fermions and Higgs bosons, all belong to **27** dimensional representations of E_6, and the low energy gauge group is a subgroup G of E_6. We have already considered in §5.3 the case where the electroweak portion of G is $SU(2)_L \times U(1)_Y \times U(1)_{Y'}$. Here we consider the case in which

$$G = SU(2)_L \times SU(2)_R \times U(1)_Y. \qquad (6.43)$$

The allowed tri-linear superfield Yukawa couplings are such that it is possible to define a possibly conserved R-parity (where as usual $R = (-1)^{2j+3B+L}$). Given this low energy group, there are still two distinct choices for how the gauge bosons of the theory are assigned R-parity. For one choice we find problems with FCNC interactions analogous to those in the conventional model just discussed, and, in particular, the W_2 will be forced to be very heavy as will many (and perhaps all but one) of the Higgs bosons. This occurs when, in first generation notation, (u, d) is a doublet under $SU(2)_R$ and thus after Cabibbo mixing W_2 can induce flavor-changing neutral currents via the usual box diagram. The second choice was considered in ref. 62. In this case the right-handed doublet partner of the u quark is the new exotic fermion discussed earlier called D. Since D has opposite R-parity from d and s it does not mix with them (unless R-parity is broken, which we shall explicitly avoid), and thus W_2 does not couple to d and s quarks and constraints on its mass from K_L–K_S mixing do not arise. This can be summarized by saying that W_2 now has opposite R-parity from the W_1 and does not yield flavor changing neutral currents involving the SM quarks and leptons, which have even R-parity. Thus, it and the Z_2 can be light. One possible drawback of the model is the fact that a small Dirac mass for the neutrino becomes rather ad hoc.

The Higgs Sector

Let us now turn to the Higgs bosons of the model. As for other superstring-motivated models, there are actually three families of Higgs in general, and as usual we imagine that we have rotated to a basis in which we need only consider the 'family' containing the neutral Higgs that acquire vacuum expectation values. For this family we shall use first family notation. Since the Higgs bosons must all belong to a **27** of E_6, it is easily seen that there are no triplet fields under either $SU(2)_L$ or $SU(2)_R$. Instead, among the $B = 0$ fields we have a bi-doublet, two Higgs doublets, and one singlet. Under the group G of eq. (6.43) the doublets transform as

$$H_L \equiv (\tfrac{1}{2}, 0, -1) \qquad H_R \equiv (0, \tfrac{1}{2}, 1), \tag{6.44}$$

where the last entry indicates the Y eigenvalue, and as in the previous subsection we have indicated a doublet field by $I_{L,R} = 1/2$. It should be noted that Y is the usual hypercharge and is not directly related to $B-L$ in this model. The superpartners of the doublet fields are indicated by

$$H_L \equiv \begin{pmatrix} H_L^0 \\ H_L^- \end{pmatrix} = \begin{pmatrix} \tilde{\nu}_E \\ \tilde{E} \end{pmatrix} \qquad H_R \equiv (H_R^0 \;\; H_R^+) = (\tilde{n} \;\; \tilde{e}^c), \tag{6.45}$$

while those of the bi-doublet and singlet are specified by

$$\phi \equiv \begin{pmatrix} \phi_1^0 & \phi_1^+ \\ \phi_2^- & \phi_2^0 \end{pmatrix} = \begin{pmatrix} \tilde{\nu}_e & \tilde{E}^c \\ \tilde{e} & \tilde{N}_E^c \end{pmatrix} \qquad N = \tilde{N}_e^c. \tag{6.46}$$

As in the previous section I_{3L} increases in going up, while I_{3R} increases in moving to the right in the above equations. Note in the superpartner correspondence we employ a notation in which all fermion fields are taken to have left-handed helicity. The R-parity assignments of the scalar fields are always opposite to those of their superpartners, and we must recall that all SM quarks and leptons have $R = +1$, while all new fermions have $R = -1$, with the possible exception of the singlet N_e^c which can be assigned either $R = +1$ or $R = -1$.

The neutral Higgs fields are thus H_L^0, H_R^0, ϕ_1^0, ϕ_2^0, and N, with R-parity values of $+1$, $+1$, -1, $+1$, and ± 1, respectively. The vacuum expectation values for these fields will be indicated in a notation that parallels that for the conventional left–right model: v_L, v_R, κ, κ', and n, respectively. To avoid breaking R-parity and, thus, destroying our ability to avoid FCNC constraints on the W_2, we clearly must avoid giving ϕ_1^0 a vacuum expectation value; *i.e.*,

$\kappa = 0$. Regarding the singlet N field, we first note that it is an E_6 singlet, and thus need not play the same role as singlet fields considered in our previous superstring-motivated models. Indeed, if it is assigned $R = -1$ it cannot be given a vacuum expectation value. If it is assigned $R = +1$, then the corresponding superfield cannot have Yukawa interactions with any of the other superfields of the model. Thus, a non-zero value for n would not play any special role, and it is simplest to set $n = 0$. We will not consider the N further.

Constraints and Phenomenology

So far, we have not discussed whether the generation of masses for the quarks and leptons leads to any FCNC mediated by one of the neutral Higgs bosons. To discuss this issue we must first note that the allowed superfield Yukawa's are such that the u quark mass is generated by κ', the d and e quark masses are generated by v_L, while the exotic D quark mass and the E mass are generated by v_R. Thus, the mass generating Lagrangian does not take the form of eq. (6.36), required when the only doublet Higgs fields appeared in the bi-doublet. Rather, the mass generation mechanisms for up and down quarks are separated from one another, and direct FCNC need not arise.

To see this in detail, consider first the Yukawa coupling responsible for the d quark masses:

$$\bar{d}_L^i (H_L^0)^k d_R^j \lambda_{ijk}, \qquad (6.47)$$

where i, j, and k are family indices.[†] Recall that we rotate to a basis where only one family of Higgs acquires non-zero VEV's, corresponding say to $k = 3$. We see immediately that diagonalization of the mass matrix in i,j-space automatically diagonalizes the down-type quark interactions, just as in the Standard Model. In the case of the bi-doublet generation of the up-quark masses, a coupling of the form

$$\left[\bar{u}_L^i (\phi_2^0)^k u_R^j + \bar{d}_L^i (\phi_1^0)^k D_R^j \right] \lambda'_{ijk} \qquad (6.48)$$

arises, where we have indicated only couplings to neutral Higgs fields. After choosing $k = 3$ corresponding to our non-zero VEV family, the u-quark interactions with the ϕ_2^0 are again diagonalized at the same time that the masses are (recall that ϕ_1^0 has no vacuum expectation value). Nonetheless, the ϕ_1^0 interaction term in eq. (6.48) might lead to FCNC problems. This is because the exotic D quark masses arise from yet a third Yukawa involving H_R^0,

$$\bar{D}_L^i (H_R^0)^k D_R^j \lambda''_{ijk}, \qquad (6.49)$$

and the diagonalizing matrices there, call them $V_D^{L,R}$, need not be the same as those for the u quark sector, denoted $V_u^{L,R}$. The second term in eq. (6.48)

[†] We use a lower case d^i vector for normal down-type quarks as distinguished from the upper case vector D^i for the exotic D quarks.

would lead to non-family-diagonal interactions of the form $\phi_1^0 \bar{d}_L V_u^{L\dagger} V_D^R D_R$ when V_u^R and V_D^R are not the same. Since the d and D quarks are not directly mixed via a non-zero VEV for the ϕ_1^0 this interaction is clearly not as dangerous as a direct family mixing between the d quarks themselves (as occurs in the conventional left–right model), but can lead to family mixing at the one-loop level. On the other hand, there is nothing to prevent $\lambda'_{ij1} \propto \lambda''_{ij1}$, so that $V_u^{L\dagger} V_D^R \propto I$. In addition, the D quarks might be very heavy so that the one-loop flavor-changing neutral currents might be very small in any case. Thus there appears to be no requirement that the mass of ϕ_1^0 be very large. Had there been, its large mass would have forced v_R to be large (assuming that Higgs potential parameters are not taken unnaturally large), and as result the W_2, Z_2 and many other Higgs masses would have had to be large, as in the conventional left–right symmetric model.

Unfortunately, this does not guarantee that there are not other FCNC processes. For instance, the same Yukawa coupling that leads to the Higgs interactions of eq. (6.48) also leads to interactions of the form

$$\bar{d}_L^i \nu^k \widetilde{D}_R^j \lambda'_{ijk}. \tag{6.50}$$

Such a term (with $k = 3$) clearly leads to $b \to s \nu_\tau \bar{\nu}_\tau$ transitions (where $k = 3$ determines the τ subscript on the ν's), via $\widetilde{D}_R$ exchange. The analogous $s \to d \nu_\tau \bar{\nu}_\tau$ process is also possible. Since the rates for these decays obviously depend on the $\widetilde{D}$ mass, which could be large, there is no current conflict with experiment. More generally, a completely satisfactory avoidance of FCNC problems requires the imposition of a number of discrete symmetries upon the theory that eliminate several potentially dangerous generation-mixing superfield Yukawas [63].

Since FCNC problems can be avoided, we must now turn to the constraints imposed upon the model by virtue of the flavor-diagonal neutral current interactions. These impose various limits on the possible values of m_{Z_2}, m_{W_2}, and the neutral Higgs field vacuum expectation values. For our purposes we need only note that for m_{W_2}/m_{W_1} to be as small as possible (~ 2.2) it is necessary to have $v_L/\kappa' \ll 1$. But, if $m_{W_2}/m_{W_1} \gtrsim 4$ then v_L can even be larger than κ'. Clearly, this situation is very different from that in the conventional left–right symmetric model.

Returning to the Higgs sector, we first note that even the most general potential for the spin-0 fields in this model is fairly simple. Because of the R-parity assignments only three parameters are required to fully specify the potential once the three non-zero VEV's are chosen (with m_{W_L} fixed), and the usual minimization conditions imposed. For details see ref. 62.

In giving mass to W_1^-, a combination of H_L^- and ϕ_1^- is absorbed leaving an orthogonal combination denoted by H_1^- with mass-squared proportional to $(\kappa'^2 + v_L^2)$, and, after giving mass to W_2^-, a combination denoted H_2^- of H_R^- and ϕ_2^- remains with mass-squared proportional to $(\kappa'^2 + v_R^2)$. Note that H_1^- is made of $\widetilde{E}$ fields, while H_2^- is made of $\widetilde{e}$ fields. Consider next the imaginary components of the four neutral bosonic fields (other than N). Two are eaten up by the Z_1 and Z_2 gauge bosons, leaving two physical spin-0 Higgs with opposite R-parity: $\sqrt{2}\mathrm{Im}\phi_1^0$ is a pure $R = -1$ mass eigenstate, while the other mass eigenstate is a mixture of the $R = +1$ states, $\sqrt{2}\mathrm{Im}\phi_2^0$, $\sqrt{2}\mathrm{Im}H_L^0$, and $\sqrt{2}\mathrm{Im}H_R^0$. Masses for both eigenstates are determined by a combination of v_R, v_L, and κ', and Higgs potential parameters. Thus, we have two pseudoscalar Higgs bosons. The real components of the four neutral scalar fields are all physical, with $\sqrt{2}\mathrm{Re}\phi_1^0$ being a pure $R = -1$ mass eigenstate degenerate with $\sqrt{2}\mathrm{Im}\phi_1^0$. (It will be convenient on occasion to use a short-hand notation in which we combine the degenerate real and imaginary component eigenstates into a single complex field ϕ_1^0.) The remaining three real components mix via a 3×3 mass matrix. Because of the restrictive form of the scalar potential, it is found that the lightest scalar must have mass less than $\sqrt{2}m_{W_1}$, even before any renormalization group analysis is performed. The bound is saturated only in the limit where $v_L = 0$. In the more likely case that the values of v_R, v_L and κ' are comparable, the bound is substantially lower.

The decays of the Higgs bosons are primarily determined by the allowed Yukawa couplings in the model. (Higgs pair decay modes are sensitive to the spin-0 field potential.) Of the two charged Higgs, H_1^-, composed of $\widetilde{E}$ fields, has $R = +1, L = 0, B = 0$ and decays much like the charged Higgs of a more conventional two-doublet model in a supersymmetric theory. Important modes will include the heaviest $d\bar{u}$ combination kinematically allowed, the $E\bar{n}$ final state, if kinematically allowed, and when allowed $\widetilde{\chi}^- \widetilde{\chi}^0$ final states with positive R-parity. The H_2^-, composed of $\widetilde{e}$ fields, has $R = -1, L = 1, B = 0$ and will primarily decay to the heaviest $D\bar{u}$, $E\nu$ or $e\nu_E$ pair allowed, or to $\widetilde{\chi}^- \widetilde{\chi}^0$ pairs with $R = -1$.

The decays of the various neutral Higgs bosons also separate according to their R-parity. Because R is conserved, the degenerate $R = -1$ scalar and pseudoscalar cannot decay to W_1W_1 (or W_2W_2) but if heavy enough could decay to W_1W_2. Their fermionic pair decay channels are determined by the allowed Yukawa interactions, and lead to a variety of modes, each containing an $R = +1$ SM quark or lepton, and an $R = -1$ exotic fermion. Examples are $\phi_1^0 \to \bar{d}D$, $\overline{E}e$. (Note that even though $\phi_1^0 \equiv \widetilde{\nu}$, these decays conserve lepton number since the D and e have lepton number $L = +1$, while the d and E have $L = 0$.) The exotic particles would in turn decay; detailed signatures have not been examined. The ϕ_1^0 Higgs could also decay to spin-0 particle pairs. But, the superpartners involved are all expected to be heavy and it is likely that such modes will be kinematically forbidden. The decays of the $R = +1$ Higgs are likely to be fairly conventional in the sense that they will be much like those of any two-Higgs-doublet model in the presence of supersymmetry. The general situation has been discussed at length in preceding sections. Of course, modes containing the D quark should be included, if kinematically allowed. A detailed examination of relative branching ratios has not, however, been performed.

Regarding production of the various Higgs bosons, the charged Higgs are best produced at an e^+e^- collider in pairs via virtual photon, Z_1 or Z_2 exchange. Of course, if they are light enough and one of the Z's is produced on-shell, they can appear in the Z decays. At a hadron collider the phenomenology of the $H_1^{\pm}$ is quite similar to that of previously discussed charged Higgs bosons. It will appear in t quark decays if $m_{H_1^{\pm}} < m_t - m_b$, and otherwise is best produced via $gb \to H_1^{\pm}t$. Mechanisms for $H_2^{\pm}$ production at a hadron collider include $D \to H_2^- u$ type decays of pair produced $D\overline{D}$ quarks, and $gu \to DH_2^+$ which could have a large cross section for large D masses that are still well below machine threshold. Production mechanisms for the $R = +1$ neutral Higgs bosons are very much the same as those for the neutral Higgs of the two-doublet models previously discussed. Assuming that the W_2 is significantly heavier than the W_1, the $W_2W_1 \to \sqrt{2}\mathrm{Re}\phi_1^0$ cross section is small, and the $R = -1$ ϕ_1^0 fields are probably best produced at an e^+e^- collider via $Z, Z^* \to \sqrt{2}\mathrm{Re}\phi_1^0 \sqrt{2}\mathrm{Im}\phi_1^0$ ($Z = Z_1, Z_2$) and at a hadron collider via $D \to d\phi_1^0$ decays (when kinematically allowed) or $gd \to D\phi_1^0$.

REFERENCES

1. W. Grimus, *Fortschr. Phys.* **36** (1988) 201.
2. T.D. Lee, *Phys. Rev.* **D8** (1973) 1226; *Phys. Rep.* **9** (1974) 143.
3. S. Weinberg, *Phys. Rev. Lett.* **37** (1976) 657.
4. G.C. Branco, *Phys. Rev.* **D22** (1980) 2901.
5. G.C. Branco, A.J. Buras, and J.M. Gerard, *Nucl. Phys.* **B259** (1985) 306.
6. P. Sikivie, *Phys. Lett.* **65B** (1976) 141.
7. A.B. Lahanas and C.E. Vayonakis, *Phys. Rev.* **D19** (1979) 2158.
8. J. Liu and L. Wolfenstein, *Nucl. Phys.* **B289** (1987) 1.
9. C.H. Albright, J. Smith, and S.-H.H. Tye, *Phys. Rev.* **D21** (1980) 711.
10. A.I. Sanda, *Phys. Rev.* **D23** (1981) 2647.
11. N. Deshpande, *Phys. Rev.* **D23** (1981) 2654.
12. Y. Dupont and T.N. Pham, *Phys. Rev.* **D28** (1983) 2169.
13. J.F. Donoghue and B.R. Holstein, *Phys. Rev.* **D32** (1985) 1152.
14. J.F. Donoghue and E. Golowich, *Phys. Rev.* **D38** (1988) 2542.
15. We would like to thank J. Donoghue for useful discussions of some of the issues in this section.
16. R. Peccei and H. Quinn, *Phys. Rev.* **D16** (1977) 1791; S. Weinberg, *Phys. Rev. Lett.* **40** (1978) 223; F. Wilczek, *Phys. Rev. Lett.* **40** (1978) 279.
17. J. E. Kim *Phys. Rep.* **149** (1987) 1.
18. M. Dine, W. Fischler and M. Srednicki, *Phys. Lett.* **104B** (1981) 199.
19. D.B. Kaplan, *Nucl. Phys.* **B260** (1985) 215; M. Srednicki, *Nucl. Phys.* **B260** (1985) 689.
20. H. Georgi, D. B. Kaplan and L. Randall, *Phys. Lett.* **169B** (1986) 73.
21. H.-Y. Cheng, *Phys. Rep.* **158** (1988) 1.

22. Y. Chikashige, R.N. Mohapatra and R. Peccei, *Phys. Lett.* **98B** (1981) 265.
23. G.B. Gelmini and M. Roncadelli, *Phys. Lett.* **99B** (1981) 411.
24. H. Georgi, S.L. Glashow and S. Nussinov, *Nucl. Phys.* **B193** (1981) 297.
25. D.S.P. Dearborn, D.N. Schramm and G. Steigman, *Phys. Rev. Lett.* **56** (1986) 26; J. Ellis and K. Olive, *Phys. Lett.* **193B** (1987) 525.
26. F. Buccella, G.B. Gelmini, A. Masiero and M. Roncadelli, *Nucl. Phys.* **B231** (1984) 493; M.C. Gonzalez-Garcia and Y. Nir, *Phys. Lett.* **232B** (1989) 383.
27. G.S. Abrams *et al.* (MARK II Collaboration), *Phys. Rev. Lett.* **63** (1989) 2173; D. Decamp *et al.* (ALEPH Collaboration), *Phys. Lett.* **231B** (1989) 519; P. Aarnio *et al.* (DELPHI Collaboration), *Phys. Lett.* **231B** (1989) 539; B. Adeva *et al.* (L3 Collaboration), *Phys. Lett.* **231B** (1989) 509; M.Z. Akrawy *et al.* (OPAL Collaboration), *Phys. Lett.* **231B** (1989) 530.
28. S. Bertolini and A. Santamaria, *Nucl. Phys.* **B310** (1988) 714; *Phys. Lett.* **213B** (1988) 487.
29. C. S. Aulakh and R. N. Mohapatra, *Phys. Lett.* **119B** (1982) 136; **121B** (1983) 147.
30. L. Hall and H. Suzuki, *Nucl. Phys.* **B231** (1984) 419; I.-H. Lee, *Phys. Lett.* **138B** (1984) 121, and *Nucl. Phys.* **B246** (1984) 120; J. Ellis, G. Gelmini, C. Jarlskog, G.G. Ross, and J.W.F. Valle, *Phys. Lett.* **150B** (1985) 142; G.G. Ross and J.W.F. Valle, *Phys. Lett.* **151B** (1985) 375; S. Dawson, *Nucl. Phys.* **B261** (1985) 297.
31. J. F. Nieves, *Phys. Lett.* **137B** (1984) 67; A. Santamaria and J.W.F. Valle, *Phys. Lett.* **195B** (1987) 423; *Phys. Rev. Lett.* **60** (1988) 397, *Phys. Rev.* **D39** (1989) 1780.
32. R. E. Shrock and M. Suzuki, *Phys. Lett.* **110B** (1982) 250; L.F. Li, Y. Liu, and L. Wolfenstein, *Phys. Lett.* **158B** (1985) 45.
33. D. Chang and W.-Y. Keung, *Phys. Lett.* **217B** (1989) 238; S. Bertolini and A. Santamaria, *Phys. Lett.* **220B** (1989) 597.
34. J.W.F. Valle, *Phys. Lett.* **131B** (1983) 87; M.C. Gonzalez-Garcia and J.W.F. Valle, *Phys. Lett.* **216B** (1989) 360.
35. S. Bertolini and A. Santamaria *Nucl. Phys.* **B315** (1989) 558.
36. J.A. Grifols and A. Mendez, *Phys. Rev.* **D22** (1980) 1725.
37. A.A. Iogansen, N.G. Ural'tsev, and V.A. Khoze, *Sov. J. Nucl. Phys.* **36** (1982) 717.
38. H. Georgi and M. Machacek, *Nucl. Phys.* **B262** (1985) 463.
39. R.S. Chivukula and H. Georgi, *Phys. Lett.* **182B** (1986) 181.
40. P. Galison, *Nucl. Phys.* **B232** (1984) 26.
41. M. S. Chanowitz and M. Golden, *Phys. Lett.* **165B** (1985) 105.
42. J.F. Gunion, R. Vega, and J. Wudka, *Phys. Rev.* **D42** (1990) 1673.
43. D. Dicus and R. Vega, *Nucl. Phys.* **B324** (1990) 533.
44. J.C. Pati and A. Salam, *Phys. Rev.* **D10** (1974) 275; R.N. Mohapatra and J.C. Pati, *Phys. Rev.* **D11** (1974) 566, 2558; G. Senjanovic and R.N. Mohapatra, *Phys. Rev.* **D12** (1974) 1502.

45. R.N. Mohapatra, in *Quarks, Leptons, and Beyond*, edited by H. Fritzsch *et al.* (Plenum Press, New York, 1985), p. 217.

46. R.N. Mohapatra and G. Senjanovic, *Phys. Rev. Lett.* **44** (1980) 912; *Phys. Rev.* **D23** (1981) 165.

47. R.N. Mohapatra and R.E. Marshak, *Phys. Rev. Lett.* **44** (1980) 1316.

48. G. Beall, M. Bander, and A. Soni, *Phys. Rev. Lett.* **48** (1982) 848.

49. P. Langacker and S. Uma Sankar, *Phys. Rev.* **D40** (1989) 1569.

50. D. Cocolicchio and G.L. Fogli, *Phys. Rev.* **D32** (1985) 3020.

51. J.F. Gunion, *et al.*, *Proceedings of the 1986 Snowmass Summer Study on the Physics of the Superconducting Super Collider*, edited by R. Donaldson and J. Marx, p. 197.

52. J.F. Gunion, A. Mendez, and F. Olness, in *Proceedings of the 1987 Madison Workshop, From Colliders to Supercolliders*, edited by V. Barger and F. Halzen (World Scientific, Singapore, 1987), p. 195.

53. J.F. Gunion, J. Grifols, A. Mendez, B. Kayser and F. Olness, *Phys. Rev.* **D40** (1989) 1546.

54. R.N. Mohapatra, G. Senjanovic and M.D. Tran, *Phys. Rev.* **D28** (1982) 546; G. Ecker, W. Grimus and H. Neufeld, *Nucl. Phys.* **B229** (1983) 421; F. Gilman and M.H. Reno, *Phys. Rev.* **D29** (1984) 937; H. Harari and M. Leurer, *Nucl. Phys.* **B233** (1984) 221.

55. U. Amaldi, A. Bohm, L.S. Durkin, P. Langacker, A. K. Mann, W.J. Marciano, *Phys. Rev.* **D36** (1987) 1385; G. Costa, J. Ellis, G.L. Fogli, D.V. Nanopoulos and F. Zwirner, *Nucl. Phys.* **B297** (1988) 244.

56. L.W. Durkin and P. Langacker, *Phys. Lett.* **166B** (1986) 436; V. Barger, N.G. Deshpande, and K. Whisnant, *Phys. Rev. Lett.* **56** (1986) 30.

57. R.N. Mohapatra, G. Senjanovic, and M.D. Tran, *Phys. Rev.* **D28** (1983) 546; F.J. Gilman and M.H. Reno, *Phys. Rev.* **D29** (1984) 937 and *Phys. Lett.* **127B** (1983) 426.

58. H. Harari, private communication.

59. F.I. Olness and M.E. Ebel, *Phys. Rev.* **D32** (1985) 1769.

60. J. Basecq and D. Wyler, *Phys. Rev.* **D39** (1989) 870.

61. J.A. Grifols, A. Mendez, and G.A. Schuler, *Mod. Phys. Lett.* **A4** (1989) 1485.

62. K.S. Babu, X.-G. H. He, and E. Ma, *Phys. Rev.* **D36** (1987) 878.

63. D. Ng and E. Ma, *Phys. Rev.* **D39** (1989) 1986.

Chapter 7

Alternatives to the Weakly-Coupled Higgs

In this book, we have focused our attention primarily on the exploration of the origin of electroweak symmetry breaking via the search for fundamental Higgs bosons. An alternative view puts the origin of the symmetry breaking in a different sector of the theory, one with new fundamental fermions that have gauge theory interactions. In this latter approach, elementary scalar bosons are completely absent. However, pseudoscalar (and scalar) bosons which are composites of the new fundamental fermions can arise (in analogy with the way pions arise in the QCD theory of quarks), and may provide an experimental signature of the new physics.

Although the composite Higgs ideas are somewhat out of fashion, they are good ideas, and in fact they suffer no particular disadvantage relative to the fundamental Higgs approaches. In low-energy supersymmetric theories, the origin of electroweak symmetry breaking is tied to the breaking of supersymmetry; however the origin of the supersymmetry-breaking is a mystery. In a technicolor theory, new gauge interactions become strong at an energy scale near 1 TeV, thereby triggering the breaking of the electroweak symmetry. Many aspects of this theory can be determined in analogy with our understanding of QCD. The mystery for technicolor is how fermions get mass.

The way in which the $SU(2)_L \times U(1)_Y$ symmetry is broken in these composite Higgs theories has been extensively studied. Gauge bosons are given mass in a rather attractive way, while fermions get mass by a brute force approach that is unconvincing and usually difficult to reconcile with experimental constraints on flavor-changing neutral currents. We will not review the theory of these approaches; instead we concentrate here on the experimental signatures and how they differ from one another and from theories with fundamental Higgs bosons. At the risk of repeating ourselves, we emphasize that whatever the answer to the origin of spontaneous symmetry breaking, a full knowledge of the experimental spectrum of spin-zero bosons will tell us what kinds of approaches are fruitful to consider. Our goal in this chapter is only to provide interested readers with a guide and introduction to this approach.

7.1   Technicolor

The technicolor approach [1–6] postulates that a new set of gauge interactions (called technicolor) acts between new, massless technifermions. The technifermions carry the ordinary quantum numbers of $SU(3)_C \times SU(2)_L \times U(1)_Y$, plus their own technicolor charges. The technicolor interaction is asymptotically free, and at an energy of order one TeV, it becomes strong. At this energy scale, one assumes that condensates of technifermions form, giving rise to a set of Goldstone bosons called technipions, in much the same way that the light pseudoscalar mesons arise in QCD. The technipions interact with the $SU(3)_C \times SU(2)_L \times U(1)_Y$ gauge bosons through technifermion loops. By the Higgs mechanism, three of the technipions become the longitudinal components of the $W^{\pm}$ and Z, and these gauge bosons acquire mass. The technicolor gauge group and technifermion representations are chosen so that the theory possesses a "custodial SU(2) symmetry" which ensures that $\rho = m_W^2 / (m_Z^2 \cos^2 \theta_W)$ remains unity [7].

This rather elegant procedure leaves the quarks and leptons massless, and further interactions [called extended technicolor (ETC)] must be postulated [8,9] to generate fermion masses. While such a method can be forced to work, it is not a compelling one, and no version naturally gives fermion masses and mixing angles without also giving possible difficulties with flavor-changing neutral currents (FCNC's); no simple equivalent of the GIM mechanism has been found. Some examples of models which attempt to control the FCNC's can be found in ref. 10. We will not concentrate on that aspect of the problem since both the theoretical and experimental aspects are in a state of flux.

After the $W^{\pm}$ and Z gain mass (which removes three of the technipions), there are generally "uneaten" technipions which remain in the physical spectrum and are experimentally detectable. The number of such technipions depends on the gauge group chosen for the technicolor interaction and on the representations chosen for the technifermions. The minimal constraints, such as the requirement that the technicolor interaction becomes strong at about a TeV, require a gauge group such as $SU(4)$ or a larger one. Since the technifermions carry ordinary electroweak and color charges, the technipions occur in multiplets of these groups. The most frequently studied model was introduced in ref. 11, and explored further in refs. 12–17. A minimal set of technipions in many models consists of a total gauge group singlet P'^0, a color singlet $Y = 0$ weak isotriplet: P^+, P^0, P^-, plus a similar set of color octets with the same electroweak quantum numbers. All of these states are light compared to the scale where technicolor becomes strong, so they behave as point-like (fundamental) spin-zero bosons.

The technipions are pseudo-Goldstone bosons; they are massless after the technicolor chiral symmetry is spontaneously broken, but obtain mass from interactions with the Standard Model gauge bosons (just as $\pi^{\pm}$ get mass from photon interactions) and also from the interactions that give rise to fermion masses (e.g., the ETC interactions). The lightest technipions are colorless;

they can only get mass from electroweak (and ETC) interactions. Thus, P'^0, P^0 and $P^\pm$ are expected to be the lightest technipions of the model and of immediate interest for phenomenology. The electroweak contribution to the technipion masses is theoretically well understood and can be reliably computed (with some dependence on the technicolor model). One finds that [9,12–15,17–19]

$$m_{P^\pm}|_{EW} \approx 5 - 14 \text{ GeV}, \qquad m_{P^0}|_{EW} = m_{P'^0}|_{EW} = 0. \qquad (7.1)$$

Unfortunately, the contributions from the extended technicolor interactions are very model-dependent, so that it is difficult to make precise predictions for the light technipion masses. In addition, P^0 and P'^0 are not necessarily mass eigenstates; in general, one would expect them to mix. (For convenience, we will neglect this mixing in the phenomenological discussion below.) However, in the above picture, it is difficult to imagine a mechanism which would yield light (colorless) technipion masses above about 40 GeV [9]. (Various model calculations invariably yield masses somewhat smaller than this generous upper limit [14,17,20].) Thus, if our present understanding of technipion masses is correct, then the light colorless technipions should be detectable or excludable at e^+e^- colliders now in operation. If a positive signal were seen, a number of further predictions could be tested at the SSC [3,6].

We now turn to the phenomenology of the light colorless technipions. As remarked above, the technipions behave like light fundamental spin-zero particles. In addition, the technipions are associated with axial vector global symmetries which are spontaneously broken when the technicolor forces become strong. As a result, all the technipions are CP-odd scalars.* Thus, the phenomenology of the $P^\pm$, P^0 and P'^0 is *indistinguishable* from that of the elementary charged and CP-odd neutral Higgs bosons of some elementary multi-Higgs model. More precisely, the interactions of the colorless technipions can always be reproduced as the low-energy limit of some elementary multi-Higgs model, in which the masses of the CP-even Higgs scalars are taken to infinity. This means that much of the phenomenology of $P^\pm$, P^0 and P'^0 is the same as the phenomenology of $H^\pm$ and A^0 studied in chapter 4. Since it is desirable to have $\rho \equiv m_W^2/(m_Z^2 \cos^2 \theta_W) = 1$, the technicolor models of interest have technipions which resemble the charged and CP-odd neutral Higgs bosons of models that contain only scalar doublets (and perhaps singlets).

---

*It is possible to construct technicolor models in which some of the pseudo-Goldstone bosons are CP-even. Consider a model in which technikaons exists. These would be analogous to technipions in precisely the same way that the K and π mesons are related in ordinary QCD. Then, if CP is conserved in the technicolor and extended technicolor interactions, the actual neutral technikaon mass eigenstates would consist of a CP-even techni-K_S and a CP-odd techni-K_L [21]. Technipions which couple to CP-even final states could also arise in extended technicolor models with CP-violation [22].

The couplings of the technipions to the gauge bosons are (nearly) model independent [3]. The relevant couplings can be obtained from the analysis of Higgs boson couplings in the two-doublet model discussed in §4.1. In particular, there are no couplings (of tree-level strength) of $P^\pm$, P^0 and P'^0 to vector boson pairs for precisely the same reasons used to explain the absence of a direct coupling of vector boson pairs to $H^\pm$ and A^0. (See Table 4.1 and the surrounding discussion.) Of course, these forbidden tree-level vertices do get generated at one-loop. In technicolor models, the new feature is that the loops contain techniquarks and technileptons (as well as the quarks and leptons). These technifermion loops dominate and are well-approximated by computing the relevant Adler-Bell-Jackiw anomalies [23] of the corresponding triangle graphs. (This calculation is precisely analogous to the famous calculation of $\pi^0 \to \gamma\gamma$.) The resulting couplings of the technipions to a pair of electroweak gauge bosons and/or gluons are given in ref. 16. Of course, since these couplings are one-loop in origin, their size is much smaller than the typical tree-level Higgs coupling. The other relevant (tree-level) interactions involve two technipions and one gauge boson. The couplings of P^+P^- to γ and Z are identical to the corresponding H^+H^- couplings. CP-invariance of the technipion-vector boson interaction rules out the coupling of γ and Z to two neutral technipions. Finally, we note that the $W^\pm P^\mp P^0$ and $W^\pm P^\mp P'^0$ couplings must satisfy:

$$g^2_{W^\pm P^\mp P^0} + g^2_{W^\pm P^\mp P'^0} = g^2_{W^\pm H^\mp A^0} , \qquad (7.2)$$

where the coupling on the right hand side above is the one from the two-doublet model of chapter 4. This relation can be derived from the requirement of tree-level unitarity of the theory. Note that the individual couplings of $W^\pm P^\mp$ to P^0 and P'^0 depend on some model-dependent mixing angle.

The couplings of the technipions to the quarks and leptons depend in detail on the structure of the extended technicolor model. As in the case of elementary Higgs models, it is necessary to guard against technipion-mediated tree-level flavor-changing neutral currents. (We have nothing to say here about other sources of flavor-changing neutral currents which have plagued extended technicolor theories and have inhibited the construction of simple phenomenologically viable models.) In analogy to the Glashow-Weinberg theorem [24] for elementary Higgs models (see §4.1), one can avoid technipion-mediated tree-level flavor-changing neutral currents by requiring that each fermion of a given electric charge gets its mass from at most one condensate of technifermions [16]. Clearly, the structure of the technipion couplings to fermions will depend in detail on the structure of the extended technicolor sector of the theory. Even with the constraint just stated above, there exists some model dependence in the form of the technipion coupling to fermions. In ref. 16, two specific models were studied which exhibit a number of interesting features:

1. The technipion-fermion coupling is CP-conserving; hence, P^0 and P'^0 have (diagonal) pseudoscalar couplings to fermion pairs.

2. The couplings of $P^\pm$ and P^0 to $f\bar{f}$ are identical to the Model I couplings of $H^\pm$ and A^0 given in eq. (4.21), with different choices for $\tan\beta$ (either $\pm\sqrt{3}$ or $\pm 1/\sqrt{3}$), depending on whether f =quark or f =lepton. In particular, $|g_{Pqq}/g_{P\ell\ell}|$ is either 3 or 1/3, corresponding to the two possible models referred to above.

3. The couplings of P^0 and P'^0 to up-type quarks (or leptons) are the same; the corresponding couplings to down-type quarks (or leptons) differ by a minus sign. This is simply a consequence of the weak-isospin properties of P^0 and P'^0.

In the specific models which lead to the above results, the factor of 3 corresponds to the number of quark colors. Clearly, some of the above details would change in models with a different extended technicolor structure. Probably, one should typically expect to find CP-violating technipion-fermion couplings in more complicated models [22]. Nevertheless, in analogy with the multi-Higgs models studies in chapter 4, we would generally expect the coupling of the technipions to fermion pairs to be proportional to the corresponding fermion masses, along with appropriate mixing angle factors. However, unlike in the case of the two-Higgs-doublet model, the above models allow for the interesting possibility that the coupling of the technipions to leptons could be unexpectedly enhanced (or suppressed) relative to the couplings to quarks. Namely, in contrast to the usual result (neglecting final state phase space factors)

$$\frac{\Gamma(A^0 \to q\bar{q})}{\Gamma(A^0 \to \ell^+\ell^-)} = \frac{3m_q^2}{m_\ell^2} \tag{7.3}$$

in the two-Higgs-doublet model, we find [16]

$$\frac{\Gamma(P^0 \to q\bar{q})}{\Gamma(P^0 \to \ell^+\ell^-)} = C\frac{3m_q^2}{m_\ell^2} \tag{7.4}$$

(and an identical result for P'^0), where $C = 9$ or $1/9$ in the two models cited above. Of course, similar features are possible in elementary multi-Higgs models of appropriate complexity.

We now survey some of the phenomenology of $P^\pm$, P^0 and P'^0. As we mentioned above, it would be very surprising if any of the colorless technipions (in particular, $P^\pm$) were as heavy as 40 GeV. Therefore the search for $Z \to P^+P^-$ at SLC and LEP should be a definitive test of conventional technicolor ideas. In particular, since supersymmetric models usually require $m_{H^\pm} > m_W$, while technicolor models seem to require $m_{P^\pm} < m_Z/2$, discovery of a charged scalar at SLC or LEP would tend to support a composite Higgs approach in preference to low energy supersymmetry, while the absence of such a discovery at SLC or LEP would be essentially fatal for the technicolor models that have been cited above. The rate for $Z \to P^+P^-$ is identical to the rate for $Z \to H^+H^-$ in the two-Higgs-doublet model discussed in §4.1.

Using eq. (4.57), we obtain the branching ratio

$$BR\left(Z \to P^+ P^-\right) \simeq 0.01 \left(1 - \frac{4m_{P^\pm}^2}{m_Z^2}\right)^{3/2}. \tag{7.5}$$

The $P^\pm$ decays may be dominantly into $b\bar{c}$, $s\bar{c}$, or $\tau\nu_\tau$ since the couplings of $P^\pm$ to fermions are probably (like those of the $H^\pm$) proportional to fermion masses. However, as discussed above, an alternative pattern of couplings (compared to the two–Higgs-doublet model of chapter 4) may arise, so a search for $P^\pm$ should allow for such a possibility. Perhaps the four jet signature or the j–j opposite $\tau\nu_\tau$ signature will be good ones.

The existence of a light charged technipion also implies that $m_t > m_{P^\pm}$, so that the semi-weak decay $t \to bP^+$ will dominate t decays when $m_t < m_W$. One consequence is that the current hadron collider limits on m_t (see §1.5) would not hold, since they depend on seeing the usual semileptonic t decay to hard e or μ. The constraints on m_t from $B^0 - \overline{B^0}$ mixing would also not apply since the contributions to the box diagram from technipion loops could be large. Note that (even if $m_t > m_W$) $t\bar{t}$ production followed by $t \to bP^+$ and $\bar{t} \to \bar{b}P^-$ will provide a copious source of $P^\pm$ at hadron colliders.

Turning next to P^0 and P'^0, the phenomenology of these technipions is similar to that of A^0 discussed in chapter 4. Because there is no electroweak contribution to the P^0 and P'^0 masses (as indicated in eq. (7.1)), we expect these neutral states to be lighter than $P^\pm$. Model calculations cited above tended to yield masses for P^0 and P'^0 no larger than a few GeV. Hence, it was expected that such neutral scalars would be observed in $\Upsilon \to \gamma P^0$ (or $\Upsilon \to \gamma P'^0$). Of course, mixing angle factors could suppress the expected rate, so the experimental absence of such a decay is not yet a definitive test of the technicolor models. Apart from Upsilon decay (and the analogous decay of toponium), it is very difficult to detect a neutral pseudoscalar. The origin of this difficulty has already been discussed in chapter 4 and alluded to above. Namely, there are no (tree-level) couplings of P^0 and P'^0 to vector boson pairs. Thus the standard methods for discovering the Higgs boson at SLC and LEP: $Z \to \phi^0 \ell^+ \ell^-$ and $e^+e^- \to Z\phi^0$, which depend on the $ZZ\phi^0$ coupling, are not appropriate for producing a neutral technipion. As discussed above, there are couplings of the neutral technipion to ZZ (and $Z\gamma$) induced at one-loop via the triangle anomaly. The largest decay rates of the Z into a technipion turns out to be the radiative decays [16]

$$BR(Z \to P^0 + \gamma) \simeq 3 \times 10^{-9}$$
$$BR(Z \to P'^0 + \gamma) \simeq 5 \times 10^{-8} \tag{7.6}$$

which are clearly unobservable. In addition, there is no technipion analog to $Z \to h^0 A^0$, since there is no light CP-even scalar state analogous to h^0 in the technicolor models cited above. Observation of any light scalar boson in Z decay or in $e^+e^- \to Z+X$ would imply the discovery of a light CP-even scalar, which would immediately falsify the conventional technicolor hypothesis.

Considering the interactions with W bosons, we note that the absence of a direct coupling[†] coupling of W^+W^- to P^0 and P'^0 implies that these states are not produced in association with a $W^\pm$ at hadron colliders. On the other hand, the decays $W^\pm \to P^\pm P^0$ and $W^\pm \to P^\pm P'^0$ can have [6] branching ratios of 1-2% (decreased only by phase space effects), so they will provide copious sources of P^0 and P'^0. This situation is quite different from the supersymmetric case, where $m_{H^\pm}$ is normally larger than m_W, so that $W^\pm \to H^\pm h^0$ is kinematically forbidden.

The couplings to gluon pairs and photon pairs are noteworthy. These arise at one-loop via the triangle anomaly as mentioned above [17,16]. The P'^0 coupling to gluon pairs exists so it can be produced via gluon-gluon fusion at a hadron collider,[‡] but P^0 does not couple to gg. Both P^0 and P'^0 have couplings to $\gamma\gamma$. From the relevant triangle anomaly, one finds:

$$\frac{\Gamma\left(P'^0 \to \gamma\gamma\right)}{\Gamma\left(P^0 \to \gamma\gamma\right)} = \frac{1}{9}, \tag{7.7}$$

so the $\gamma\gamma$ mode may be most useful for P^0. Indeed, it can be several times larger for P^0 than for a Standard Model ϕ^0 or a supersymmetric h^0, so it could be a valuable mode for finding P^0 at a hadron collider, particularly for larger values of m_{P^0}. In the model of ref. 11,

$$\Gamma\left(P^0 \to \gamma\gamma\right) = \frac{\alpha^2}{3\pi^3}\frac{m_{P^0}^3}{F_\pi^2}, \tag{7.8}$$

where $F_\pi \simeq 125$ GeV is the technipion decay constant which governs the W and Z boson masses.

If a new light spin-zero particle, h, were to be discovered, it will be important to determine whether this particle can be a technipion. Various ways to distinguish light technipions from the Standard Model Higgs boson have been discussed in the literature [25–28]. Invariably, such methods involve trying to prove that the observed spin-zero particle is a CP-odd boson. We have already noted above that the observation of a ZZh coupling would be very strong evidence against a technipion interpretation. One can also attempt to measure the P quantum number via the interaction of h with fermions. Whether a given approach can work depends on the masses of the scalars and fermions involved. For example, the decay of a scalar to $f\bar{f}$ is p-wave whereas the decay of a pseudoscalar is s-wave, and as a result the associated phase space factors differ. Thus the ratios of branching ratios can be very different

---

[†]In fact, the one loop couplings of W^+W^- to P^0 and P'^0 via the triangle anomaly also vanish [16].

[‡]Detailed production cross sections for all the technipions at $\bar{p}p$ colliders have been given in ref. 6. It will not be easy to discover P^0, P'^0, and $P^\pm$ at a hadron collider. However, sufficiently many are produced, either via gluon-gluon fusion or by real or virtual W-decay so that if they are found elsewhere, they could probably be studied at a hadron collider.

if $4m_f^2/m_h^2$ is not too small [29]. One can also attempt to measure various angular correlations in the production and decay process. It is important to reiterate that no test exists in the low-energy domain which can distinguish a technipion from a light neutral CP-odd (or charged) elementary Higgs boson. Definitive proof of a technicolor structure requires one to probe energies of order the electroweak scale. Nevertheless, we have already indicated certain features of the technipion-fermion interaction [see eq. (7.4) above] which would seem odd in the context of an elementary multi-Higgs model, but are perhaps more natural in a technicolor approach.

Finally, it is important to inject a word of caution. A great deal of attention in this section has been given to the properties of the light colorless technipions. We claimed that either the absence of a light CP-odd charged scalar or the detection of a light CP-even scalar in Z decay would be fatal to the technicolor approach described above. This remark should be qualified in two ways. First, it is possible that the structure of the technicolor model is such that no light colorless pseudo-Goldstone bosons exist. Although we are unaware of a realistic model of this kind, we know of no theorem which rules out such a possibility. Second, variants of the technicolor model do exist which contain a light CP-even scalar! One can construct models in which the spontaneous breaking of chiral symmetry also leads to a spontaneous breaking of an approximate scale symmetry. Such models contain a light pseudo-Goldstone boson of the spontaneously broken scale symmetry, called a dilaton. The phenomenology of the dilaton is quite similar to that of a CP-even Higgs scalar, although there are some distinguishing features due to the Goldstone nature of the dilaton. For further discussion and additional references, see ref. 30.

So far, we have concentrated on the light colorless technipions of the model. However, as noted above, discovery of such states can never be an unambiguous confirmation of the technicolor approach. To confirm the technicolor scenario, we must probe new features of the theory which begin to emerge at the electroweak scale. Recall that technicolor models generally possess multiplets of *colored* technipions, whose discovery would unambiguously signal new nontrivial physics beyond the Standard Model. Typically, one finds color octet technipions with various electroweak quantum numbers: an electroweak singlet (called η_T or $P_8'^0$), and a $Y = 0$ weak isotriplet ($P_8^{\pm}$ and P_8^0). The masses of these states have been estimated to be around 250 GeV [12,13,15]. The dominant source of the mass of the octet technipions derives from the strong color interactions. (The electroweak contributions can also be computed, and are found to be small.) In addition, there is a model-dependent contribution to the mass from ETC interactions. However, the various contributions add quadratically, so the ETC contribution to the mass should be unimportant. The phenomenology of the color-octet technipions is studied in detail in ref. 6, where cross sections are given and several earlier studies are reviewed. Because they are color octets, their production cross sections are large. The various decay mode branching ratios of the color octet technipions are conveniently summarized in ref. 17. Their dominant decays are to $t\bar{t}$ and gg, with a total width of order 1 GeV. Thus, detection will depend very much on what resolution can be obtained for the invariant masses $M(t\bar{t})$ and $M(jj)$. An $\eta_T \rightarrow g\gamma$ mode with a branching ratio of order 10^{-3} and gZ decay with a branching ratio of order 1% provide rare decays whose signature may be much clearer than the dominant decay modes. The cross section for η_T is of order 10^{-35} cm^2 at the Tevatron collider, so with an integrated luminosity above 10^{38} cm^{-2} the Tevatron becomes a good machine to use for η_T searches.

There are other states in the technicolor zoo that may be phenomenologically relevant. For example, additional narrow states that have leptoquark quantum numbers, and vector bosons (*e.g.*, the "techni-rho" meson) with substantial widths and masses in the 800–1000 GeV range, are expected. If any of the lighter states are found in the next few years, these additional states would eventually provide important information to distinguish between composite and fundamental Higgs approaches. In addition, the large amount of information that could be obtained on the masses, production cross sections, and decay branching ratios of the various states (such as those described above) would help distinguish among various approaches to composite Higgs theories.

7.2 Walking Technicolor

Recently, the dynamics of technicolor has been reexamined in an approach known as "walking technicolor". This name derives from the idea that models can be constructed in which the technicolor gauge coupling runs much more slowly (than is typical for an asymptotically free coupling) above the scale of electroweak interactions. This allows the extended technicolor scale to be much larger than in standard extended technicolor models, thereby suppressing flavor-changing neutral currents while not affecting the size of fermion masses which can be generated. Stimulated by a paper of Holdom [31], a number of authors have examined the possible dynamics of such a theory [32–38] . Unfortunately, no approach has advanced to the point where it can make even tentative predictions without qualifications. For example, in walking technicolor, the technipion masses are enhanced by a factor such as $\ln(m_{\mathrm{ETC}}^2/m_W^2)$. Such a factor arises in walking technicolor theories from loop contributions to the technipion masses that are cut off at m_{ETC} (or some other intermediate scale relevant for the generation of fermion masses), in contrast to the usual running technicolor theories, where the loop contributions are cut off at a scale, m_{TC}, characteristic of the technicolor theory (presumably below 1 TeV). Since $m_{\mathrm{ETC}} \gg m_{\mathrm{TC}}$, walking technicolor models permit technipions with masses that can be significantly larger than the corresponding states of the usual technicolor models. However, some walking technicolor approaches do exist which give an interesting spectrum of relatively light pseudo-Goldstone bosons and vector boson states [39]. Clearly, further theoretical work along these lines is needed before the phenomenology of these models can be understood in detail.

7.3   Composite Higgs Bosons

A more general approach to constructing composite Higgs bosons has been taken by Kaplan, Georgi and collaborators [40]. Their models possess a number of features in common with technicolor. Again, one introduces a new gauge force ("ultracolor"), and a set of new massless ultrafermions. The resulting global symmetry of the model is broken dynamically to a smaller subgroup by ultrafermion condensates at an energy scale where the ultracolor force becomes strong. Once again, there are pseudo-Goldstone bosons corresponding to the generators of the broken global symmetry. However, the new feature here is that the electroweak symmetry remains unbroken at this stage. This is in contrast to technicolor models in which the electroweak gauge symmetry is broken when condensates of technifermions form. The pseudo-Goldstone bosons include both CP-even and CP-odd scalars and are composites of the ultrafermions. In this approach, the electroweak symmetry is broken when a pseudo-Goldstone boson with the same quantum numbers as those of the Standard Model Higgs boson acquires a vacuum expectation value. This will occur because a Higgs potential can develop when electroweak gauge interactions are taken into account. [To build a successful model, one must expand the low-energy electroweak gauge group beyond SU(2) × U(1).] Because the Higgs bosons of the model are pseudo-Goldstone bosons, their masses are calculable, at least in part. In models which have been constructed, one typically finds $m_H \sim m_W$.

As in technicolor, one must introduce additional (extended ultracolor) interactions in order to generate fermion masses. However, these models are more flexible than the technicolor models discussed in §7.1. In particular, one can generate quark and lepton masses of an appropriate size, and at the same time avoid large flavor-changing neutral currents. In addition, colorless pseudo-Goldstone bosons can also be heavier than in traditional technicolor models, with masses of order m_W or less. The phenomenology of these scalars has yet to be explored in detail. (It may be possible to construct models where no pseudo-Goldstone states exist [41], but no model has been constructed so far where that happens and where fermions get masses in a satisfactory manner.)

Clearly the theory of composite Higgs bosons should be regarded as a viable alternative to the approach of the Standard Model. The theoretical questions involved are difficult and challenging. Perhaps major progress in the alternative approaches discussed in this chapter will come when the first direct experimental results associated with the origin of electroweak symmetry breaking begin to appear.

REFERENCES

1. L. Susskind, *Phys. Rev.* **D20** (1979) 2619; S. Weinberg, *Phys. Rev.* **D19** (1979) 1277.
2. E. Farhi and L. Susskind, *Phys. Rep.* **74** (1981) 277.
3. K. Lane, in *Proceedings of the 1982 DPF Summer Study on Elementary Particle Physics and Future Facilities*, edited by R. Donaldson, R. Gustafson, and F. Paige (Fermilab, Batavia, Illinois, 1982), p. 222.
4. G.L. Kane in *Gauge Theories in High Energy Physics*, ed. by M.K. Gaillard and R. Stora, (North-Holland, Amsterdam, 1983), p. 416.
5. R.K. Kaul, *Rev. Mod. Phys.* **55** (1983) 449.
6. E. Eichten, I. Hinchliffe, K. Lane and C. Quigg, *Phys. Rev.* **D34** (1986) 1547.
7. P. Sikivie, L. Susskind, M. Voloshin, and V. Zakharov, *Nucl. Phys.* **B173** (1980) 189.
8. S. Dimopoulos and L. Susskind, *Nucl. Phys.* **B155**(1979) 237.
9. E. Eichten, and K.D. Lane, *Phys. Lett.* **90B** (1980) 125.
10. S. Dimopoulos and H. Georgi, *Phys. Lett.* **127B** (1983) 101; B. Holdom, *Phys. Lett.* **143B** (1984) 227; A.E. Nelson, *Phys. Rev.* **D38** (1988) 2875; K.-I. Aoki, M. Bando, H. Mino, T. Nonoyama, H. So and K. Yamawaki, *Prog. Theor. Phys.* **82** (1989) 388.
11. E. Farhi and L. Susskind, *Phys. Rev.* **D20** (1979) 3404.
12. S. Dimopoulos, *Nucl. Phys.* **B168** (1980) 69.
13. M. Peskin, *Nucl. Phys.* **B175** (1980) 197.
14. S. Dimopoulos, S. Raby and P. Sikivie, *Nucl. Phys.* **B176** (1980) 449.
15. J. Preskill, *Nucl. Phys.* **B177** (1981) 21.
16. J. Ellis, M.K. Gaillard, D.V. Nanopoulos, and P. Sikivie, *Nucl. Phys.* **B182** (1981) 529.
17. S. Dimopoulos, S. Raby, and G.L. Kane, *Nucl. Phys.* **B181** (1981) 77.
18. S. Chadha and M. Peskin, *Nucl. Phys.* **B185** (1981) 61; **B187** (1981) 541.
19. V. Baluni, *Ann. Phys. (N.Y.)* **165** (1985) 148.
20. P. Binetruy, S. Chadha and P. Sikivie, *Phys. Lett.* **107B** (1981) 425; *Nucl. Phys.* **B207** (1982) 505.
21. H.E. Haber, SLAC-PUB-3193 (1983), unpublished.
22. E. Eichten, K. Lane and J. Preskill, *Phys. Rev. Lett.* **45** (1980) 225; A. Manohar, *Phys. Lett.* **113B** (1982) 253; A.J. Buras, S. Dawson and A.N. Schellekens, *Phys. Rev.* **D27** (1983) 1171; W. Goldstein, *Nucl. Phys.* **B213** (1983) 477; **B229** (1983) 157.
23. J.S. Bell and R. Jackiw, *Nuovo Cim.* **60A** (1969) 47; S.L. Adler, *Phys. Rev.* **117** (1969) 2426.
24. S. Glashow and S. Weinberg, *Phys. Rev.* **D15** (1977) 1958.
25. G.J. Gounaris and A. Nicolaidis, *Phys. Lett.* **102B** (1981) 144; *Phys. Lett.* **109B** (1982) 221.
26. H.E. Haber and G.L. Kane, *Nucl. Phys.* **B250** (1985) 716.

27. J. Pantaleone, *Phys. Lett.* **172B** (1986) 261.
28. C.A. Nelson, *Phys. Rev.* D30 (1984) 1937 [E: **D32** (1985) 1848]; J.R. Dell'Aquila and C.A. Nelson, *Phys. Rev.* **D33** (1986) 80, 93; *Nucl. Phys.* **B320** (1989) 86.
29. H.E. Haber, Iraj Kani, G.L. Kane, and M. Quiros, *Nucl. Phys.* **B283** (1987) 111.
30. W.A. Bardeen, C.N. Leung, and S.T. Love, *Phys. Rev. Lett.* **56** (1986) 1230; T.E. Clark, C.N. Leung, and S.T. Love, *Phys. ReV.* **D35** (1987) 997.
31. B. Holdom, *Phys. Lett.* **150B** (1985) 301.
32. T. Akiba and T. Yanagida, *Phys. Lett.* **169B** (1986) 432.
33. M. Bando, K. Matumoto, and K. Yamawaki, *Phys. Lett.* **178B** (1986) 308.
34. T. Appelquist, D. Karabali, and L.C.R. Wijewardhana, *Phys. Rev. Lett.* **57** (1986) 957.
35. M. Bando, T. Morozumi, H. So, and K. Yamawaki, *Phys. Rev. Lett.* **59** (1987) 389.
36. S. King, *Phys. Lett.* **184B** (1987) 49.
37. B. Holdom, *Phys. Lett.* **198B** (1987) 535.
38. T. Appelquist and L.C.R. Wijewardhana, *Phys. Rev.* **D35** (1987) 774; *Phys. Rev.* **D36** (1987) 568.
39. B. Holdom, private communication.
40. D.B. Kaplan, H. Georgi and S. Dimopoulos, *Phys. Lett.* **136B** (1984) 183; D.B. Kaplan and H. Georgi, *Phys. Lett.* **136B** (1984) 187, *Phys. Lett.* **143B** (1984) 152, and *Phys. Lett.* **145B** (1984) 216; M.J. Dugan, H. Georgi and D.B. Kaplan, *Nucl. Phys.* **B254** (1985) 299; R.S. Chivukula, and H. Georgi, *Phys. Lett.* **188B** (1987) 99 and *Phys. Rev.* **D36** (1987) 2102, and further references in these papers.
41. This possibility has been emphasized by H. Georgi.

Appendix A

Higgs Boson Feynman Rules in the Minimal Supersymmetric Model

The conventions employed in this book are those used in ref. 1, and can be summarized as follows. We use the Bjorken and Drell metric and conventions for γ-matrices. Our covariant derivative acting on an $SU(2)_L$ weak-doublet field with hypercharge Y is given by

$$D_\mu = \partial_\mu + \tfrac{1}{2}ig\tau^a W_\mu^a + \tfrac{1}{2}ig'Y B_\mu \qquad (A.1)$$

where τ^a are the usual Pauli matrices ($\mathrm{Tr}\,\tau^a\tau^b = 2\delta^{ab}$), and the electric charge operator is given by $Q = \tfrac{1}{2}(\tau^3 + Y)$. We define $e = g\sin\theta_W = g'\cos\theta_W$; in our conventions $e > 0$. The Z boson and photon fields are defined by

$$\begin{aligned} A &= W^3\sin\theta_W + B\cos\theta_W \\ Z &= W^3\cos\theta_W - B\sin\theta_W\,, \end{aligned} \qquad (A.2)$$

where W^3 is the neutral member of the $SU(2)_L$ triplet of gauge fields, and B is the $U(1)_Y$ gauge field. The physical Higgs fields are defined in eqs. (4.12), (4.14), and (4.16), and the (unphysical) Goldstone boson fields are defined in eqs. (4.11) and (4.13).

Feynman rules for the interaction of Higgs bosons in the minimal supersymmetric extension of the Standard Model in the unitary gauge have been given in ref. 2. In this appendix, we give the Feynman rules appropriate for the R_ξ-gauge (with the choice $\xi = 1$) in which interactions involving unphysical particles, namely the Goldstone fields and Faddeev-Popov ghosts, must be taken into account.* This will certainly cause a proliferation of Feynman

---

*For a pedagogical introduction to the R-gauge, see ref. 3. A partial compilation of R-gauge Feynman rules in two-Higgs-doublet models can be found in refs. 4, 5 and 6.

diagrams for a given physical process. Nevertheless, there are a number of advantages to having this enlarged set of rules. First, one can use the rules involving the Goldstone fields to check certain calculations using the Equivalence Theorem. This theorem states that the S-matrix amplitudes involving external longitudinally polarized gauge bosons (V) may be evaluated in the R-gauge by substituting the corresponding Goldstone bosons as external particles; the result thus obtained is accurate to leading order in m_V/E_V, where E_V is the vector boson energy [9]. (See §2.6 for further discussion.) Second, calculations in the R_ξ-gauge contain the gauge parameter ξ in intermediate stages of the calculation. However, any physical S-matrix element must be independent of ξ, and this provides a useful check of some calculations. Finally, even if one foregoes the gauge invariance check just mentioned, the choice of $\xi = 1$ (the 't Hooft gauge) is particularly convenient since the massive gauge boson propagator (which is ξ-dependent) takes on its simplest form in this gauge. In particular, loop calculations involving internal massive vector boson loops are much easier (and less prone to error) in the 't Hooft gauge than in the unitary gauge; this more than makes up for the additional work generated by additional diagrams involving unphysical particles.

The 't Hooft gauge ($\xi = 1$) corresponds to the following gauge-fixing term

$$\mathcal{L}_{\text{g.f.}} = -\tfrac{1}{2}(\partial_\mu A^\mu)^2 - \tfrac{1}{2}(\partial_\mu Z^\mu + m_Z G^0)^2 - |\partial_\mu W^{+\mu} + i m_W G^+|^2 , \quad \text{(A.3)}$$

where $G^\pm$ and G^0 are the unphysical Goldstone bosons (which are eaten by the $W^\pm$ and Z^0). In the $\xi = 1$ gauge the masses of $G^\pm$ and G^0 (which appear in the propagators) are m_W and m_Z, respectively. When one computes the quadratic part of the electroweak Lagrangian, before including $\mathcal{L}_{\text{g.f.}}$ (but keeping the unphysical Goldstone fields), one finds the following non-diagonal pieces:

$$i m_W (W_\mu^+ \partial^\mu G^- - W_\mu^- \partial^\mu G^+) - m_Z Z_\mu \partial^\mu G^0 . \quad \text{(A.4)}$$

These pieces are precisely canceled by $\mathcal{L}_{\text{g.f.}}$ (after an integration by parts), thereby eliminating mixing between the gauge bosons and the unphysical Goldstone fields. If we examine the trilinear interaction terms involving two gauge bosons and a single Goldstone field [see eqs. (6.6)–(6.7)], we find

$$\mathcal{L}_{GVV} = m_W \left(e A^\mu - \frac{g \sin^2 \theta_W}{\cos \theta_W} Z^\mu \right) (G^- W_\mu^+ + G^+ W_\mu^-) . \quad \text{(A.5)}$$

The vertices contained in eq. (A.5) play an important role in the R-gauge calculations of $\phi^0 \to \gamma\gamma$, $\phi^0 \to Z\gamma$, $H^+ \to W^+\gamma$, and $H^+ \to W^+ Z$. However, there is a slight variation of the gauge choice which leads to remarkable simplifications. Gavela et al. [10] observed that the choice of the nonlinear R-gauge, first suggested by Fujikawa [11], provided significant simplifications in the calculation of $\phi^0 \to \gamma\gamma$.[†] In the nonlinear R-gauge, the following modification of the gauge-fixing term is used (for $\xi = 1$)

$$\mathcal{L}_{g.f.} = -\tfrac{1}{2}(\partial_\mu A^\mu)^2 - \tfrac{1}{2}(\partial_\mu Z^\mu + m_Z G^0)^2 - |(\partial_\mu + ig W_\mu^3)W^{+\mu} + im_W G^+|^2 . \tag{A.6}$$

There are two basic advantages of the nonlinear R-gauge. First, if one combines the new trilinear terms of eq. (A.6) with the GVV-interaction terms given in eq. (A.5), one finds

$$\mathcal{L}_{GVV}^{nonlinear} = -gm_Z Z^\mu(G^- W_\mu^+ + G^+ W_\mu^-) . \tag{A.7}$$

Note that the $\gamma W^+ G^-$ vertex has *vanished*! For this reason, one-loop calculations involving external photons are simpler in this gauge. However, a $Z W^+ G^-$ vertex still remains [eq. (A.7)]. In fact, by a suitable generalization of the nonlinear gauge condition, one can choose to have either one or both of the $\gamma W^+ G^-$ and $Z W^+ G^-$ vertices absent [12]! A second advantage of the nonlinear R-gauge is that the Z–γ transition is purely transverse. (This is not true in the usual R-gauge.) As a result, all Z–γ mixing on external lines is automatically zero, which greatly simplifies the calculation of processes with an external γ and/or Z.

One does pay a price for using the nonlinear gauge. First, the Faddeev-Popov ghost Lagrangian must be appropriately modified; in particular, a new quartic coupling involving two Faddeev-Popov ghost fields and two photons is introduced. This is discussed in detail in ref. 10. This is only a minor nuisance in the calculation of $\phi \to \gamma\gamma$; and in charged Higgs decay at one loop no graphs involving Faddeev-Popov ghosts contribute. Second, the three and four gauge boson vertices are modified. (The modified rules explicitly involve the gauge parameter ξ, as shown in ref. 10. Once again, it is easiest to work in the gauge where $\xi = 1$.) However, the simplification in the nonlinear R-gauge computation more than makes up for the necessity for using slightly more complicated Feynman rules for the three and four gauge boson vertices. Finally, we note that the discussion above applies to all electroweak models based on SU(2) × U(1), independent of the number of Higgs multiplets. This is true because the Feynman rules involving the Goldstone fields do not depend on the choice of the Higgs structure of the model. The reader is referred to refs. 10, 12, 16, and 17 for further discussion of the nonlinear R-gauge Feynman rules and their applications.

Let us now turn to the minimal supersymmetric extension of the Standard Model, with exactly two Higgs doublet fields with hypercharge $Y = -1$ and $Y = +1$ respectively, and no Higgs singlets. We present here a complete list

---

[†]The nonlinear gauge condition has also been exploited in ref. 12 for the calculation of neutral Higgs boson decay to two photons, in ref. 13 for the calculation of the flavor-changing electromagnetic vertex, in refs. 14 and 15 for the calculation of the charge radius of the neutrino, in ref. 16 for neutralino radiative decay and in ref. 17 for one-loop decays of the charged Higgs boson.

of the (linear) R-gauge Feynman rules with $\xi = 1$, corresponding to the gauge fixing term of eq. (A.3). (Appropriate changes to the rules in the nonlinear R-gauge can be found in the references quoted above.) In this model, the physical Higgs fields $H^\pm$, h^0, H^0, A^0 and the Goldstone fields G^0 and $G^\pm$ are defined in eq. (4.72). Of course, the unitary gauge rules are that subset of the R-gauge rules for interactions that involve physical degrees of freedom only. We present most of the Feynman vertices in a series of figures; quartic interactions among the various Higgs bosons are merely listed. We will only present Feynman rules for vertices which explicitly involve the Higgs bosons of the two doublet model. For the Feynman rules involving the other Standard Model particles, we recommend that the reader consult ref. 18, in which a complete set of R_ξ-gauge Feynman rules (in the above conventions) is given. For example, the conventions implicit in eqs. (A.1) and (A.2) yield the following Feynman rules:

$$W f \bar{f}' \qquad\qquad \frac{-ig}{2\sqrt{2}}\, \gamma^\mu (1 - \gamma_5)$$

$$\gamma f \bar{f} \qquad\qquad -i e e_f \gamma^\mu$$

$$Z f \bar{f} \qquad\qquad \frac{-ig}{2 \cos \theta_W} \gamma^\mu [(T_f^{3L} - e_f \sin^2 \theta_W)(1 - \gamma_5)$$
$$-e_f \sin^2 \theta_W (1 + \gamma_5)]$$

$$W_\mu^+(p_+) W_\nu^-(p_-) V_\alpha^0(p_0) \qquad\qquad i g_V [g^{\mu\nu}(p_+ - p_-)^\alpha + g^{\nu\alpha}(p_- - p_0)^\mu$$
$$+ g^{\alpha\mu}(p_0 - p_+)^\nu],$$

with charge e_f in units of e and 3rd component of weak isospin $T_f^{3L} = \pm\frac{1}{2}$, and $V^0 = \gamma$ or Z with

$$g_V = \begin{cases} e, & V^0 = \gamma \\ g \cos \theta_W, & V^0 = Z. \end{cases} \qquad (A.8)$$

Note that in the $W^+ W^- V^0$ Feynman rule given above, all momenta are assumed to point into the vertex. The reader should be cautioned that the $W^+ W^- \gamma$ and $W^+ W^- Z^0$ Feynman rules given in ref. 18 have the wrong sign. (We are also aware of some errors in the four-scalar vertices involving Goldstone bosons; these will be mentioned below when we discuss how to obtain the rules for the minimal one-Higgs-doublet model as a limiting case of the rules given below.) Additional Feynman rules involving supersymmetric particles can be found in refs. 19, 20 and 2.

In the Feynman rules which follow, there appear numerous "angle factors", which depend on α and β. In the supersymmetric two-Higgs doublet model, it is convenient to choose phase conventions for the Higgs fields in which $-\pi/2 \le \alpha \le 0$ and $0 \le \beta \le \pi/2$. One can then obtain expressions for the various angle factors in terms of the physical Higgs masses in the supersymmetric model (but not in the most general two-Higgs-doublet model) since supersymmetry imposes a number of relations among the Higgs parameters. We summarize the most useful relations (taken from Appendix A of ref. 21) below.

First, it is convenient to rescale all mass variables and define

$$r_H \equiv \frac{m_{H^0}^2}{m_Z^2}, \quad r_h \equiv \frac{m_{h^0}^2}{m_Z^2}, \quad r_A \equiv \frac{m_{A^0}^2}{m_Z^2}, \quad r_\pm \equiv \frac{m_{H^\pm}^2}{m_Z^2}, \quad r_W \equiv \frac{m_W^2}{m_Z^2}. \quad (A.9)$$

From eq. (4.75), it follows that

$$r_H \geq 1, \quad r_h \leq 1, \quad r_A = r_H + r_h - 1, \quad r_\pm = r_A + r_W. \quad (A.10)$$

It must be emphasized that all Higgs boson masses appearing in the formulae given in this Appendix are *tree-level* masses. Given any two Higgs masses (we choose r_H and r_h), all angle factors given below are determined up to a sign. The sign is uniquely fixed once we specify whether $\tan\beta$ is larger or smaller than 1. Thus, we introduce the notation

$$\epsilon = \text{sign}(v_2 - v_1) \quad (A.11)$$

i.e., $\epsilon = +1$ for $\tan\beta > 1$ and vice versa. (For the borderline case of $\tan\beta = 1$, $r_h = 0$, in which case none of the expressions which are nonzero depend on ϵ.) We now give the expressions for the various angle factors which appear in the Feynman rules

$$\sin(\beta + \alpha) = \epsilon\sqrt{\frac{r_h(r_H - 1)}{r_H - r_h}}$$

$$\cos(\beta - \alpha) = -\epsilon\sqrt{\frac{r_h(1 - r_h)}{(r_H - r_h)(r_H + r_h - 1)}}$$

$$\sin(\beta - \alpha) = \sqrt{\frac{r_H(r_H - 1)}{(r_H - r_h)(r_H + r_h - 1)}}$$

$$\cos(\beta + \alpha) = \sqrt{\frac{r_H(1 - r_h)}{r_H - r_h}}$$

$$\cos 2\beta = -\epsilon\sqrt{\frac{r_h r_H}{r_H + r_h - 1}} \qquad (A.12)$$

$$\sin 2\beta = \sqrt{\frac{(r_H - 1)(1 - r_h)}{r_H + r_h - 1}}$$

$$\sin 2\alpha = -\sin 2\beta\left(\frac{r_H + r_h}{r_H - r_h}\right)$$

$$\cos 2\alpha = -\cos 2\beta\left(\frac{r_H + r_h - 2}{r_H - r_h}\right).$$

All square roots in the above equation are to be taken positive. Note that the following relations are always true

$$\sin 2\alpha \leq 0$$
$$\sin 2\beta \geq 0$$
$$\sin(\beta - \alpha) \geq 0$$
$$\cos(\beta + \alpha) \geq 0.$$

(A.13)

Finally, we provide an algorithm which, when applied to the supersymmetric two-Higgs doublet model rules, successfully reproduces the Feynman rules of the minimal (nonsupersymmetric) model with one physical Higgs doublet. If we take the limit $m_{A^0} \to \infty$, then by eq. (A.10), both $m_{H^\pm}$, $m_{H^0} \to \infty$. As a result, $H^\pm$, H^0 and A^0 decouple from the theory, and only h^0 remains. Furthermore, from eq. (A.12), we find that in this limit,

$$\sin 2\beta = \cos(\beta + \alpha) = -\sin 2\alpha = \sqrt{1 - \frac{m_{h^0}^2}{m_Z^2}},$$

$$\cos 2\alpha = \sin(\beta + \alpha) = -\cos 2\beta = \frac{\epsilon m_{h^0}}{m_Z},$$

(A.14)

$$\sin(\beta - \alpha) = 1, \qquad \cos(\beta - \alpha) = 0.$$

In addition, these relations imply that

$$\cos \alpha = \sin \beta, \qquad \sin \alpha = -\cos \beta.$$

(A.15)

If we insert these expressions into the Feynman rules below, and delete all vertices containing $H^\pm$, H^0, A^0, and/or any supersymmetric particles and set $h^0 = \phi^0$, we obtain precisely the Standard Model Feynman rules for the minimal Higgs boson. Note that although the parameter ϵ (which is a remnant of the two-Higgs-doublet model) appears in eq. (A.14), one finds that in the above procedure, only $\epsilon^2 = 1$ appears. We have compared the resulting Standard Model rules with those of ref. 18, and we note the following errors: the $ZW^\pm G^\mp \phi^0$ and $ZW^\pm G^\mp G^0$ vertices given in ref. 18 are incorrect; the factor $\frac{1}{2}\cos 2\theta_W - 1$ should be replaced by $-\sin^2 \theta_W$.

In the Feynman rules which follow, a directed arrow on a line indicates the flow of momentum and/or electric charge as shown in the graph. In cases with no arrows displayed, any choice of electric charges consistent with charge conservation is allowed and has the same Feynman rule. Finally, in cases where a change of sign of the charge of some of the particles results in a change of sign of the Feynman rule, one must associate the corresponding upper (and lower) signs if indicated.

Although the Higgs boson Feynman rules we now present are specifically for the minimal supersymmetric extension of the Standard Model (MSSM), some of the rules are, in fact, more general. Those rules which are also valid for *any* general CP-conserving (non-supersymmetric) two-Higgs-doublet model are: vertices with one or two gauge bosons or Faddeev-Popov ghosts and vertices involving two fermions and a Goldstone boson. In addition, the rules for the $G^{\pm}H^{\mp}h$ vertices ($h = H^0, h^0$ or A^0) can be written in a rather simple form (in which the corresponding vertex rules are proportional to $m_{H^{\pm}}^2 - m_h^2$) such that they are also valid in the more general two-Higgs doublet model. All other vertices are model dependent. The couplings of fermions to physical Higgs bosons given below correspond to the Model II interaction given by eq. (4.22). A second possible Higgs fermion interaction is of the Model I form shown in eq. (4.21). Other possible choices can also be made. The three and four scalar vertices involving physical Higgs and/or Goldstone bosons are generally model dependent functions of the λ_i which appear in the Higgs potential of eq. (4.8). Only in the minimal supersymmetric model, where the λ_i are greatly constrained do the three and four scalar vertices take on the relatively simple forms shown below. The exceptional case of the $G^{\pm}H^{\mp}h$ vertices, which can be written in a simple model-independent form was noted above. For the reader's convenience, *both* the model-dependent (MSSM) and model-independent form of the $G^{\pm}H^{\mp}h$ rules are displayed below.

For convenience, we provide a table of contents which displays our organization of the R_{ξ}-gauge Feynman rules (for $\xi = 1$) for the supersymmetric two-Higgs-doublet model. In the descriptive titles below, the term "Higgs boson" refers either to a physical Higgs particle or an unphysical Goldstone boson.

Higgs Boson Feynman Rules in the MSSM

A.1   Feynman Rules with One Gauge Boson and Two Higgs Bosons

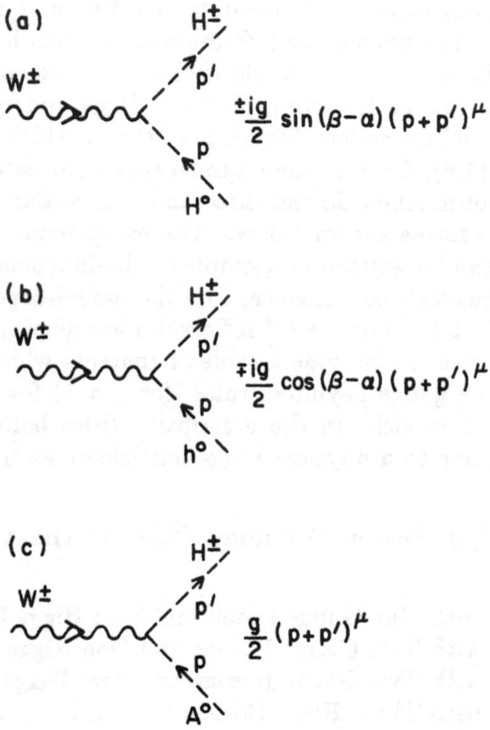

(a)

$H^{\pm}$

$W^{\pm}$ p'

$\pm \dfrac{ig}{2} \sin(\beta-\alpha)(p+p')^{\mu}$

p

H^{0}

(b)

$H^{\pm}$

$W^{\pm}$ p'

$\mp \dfrac{ig}{2} \cos(\beta-\alpha)(p+p')^{\mu}$

p

h^{0}

(c)

$H^{\pm}$

$W^{\pm}$ p'

$\dfrac{g}{2}(p+p')^{\mu}$

p

A^{0}

Figure A.1 Feynman rules for vertices with $W^{\pm}$, $H^{\pm}$, and a neutral Higgs boson. (There is no $W^{\pm}H^{\mp}G^{0}$ vertex.) The direction of momentum is indicated.

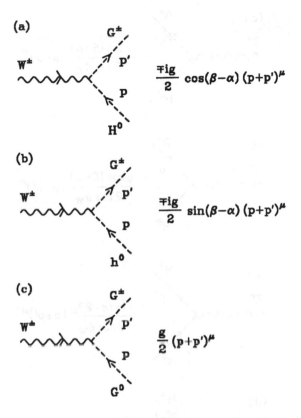

(a)

$$\frac{\mp ig}{2} \cos(\beta-\alpha)\,(p+p')^{\mu}$$

(b)

$$\frac{\mp ig}{2} \sin(\beta-\alpha)\,(p+p')^{\mu}$$

(c)

$$\frac{g}{2}\,(p+p')^{\mu}$$

Figure A.2 Feynman rules for vertices with $W^{\pm}$, $G^{\pm}$, and a neutral Higgs or Goldstone boson. The direction of momentum is indicated.

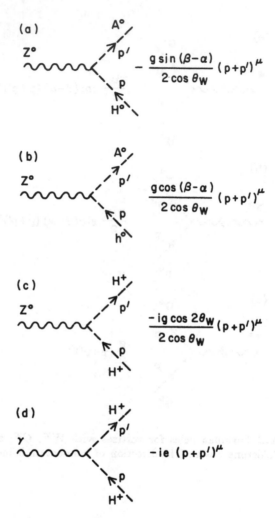

Figure A.3 Feynman rules for vertices with a Z or γ and two neutral or charged Higgs bosons. The direction of momentum is indicated.

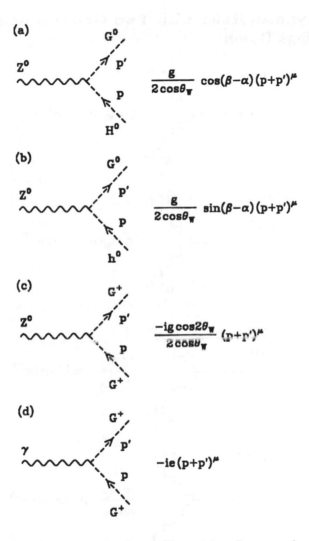

Figure A.4 Feynman rules for vertices with a Z or γ and one or two Goldstone bosons. The direction of momentum is indicated.

A.2   Feynman Rules with Two Gauge Bosons and One Higgs Boson

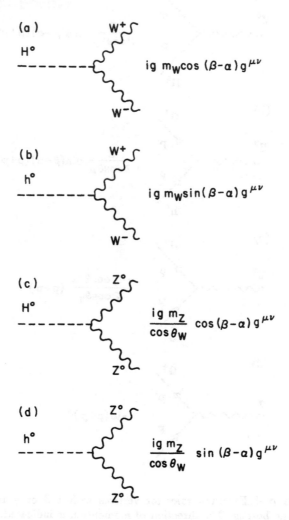

Figure A.5 Feynman rules for vertices with two vector bosons and a neutral Higgs boson.

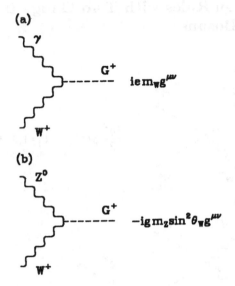

Figure A.6 Feynman rules for vertices with a γ or Z, a W^+ and a G^+.

A.3 Feynman Rules with Two Gauge Bosons and Two Higgs Bosons

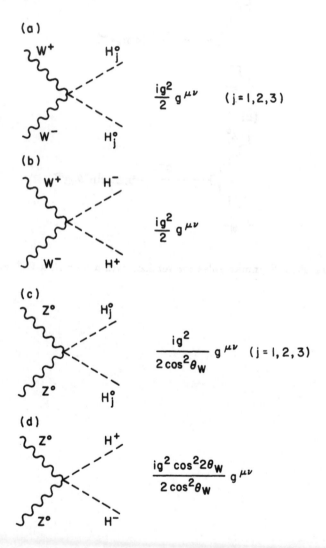

(a)

$$\frac{ig^2}{2} g^{\mu\nu} \qquad (j=1,2,3)$$

(b)

$$\frac{ig^2}{2} g^{\mu\nu}$$

(c)

$$\frac{ig^2}{2\cos^2\theta_W} g^{\mu\nu} \qquad (j=1,2,3)$$

(d)

$$\frac{ig^2 \cos^2 2\theta_W}{2\cos^2\theta_W} g^{\mu\nu}$$

Figure A.7 Feynman rules for vertices with two vector bosons and two Higgs bosons. The index values $j = 1, 2, 3$ correspond to H^0, h^0 and A^0, respectively. The corresponding vertices involving Goldstone bosons are given in Fig. A.10.

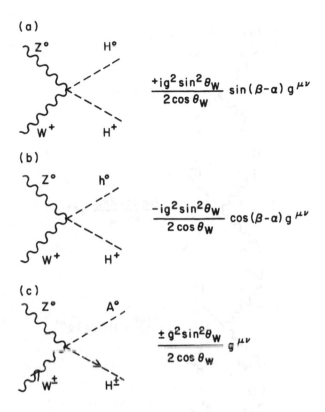

(a)

$$+\frac{ig^2\sin^2\theta_W}{2\cos\theta_W}\sin(\beta-\alpha)\,g^{\mu\nu}$$

(b)

$$-\frac{ig^2\sin^2\theta_W}{2\cos\theta_W}\cos(\beta-\alpha)\,g^{\mu\nu}$$

(c)

$$\pm\frac{g^2\sin^2\theta_W}{2\cos\theta_W}\,g^{\mu\nu}$$

Figure A.8 Feynman rules for vertices with a $W^\pm$, a Z, a $H^\pm$, and a neutral Higgs boson. The direction of momentum is indicated. The corresponding vertices involving Goldstone bosons are given in Fig. A.11.

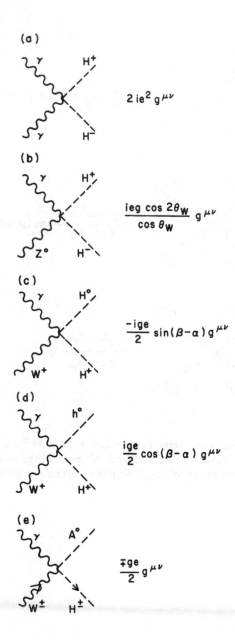

Figure A.9 Feynman rules for vertices with a γ, a second gauge boson, and two Higgs bosons. The direction of momentum is indicated. The corresponding vertices involving Goldstone bosons are given in Fig. A.12.

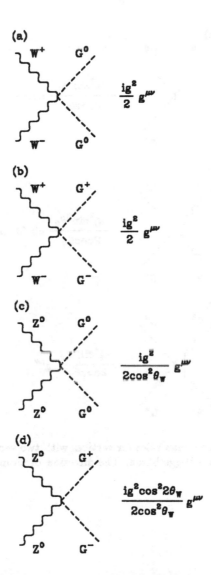

Figure A.10 Feynman rules for vertices with two vector bosons and two Goldstone bosons.

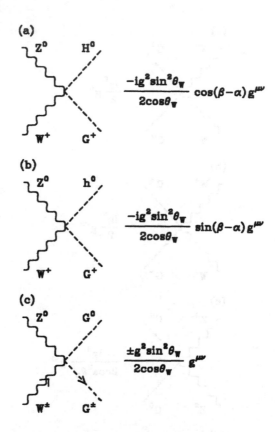

Figure A.11 Feynman rules for vertices with two vector bosons, a Goldstone boson, and a Higgs boson. The direction of momentum is indicated.

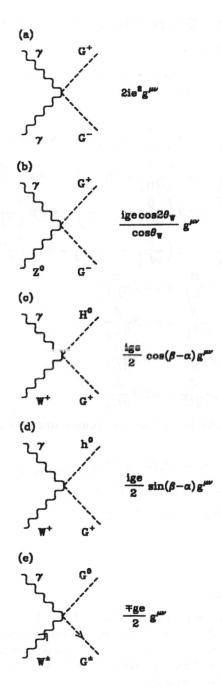

(a)

$$2ie^2 g^{\mu\nu}$$

(b)

$$\frac{ige\cos2\theta_W}{\cos\theta_W} g^{\mu\nu}$$

(c)

$$\frac{ige}{2} \cos(\beta-\alpha) g^{\mu\nu}$$

(d)

$$\frac{ige}{2} \sin(\beta-\alpha) g^{\mu\nu}$$

(e)

$$\frac{\mp ge}{2} g^{\mu\nu}$$

Figure A.12 Feynman rules for four-point vertices with one or two γ's and one or two Goldstone bosons. The direction of momentum is indicated.

A.4   Cubic and Quartic Higgs Scalar Interactions

Method of Calculation

We begin with the scalar potential for the minimal supersymmetric model as given in eq. (4.71). Using eq. (4.72) we see that it is convenient to introduce differential operators

$$D_{H^0} = \frac{1}{\sqrt{2}}\left[\cos\alpha \left(\frac{\partial}{\partial H_1^1} + \frac{\partial}{\partial H_1^{1*}}\right) + \sin\alpha \left(\frac{\partial}{\partial H_2^2} + \frac{\partial}{\partial H_2^{2*}}\right)\right]$$

$$D_{h^0} = \frac{1}{\sqrt{2}}\left[-\sin\alpha \left(\frac{\partial}{\partial H_1^1} + \frac{\partial}{\partial H_1^{1*}}\right) + \cos\alpha \left(\frac{\partial}{\partial H_2^2} + \frac{\partial}{\partial H_2^{2*}}\right)\right]$$

$$D_{A^0} = i\frac{1}{\sqrt{2}}\left[\sin\beta \left(\frac{\partial}{\partial H_1^1} - \frac{\partial}{\partial H_1^{1*}}\right) + \cos\beta \left(\frac{\partial}{\partial H_2^2} - \frac{\partial}{\partial H_2^{2*}}\right)\right]$$

$$D_{G^0} = i\frac{1}{\sqrt{2}}\left[-\cos\beta \left(\frac{\partial}{\partial H_1^1} - \frac{\partial}{\partial H_1^{1*}}\right) + \sin\beta \left(\frac{\partial}{\partial H_2^2} - \frac{\partial}{\partial H_2^{2*}}\right)\right]$$

$$D_{H^-} = \sin\beta\frac{\partial}{\partial H_1^2} + \cos\beta\frac{\partial}{\partial H_2^{1*}}$$

$$D_{H^+} = \sin\beta\frac{\partial}{\partial H_1^{2*}} + \cos\beta\frac{\partial}{\partial H_2^1}$$

$$D_{G^-} = -\cos\beta\frac{\partial}{\partial H_1^2} + \sin\beta\frac{\partial}{\partial H_2^{1*}}$$

$$D_{G^+} = -\cos\beta\frac{\partial}{\partial H_1^{2*}} + \sin\beta\frac{\partial}{\partial H_2^1}$$

and that we can easily find three- and four-scalar interactions from the formulae

$$g_{abc} = -i\, D_a D_b D_c \, V|_{H_1^1 = v_1, H_2^2 = v_2, H_1^2 = H_2^1 = 0}$$

$$g_{abcd} = -i\, D_a D_b D_c D_d \, V$$

where $a, b, c, d = H^0, h^0, A^0, G^0, H^\pm, G^\pm$.

Cubic Interactions

We present here all the Feynman rules for the cubic self interactions among the Higgs fields of the model, including Goldstone modes, in the R-gauge. We note that all particles are treated as incoming, except in a very few cases where the momentum directions are indicated. The Feynman rules are given graphically in a series of figures.

(a)

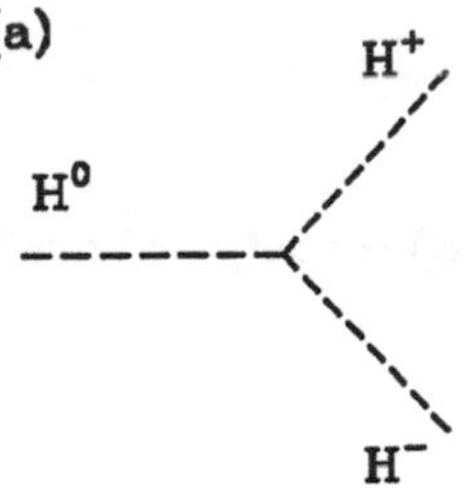

$$-ig\left[m_W\cos(\beta-\alpha) - \frac{m_Z}{2\cos\theta_W}\cos2\beta\,\cos(\beta+\alpha)\right]$$

(b)

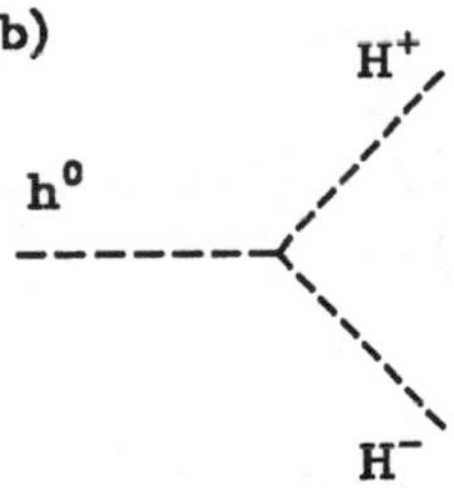

$$-ig\left[m_W\sin(\beta-\alpha) + \frac{m_Z}{2\cos\theta_W}\cos2\beta\,\sin(\beta+\alpha)\right]$$

(c)

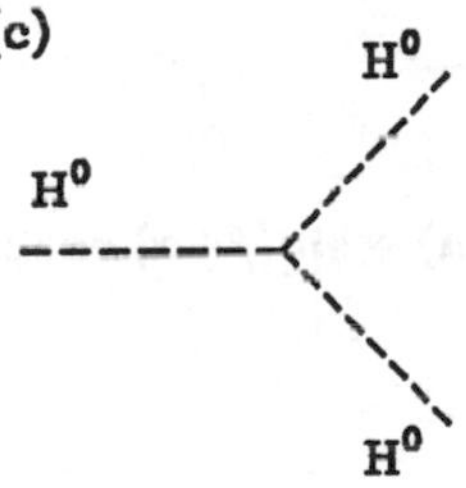

$$\frac{-3igm_Z}{2\cos\theta_W}\cos2\alpha\,\cos(\beta+\alpha)$$

(d)

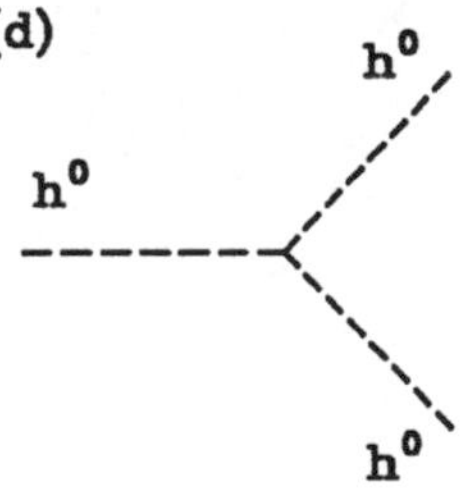

$$\frac{-3igm_Z}{2\cos\theta_W}\cos2\alpha\,\sin(\beta+\alpha)$$

Figure A.13 Feynman rules for vertices with three Higgs bosons.

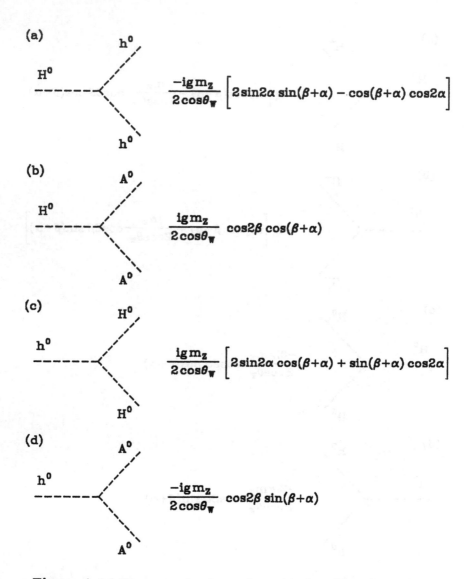

Figure A.14 Feynman rules for vertices with three Higgs bosons, continued.

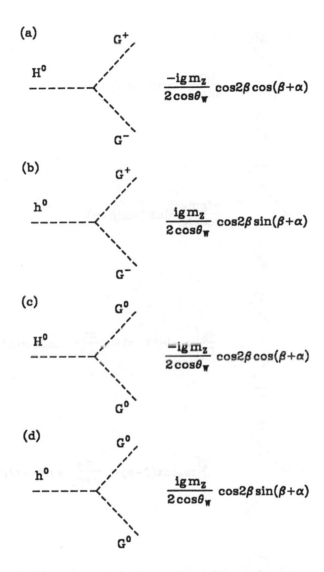

Figure A.15 Feynman rules for vertices with one Higgs boson and two Goldstone bosons.

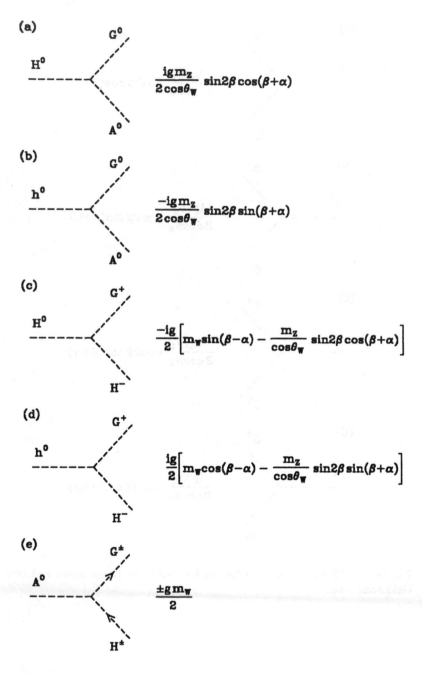

Figure A.16 Feynman rules for vertices with two Higgs bosons and one Goldstone boson.

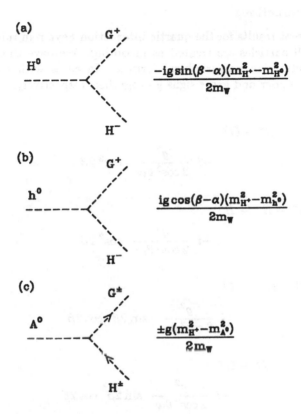

Figure A.17 An alternative form for the Feynman rules with two Higgs bosons and one charged Goldstone boson. The rules of fig. A.16c-e have been rewritten in a more symmetrical form using eqs. (A.10) and (A.12). In contrast to the rules given in fig. A.16, the $G^{\pm}H^{\mp}h$ vertex rules ($h = H^0, h^0$ or A^0) in the above form are also valid in the more general (non-supersymmetric) two doublet Higgs model.

Quartic Interactions

Here we present results for the quartic interaction Feynman rules in R-gauge. As above, all particles are treated as incoming. We have chosen to present these interactions in non-graphical format. In cases with multiple charge superscripts, upper and lower signs go together, respectively.

$G^- - G^+ - G^- - G^+$

$$-i \; \frac{g^2}{2\cos^2\theta_W} \; \cos^2 2\beta$$

$H^- - H^+ - H^- - H^+$

$$-i \; \frac{g^2}{2\cos^2\theta_W} \; \cos^2 2\beta$$

$G^- - G^+ - G^\mp - H^\pm$

$$i \; \frac{g^2}{2\cos^2\theta_W} \; \sin 2\beta \; \cos 2\beta$$

$H^- - H^+ - H^\mp - G^\pm$

$$-i \; \frac{g^2}{2\cos^2\theta_W} \; \sin 2\beta \; \cos 2\beta$$

$H^\mp - H^\mp - G^\pm - G^\pm$

$$-i \; \frac{g^2}{2\cos^2\theta_W} \; \sin^2 2\beta$$

$H^- - H^+ - G^- - G^+$

$$i \; \frac{g^2}{4\cos^2\theta_W} \; [\cos^2 2\beta - \sin^2 2\beta]$$

$H^0 - H^0 - G^- - G^+$

$$-i \; \frac{g^2}{4} \; [1 - \sin 2\beta \sin 2\alpha + \tan^2\theta_W \cos 2\beta \cos 2\alpha]$$

$H^0 - H^0 - H^- - H^+$

$$-i \; \frac{g^2}{4} \; [1 + \sin 2\beta \sin 2\alpha - \tan^2\theta_W \cos 2\beta \cos 2\alpha]$$

$H^0 - H^0 - H^\mp - G^\pm$

$$i \, \frac{g^2}{4} \left[\cos 2\beta \sin 2\alpha + \tan^2 \theta_W \sin 2\beta \cos 2\alpha\right]$$

$h^0 - h^0 - G^- - G^+$

$$-i \, \frac{g^2}{4} \left[1 + \sin 2\beta \sin 2\alpha - \tan^2 \theta_W \cos 2\beta \cos 2\alpha\right]$$

$h^0 - h^0 - H^- - H^+$

$$-i \, \frac{g^2}{4} \left[1 - \sin 2\beta \sin 2\alpha + \tan^2 \theta_W \cos 2\beta \cos 2\alpha\right]$$

$h^0 - h^0 - H^\mp - G^\pm$

$$-i \, \frac{g^2}{4} \left[\cos 2\beta \sin 2\alpha + \tan^2 \theta_W \sin 2\beta \cos 2\alpha\right]$$

$H^0 - h^0 - G^- - G^+$

$$i \, \frac{g^2}{4} \left[\sin 2\beta \cos 2\alpha + \tan^2 \theta_W \cos 2\beta \sin 2\alpha\right]$$

$H^0 - h^0 - H^- - H^+$

$$-i \, \frac{g^2}{4} \left[\sin 2\beta \cos 2\alpha + \tan^2 \theta_W \cos 2\beta \sin 2\alpha\right]$$

$H^0 - h^0 - H^\mp - G^\pm$

$$i \, \frac{g^2}{4} \left[\cos 2\beta \cos 2\alpha - \tan^2 \theta_W \sin 2\beta \sin 2\alpha\right]$$

$A^0 - A^0 - G^- - G^+$

$$-i \, \frac{g^2}{4} \left[1 + \sin^2 2\beta - \tan^2 \theta_W \cos^2 2\beta\right]$$

$A^0 - A^0 - H^- - H^+$

$$-i \, \frac{g^2}{4 \cos^2 \theta_W} \cos^2 2\beta$$

$A^0 - A^0 - H^\mp - G^\pm$

$$-i \, \frac{g^2}{4 \cos^2 \theta_W} \sin 2\beta \, \cos 2\beta$$

$$G^0 - G^0 - G^- - G^+$$

$$-i \ \frac{g^2}{4\cos^2\theta_W} \ \cos^2 2\beta$$

$$G^0 - G^0 - H^- - H^+$$

$$-i \ \frac{g^2}{4} \left[1 + \sin^2 2\beta - \tan^2\theta_W \cos^2 2\beta \right]$$

$$G^0 - G^0 - H^\mp - G^\pm$$

$$i \ \frac{g^2}{4\cos^2\theta_W} \ \sin 2\beta \ \cos 2\beta$$

$$A^0 - G^0 - G^- - G^+$$

$$i \ \frac{g^2}{4\cos^2\theta_W} \ \sin 2\beta \ \cos 2\beta$$

$$A^0 - G^0 - H^- - H^+$$

$$-i \ \frac{g^2}{4\cos^2\theta_W} \ \sin 2\beta \ \cos 2\beta$$

$$A^0 - G^0 - H^\mp - G^\pm$$

$$i \ \frac{g^2}{4} \left[\cos^2 2\beta - \tan^2\theta_W \sin^2 2\beta \right]$$

$$H^0 - G^0 - H^\mp - G^\pm$$

$$\mp \frac{g^2}{4} \ \sin(\beta - \alpha)$$

$$h^0 - G^0 - H^\mp - G^\pm$$

$$\pm \frac{g^2}{4} \ \cos(\beta - \alpha)$$

$$H^0 - A^0 - H^\mp - G^\pm$$

$$\mp \frac{g^2}{4} \ \cos(\beta - \alpha)$$

$$h^0 - A^0 - H^\mp - G^\pm$$

$$\mp \frac{g^2}{4} \ \sin(\beta - \alpha)$$

$G^0 - G^0 - G^0 - G^0$

$$-i \; \frac{3g^2}{4\cos^2\theta_W} \; \cos^2 2\beta$$

$H^0 - H^0 - H^0 - H^0$

$$-i \; \frac{3g^2}{4\cos^2\theta_W} \; \cos^2 2\alpha$$

$h^0 - h^0 - h^0 - h^0$

$$-i \; \frac{3g^2}{4\cos^2\theta_W} \; \cos^2 2\alpha$$

$A^0 - A^0 - A^0 - A^0$

$$-i \; \frac{3g^2}{4\cos^2\theta_W} \; \cos^2 2\beta$$

$h^0 - H^0 - H^0 - H^0$

$$i \; \frac{3g^2}{4\cos^2\theta_W} \; \sin 2\alpha \; \cos 2\alpha$$

$H^0 - h^0 - h^0 - h^0$

$$-i \; \frac{3g^2}{4\cos^2\theta_W} \; \sin 2\alpha \; \cos 2\alpha$$

$G^0 - A^0 - A^0 - A^0$

$$-i \; \frac{3g^2}{4\cos^2\theta_W} \; \sin 2\beta \; \cos 2\beta$$

$A^0 - G^0 - G^0 - G^0$

$$i \; \frac{3g^2}{4\cos^2\theta_W} \; \sin 2\beta \; \cos 2\beta$$

$H^0 - H^0 - h^0 - h^0$

$$-i \; \frac{g^2}{4\cos^2\theta_W} \; [3\sin^2 2\alpha - 1]$$

$A^0 - A^0 - G^0 - G^0$

$$-i \; \frac{g^2}{4\cos^2\theta_W} \; [3\sin^2 2\beta - 1]$$

$H^0 - H^0 - G^0 - G^0$

$$-i \; \frac{g^2}{4 \cos^2 \theta_W} \; \cos 2\beta \; \cos 2\alpha$$

$h^0 - h^0 - G^0 - G^0$

$$i \; \frac{g^2}{4 \cos^2 \theta_W} \; \cos 2\beta \; \cos 2\alpha$$

$H^0 - h^0 - G^0 - G^0$

$$i \; \frac{g^2}{4 \cos^2 \theta_W} \; \cos 2\beta \; \sin 2\alpha$$

$H^0 - H^0 - A^0 - G^0$

$$i \; \frac{g^2}{4 \cos^2 \theta_W} \; \sin 2\beta \; \cos 2\alpha$$

$h^0 - h^0 - A^0 - G^0$

$$-i \; \frac{g^2}{4 \cos^2 \theta_W} \; \sin 2\beta \; \cos 2\alpha$$

$H^0 - h^0 - A^0 - G^0$

$$-i \; \frac{g^2}{4 \cos^2 \theta_W} \; \sin 2\beta \; \sin 2\alpha$$

$H^0 - H^0 - A^0 - A^0$

$$i \; \frac{g^2}{4 \cos^2 \theta_W} \; \cos 2\beta \; \cos 2\alpha$$

$h^0 - h^0 - A^0 - A^0$

$$-i \; \frac{g^2}{4 \cos^2 \theta_W} \; \cos 2\beta \; \cos 2\alpha$$

$H^0 - h^0 - A^0 - A^0$

$$-i \; \frac{g^2}{4 \cos^2 \theta_W} \; \cos 2\beta \; \sin 2\alpha$$

A.5   Higgs Couplings to Quark (or Lepton) Pairs

We present here all the Feynman rules for the interactions of Higgs and Goldstone bosons with one generation of quarks. We assume Model II type couplings as defined in eq. (4.22). To obtain the corresponding interactions for
leptons, make the following substitutions: $u \to \nu$ and $d \to e$. The neutral
scalar interactions are flavor conserving, so they apply universally to any of
the three generations of quarks and leptons (assuming that the appropriate
fermion mass is used). The couplings of the charged Higgs and charged Goldstone bosons are shown for the case of one generation. In the case of three
generations, the fermion masses must be replaced by the corresponding mass
matrices and a factor of the Kobayashi-Maskawa mixing matrix K must be
inserted as shown in eqs. (4.22) and (4.23), respectively.

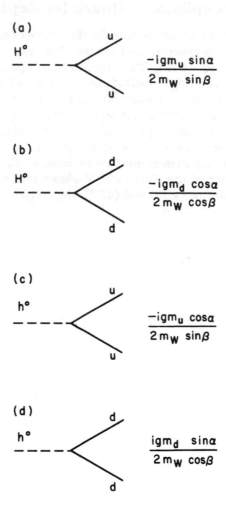

Figure A.18 Feynman rules for Higgs–quark–quark vertices.

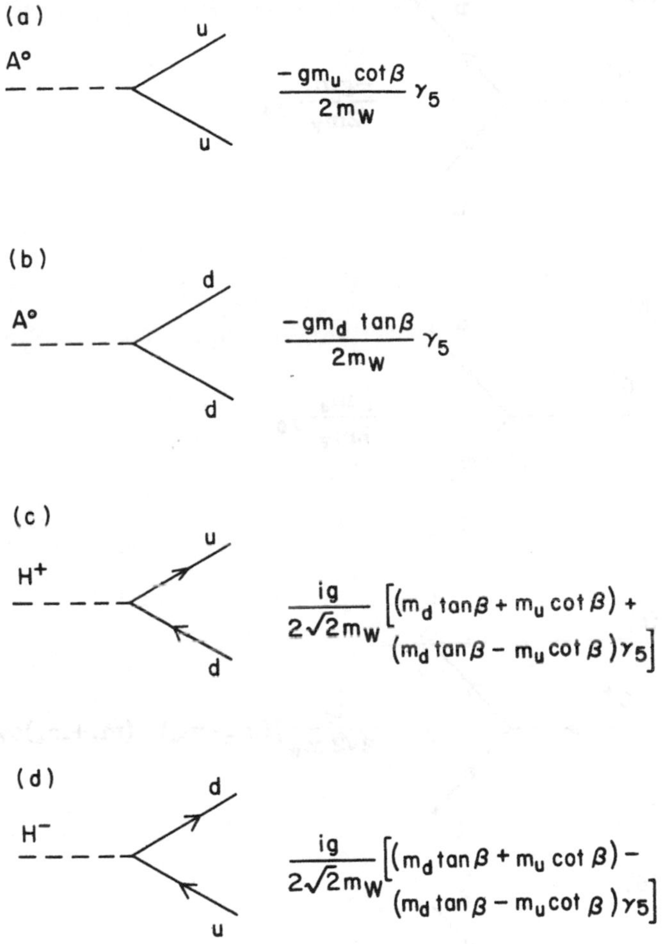
(a)
A^0
u
u
$\frac{-gm_u \cot\beta}{2m_W}\gamma_5$
(b)
A^0
d
d
$\frac{-gm_d \tan\beta}{2m_W}\gamma_5$
(c)
H^+
u
d
$\frac{ig}{2\sqrt{2}m_W}\left[(m_d \tan\beta + m_u \cot\beta) + (m_d \tan\beta - m_u \cot\beta)\gamma_5\right]$
(d)
H^-
d
u
$\frac{ig}{2\sqrt{2}m_W}\left[(m_d \tan\beta + m_u \cot\beta) - (m_d \tan\beta - m_u \cot\beta)\gamma_5\right]$

Figure A.19 Feynman rules for Higgs–quark–quark vertices, continued.

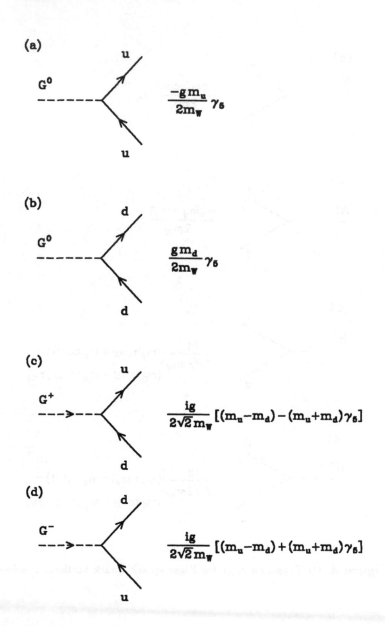

Figure A.20 Feynman rules for Goldstone–boson–quark–quark vertices.

A.6   Faddeev-Popov Ghost Vertices

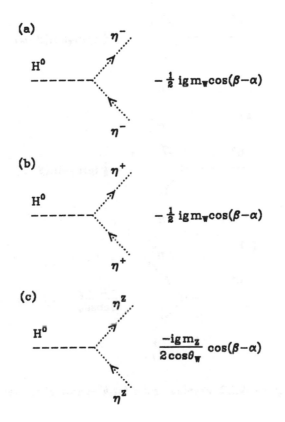

(a) $-\frac{1}{2}\,ig\,m_W\cos(\beta-\alpha)$

(b) $-\frac{1}{2}\,ig\,m_W\cos(\beta-\alpha)$

(c) $\dfrac{-ig\,m_Z}{2\cos\theta_W}\,\cos(\beta-\alpha)$

Figure A.21 Feynman rules for H^0–ghost–ghost vertices.

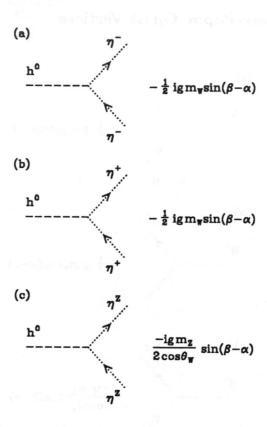

Figure A.22 Feynman rules for h^0-ghost-ghost vertices.

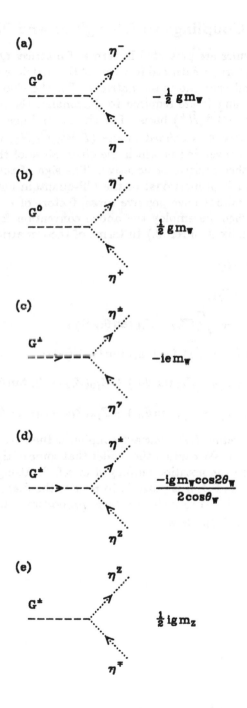

Figure A.23 Feynman rules for G–ghost–ghost vertices.

A.7   Higgs Couplings to Charginos and Neutralinos

These Feynman rules are presented in terms of matrices Q, S, Q'^L, Q'^R, and Q'', S'' which, in turn, are defined in terms of the matrices used to diagonalize the neutralino and chargino mass matrices. For the charginos, two unitary 2×2 matrices U and V are required to diagonalize the mass matrix. They are defined in the $(\widetilde{W}^+, \widetilde{H}^+)$ basis. For the neutralinos, the 4×4 unitary diagonalizing matrix Z is defined in the $(\widetilde{B}, \widetilde{W}_3, \widetilde{H}_1, \widetilde{H}_2)$ basis. The matrix Z is defined in a convention in which the eigenvalues of the neutralino mass matrix can be either positive or negative. The sign of the i'th eigenvalue is denoted by ϵ_i and is proportional to the CP-quantum number of $\widetilde{\chi}_i^0$. Since the physical neutralinos have positive mass, factors of ϵ_i will appear in the Feynman rules when we employ the above convention for Z. (For further details, see Appendix A of ref. 2.) In terms of these matrices we have

$$
\begin{aligned}
Q_{ij} &= \sqrt{\tfrac{1}{2}} V_{i1} U_{j2}, \\[4pt]
S_{ij} &= \sqrt{\tfrac{1}{2}} V_{i2} U_{j1}, \\[4pt]
Q'^L_{ij} &= Z_{i4} V_{j1} + \sqrt{\tfrac{1}{2}}(Z_{i2} + Z_{i1} \tan\theta_W) V_{j2}, \\[4pt]
Q'^R_{ij} &= \Big[Z_{i3} U_{j1} - \sqrt{\tfrac{1}{2}}(Z_{i2} + Z_{i1} \tan\theta_W) U_{j2} \Big] \epsilon_i, \\[4pt]
Q''_{ij} &= \tfrac{1}{2} \Big[Z_{i3}(Z_{j2} - Z_{j1}\tan\theta_W) + Z_{j3}(Z_{i2} - Z_{i1}\tan\theta_W) \Big] \epsilon_i, \\[4pt]
S''_{ij} &= \tfrac{1}{2} \Big[Z_{i4}(Z_{j2} - Z_{j1}\tan\theta_W) + Z_{j4}(Z_{i2} - Z_{i1}\tan\theta_W) \Big] \epsilon_i.
\end{aligned}
\tag{A.16}
$$

Note that if we assume CP-invariant couplings, then U, V, and Z are real orthogonal matrices. We caution the reader that some of the definitions above differ slightly from the notation employed in ref. 2. Using various identities satisfied by U, V, and Z, it is possible to rewrite the diagonal Feynman rules (coupling Higgs to $\widetilde{\chi}_i \widetilde{\chi}_i$) to include a term proportional to $M_{\widetilde{\chi}_i}/m_W$. Rules in this form can be found in ref. 2.

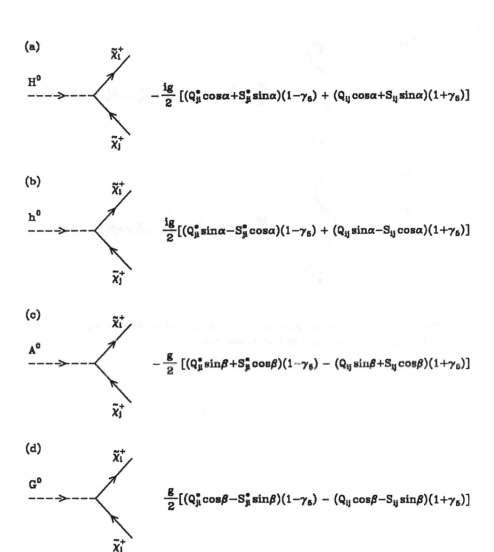

(a)

$$-\frac{ig}{2}\left[(Q_{ji}^{*}\cos\alpha+S_{ji}^{*}\sin\alpha)(1-\gamma_{5}) + (Q_{ij}\cos\alpha+S_{ij}\sin\alpha)(1+\gamma_{5})\right]$$

(b)

$$\frac{ig}{2}\left[(Q_{ji}^{*}\sin\alpha-S_{ji}^{*}\cos\alpha)(1-\gamma_{5}) + (Q_{ij}\sin\alpha-S_{ij}\cos\alpha)(1+\gamma_{5})\right]$$

(c)

$$-\frac{g}{2}\left[(Q_{ji}^{*}\sin\beta+S_{ji}^{*}\cos\beta)(1-\gamma_{5}) - (Q_{ij}\sin\beta+S_{ij}\cos\beta)(1+\gamma_{5})\right]$$

(d)

$$\frac{g}{2}\left[(Q_{ji}^{*}\cos\beta-S_{ji}^{*}\sin\beta)(1-\gamma_{5}) - (Q_{ij}\cos\beta-S_{ij}\sin\beta)(1+\gamma_{5})\right]$$

Figure A.24 Feynman rules for the couplings of charginos to a neutral Higgs or Goldstone boson.

(a)

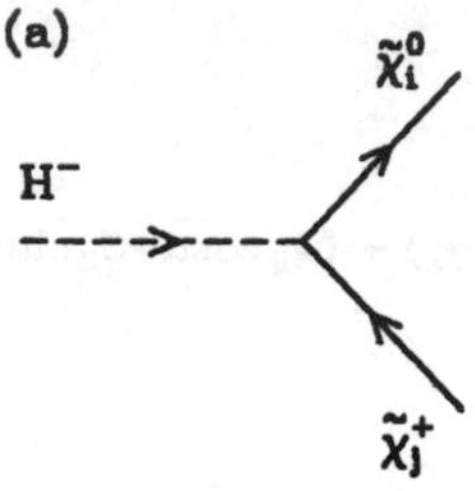

$$\tilde{\chi}_i^0$$

$$H^-$$

$$-\frac{ig}{2}\left[Q_{ij}'^L \cos\beta(1-\gamma_5) + Q_{ij}'^R \sin\beta(1+\gamma_5)\right]$$

$$\tilde{\chi}_j^+$$

(b)

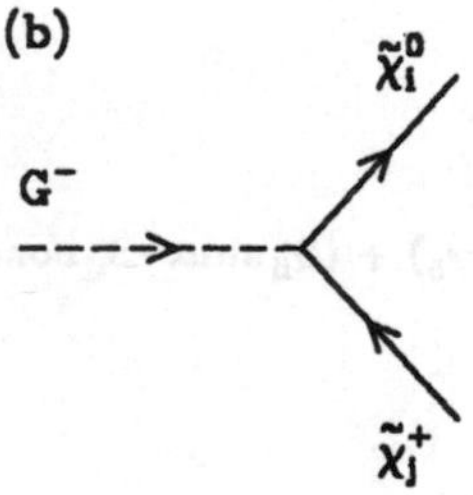

$$\tilde{\chi}_i^0$$

$$G^-$$

$$-\frac{ig}{2}\left[Q_{ij}'^L \sin\beta(1-\gamma_5) - Q_{ij}'^R \cos\beta(1+\gamma_5)\right]$$

$$\tilde{\chi}_j^+$$

Figure A.25 Feynman rules for the coupling of a chargino and a neutralino to a charged Higgs or Goldstone boson.

(a)

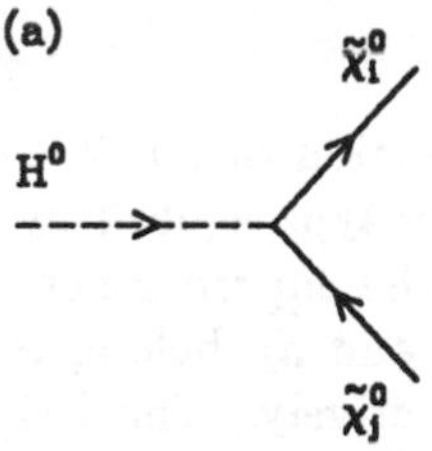

$\tilde{\chi}_i^0$

H^0

$$-\frac{ig}{2}\left[(Q_{ji}''^{*}\cos\alpha - S_{ji}''^{*}\sin\alpha)(1-\gamma_5) + (Q_{ij}''\cos\alpha - S_{ij}''\sin\alpha)(1+\gamma_5)\right]$$

$\tilde{\chi}_j^0$

(b)

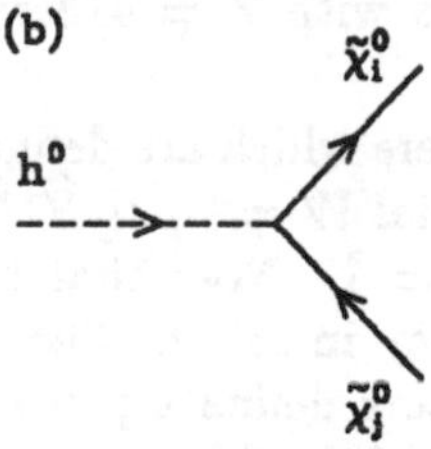

$\tilde{\chi}_i^0$

h^0

$$\frac{ig}{2}\left[(Q_{ji}''^{*}\sin\alpha + S_{ji}''^{*}\cos\alpha)(1-\gamma_5) + (Q_{ij}''\sin\alpha + S_{ij}''\cos\alpha)(1+\gamma_5)\right]$$

$\tilde{\chi}_j^0$

(c)

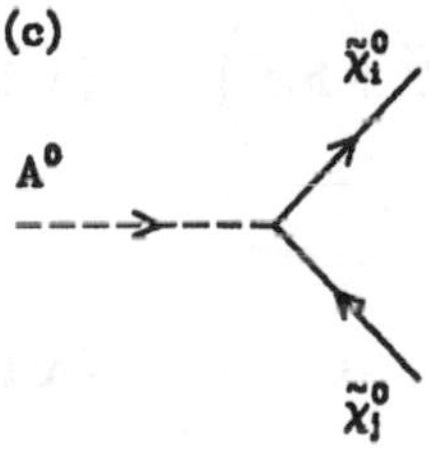

$\tilde{\chi}_i^0$

A^0

$$-\frac{g}{2}\left[(Q_{ji}''^{*}\sin\beta - S_{ji}''^{*}\cos\beta)(1-\gamma_5) - (Q_{ij}''\sin\beta - S_{ij}''\cos\beta)(1+\gamma_5)\right]$$

$\tilde{\chi}_j^0$

(d)

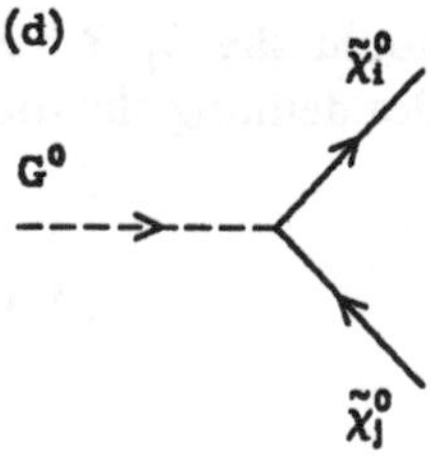

$\tilde{\chi}_i^0$

G^0

$$\frac{g}{2}\left[(Q_{ji}''^{*}\cos\beta + S_{ji}''^{*}\sin\beta)(1-\gamma_5) - (Q_{ij}''\cos\beta + S_{ij}''\sin\beta)(1+\gamma_5)\right]$$

$\tilde{\chi}_j^0$

Figure A.26 Feynman rules for the coupling of neutralinos to a neutral Higgs or Goldstone boson.

A.8 Trilinear Higgs Couplings to Squark (or Slepton) Pairs

These Feynman rules are presented in terms of the notation of ref. 2. We denote a generic up-type squark by $\tilde{u}$ and generic down-type squark by $\tilde{d}$. The L and R subscripts indicate that the squarks are the superpartners of left and right handed quarks. In our notation, the $\tilde{u}_L$ and $\tilde{d}_L$ belong to a $SU(2)_L$ doublet and have $T^{3L} = +1/2$ and $-1/2$, respectively. The $U(1)_Y$ quantum number of both doublet members is $Y = 1/3$. The charge formula is $Q = T^{3L} + Y/2$. The $\tilde{u}_R$ and $\tilde{d}_R$ are $SU(2)_L$ singlets with $Y = 4/3$ and $Y = -2/3$, respectively.

The Feynman rules depend on a number of parameters which are defined below. The μ parameter enters through the superpotential $W = -\mu\epsilon_{ij}\hat{H}_1^i\hat{H}_2^j$ [see eq. (4.65)]. In our notation, $\epsilon_{ij} = -\epsilon_{ji}$ with $\epsilon_{12} = 1$. Note that the choice of the sign of μ above corrects a sign inconsistency in ref. 2. The A-parameters are denoted by A_d and A_u. These parameters define a piece of the most general soft supersymmetry-breaking potential of the form

$$V_{\text{soft}} = \frac{g}{\sqrt{2}m_W}\left(\epsilon_{ij}\frac{m_d A_d}{\cos\beta}H_1^i\tilde{Q}^j\tilde{D} - \epsilon_{ij}\frac{m_u A_u}{\sin\beta}H_2^i\tilde{Q}^j\tilde{U} + \text{h.c.}\right), \quad (A.17)$$

where m_u and m_d are the *quark* masses, and

$$\tilde{Q}^i = \begin{pmatrix}\tilde{u}_L \\ \tilde{d}_L\end{pmatrix}, \quad \tilde{U}^* = \tilde{u}_R, \quad \tilde{D}^* = \tilde{d}_R. \quad (A.18)$$

When these parameters are non-zero, the L and R basis states are not mass eigenstates. Rules for conversion to mass eigenstates, should the L, R basis states mix, can be easily given in terms of the mixing angles defining the mass eigenstates

$$\tilde{q}_1 = \tilde{q}_L\cos\phi_q + \tilde{q}_R\sin\phi_q,$$
$$\tilde{q}_2 = -\tilde{q}_L\sin\phi_q + \tilde{q}_R\cos\phi_q. \quad (A.19)$$

In order to shorten our notation let us define

$$c_q = \cos\phi_q \qquad s_q = \sin\phi_q, \quad (A.20)$$

with similar definitions for a second quark type q'. The Feynman rule for a vertex of the type $X\tilde{q}_i\tilde{q}'_j$, where $i, j = 1, 2$ correspond to mass eigenstates, may be obtained from those for the vertices $X\tilde{q}_\alpha\tilde{q}'_\beta$, where $\alpha, \beta = L, R$, by the relation

$$V(X\tilde{q}_i\tilde{q}'_j) = \sum_{\alpha,\beta=L,R} T_{ij\alpha\beta}V(X\tilde{q}_\alpha\tilde{q}'_\beta), \quad (A.21)$$

where the conversion coefficients, T, appear in table A.1.

Table A.1

Table of conversion coefficients $T_{ij\alpha\beta}$ for use in eq. (A.21).

i,j \ $\alpha\beta$	L, L	R, R	L, R	R, L
1,1	$c_q c_{q'}$	$s_q s_{q'}$	$c_q s_{q'}$	$s_q c_{q'}$
1,2	$-c_q s_{q'}$	$s_q c_{q'}$	$c_q c_{q'}$	$-s_q s_{q'}$
2,1	$-s_q c_{q'}$	$c_q s_{q'}$	$-s_q s_{q'}$	$c_q c_{q'}$
2,2	$s_q s_{q'}$	$c_q c_{q'}$	$-s_q c_{q'}$	$-c_q s_{q'}$

All the rules that follow are given for the case of one generation of squarks. The generalization to the case of three generations can be found in Appendix B of ref. 2. The corresponding rules for sleptons are obtained by making the substitutions $\tilde{u} \to \tilde{\nu}$ and $\tilde{d} \to \tilde{e}$ (and omitting all rules with $\tilde{\nu}_R$).

(a)

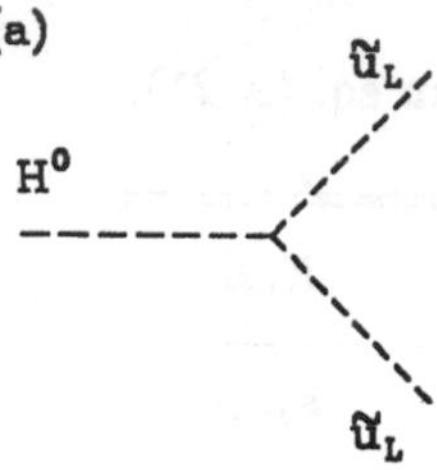

$$\frac{-igm_Z}{\cos\theta_W}\left(\tfrac{1}{2} - e_u\sin^2\theta_W\right)\cos(\alpha+\beta) - \frac{igm_u^2}{m_W\sin\beta}\sin\alpha$$

(b)

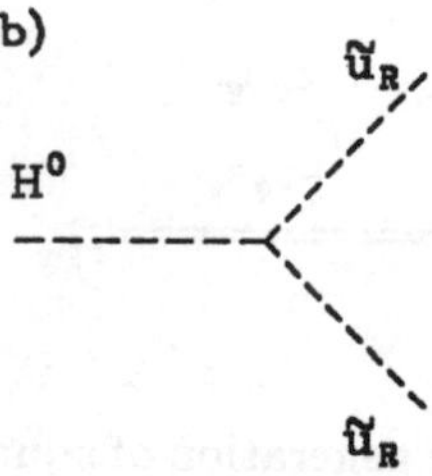

$$\frac{-igm_Z}{\cos\theta_W}e_u\sin^2\theta_W\cos(\alpha+\beta) - \frac{igm_u^2}{m_W\sin\beta}\sin\alpha$$

(c)

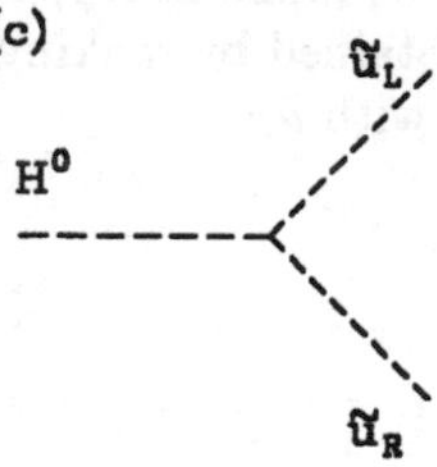

$$\frac{igm_u}{2m_W\sin\beta}\left(\mu\cos\alpha - A_u\sin\alpha\right)$$

Figure A.27 Feynman rules for H^0 couplings to up-type squarks.

(a)

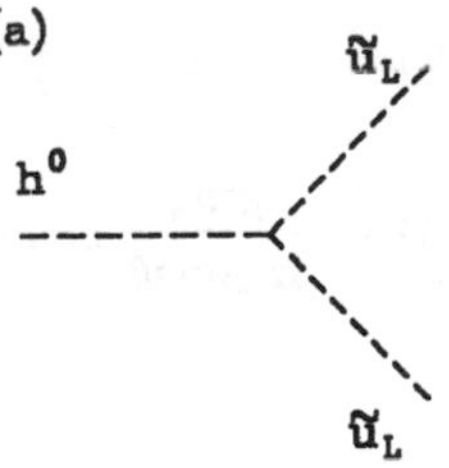

$$\frac{igm_Z}{\cos\theta_W}\left(\tfrac{1}{2}-e_u\sin^2\theta_W\right)\sin(\alpha+\beta)-\frac{igm_u^2}{m_W\sin\beta}\cos\alpha$$

(b)

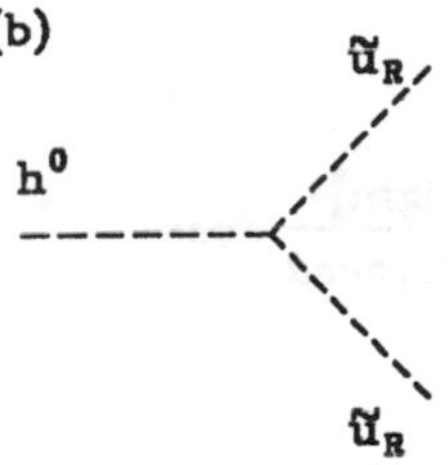

$$\frac{igm_Z}{\cos\theta_W}e_u\sin^2\theta_W\sin(\alpha+\beta)-\frac{igm_u^2}{m_W\sin\beta}\cos\alpha$$

(c)

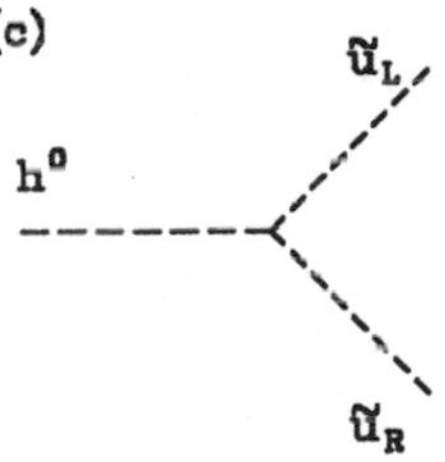

$$\frac{-igm_u}{2m_W\sin\beta}(\mu\sin\alpha+A_u\cos\alpha)$$

Figure A.28 Feynman rules for h^0 couplings to up-type squarks.

(a)

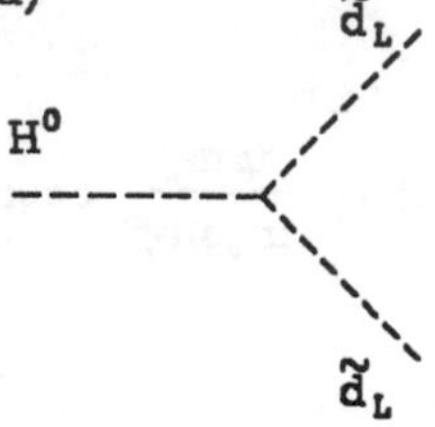

$$\frac{igm_Z}{\cos\theta_W} \left(\tfrac{1}{2} + e_d\sin^2\theta_W\right)\cos(\alpha+\beta) - \frac{igm_d^2}{m_W\cos\beta}\cos\alpha$$

(b)

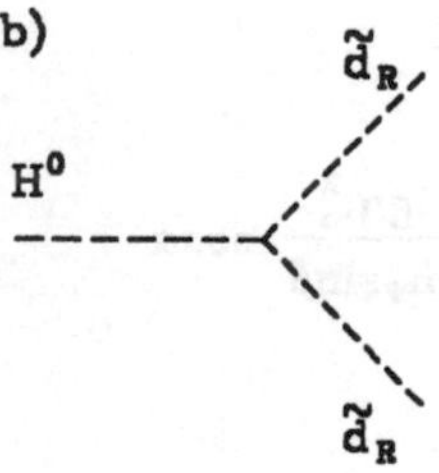

$$\frac{-igm_Z}{\cos\theta_W} e_d\sin^2\theta_W\cos(\alpha+\beta) - \frac{igm_d^2}{m_W\cos\beta}\cos\alpha$$

(c)

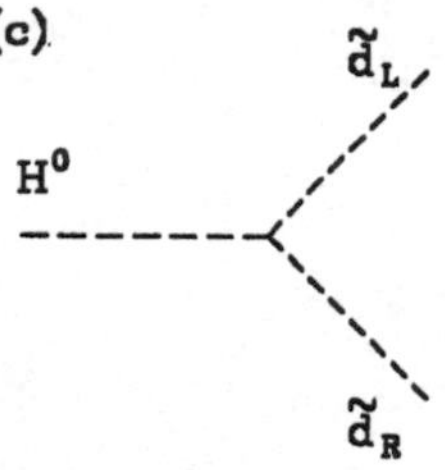

$$\frac{igm_d}{2m_W\cos\beta} \left(\mu\sin\alpha - A_d\cos\alpha\right)$$

Figure A.29 Feynman rules H^0 couplings to down-type squarks.

(a)

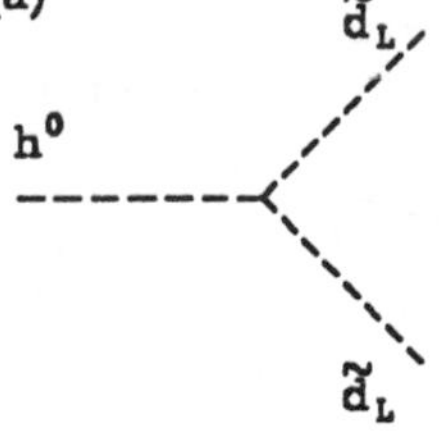

$$\frac{-igm_Z}{\cos\theta_W}\left(\tfrac{1}{2}+e_d\sin^2\theta_W\right)\sin(\alpha+\beta)+\frac{igm_d^2}{m_W\cos\beta}\sin\alpha$$

(b)

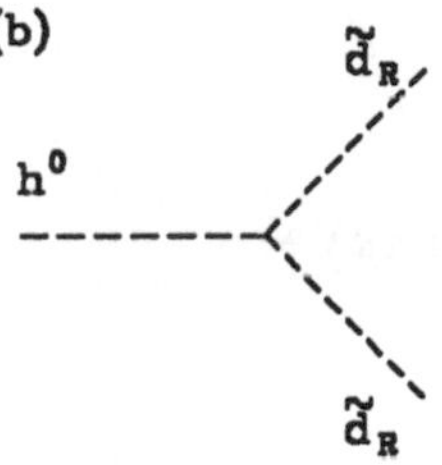

$$\frac{igm_Z}{\cos\theta_W}e_d\sin^2\theta_W\sin(\alpha+\beta)+\frac{igm_d^2}{m_W\cos\beta}\sin\alpha$$

(c)

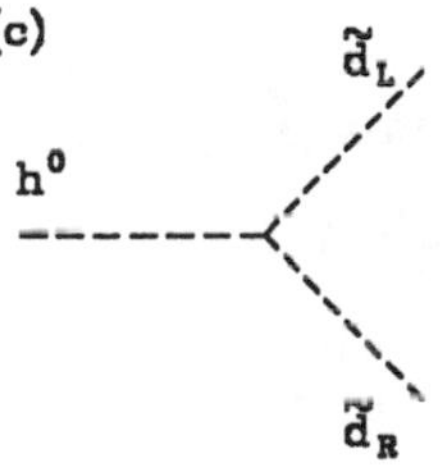

$$\frac{igm_d}{2m_W\cos\beta}\left(\mu\cos\alpha+A_d\sin\alpha\right)$$

Figure A.30 Feynman rules h^0 couplings to down-type squarks.

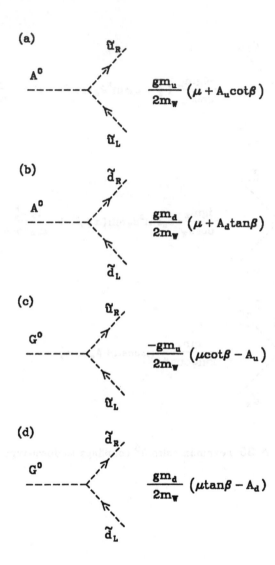

Figure A.31 Feynman rules for squark couplings to a neutral pseudoscalar Higgs or Goldstone boson. The direction of momentum is indicated. To obtain the appropriate rules in the $\tilde{q}_1 - \tilde{q}_2$ basis, simply replace L with 1 and R with 2.

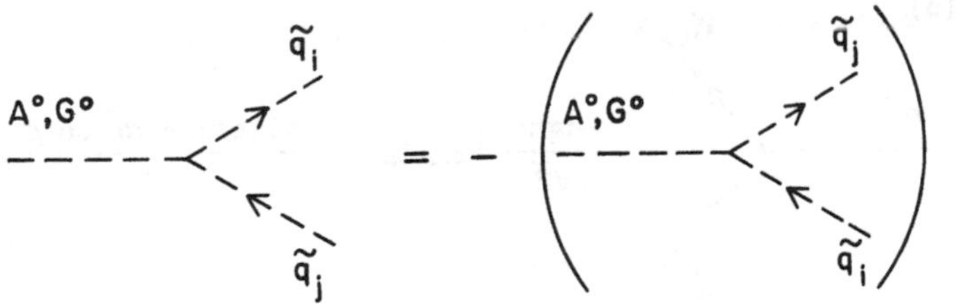

Figure A.32 Rule for changing momenta directions in fig. A.31. The indices i, j refer to either the $L - R$ or $1 - 2$ bases. Note that this rule implies that for $i = j$ the vertex vanishes.

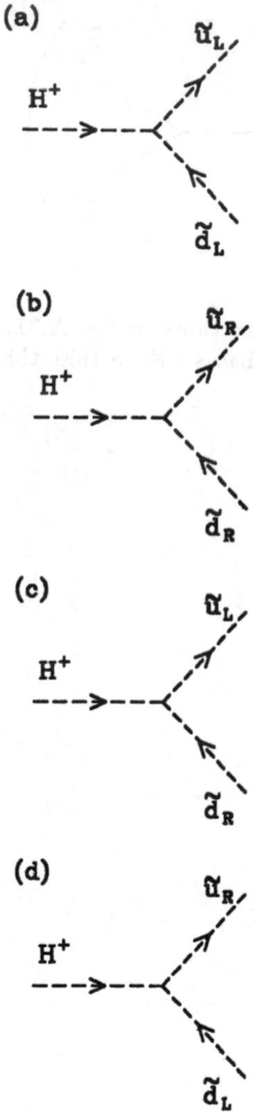

(a)

$$\frac{-igm_W}{\sqrt{2}} \left[\sin 2\beta - \frac{m_d^2 \tan\beta + m_u^2 \cot\beta}{m_W^2} \right]$$

(b)

$$\frac{igm_u m_d}{\sqrt{2} m_W} (\tan\beta + \cot\beta)$$

(c)

$$\frac{igm_d}{\sqrt{2} m_W} (\mu + A_d \tan\beta)$$

(d)

$$\frac{igm_u}{\sqrt{2} m_W} (\mu + A_u \cot\beta)$$

Figure A.33 Feynman rules for charged Higgs couplings to squarks.

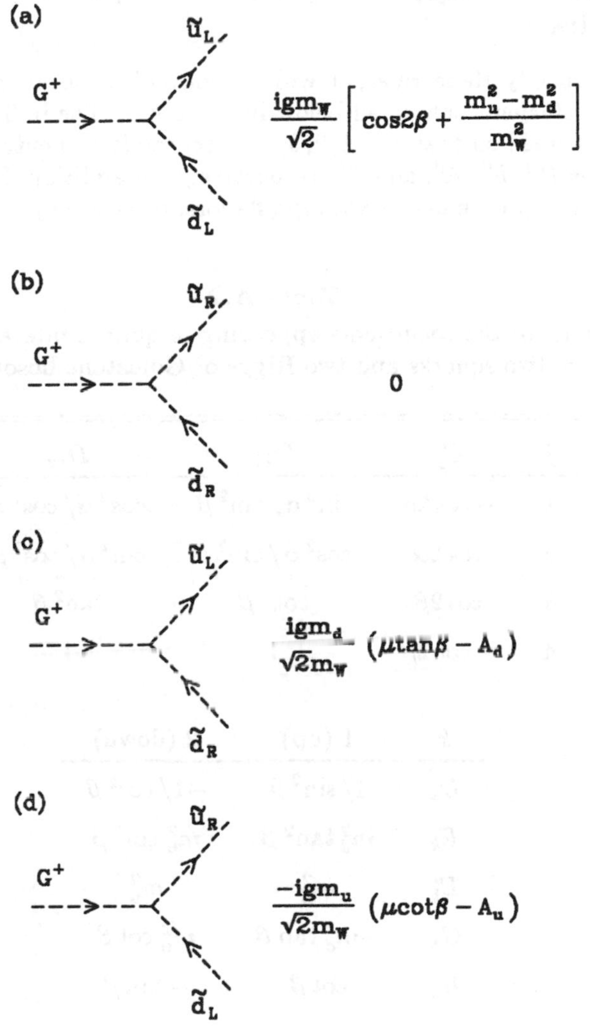

(a)

$\tilde{u}_L$

G^+

$\tilde{d}_L$

$$\frac{igm_W}{\sqrt{2}} \left[\cos2\beta + \frac{m_u^2 - m_d^2}{m_W^2} \right]$$

(b)

$\tilde{u}_R$

G^+

$\tilde{d}_R$

$$0$$

(c)

$\tilde{u}_L$

G^+

$\tilde{d}_R$

$$\frac{igm_d}{\sqrt{2}m_W} \left(\mu\tan\beta - A_d \right)$$

(d)

$\tilde{u}_R$

G^+

$\tilde{d}_L$

$$\frac{-igm_u}{\sqrt{2}m_W} \left(\mu\cot\beta - A_u \right)$$

Figure A.34 Feynman rules for charged Goldstone boson couplings to squarks.

A.9   Quartic Higgs Couplings to Squark (or Slepton) Pairs

In order to specify these rules, it will be convenient to define a number of subsidiary coefficients tabulated below in table A.2. The indices on the coefficients are defined so that $j = 1, 2, 3, 4$ corresponds to neutral Higgs (Goldstone) bosons H^0, h^0, A^0, and G^0, respectively. In addition, for D_{jk}, $k = 1, 2$ corresponds to up-type and down-type flavors, respectively.

Table A.2

Table for the coefficients appearing in quartic interactions
of two squarks and two Higgs or Goldstone bosons.

j	C_j	D_{j1}	D_{j2}
1	$-\cos 2\alpha$	$\sin^2 \alpha / \sin^2 \beta$	$\cos^2 \alpha / \cos^2 \beta$
2	$\cos 2\alpha$	$\cos^2 \alpha / \sin^2 \beta$	$\sin^2 \alpha / \cos^2 \beta$
3	$\cos 2\beta$	$\cot^2 \beta$	$\tan^2 \beta$
4	$-\cos 2\beta$	1	1

k	1 (up)	2 (down)
D_k	$1/\sin^2 \beta$	$-1/\cos^2 \beta$
E_k	$m_d^2 \tan^2 \beta$	$m_u^2 \cot^2 \beta$
E_k'	m_d^2	m_u^2
G_k	$-m_d^2 \tan \beta$	$m_u^2 \cot \beta$
H_k	$\cot \beta$	$-\tan \beta$

Table A.2, cont.

j	R_j	S_j	T_j	V_j
1	$\sin(\beta + \alpha)$	$\sin\alpha\cos\beta/\sin^2\beta$	$\cos\alpha\sin\beta/\cos^2\beta$	$\cos(\beta - \alpha)$
2	$\cos(\beta + \alpha)$	$\cos\alpha\cos\beta/\sin^2\beta$	$-\sin\alpha\sin\beta/\cos^2\beta$	$\sin(\beta - \alpha)$
3	$i\cos 2\beta$	$i\cot^2\beta$	$-i\tan^2\beta$	0
4	$i\sin 2\beta$	$i\cot\beta$	$i\tan\beta$	i

j	R'_j	S'_j	T'_j	V'_j
1	$-\cos(\beta + \alpha)$	$\sin\alpha/\sin\beta$	$-\cos\alpha/\cos\beta$	$\sin(\beta - \alpha)$
2	$\sin(\beta + \alpha)$	$\cos\alpha/\sin\beta$	$\sin\alpha/\cos\beta$	$-\cos(\beta - \alpha)$
3	$i\sin 2\beta$	$i\cot\beta$	$i\tan\beta$	$-i$
4	$-i\cos 2\beta$	$+i$	$-i$	0

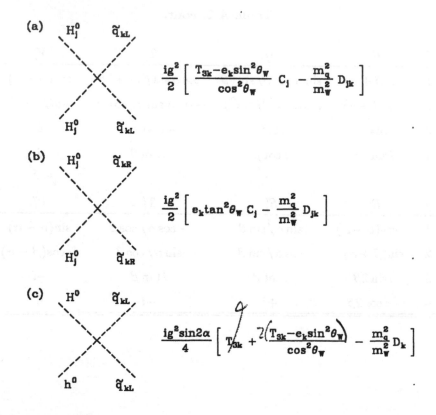

Figure A.35 Feynman rules for quartic vertices with two neutral Higgs or Goldstone bosons and two squarks. The index $j = 1, \ldots, 4$ corresponds to H^0, h^0, A^0 and G^0 respectively. C_j, D_{jk} and D_k are defined in table A.2. Note that there are no $A^0 H^0 \widetilde{q} \widetilde{q}$, $A^0 h^0 \widetilde{q} \widetilde{q}$, $G^0 H^0 \widetilde{q} \widetilde{q}$, or $G^0 h^0 \widetilde{q} \widetilde{q}$ vertices due to CP conservation.

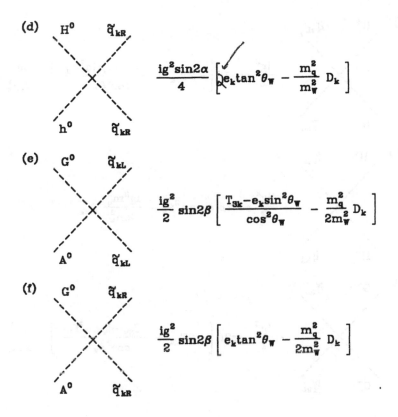

(d) H^0 $\tilde{q}_{kR}$

h^0 $\tilde{q}_{kR}$

$$\frac{ig^2 \sin 2\alpha}{4} \left[e_k \tan^2\theta_W - \frac{m_q^2}{m_W^2} D_k \right]$$

(e) G^0 $\tilde{q}_{kL}$

A^0 $\tilde{q}_{kL}$

$$\frac{ig^2}{2} \sin 2\beta \left[\frac{T_{3k} - e_k \sin^2\theta_W}{\cos^2\theta_W} - \frac{m_q^2}{2m_W^2} D_k \right]$$

(f) G^0 $\tilde{q}_{kR}$

A^0 $\tilde{q}_{kR}$

$$\frac{ig^2}{2} \sin 2\beta \left[e_k \tan^2\theta_W - \frac{m_q^2}{2m_W^2} D_k \right]$$

Figure A.35, continued.

(a)

$$\frac{ig^2}{2}\cos2\beta\left[-2T_{3k} + \frac{T_{3k}-e_k\sin^2\theta_W}{\cos^2\theta_W}\right] - \frac{ig^2}{2m_W^2}E_k$$

(b)

$$\frac{ig^2}{2}\cos2\beta\;e_k\tan^2\theta_W - \frac{ig^2m_q^2}{2m_W^2}D_{3k}$$

(c)

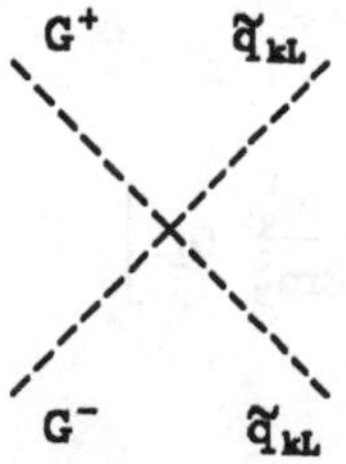

$$-\frac{ig^2}{2}\cos2\beta\left[-2T_{3k} + \frac{T_{3k}-e_k\sin^2\theta_W}{\cos^2\theta_W}\right] - \frac{ig^2}{2m_W^2}E_k'$$

Figure A.36 Feynman rules for quartic vertices with two charged Higgs or Goldstone bosons and two squarks. D_{jk} $(j=3)$, E_k, E_k', G_k, and H_k are defined in table A.2.

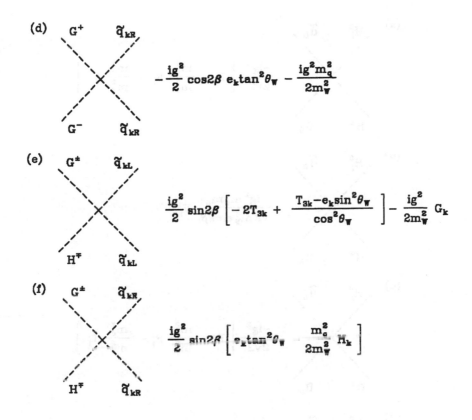

(d)

G^+ $\tilde{q}_{kR}$

G^- $\tilde{q}_{kR}$

$$-\frac{ig^2}{2}\cos2\beta\ e_k\tan^2\theta_W - \frac{ig^2m_q^2}{2m_W^2}$$

(e)

$G^\pm$ $\tilde{q}_{kL}$

$H^\mp$ $\tilde{q}_{kL}$

$$\frac{ig^2}{2}\sin2\beta\left[-2T_{3k} + \frac{T_{3k}-e_k\sin^2\theta_W}{\cos^2\theta_W}\right] - \frac{ig^2}{2m_W^2}G_k$$

(f)

$G^\pm$ $\tilde{q}_{kR}$

$H^\mp$ $\tilde{q}_{kR}$

$$\frac{ig^2}{2}\sin2\beta\left[e_k\tan^2\theta_W \quad \frac{m_q^2}{2m_W^2}\ H_k\right]$$

Figure A.36, continued.

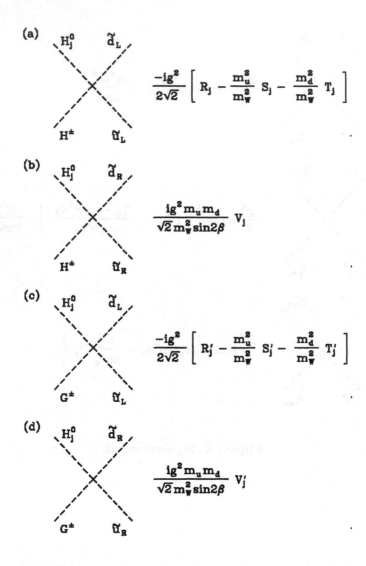

Figure A.37 Feynman rules for quartic vertices with one charged and one neutral Higgs or Goldstone boson and two squarks. The index $j = 1, \ldots, 4$ corresponds to H^0, h^0, A^0 and G^0 respectively. See table A.2 for the auxiliary coefficients. Note that there is no $A^0 H^- \tilde{d}_R \tilde{u}_R$ or $G^0 H^- \tilde{d}_R \tilde{u}_R$ vertex.

REFERENCES

1. I.J.R. Aitchison and A.J.G. Hey, *Gauge Theories in Particle Physics* (Adam Hilger, Bristol, 1982).
2. J.F. Gunion and H.E. Haber, *Nucl. Phys.* **B272** (1986) 1.
3. T.-P. Cheng and L.-F. Li, *Gauge Theory of Elementary Particle Physics* (Oxford University Press, Oxford, 1984).
4. W. Hollik, *Z. Phys.* **C32** (1986) 291; *Z. Phys.* **C37** (1988) 569.
5. S. Bertolini, *Nucl. Phys.* **B272** (1986) 77.
6. P. Kalyniak, R. Bates and J.N. Ng *Phys. Rev.* **D33** (1986) 755; R. Bates, J.N. Ng, and P. Kalyniak, *Phys. Rev.* **D34** (1986) 172.
7. J.S. Bell, *Nucl. Phys.* **B60** (1973) 427; C.H. Llewellyn Smith, *Phys. Lett.* **46B** (1973) 233; J.M. Cornwall, D.N. Levin, and G. Tiktopoulos, *Phys. Rev. Lett.* **30** (1973) 1268, and *Phys. Rev.* **D10** (1974) 1145.
8. B. Cox and F.J. Gilman, *Proceedings of the 1984 Snowmass Summer Study on the Design and Utilization of the Superconducting Super Collider*, edited by R. Donaldson and J. G. Morfin, p. 87.
9. J. Cornwall *et al.*, in ref. 7; C. Vayonakis, *Lett. Nuovo Cim.* **17** (1976) 383; M. Chanowitz and M. Gaillard, in ref. 8; and G.J. Gounaris, R. Kogerler, and H. Neufeld, *Phys. Rev.* **D34** (1986) 3257.
10. M.B. Gavela, G. Girardi, C. Malleville and P. Sorba, *Nucl. Phys.* **B193** (1981) 257.
11. K. Fujikawa, *Phys. Rev.* **D7** (1973) 393.
12. G. Keller and D. Wyler, *Nucl. Phys.* **B274** (1986) 410.
13. N. Deshpande and M. Nazerimonfared, *Nucl. Phys.* **B213** (1983) 390.
14. N.M. Monyonko, J.H. Reid and A. Sen, *Phys. Lett.* **136B** (1984) 265.
15. N.M. Monyonko and J.H. Reid, *Prog. Theor. Phys.* **73** (1984) 734.
16. H.E. Haber and D. Wyler, *Nucl. Phys.* **B323** (1989) 267.
17. M. Capdequi Peyranere, H.E. Haber and P. Irulegui, *Phys. Rev.* **D44** (1991) in press.
18. R. Bailin and A. Love, *Introduction to Gauge Field Theory* (Adam Hilger, Bristol, 1986).
19. H.E. Haber and G.L. Kane, *Phys. Rep.* **117C** (1985) 75.
20. J. Rosiek, *Phys. Rev.* **D41** (1990) 3464.
21. J.F. Gunion and H.E. Haber, *Nucl. Phys.* **B278** (1986) 449.

Appendix B

Formulae for Higgs Boson Decay Widths for Two-Body Channels in the Minimal Supersymmetric Model

For the reader's convenience, we list below the explicit expressions for the partial widths of all Higgs boson decay channels involving both non-supersymmetric and supersymmetric final states. These have been derived using the Feynman rules given in Appendix A.

The decay widths for channels involving Standard Model quark and lepton pairs are well-known. Using the notation of the third family fermions,

$$
\Gamma(H^+ \to t\bar{b}) = \frac{3g^2 \lambda^{1/2}}{32\pi m_W^2 m_{H+}^3}
$$
$$
\times \left[(m_{H+}^2 - m_b^2 - m_t^2)(m_b^2 \tan^2 \beta + m_t^2 \cot^2 \beta) - 4m_b^2 m_t^2 \right]
\tag{B.1}
$$

$$
\Gamma(h \to t\bar{t}) = \frac{3g^2 m_t^2 d_h^2 m_h}{32\pi m_W^2 \sin^2 \beta} \left(1 - \frac{4m_t^2}{m_h^2} \right)^p ,
\tag{B.2}
$$

$$
\Gamma(h \to b\bar{b}) = \frac{3g^2 m_b^2 e_h^2 m_h}{32\pi m_W^2 \cos^2 \beta} \left(1 - \frac{4m_b^2}{m_h^2} \right)^p ,
\tag{B.3}
$$

where d_h and e_h are given by

$$
d_h = \begin{cases} -\sin \alpha, & h = H^0 \\ \cos \alpha, & h = h^0 \\ \cos \beta, & h = A^0 \end{cases} ,
\tag{B.4}
$$

$$
e_h = \begin{cases} \cos \alpha, & h = H^0 \\ \sin \alpha, & h = h^0 \\ -\sin \beta, & h = A^0 \end{cases} ,
\tag{B.5}
$$

respectively, and the power p is

$$p = \begin{cases} 3/2, & h = h^0, H^0 \\ 1/2, & h = A^0 \end{cases}. \tag{B.6}$$

The factor $\lambda^{1/2}$ is the usual kinematic factor

$$\lambda^{1/2} \equiv [(m_1^2 + m_2^2 - m_3^2)^2 - 4m_1^2 m_2^2]^{1/2}, \tag{B.7}$$

where m_1, m_2 and m_3 are the masses of the three particles involved in the decay. Eqs. (B.2) and (B.3) are valid for any up-type or down-type quark. For leptonic final states, the same three equations apply if one removes the color factor of 3.

The only other possible tree-level two-body decay modes of the Higgs bosons into non-supersymmetric final states involve gauge bosons and/or other Higgs bosons. First, consider the decays into a pair of (massive) vector bosons. There are no tree level couplings of vector boson pairs to H^+ and A^0. Furthermore, h^0 is lighter than the Z and so it has no vector boson decays. The only remaining Higgs boson to consider is H^0. The relevant decay rates are listed below

$$\Gamma(H^0 \to W^+ W^-) =$$

$$\frac{g^2(m_{H^0}^4 - 4m_{H^0}^2 m_W^2 + 12m_W^4) \cos^2(\beta - \alpha)}{64\pi m_W^2 m_{H^0}} \left(1 - \frac{4m_W^2}{m_{H^0}^2}\right)^{1/2}, \tag{B.8}$$

$$\Gamma(H^0 \to ZZ) =$$

$$\frac{g^2(m_{H^0}^4 - 4m_{H^0}^2 m_Z^2 + 12m_Z^4) \cos^2(\beta - \alpha)}{128\pi m_Z^2 m_{H^0} \cos^2 \theta_W} \left(1 - \frac{4m_Z^2}{m_{H^0}^2}\right)^{1/2}, \tag{B.9}$$

Note that the suppression of the H^0 width into vector boson pairs with respect to the minimal Higgs model is due to the factor $\cos^2(\beta - \alpha)$. As shown in fig. 4.7, this factor is less than about 10^{-2} over the entire region of parameter space where $m_{H^0} \geq 2m_W$, and decreases by a factor proportional to $1/m_{H^0}^2$ as m_{H^0} gets large.

Second, consider Higgs decays to a final state consisting of one Higgs boson and one gauge boson. Taking the Higgs boson mass relations into account, only the following two decays may be kinematically allowed

$$\Gamma(H^+ \to W^+ h^0) = \frac{g^2 \lambda^{3/2} \cos^2(\beta - \alpha)}{64\pi m_W^2 m_{H^+}^3} \tag{B.10}$$

$$\Gamma(A^0 \to Z h^0) = \frac{g^2 \lambda^{3/2} \cos^2(\beta - \alpha)}{64\pi m_Z^2 m_{A^0}^3 \cos^2 \theta_W} \tag{B.11}$$

In the above formulae, one should use the appropriate kinematic factor λ [see eq. (B.7)]. Note that both decays are again suppressed by a factor of $\cos^2(\beta - \alpha)$.

Finally, consider Higgs decay to a two-Higgs-boson final state. Once again the Higgs boson mass relations imply that only two decays, $H^0 \rightarrow h^0 h^0, A^0 A^0$ may be kinematically allowed. The corresponding partial widths are

$$\Gamma(H^0 \rightarrow hh) = \frac{g^2 m_Z^2 f_h^2}{128\pi m_{H^0} \cos^2 \theta_W} \left(1 - \frac{4m_h^2}{m_{H^0}^2}\right)^{1/2}, \qquad (B.12)$$

where f_h consists of the following mixing angle factors

$$f_h = \begin{cases} \cos 2\alpha \cos(\beta + \alpha) - 2\sin 2\alpha \sin(\beta + \alpha), & h = h^0 \\ \cos 2\beta \cos(\beta + \alpha), & h = A^0 \end{cases}. \qquad (B.13)$$

We now present formulae for the partial widths into charginos ($\widetilde{\chi}_i^+$) and neutralinos ($\widetilde{\chi}_i^0$). In the minimal supersymmetric model, there are two charginos and four neutralinos.

First, we give the decay rate for the neutral Higgs bosons

$$\Gamma(h \rightarrow \widetilde{\chi}_i \widetilde{\chi}_j) =$$
$$\frac{g^2 \lambda^{1/2} \left[(F_{ijh}^2 + F_{jih}^2)(m_h^2 - M_i^2 - M_j^2) - 4F_{ijh} F_{jih} \epsilon_i \epsilon_j \eta_h M_i M_j\right]}{16\pi m_h^3 (1 + \delta(i,j))}$$
$$(B.14)$$

where the factor of η_h is equal to 1 for the scalar Higgses ($h = h^0, H^0$), and is equal to -1 for the pseudoscalar Higgs ($h = A^0$). We use the general notation $\widetilde{\chi}_i$ to denote either a neutralino or chargino; its (positive) mass will be denoted by M_i. The factor $\delta(i,j)$ is inserted only when there are two identical Majorana neutralinos in the final state. In that case, $\delta(i,j) = 1$, otherwise, it is equal to 0. The kinematical factor λ is given by

$$\lambda = (M_i^2 + M_j^2 - m_h^2)^2 - 4M_i^2 M_j^2. \qquad (B.15)$$

The factor ϵ_i stands for the sign of the neutralino mass. When the neutralino mass matrix is diagonalized, we allow the sign of the ith eigenvalue (ϵ_i) to be either positive or negative. For the chargino, it is trivial to insure positive mass eigenvalues (by appropriate choice of the diagonalizing matrices), so we simply set $\epsilon = 1$ for all chargino states. The factors F_{ijh} are given in terms of the diagonalizing matrix elements for the charginos and neutralinos. For the charginos, two 2×2 matrices U and V are required to diagonalize the mass matrix. For the neutralinos, the diagonalizing matrix Z is defined in the ($\widetilde{B}, \widetilde{W}_3, \widetilde{H}_1, \widetilde{H}_2$)–basis (where H_2 is the doublet that couples to the top-quark). We assume CP-invariant couplings, in which case U, V and Z

are real orthogonal matrices. (Further discussion can be found in Appendix A of ref. 1.) The factors F_{ijh} appearing in eq. (B.14), corresponding to two possible types of decay modes, are given below

1. For $h \to \tilde{\chi}_i^+ \tilde{\chi}_j^-$

$$F_{ijh} = \frac{1}{\sqrt{2}}[e_h V_{i1} U_{j2} - d_h V_{i2} U_{j1}]. \qquad (B.16)$$

2. For $h \to \tilde{\chi}_i^0 \tilde{\chi}_j^0$

$$F_{ijh} = \frac{e_h}{2}\left[Z_{i3}Z_{j2} + Z_{j3}Z_{i2} - \tan\theta_W\left(Z_{i3}Z_{j1} + Z_{j3}Z_{i1}\right)\right]$$
$$+\frac{d_h}{2}\left[Z_{i4}Z_{j2} + Z_{j4}Z_{i2} - \tan\theta_W\left(Z_{i4}Z_{j1} + Z_{j4}Z_{i1}\right)\right], \qquad (B.17)$$

where d_h and e_h are is given in eqs. (B.4) and (B.5).

For the decay of the charged Higgs boson, we have

$$\Gamma(H^+ \to \tilde{\chi}_i^+ \tilde{\chi}_j^0) = \frac{g^2\lambda^{1/2}\left[(F_L^2 + F_R^2)(m_{H^+}^2 - M_i^2 - M_j^2) - 4F_L F_R \epsilon_j M_i M_j\right]}{16\pi m_{H^+}^3}, \qquad (B.18)$$

where F_L and F_R are given by

$$F_L = \cos\beta \left[Z_{j4}V_{i1} + \frac{1}{\sqrt{2}}(Z_{j2} + Z_{j1}\tan\theta_W)V_{i2}\right] \qquad (B.19)$$

$$F_R = \sin\beta \left[Z_{j3}U_{i1} - \frac{1}{\sqrt{2}}(Z_{j2} + Z_{j1}\tan\theta_W)U_{i2}\right]. \qquad (B.20)$$

Note that all the rates for Higgs decay into chargino/neutralino pairs are invariant under the interchange of v_1 and v_2 (*i.e.* $\tan\beta \to 1/\tan\beta$). This is proved by observing that under this interchange all neutralino and chargino masses (including the sign), and the mixing elements Z_{i1} and Z_{i2}, remain unchanged. In addition, $\cos\alpha \leftrightarrow \sin\alpha$, $U_{ij} \leftrightarrow V_{ij}$, $Z_{i3} \to -Z_{i4}$ and $Z_{i4} \to -Z_{i3}$. It follows that under the interchange of v_1 and v_2, $F_{ijh} \leftrightarrow \pm F_{jih}$ [in eqs. (B.16) and (B.17)] and $F_L \leftrightarrow -F_R$ [in eqs. (B.18)–(B.20)]. Thus, all width formulas are invariant under $\tan\beta \to 1/\tan\beta$, as claimed above.

Finally, we give the decay rate for Higgs boson decay to a pair of squarks (or sleptons). Using the Feynman rules given in Appendix A (§A.8), we can read off the appropriate couplings, $g_{h\tilde{f}_i\tilde{f}_j}$. (Note that this coupling has units of mass.) Then, for any neutral or charged Higgs boson, h,

$$\Gamma(h \to \tilde{f}_i \overline{\tilde{f}_j}) = \frac{N_c g_{h\tilde{f}_i\tilde{f}_j}^2 \lambda^{1/2}}{16\pi m_h^3}, \qquad (B.21)$$

where $N_c = 3$ ($N_c = 1$) for squark (slepton) final states, and λ is the usual kinematic factor [see eq. (B.15)]. If $i \neq j$, then one should multiply the above width by 2 to include the $\overline{\tilde{f}_i}\tilde{f}_j$ final state.

REFERENCES

1. J.F. Gunion and H.E. Haber, *Nucl. Phys.* **B272** (1986) 1.

Appendix C

Formulae for Neutral Higgs Boson Decays to $\gamma\gamma$ and $Z\gamma$ in the Two-Doublet and the Minimal Supersymmetric Model

In this appendix we give the complete formulae for the $\gamma\gamma$ decay widths of the neutral Higgs bosons in both the two-doublet model [with Model II Higgs–fermion couplings defined in eq. (4.22)] and in the two-doublet minimal supersymmetric model (MSSM) including superpartner loops. The formulae presented here are based on ref. 1, but we make use of the notation of §2.1 for the auxiliary functions. For the $Z\gamma$ decay mode we will confine ourselves to only the two-doublet extension, and will not give explicit formulae for the superpartner loops. The latter can be found in ref. 2. For completeness, we note that expressions exist in the literature for the $h\gamma\gamma$ and $hZ\gamma$ couplings, in which one particle (either γ or Z) is off-shell. These expressions can be found in ref. 3 (for $h = \phi^0$), and in ref. 2 (for $h = H^0$, h^0, or A^0 in the MSSM).

$\gamma\gamma$ Final State

The $\gamma\gamma$ widths are obtained by a relatively straightforward extension of eqs. (2.16)–(2.20). We have

$$\Gamma\left(\phi^0 \to \gamma\gamma\right) = \frac{\alpha^2 g^2}{1024\pi^3} \frac{m_h^3}{m_W^2} \left| \sum_i I_h^i \right|^2, \qquad \text{(C.1)}$$

where i runs over all the loop contributions appropriate to the particular model being considered. In order to include the pseudoscalar A^0 in a convenient fashion it is necessary to define a generalization of the function $F_{1/2}$

defined in eq. (2.17). We write:

$$F_{1/2}^h(\tau) = -2\tau\left[\xi^h + (1 - \tau\xi^h)f(\tau)\right],\qquad\text{(C.2)}$$

where $f(\tau)$ is defined in eq. (2.19) and

$$\xi^h = \begin{cases} 1 & h^0 \\ 1 & H^0 \\ 0 & A^0 \end{cases}.\qquad\text{(C.3)}$$

In the $\tau \to \infty$ limit, $F_{1/2}^h(\tau) \to -4/3$ for $h = H^0, h^0$ [as in eq. (2.21)] but $F_{1/2}^{A^0} \to 2$. We now list the contributions deriving from quark (or lepton), W, charged Higgs, squark (or slepton), and chargino loops, respectively. Only the first three are to be included in a non-supersymmetric extension of the Standard Model.* Using F_0 and F_1 as defined in eq. (2.17) and $F_{1/2}^h$ defined in eq. (C.2), we have

$$
\begin{aligned}
I_h^f &= N_{cf}e_f^2 R_f^h F_{1/2}^h(\tau_f) \\
I_h^W &= R_W^h F_1(\tau_W) \\
I_h^{H\pm} &= R_{H\pm}^h F_0(\tau_{H\pm})\frac{m_W^2}{m_{H\pm}^2} \\
I_h^{\tilde{f}} &= N_{cf}e_f^2 R_{\tilde{f}}^h F_0(\tau_{\tilde{f}})\frac{m_Z^2}{m_{\tilde{f}}^2} \\
I_h^{\tilde{\chi}\pm} &= R_{\tilde{\chi}\pm}^h F_{1/2}^h(\tau_{\tilde{\chi}\pm})\frac{m_W}{m_{\tilde{\chi}\pm}},
\end{aligned}
\qquad\text{(C.4)}
$$

where $\tau_i = 4m_i^2/m_h^2$, $N_c = 3$ for quarks and squarks ($N_c = 1$ for leptons and sleptons), and the R_i^h are defined as follows

$$R_{u,c,t}^h = \begin{cases} \frac{\cos\alpha}{\sin\beta} & h^0 \\ \frac{\sin\alpha}{\sin\beta} & H^0 \\ \cot\beta & A^0 \end{cases}, \quad R_{d,s,b,e,\mu,\tau}^h = \begin{cases} \frac{-\sin\alpha}{\cos\beta} & h^0 \\ \frac{\cos\alpha}{\cos\beta} & H^0 \\ \tan\beta & A^0 \end{cases},\qquad\text{(C.5)}$$

$$R_W^h = \begin{cases} \sin(\beta - \alpha) & h^0 \\ \cos(\beta - \alpha) & H^0 \\ 0 & A^0 \end{cases}, \quad R_{H\pm}^h = \begin{cases} \sin(\beta - \alpha) + \frac{\cos 2\beta \sin(\beta+\alpha)}{2\cos^2\theta_W} & h^0 \\ \cos(\beta - \alpha) - \frac{\cos 2\beta \cos(\beta+\alpha)}{2\cos^2\theta_W} & H^0 \\ 0 & A^0 \end{cases},$$

$$\text{(C.6)}$$

_____

*For definiteness, we quote the charged Higgs contribution based on the H^+H^-h coupling of the MSSM. In a non-supersymmetric two-doublet model, one must alter the expression for $R_{H\pm}^h$ [eq. (C.6)] appropriately to reflect a more general H^+H^-h coupling.

$$
R^h_{\tilde{f}_{L,R}} = \begin{cases} \dfrac{m_f^2}{m_Z^2} R_f^h \mp (T_f^{3L} - e_f \sin^2\theta_W)\sin(\beta+\alpha) & h^0 \\[2ex] \dfrac{m_f^2}{m_Z^2} R_f^h \pm (T_f^{3L} - e_f \sin^2\theta_W)\cos(\beta+\alpha) & H^0 \\[2ex] 0 & A^0 \end{cases} \tag{C.7}
$$

with $T_f^{3L} = \pm 1/2$ for $\tilde{f}_L$ and 0 for $\tilde{f}_R$, and the upper signs in eq. (C.7) taken for $\tilde{f}_L$ and the lower signs taken for $\tilde{f}_R$. In addition,

$$
R_{\tilde{\chi}_i^\pm} = 2 \begin{cases} S_{ii}\cos\alpha - Q_{ii}\sin\alpha & h^0 \\[1ex] S_{ii}\sin\alpha + Q_{ii}\cos\alpha & H^0 \\[1ex] -S_{ii}\cos\beta - Q_{ii}\sin\beta & A^0 \end{cases}, \tag{C.8}
$$

where, in the minimal supersymmetry model, $i = 1, 2$ runs over the two chargino mass eigenstates, and the S_{ii} and Q_{ii} are defined in eq. (A.16). In terms of the matrices U and V required to diagonalize the chargino mass matrix, discussed briefly in Appendix A (see §A.7), the S and Q matrices are given by

$$
S_{ij} = \sqrt{\tfrac{1}{2}}\, V_{i2}U_{j1}, \quad Q_{ij} = \sqrt{\tfrac{1}{2}}\, V_{i1}U_{j2}. \tag{C.9}
$$

For more details see ref. 4.

$Z\gamma$ Final State

The modifications to the results of §2.1, as given in eq. (2.22), are easily stated for the two-doublet extension of the Standard Model. We write

$$
\Gamma(h \to Z\gamma) = \frac{1}{32\pi} m_h^3 \left(1 - \frac{m_Z^2}{m_h^2}\right)^3 |\mathcal{A}^h|^2, \tag{C.10}
$$

where

$$
\mathcal{A}^h = \frac{\alpha g}{4\pi m_W}\left(A_F^h + A_W^h + A_{H\pm}^h\right), \tag{C.11}
$$

with

$$
A_F^h = \sum_f N_{cf} R_f^h \frac{-2e_f(T_f^{3L} - 2e_f \sin^2\theta_W)}{\sin\theta_W \cos\theta_W}\left[\xi^h I_1(\tau_f, \lambda_f) - I_2(\tau_f, \lambda_f)\right], \tag{C.12}
$$

$$
A_W^h = -R_W^h \cot\theta_W \left\{ 4(3 - \tan^2\theta_W)I_2(\tau_W, \lambda_W) + \left[\left(1 + \frac{2}{\tau_W}\right)\tan^2\theta_W - \left(5 + \frac{2}{\tau_W}\right)\right]I_1(\tau_W, \lambda_W)\right\}, \tag{C.13}
$$

and

$$A^h_{H\pm} = R^h_{H\pm} \frac{1 - 2\sin^2\theta_W}{\cos\theta_W \sin\theta_W} I_1\left(\tau_{H\pm}, \lambda_{H\pm}\right) \frac{m^2_W}{m^2_{H\pm}}, \tag{C.14}$$

where τ_i and λ_i are defined in analogy to eq. (2.23) by

$$\tau_i = \frac{4m^2_i}{m^2_h}, \quad \lambda_i = \frac{4m^2_i}{m^2_Z}, \tag{C.15}$$

ξ^h is defined in eq. (C.3), the various R^h_i $(i = f, W, H^\pm)$ are defined in eqs. (C.5) and (C.6), and I_1 and I_2 are defined in eq. (2.24).

It is perhaps useful to note the following reductions that allow one to relate the $Z\gamma$ width expressions to those for $\gamma\gamma$. The limit of interest is when the λ_i's become infinite, $i.e.$ when $m_Z \rightarrow 0$. We have

$$\begin{aligned} I_1(\tau, \lambda) &\rightarrow -\tfrac{1}{2}F_0(\tau) \\ \xi^h I_1(\tau, \lambda) - I_2(\tau, \lambda) &\rightarrow \tfrac{1}{4}F^h_{1/2}(\tau). \end{aligned} \tag{C.16}$$

Taking these limits, and replacing the vector Z couplings appearing in the $Z\gamma$ expressions by the appropriate charge coupling yields the $\gamma\gamma$ width expressions in the case of fermion-loop and charged-Higgs-loop contributions.

REFERENCES

1. J.F. Gunion, G. Gamberini, and S.F. Novaes, *Phys. Rev.* **D38** (1988) 3481.
2. T.J. Weiler and T.-C. Yuan, *Nucl. Phys.* **B318** (1989) 337.
3. A. Barroso, J. Pulido and J.C. Romão, *Nucl. Phys.* **B267** (1986) 509; J.C. Romão and A. Barroso, *Nucl. Phys.* **B272** (1986) 693.
4. J.F. Gunion and H.E. Haber, *Nucl. Phys.* **B272** (1986) 1.

Index